Genetics, Genomics and Breeding of Crops

Genetics, Genomics and Breeding of Crops

Edited by Craig Steward

SYRAWOOD
PUBLISHING HOUSE
New York

Published by Syrawood Publishing House,
750 Third Avenue, 9th Floor,
New York, NY 10017, USA
www.syrawoodpublishinghouse.com

Genetics, Genomics and Breeding of Crops
Edited by Craig Steward

International Standard Book Number: 978-1-68286-510-1 (Hardback)

Cataloging-in-Publication Data

Genetics, genomics and breeding of crops / edited by Craig Steward.
p. cm.
Includes bibliographical references and index.
ISBN 978-1-68286-510-1
1. Genetics. 2. Genomics. 3. Plant breeding. 4. Crops. I. Steward, Craig.
QH430 .G46 2018
576.5--dc23

Printed in China.

TABLE OF CONTENTS

PREFACE

Conventional plant breeding techniques have existed since prehistoric times. As times have evolved and new techniques and advancements in genetics and molecular biology have been developed, plant breeding is now carried out by using molecular genetic tools. Techniques may vary from promoting desirable characteristics in plants to increasing crop yield. This book provides comprehensive insights into the field of crop breeding. It aims to shed light on some of the unexplored aspects of genetics and breeding of crops and the recent researches in this field. It is an essential guide for both academicians and those who wish to pursue this discipline further.

This book unites the global concepts and researches in an organized manner for a comprehensive understanding of the subject. It is a ripe text for all researchers, students, scientists or anyone else who is interested in acquiring a better knowledge of this dynamic field.

I extend my sincere thanks to the contributors for such eloquent research chapters. Finally, I thank my family for being a source of support and help.

Editor

1

Genome-Wide SNP Detection, Validation, and Development of an 8K SNP Array for Apple

David Chagné[1], Ross N. Crowhurst[2], Michela Troggio[3], Mark W. Davey[4], Barbara Gilmore[5], Cindy Lawley[6], Stijn Vanderzande[4], Roger P. Hellens[2], Satish Kumar[7], Alessandro Cestaro[3], Riccardo Velasco[3], Dorrie Main[8], Jasper D. Rees[9], Amy Iezzoni[10], Todd Mockler[11], Larry Wilhelm[12], Eric Van de Weg[13], Susan E. Gardiner[1], Nahla Bassil[5], Cameron Peace[8]*

1 Plant and Food Research, Palmerston North Research Centre, Palmerston North, New Zealand, **2** Plant and Food Research, Mount Albert Research Centre, Auckland, New Zealand, **3** IASMA Research and Innovation Centre, Foundation Edmund Mach, San Michele all'Adige, Trento, Italy, **4** Laboratory for Fruit Breeding and Biotechnology, Department of Biosystems, Katholieke Universiteit Leuven, Heverlee, Leuven, Belgium, **5** USDA-ARS, National Clonal Germplasm Repository, Corvallis, Oregon, United States of America, **6** Illumina Inc., Hayward, California, United States of America, **7** Plant and Food Research, Hawke's Bay Research Centre, Havelock North, New Zealand, **8** Department of Horticulture and Landscape Architecture, Washington State University, Pullman, Washington, United States of America, **9** Agricultural Research Council, Onderstepoort, South Africa, **10** Department of Horticulture, Michigan State University, East Lansing, Michigan, United States of America, **11** The Donald Danforth Plant Science Center, St. Louis, Missouri, United States of America, **12** Oregon Health Sciences University, Portland, Oregon, United States of America, **13** Plant Breeding, Wageningen University and Research Centre, Wageningen, The Netherlands

Abstract

As high-throughput genetic marker screening systems are essential for a range of genetics studies and plant breeding applications, the International RosBREED SNP Consortium (IRSC) has utilized the Illumina Infinium® II system to develop a medium- to high-throughput SNP screening tool for genome-wide evaluation of allelic variation in apple (*Malus ×domestica*) breeding germplasm. For genome-wide SNP discovery, 27 apple cultivars were chosen to represent worldwide breeding germplasm and re-sequenced at low coverage with the Illumina Genome Analyzer II. Following alignment of these sequences to the whole genome sequence of 'Golden Delicious', SNPs were identified using *SoapSNP*. A total of 2,113,120 SNPs were detected, corresponding to one SNP to every 288 bp of the genome. The Illumina GoldenGate® assay was then used to validate a subset of 144 SNPs with a range of characteristics, using a set of 160 apple accessions. This validation assay enabled fine-tuning of the final subset of SNPs for the Illumina Infinium® II system. The set of stringent filtering criteria developed allowed choice of a set of SNPs that not only exhibited an even distribution across the apple genome and a range of minor allele frequencies to ensure utility across germplasm, but also were located in putative exonic regions to maximize genotyping success rate. A total of 7867 apple SNPs was established for the IRSC apple 8K SNP array v1, of which 5554 were polymorphic after evaluation in segregating families and a germplasm collection. This publicly available genomics resource will provide an unprecedented resolution of SNP haplotypes, which will enable marker-locus-trait association discovery, description of the genetic architecture of quantitative traits, investigation of genetic variation (neutral and functional), and genomic selection in apple.

Editor: Mohammed Bendahmane, Ecole Normale Superieure, France

Funding: This work was partially funded by the U.S. Department of Agriculture's National Institute of Food and Agriculture–Specialty Crop Research Initiative project, "RosBREED: Enabling marker-assisted breeding in Rosaceae" (2009-51181-05808). Apple genomics at Plant and Food Research is partially supported by the New Zealand Ministry of Science and Innovation project C06X0812 "Exploiting Opportunities from Horticultural Genomics", PREVAR Ltd, as well as Plant and Food Research's Pipfruit Internal Investment Project. Apple genome research at Istituto Agrarie San Michele Adige-Fundazione Edmund Mach is supported by the research office of the Provincia autonoma di Trento. The funders had no role in study design, data collection and analysis, decision to publish, or preparation of the manuscript. The funders had no role in study design, data collection and analysis, decision to publish, or preparation of the manuscript.

Competing Interests: DC, RPH, SK, and SEG received funding from PREVAR Ltd and The New Zealand Institute for Plant and Food Research Limited (Plant and Food Research). There are no patents, products in development or marketed products to declare. DC, RNC, RPH, SK, and SEG belong to Plant and Food Research, a New Zealand government-owned crown-research institute. CL belongs to Illumina Inc. There are no patents, products in development or marketed products to declare.

* E-mail: cpeace@wsu.edu

Introduction

Understanding the links between phenotypic variations and their underlying DNA variation is the major challenge facing plant geneticists today. Recent advances in genomics technologies, including highly parallel sequencing and genetic marker methods for genome-wide assays of allelic variation, have now made high-resolution genetic characterization of crop germplasm feasible. The use of genomics tools has tremendous potential to assist Rosaceae crop breeders to produce significant genetic gains for consumer and grower traits more precisely and efficiently, as well as to improve understanding of the genetic architecture of agronomic characters [1]. Genetic marker-based strategies such as QTL interval mapping, pedigree-based analysis, association mapping, and genomic selection are now powerfully enabled by genomics technologies. Although genetic markers such as Simple

Sequence Repeats (SSRs) [2,3,4] and Single Nucleotide Polymorphisms (SNPs) [5,6,7] have been developed for apple, their density across the genome is not sufficient for fine dissection of functional genetic variation. We focus here on the development of a medium to high-throughput multiplexed SNP assay tool based on the Illumina Infinium® assay for genome-wide evaluation of allelic variation in apple breeding germplasm. This product of international research community collaboration is intended to benefit apple breeders directly, and consequently growers and consumers who demand better quality pome fruit that is produced more sustainably.

SNP marker development for a genome screening tool is based on three steps: detection, validation, and final selection. The objectives and throughput required are different for each step. The detection step requires identification of a large pool of SNPs in the crop and typically involves non-targeted techniques such as re-sequencing, or High Resolution Melting (HRM) [8], of a few representative yet diverse individuals. This maximizes opportunities for detecting SNPs, while ensuring that the SNPs will be suitable for their final application for screening breeding germplasm. While sequencing has historically been carried out using the Sanger method, high-throughput sequencing-by-synthesis techniques have now made it possible to re-sequence entire genomes affordably [9]. Although not strictly necessary, validation is desirable to inform SNP assay development, in order to maximize the number of functional polymorphic markers in the final genetic analysis. Because of the current prohibitive cost of validating every detected SNP, validation is often based on a subset of the detected SNPs that are screened over an informative set of individuals. If a range of SNP filtering parameter values are included in the SNPs used for the validation step, suitable parameter thresholds can then be established for those SNPs to be selected for the final assay. For development of genome-wide SNP assays, adequate genome coverage by the final set chosen is critical [10]. In crops with whole genome sequences and saturated genetic maps, such as apple [11], genome coverage can be on a physical and/or a genetic basis, as desired. When such genomics resources are lacking, the assumption may be made that random distribution of a large number of SNPs will achieve adequate coverage, with the caveat that certain genomic regions will be over- or under-represented depending on the SNP detection approach adopted.

Once SNP sets have been developed, screening can be carried out with a range of highly parallel techniques on segregating germplasm sets according to the research need or breeding application. Multiplexing and automation have enhanced the efficiency of SNP genotyping enormously and several high-throughput platforms are now available for the genotyping of a variable number of samples for one to up to one million SNPs in parallel. They include, but are not limited to, TaqMan® from Applied Biosystems and array-based technologies from Illumina (GoldenGate® and Infinium®) or Affymetrix [10,12,13].

We used next-generation sequencing (NGS) to detect SNPs covering the apple genome. This effort involved re-sequencing a small set of cultivars, ancestors, and founders, chosen to represent the pedigrees of worldwide apple breeding programs by RosBREED, a consortium established to enable marker-assisted breeding for Rosaceae crops (www.rosbreed.org; [14]). We validated the SNPs detected and determined adequate filtering parameter values, using the Illumina GoldenGate® assay to screen a larger set of accessions from the international apple breeding germplasm. Based on results of the GoldenGate® assay, we further refined SNP-filtering criteria to enable the development of an 8K Infinium® II array and evaluated it, using a segregating population of apple seedlings.

Materials and Methods

The workflow and design parameters described below are summarized in Figure 1.

Whole genome re-sequencing of apple breeding accessions

A SNP detection panel of 27 apple germplasm accessions was chosen for whole-genome, low coverage re-sequencing. The accessions were founders, intermediate ancestors, or important breeding parents used extensively in apple breeding programs worldwide. The complete set maximized coverage of the genetic background of the world's cultivated apple. Individuals in the set were *Malus×domestica* 'Braeburn', Co-op 15, 'Cox's Orange Pippin', 'CrimsonCrisp', 'Cripps Pink', 'Delicious' (both the original form and the derived mutant 'Red Delicious'), 'Dolgo', 'Duchess of Oldenburg', $F_2$26829-2-2, 'Frostbite', 'Fuji', 'Geneva', 'Golden Delicious', 'Granny Smith', 'Haralson', 'Honeycrisp', 'Idared', 'James Grieve', 'Jonathan', 'McIntosh', 'Ralls Janet', 'Red Dougherty', 'Rome Beauty', 'Splendour', 'Zestar', and *Malus sieversii* PI613981. Each accession was sequenced using one lane of Illumina GA II (except for 'Cox's Orange Pippin' that was sequenced twice) with 75 to 80 cycles per read and small insert paired-end or single-end sequencing as specified in Table 1. The raw sequenced data were retrieved each apple accession and then aligned separately to the reference genome of 'Golden Delicious' [11] using *Soap2* [15].

Detection and filtering of SNPs

SNPs were detected using *SoapSNP* (http://soap.genomics.org.cn/soapsnp.html) essentially as described by [16], but with the following modifications (Stage 1 filtering): *SoapSNP* calls were discarded if (1) the SNP call was within the first or last 60 bases of a linkage group (LG) sequence, (2) the reference genome base was ambiguous at the SNP position or within 60 bases of it, (3) another SNP was detected within 50 bases, (4) the average copy number of the SNP flanking region was more than two, (5) the Illumina quality score of either allele was less than 20, (6) the number of reads supporting either allele was less than two reads per accession for at least half the accessions, (7) the number of reads for either allele across all accessions was greater than the average read depth of all SNPs plus three standard deviations, and (8) if the call was an A/T or C/G transversion. This filtration yielded "Stage 1 SNPs".

Stage 1 SNPs were then subjected to a Stage 2 filter, whereby SNPs not located in a predicted exonic region were discarded. For each LG, the exon space was defined by mapping all gene models [11] and 396,643 cDNA sequences from the Plant & Food Research *Malus* Sequence Database libraries [17] ('AAAA', 'AABA', 'AACA', 'AADA', 'AAEA', 'AAMA', 'AAFA', 'AAFB', 'AAGA', 'AAHA', 'AAIA', 'AAJA', 'AAKA', 'AALA', 'AALB', 'AANA', 'AAOA', 'AAPA', 'AAQA', 'AARA', 'AASA', 'AAUA', 'AAVB', 'AVBC', 'AAWA', 'AAXA', 'AAYA', 'AAZA', 'AATA', 'ABAA', 'ABBA', 'AASB', 'AASC', 'ABBB', 'ABAB', 'ABDA', 'ABCA', 'ABEA', 'ABHA', 'ABKA', 'AGDQ', 'ABPB', 'ABLB', 'ABMA', 'ABPA', 'ABNA', 'ABNB', 'ABLC', 'ABLA', 'ABQA', 'ABRA', 'ABSA', 'ABTA', 'ABUA', 'ABVA', 'ABWA', 'ABXA', 'ABYA', 'ABZA', 'ACAA', 'ACBA', 'AELA', 'AEPA', 'AEAA', 'AENA', 'AFBB', 'AFBA', 'AFBC', 'AGBM', 'AGBN', 'AGBO', 'AGBP', 'AGBQ', 'AGBR', 'AGBS', 'AGBT', 'AGBU', 'AGBV', 'AGBW', 'AGCD', 'AGCE', 'AGCI', 'AGCJ', 'AGCK', 'AGCL', 'AGCM', 'AGCN', 'AGCP', 'AGCQ', 'AGCR', 'AGCS', 'AGCT', 'AGCU', 'AGCV', 'AGCW', 'AGCX', 'AGCY', 'AGCZ', 'AGDA', 'AGDB', 'AGDC', 'AGDD', 'AGDE', 'AGDF', 'AGDG', 'AGDH', 'AGDK', 'AGDM', 'AGDN', 'AGDO',

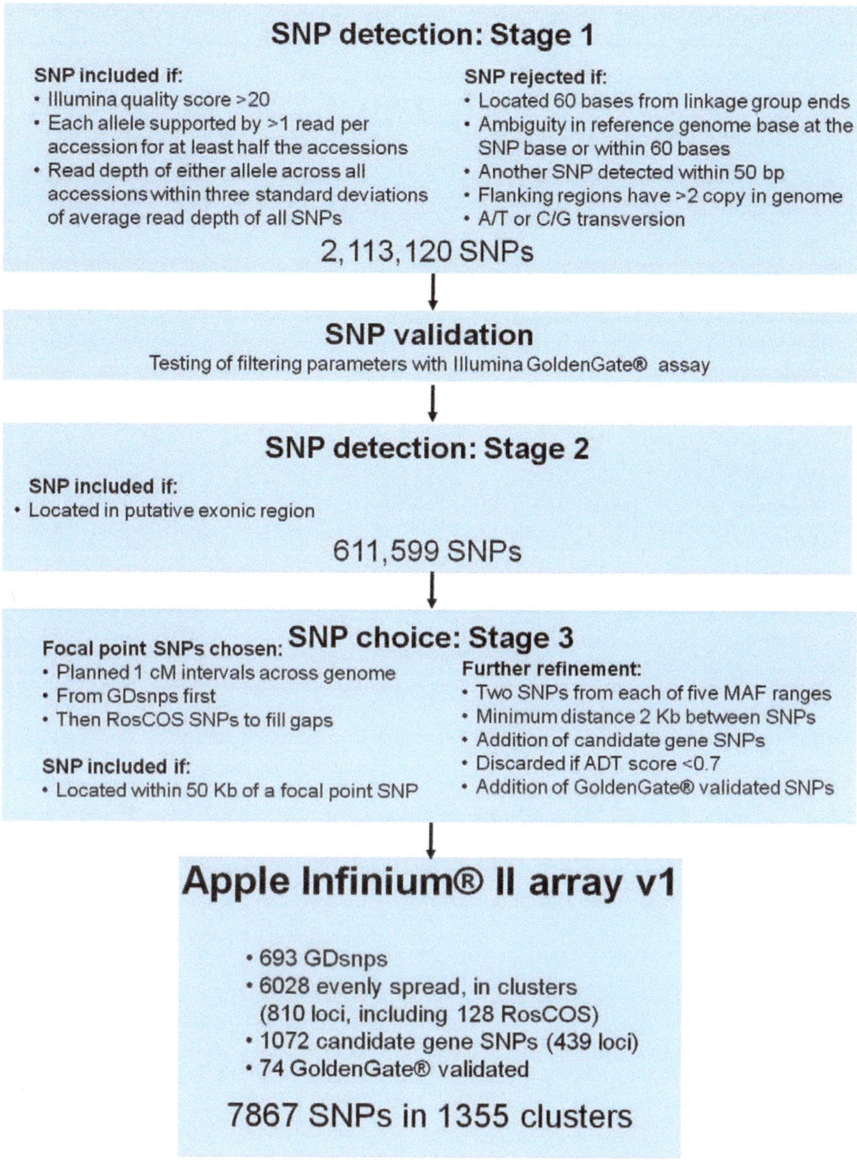

Figure 1. Workflow for single nucleotide polymorphism (SNP) detection, validation, and final choice employed for development of the IRSC apple 8K SNP array v1. GDsnp: 'Golden Delicious'-validated SNP; RosCOS: Rosaceae Conserved Orthologous Set; MAF: Minor Allele Frequency; ADT: Assay Design Tool.

'AGDP', 'AGDL', 'AGDR', 'AGDS', 'AGDW', 'AHAA', 'AOFA', 'AVBB', 'ASYA', 'AYFB', AYFA') to the reference genome using *gmap* [18] with command line options -B 2 -K 1000 -L 50000 -f 2 -O.

SNP validation with the GoldenGate® assay

A subset of 144 SNPs was chosen to validate the efficiency of SNP detection and fine-tune the filtering parameters (Figure 2). The initial selection was of 100 Stage 1 SNPs evenly spread across the apple genome, i.e., the 17 LGs representing the 17 haploid chromosomes of apple, according to the primary assembly of 'Golden Delicious' [11]. These included a SNP located in the first 200 kb from each end of each LG. SNPs between these were evenly spaced along the LG every 6-8 Mb, depending on LG length. Spacing across all LGs averaged 7.26 Mb (standard deviation of 0.66 Mb), with a maximum average of 8.01 Mb (±0.07 Mb) for LG2 and a minimum average of 5.84 Mb (±0.36 Mb) for LG16. Twenty more SNPs were targeted to a 1.4 Mb region near the top of LG16, spanning a major trait locus associated with malic acid content in fruit (*Ma* locus; [19]). These *Ma* locus SNPs were spaced an average of 75 Kb apart (ranging from 5 Kb to 186 Kb). It was planned that approximately 20% of the 144 SNPs would be accession-specific, i.e., their minor allele would be detected in only one re-sequenced accession of the detection panel. As only four of the evenly spread SNPs met this criterion, 24 additional SNPs were chosen as accession-specific. These latter SNPs were spread over the genome at 1–2 per LG but otherwise randomly distributed on each LG. Approximately 20% of the 144 SNPs were allocated to meet the criterion of being within exons of candidate genes for fruit quality and plant architectural traits. These included 24 of the evenly spread SNPs, four of the *Ma* locus SNPs, and one of the accession-specific SNPs,

Table 1. Apple cultivars used for low coverage re-sequencing and subsequent single nucleotide polymorphism detection.

Accession	Number of reads	Type of reads	Estimated genome coverage (X)	Location of sequencing
'Braeburn'	34,699,040	Paired-end	6.9	PFR
'Co-op 15'	51,864,610	Single-end	5.2	ARC
'Cox's Orange Pippin'	35,120,961	Paired-end	7.0	PFR & ARC
'Crimson Crisp'	6,128,575	Paired-end	1.3	RosBREED
'Cripps Pink'	56,738,286	Single-end	6.1	ARC
'Delicious'	17,995,052	Paired-end	1.9	RosBREED
'Dolgo'	59,863,282	Single-end	6.4	ARC
'Duchess of Oldenburg'	17,762,643	Paired-end	3.8	RosBREED
$F_2$26829-2-2	31,580,732	Paired-end	6.7	RosBREED
'Frostbite'	44,740,552	Single-end	4.5	ARC
'Fuji'	44,545,764	Single-end	4.8	ARC
'Geneva'	31,977,525	Paired-end	6.4	PFR
'Golden Delicious'	22,101,159	Paired-end	4.7	RosBREED
'Granny Smith'	36,112,204	Paired-end	7.2	PFR
'Haralson'	15,499,148	Paired-end	3.3	RosBREED
'Honeycrisp'	20,887,451	Paired-end	4.5	RosBREED
'Idared'	33,479,444	Paired-end	6.7	PFR
'James Grieve'	26,005,304	Paired-end	5.2	PFR
'Jonathan'	18,812,375	Paired-end	4.0	RosBREED
'McIntosh'	59,934,507	Paired-end	12.8	RosBREED
Malus sieversii PI613981	23,500,375	Paired-end	4.7	USDA-ARS
'Ralls Janet'	35,485,026	Paired-end	7.1	PFR
'Red Delicious'	25,993,982	Paired-end	5.2	PFR
'Red Dougherty'	32,718,666	Paired-end	6.5	PFR
'Rome Beauty'	23,220,046	Paired-end	5.0	RosBREED
'Splendour'	36,353,918	Paired-end	7.3	PFR
'Zestar'	54,404,792	Single-end	5.4	ARC

An approximate genome coverage was estimated for each accession using a genome size of 750 Mb. The source of cultivars and numbers of Illumina GA II reads obtained are indicated. ARC: Agricultural Research Council, South Africa; PFR: The New Zealand Institute for Plant & Food Research Ltd; RosBREED: U.S.-based international project.

totaling 27 exonic candidate gene SNPs. The 144 SNPs also deliberately included a wide range of minor allele frequencies (MAFs) and a wide range of Illumina Assay Design Tool (ADT) scores. Approximately 5% of SNPs were chosen from those previously validated in 'Golden Delicious' (GDsnp) using the SNPlex technique [11] and these included seven evenly spread SNPs and one *Ma* locus SNP, to serve as positive controls.

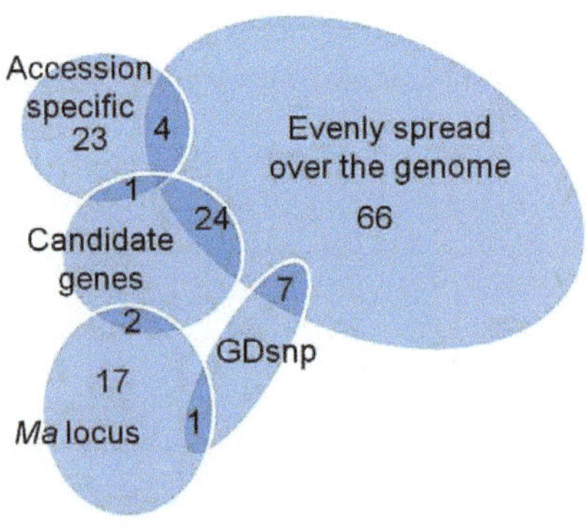

Figure 2. Classification of the 144 apple single nucleotide polymorphisms (SNPs) used for validation using the Illumina GoldenGate® assay.

To test the above SNP parameters for their effects on genotyping efficiency, a validation panel of 160 apple accessions (Table S1) was screened with the 144 SNP subset, using the GoldenGate® assay. Individuals in the validation panel are founders, intermediate ancestors, and important breeding parents of modern apple cultivars, forming a complex pedigree structure linking much of the world's cultivated apple crop and incorporating the 27 accessions of the SNP detection panel. The validation panel also included bin-mapping sets of approximately eight seedlings each for: 'Golden Delicious'×'Anna' [20], 'Malling 9'×'Robusta 5' [21], 'Prima'×'Fiesta' [19], 'Royal Gala'×'Braeburn' [11], and 'Telamon'×'Braeburn' [22]. Genomic DNA was purified from each accession using the E-Z 96 Tissue DNA Kit (Omega Bio-Tek, Inc., Norcross, USA). DNA was quantitated with the Quant-iT™ PicoGreen® Assay (Invitrogen, Carlsbad, USA), using the Victor multiplate reader (Perkin Elmer Inc., San Jose, USA). Concentrations were adjusted to a minimum of 50 ng/μl in 5 μl aliquots and were submitted to the Research

Technology Support Facility at Michigan State University (http://rtsf.msu.edu/illumina-beadxpress-reader-system), where the GoldenGate® assay was performed following the manufacturer's protocol (Illumina Inc., San Diego, USA). Following amplification, the PCR products were hybridized to VeraCode microbeads via the address sequence for detection on a VeraCode BeadXpress Reader. SNP genotypes were scored with the Genotyping Module of the GenomeStudio Data Analysis software (Illumina Inc.).

SNP final choice for 8K Infinium® II array

A clustering strategy was devised that would evenly span the genetic map of apple with clusters of exonic SNPs, in order to provide a final SNP genome scan with the capability of determining SNP haplotypes at distinct loci. The design featured focal points at approximately 1 cM intervals, with 4–10 SNPs clustered at each focal point. Choice of focal points began with GDsnps [11] according to their location in the IASMA-FEM 'Golden Delicious'×'Scarlet' reference genetic map (http://genomics.research.iasma.it/cgi-bin/cmap/viewer), followed by Rosaceae Conserved Orthologous Set (RosCOS; [23]) loci to fill genetic gaps between GDsnps, according to physical map location [24]. Chromosome ends, the first 200 Kb of each LG according to the apple draft genome pseudo-chromosomes [11], were also targeted. SNPs not located within ±50 Kb of a focal point SNP were then discarded. Within this pool of candidates, SNPs were binned according to MAF (bins of 0.1, 0.2, 0.3, 0.4, and 0.5 corresponding to MAF ranges of 0.01–0.1, 0.101–0.2, and so on). Two SNPs were chosen from each bin for each cluster, to give a total of up to ten SNPs per cluster (including the focal point SNP). To minimize detection of redundant haplotypes within clusters in subsequent genotyping, the minimum distance between SNPs was set at 2 Kb. Additional SNP clusters were developed from candidate genes for fruit quality, tree architecture, and flowering that were chosen using the available literature [25,26,27]. The best hits from a BLAST search of these candidate genes against the apple "Consensus CDS Peptides", available from the GDR website (http://www.rosaceae.org), created a list of homologous apple sequences that were then located in the apple genome and became new focal points. Stage 2 SNPs identified in these genes were subjected to Stage 3 filtering criteria, except that the "2 Kb minimum distance" rule was relaxed so that multiple SNPs could be chosen within a gene to form a cluster. Instead, up to four SNPs within each candidate gene were manually chosen to lie as far from each other as possible. Finally, SNPs with an ADT score below 0.7 were discarded.

SNP array evaluation and cluster file development

A set of populations from various crosses, as well as accessions from the apple germplasm, were used to evaluate the Apple 8K Infinium® II array. This set included a 'Royal Gala'×'Granny Smith' F_1 population of 186 seedlings [28], seven controlled F_1 crosses that are used as a training population for genomic selection at Plant & Food Research and comprise 1313 individuals [29], and a set of 117 accessions from the Plant & Food Research germplasm collection (S. Kumar, unpublished). Genomic DNA (gDNA) was extracted using the NucleoSpin® Plant II kit (Macherey-Nagel GmbH & Co KG, Düren, Germany), and quantitated using the Quant-iT™ PicoGreen® Assay (Invitrogen). 'Royal Gala', 'Granny Smith', and 'Golden Delicious' were used as controls. Two hundred nanograms of gDNA were used as template for the reaction, following the manufacturer's instructions. SNP genotypes were scored with the Genotyping Module of the GenomeStudio Data Analysis software (Illumina Inc., San Diego, CA). Individuals with low SNP call quality (*p50GC*<0.54), as well as seedlings putatively resulting from an unintended pollination, were removed from the analysis. SNPs with a *GenTrain* score >0.6 were retained and those with scores ranging between 0.3 and 0.6 were visually checked for accuracy of the SNP calling. Clusters were manually edited when the parent-offspring segregation was not correct, or when the number of missing genotypes was greater than 20.

Two trios with both parents and one seedling were used to test the usefulness of SNP clusters for identifying haplotypes: 'Royal Gala'×'Braeburn'−>'Scifresh' and '(Royal) Gala'×'Splendour'−>'Sciros'. SNPs were coded using A and B alleles and haplotypes were inferred using FlexQTL™ (www.flexqtl.nl) for each cluster of SNPs.

Results

NGS re-sequencing of apple breeding accessions and SNP detection

A total of 67 Gb of DNA sequence from 898 million 75-base reads were generated for the 27 *Malus* accessions (Table 1). The total number of reads per accession ranged from six to 60 million, for 'Crimson Crisp' and 'McIntosh', respectively. The difference was due to an increase in the cluster density and resulting sequencing yield of the Illumina GA instrument over the period that sequencing was performed. A total of 10,915,756 SNPs was identified using *SoapSNP*, of which 2,113,120 (19.5%) passed the filtering criteria for Stage 1 detection (Table 2, Figure 1). The average SNP frequency was one per 288 bp. Of these SNPs, 611,599 (28.9%) were predicted to be located in exonic regions and passed the Stage 2 filter (Table 2, Figure 1). When the SNP calls based on re-sequencing data for the two independently sequenced 'Cox's Orange Pippin' samples were compared, an average of 21.3% of detected SNPs gave different genotypes between the two samples.

SNP validation

In the GoldenGate® validation assay, 148 apple accessions in the test panel gave good quality scores (call rate >0.8 and 10%GC Score >0.5) and 12 accessions failed because of poor DNA quality. Seventy-three (50.7%) of SNPs were polymorphic, 46 (31.9%) had failed reactions, and 25 (17.3%) were monomorphic (MAF <0.05 or A/B frequency <0.1) (Table 3). All eight GDsnps were polymorphic. Evenly spread SNPs and SNPs at the *Ma* locus had similar success rates to the overall set (54% and 55% polymorphic SNPs, respectively). SNPs located within candidate genes were generally more successful than the overall set (63% polymorphic). Of the 28 SNPs chosen on the basis of their accession specificity, 14 (50%) had a MAF <0.05, of which 11 were not monomorphic. The Illumina ADT score calculated for each SNP did not significantly influence the success of the SNPs (Table 3).

SNP final choice

The third stage of filtering, which involved designating focal points at approximate 1 cM intervals over the apple genome and choosing SNPs in 100 Kb windows around them (Figure 3), was initially based on 712 suitable GDsnp markers. These markers left 107 gaps in the genome greater than 3 cM and these gaps were filled using two SSRs and 128 RosCOS as additional focal points, giving a total of 842 focal points. By this stage, 6074 SNPs had been chosen in addition to the initial 712 GDsnps. Next, a further 5528 SNPs were identified in candidate genes and narrowed down to 1652 SNPs after removing putative redundancies. Filtering of the 8438 SNPs chosen by this stage, using the ADT score, reduced the pool to 7793 SNPs: 693 GDsnps, 6028 SNPs around focal points, and 1072 SNPs within candidate genes (Figure 1). Finally,

Table 2. Single nucleotide polymorphisms (SNPs) detected across the apple genome by re-sequencing 27 apple accessions.

Chromosome	Sequence used for SNP detection (bp)	SNPs examined from *SoapSNP*	SNPs passing "Stage 1" filtering	Average distance between SNPs (bp)	Exonic SNPs
1	36,084,648	556,642	102,825	350	31,678
2	40,172,783	793,583	144,715	277	42,496
3	39,907,579	700,701	134,595	296	36,005
4	25,411,901	474,397	91,212	278	27,540
5	37,603,833	661,644	134,672	279	38,400
6	30,670,413	514,972	97,014	316	29,450
7	31,181,013	546,126	117,652	265	33,155
8	35,800,717	625,020	114,173	313	31,946
9	37,514,065	737,124	146,472	256	41,935
10	38,388,612	687,313	146,102	262	43,522
11	40,097,013	751,789	146,718	273	42,681
12	36,276,268	664,326	121,739	297	35,456
13	39,686,055	720,281	158,912	249	43,699
14	34,156,235	607,215	114,259	298	32,712
15	55,775,419	919,506	159,368	349	46,218
16	23,462,870	421,197	85,530	274	24,357
17	27,122,502	533,920	97,162	279	30,349
Total	609,311,926	10,915,756	2,113,120	288	611,599

the 74 validated SNPs from the GoldenGate® assay were included, to provide a grand total of 7867 SNPs (Table S2) in 1355 clusters for construction of the final apple Infinium® II SNP array, officially named the International RosBREED SNP Consortium (IRSC) apple 8K SNP array v1.

The IRSC apple 8K SNP array v1

The 7867 SNPs were uniformly represented on the 17 apple LGs (Table 4). The number of SNP clusters ranged from 64 to 113 on LGs 6 and 2, respectively, with an average of 5.8 SNPs per cluster. The average physical distance between SNP clusters ranged from one cluster every 316.2 kb to 538.3 kb on LG 16 and 3, respectively. The overall cluster density on the 'Golden Delicious'×'Scarlet' reference genetic map was one cluster every centiMorgan, with only five gaps between clusters larger than 10 cM.

IRSC apple 8K SNP array v1 evaluation

Evaluation of the IRSC apple Infinium® II 8K array using 1619 individuals, including individual accessions and segregating populations, yielded 7692 successful beadtypes (97.7%) of which 5554 (72.2%) were polymorphic (Table 4). The remaining 2138 SNPs (27.8%) exhibited poor quality genotype clustering or were monomorphic. Numbers of polymorphic SNPs per LG ranged

Table 3. Results from a GoldenGate® assay of 144 single nucleotide polymorphisms (SNPs) screened over 160 apple accessions (Table S1).

		Proportion of SNPs			
SNP type	Total	Failed	Mono-morphic	Polymorphic (MAF<0.05)	Polymorphic (MAF>0.05)
Evenly spread	100	0.32	0.08	0.04	0.56
Accession-specific	28	0.25	0.11	0.39	0.25
Candidate genes	27	0.22	0.11	0.04	0.63
Ma locus	20	0.40	0.00	0.05	0.55
GDsnp	8	0.00	0.00	0.00	1.00
ADT score <0.8	40	0.35	0.00	0.05	0.60
ADT score 0.8–0.9	35	0.40	0.06	0.03	0.51
ADT score <0.9	69	0.26	0.12	0.17	0.45
Mean ADT score (s.d.)	0.85 (0.13)	0.83 (0.12)	0.94 (0.04)	0.92 (0.10)	0.83 (0.14)
Total	144	0.32	0.07	0.10	0.51

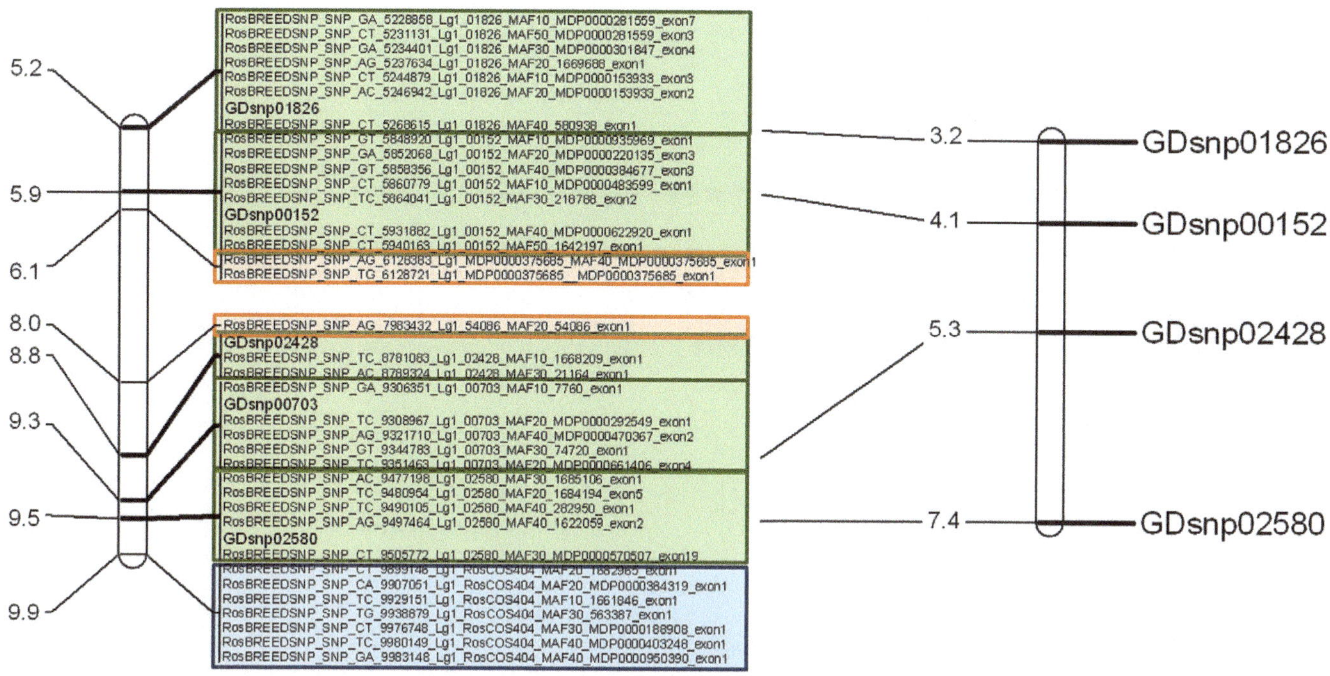

Figure 3. Detailed view of a genomic region at the top of Linkage Group 1 showing SNPs chosen for the International RosBREED SNP Consortium (IRSC) apple 8K SNP array v1. Physical map locations of SNPs (left; in megabases) are compared with known genetic map locations of SNP markers developed from 'Golden Delicious' (GDsnp in centiMorgans). SNP clusters around focal points are boxed, with green boxes denoting GDsnps, blue boxes denoting Rosaceae Conserved Orthologous Set markers (RosCOS), and orange boxes denoting candidate genes.

between 519 and 215 on LG2 and LG7, respectively and the number of clusters per LG varied according to LG genetic length and ranged from 62 to 108. Polymorphic SNPs were observed for 1190 (93.7%) of the attempted clusters. The GenomeStudio® *GenTrain* scores for the polymorphic SNPs ranged from 0.35 to 0.92 and the call frequency ranged from 0.73 to 1. In total, 1061 polymorphic SNPs had a very low MAF of between 0 and 0.05, while 3371 polymorphic SNPs had MAF>0.05 with a call frequency >0.95 and reliability score (50% GC) >0.4. Of the previously validated GDsnps and GoldenGate®-validated SNPs, 629 (90.7%) and 55 (74.3%), respectively, were polymorphic. In the 'Royal Gala'×'Granny Smith' segregating population, 2810 SNPs in 994 clusters were polymorphic. Of these, 2223 SNPs (79.2%) were heterozygous for one parent and homozygous for the other (pseudo-testcross configuration) and the remaining 587 (20.8%) were heterozygous for both parents. Polymorphic pseudo-testcross SNPs in this population were successfully used as independent markers for linkage map construction (Figure 4). The average distance between markers was 0.88 cM and 0.91 cM for 'Royal Gala' and 'Granny Smith', respectively, and the largest gap between informative clusters was 24.3 cM for 'Royal Gala' and 25 cM for 'Granny Smith'. However, the marker density could be increased in these large gaps by adding markers heterozygous for both parents.

The small region of LG1 presented in Figure 3 spanned 25 polymorphic markers in two trios with both parents and one progeny (Figure 5): 'Royal Gala'×'Braeburn'−>'Scifresh' and '(Royal) Gala'×'Splendour'−>'Sciros'. The haplotypes inherited from each parent could be inferred for the eight SNP clusters, with three of the clusters having four haplotypes. One putative recombination event was detected in the '(Royal) Gala'×'Splendour'−>'Sciros' trio between the second last and last two clusters.

Discussion

A SNP-based strategy for apple genome scanning

An international community of apple genomics, genetics, and breeding researchers recently united with the goal of developing high-resolution genome scanning capability for germplasm of cultivated apple. The resulting medium to high-throughput multiplexed SNP assay presented and evaluated here is the culmination of a collective technical effort and shared resources that would have been beyond the means of any single entity. This publicly available genomics resource will provide unprecedented resolution for discovery of marker-locus-trait association discovery, for description of the genetic architecture of quantitative traits, for investigation of genetic variation (neutral and functional), and will enable genomic selection in apple. To ensure maximum utility, the apple SNP array involved several key design features.

To ensure that genome scans using the SNP array relate directly to the draft apple genome sequence of [11], priority was given to including on the array previously validated and genetically mapped SNPs from 'Golden Delicious' that had been used to anchor the genetic and physical maps of apple. GDsnps also provided internal controls for validation of SNPs with the GoldenGate® assay. The series of previously validated GDsnps [7,11] as a whole performed better than the new series of SNPs in both the GoldenGate® assay and the apple Infinium® 8K array evaluations (100% and 90.7% success rates, respectively). This result underscores that prior validation is useful when developing a high-throughput SNP array, to minimize the proportion of attempted SNPs that are not informative. Future SNP arrays for apple will be firmly based on the SNPs validated here with the first apple Infinium® array.

Genetic and physical distribution of SNPs across the apple genome was also a strong consideration in array design. The whole

Table 4. Apple Infinium® II array v1 content and evaluation in the Plant & Food Research dataset. The number of attempted SNPs and clusters is indicated per linkage group (LG), as well as their density based on the apple genome assembly in kilobases (Kb) and the 'Golden Delicious' genetic map (cM).

	Attempted SNPs on IRSC Apple Infinium II array v1				Array evaluation in the Plant & Food Research dataset						
LG	No. of SNPs	No. of clusters	Average physical distance between focal points (kb)	Average genetic distance between focal points (cM)	No. of successful beadtypes	No. of poly-morphic SNPs	% poly-morphic SNPs	No. of poly-morphic SNPs with MAF>0.05	MAF>0.05, 50 GC>0.4, call rate >0.95	No. of poly-morphic clusters	% poly-morphic clusters
1	443	81	443.7	1.10	434	299	68.9	250	180	75	92.6
2	695	113	355.1	0.75	684	519	75.9	436	313	108	95.6
3	499	74	538.3	1.19	487	350	71.9	288	212	72	97.3
4	392	65	388.7	1.00	386	292	75.6	235	201	63	96.9
5	492	82	456.9	1.10	486	360	74.1	308	239	76	92.7
6	351	64	473.2	1.17	340	233	68.5	194	117	63	98.4
7	345	71	438.5	0.83	340	215	63.2	181	129	62	87.3
8	405	68	519.9	1.07	399	276	69.2	229	182	62	91.2
9	490	80	449.2	0.91	477	364	76.3	311	232	76	95.0
10	554	83	446.6	1.12	531	377	71.0	271	224	79	95.2
11	466	80	500.7	0.97	456	322	70.6	277	177	73	91.3
12	469	85	426.5	0.87	459	338	73.6	258	210	81	95.3
13	433	81	489.6	0.94	423	299	70.7	229	178	76	93.8
14	388	72	472.2	1.02	374	266	71.1	215	151	64	88.9
15	629	106	525.6	1.09	621	440	70.9	365	253	101	95.3
16	355	74	316.2	0.76	347	271	78.1	215	158	69	93.2
17	461	76	356.7	0.99	448	333	74.3	231	215	70	92.1
Total	7867	1355	446.2	1.00	7692	5554	72.2	4493	3371	1270	93.7

The Apple Infinium® array was evaluated by screening 1398 progeny in 8 segregating populations and 117 individuals from the Plant & Food Research apple germplasm collection.

genome sequence of apple [11] was a critical resource that greatly increased the likelihood that the final array would cover the majority of the apple genome. Organizing each SNP according to genetic and physical location enabled efficient genome saturation, helped to avoid redundancy and spanned gaps that would have occurred if a collection of random SNPs had been employed. Attention was given to ends of chromosomes; however, not beyond the edges of a genetically defined linkage group. Our genetic positioning of focal points relied on the assembly of the apple genome sequence into linkage groups by [11] that involved anchoring of metacontigs to a genetic map comprising 1643 SNP loci. Errors in this assembly will reduce the effectiveness of the IRSC apple 8K SNP array v1 to span the apple genome genetically. As the number of high quality (MAF>0.05, *p50GC* >0.4, call rate >0.95) polymorphic markers provided by the array developed here is greater than that was used for the genome assembly (3371 vs. 1643), the new set of SNPs from the array will be useful for improvement of the apple genome assembly.

A haplotype-targeting strategy was adopted to maximize information gained using the SNP array and to decrease computation time for mapping software. Clusters of multiple SNPs were spaced at 1 cM intervals (equivalent to an average of 446.2 kb, [11]) rather than evenly spreading independent loci (approximately every 0.17 cM or 75 Kb). As bi-allelic SNPs are often less informative than co-dominant multi-allelic markers such as SSRs, we clustered four to 10 SNP markers within a small, essentially non-recombining, physical distance, i.e., ±50 Kb around focal points. Individual SNPs within clusters were spaced at a minimum of 2 Kb to reduce redundancy of haplotype capture. This strategy will enable combination of the information from individual SNPs into haplotypes that will provide fully informative and easy to handle multi-allelic markers and capture all unique haplotypes within genotyped germplasm. A preliminary analysis using two parent-child trios indicated that our clustering strategy is capable of facilitating the identification of haplotypes within and among SNP clusters. Nevertheless, it was recently observed that the 'Golden Delicious' genome assembly contains some erroneous gene position assignments as reported recently for a set of candidate genes for flowering [30] and RosCOS markers [31]. Therefore, it is advisable to perform independent linkage analysis of polymorphic SNPs to ensure that they locate in the expected cluster, to avoid haplotype construction using SNPs that are not truly physically linked. Use of a consensus map of fully informative markers is more powerful for QTL analysis than partially informative markers, as all alleles of each marker locus can be simultaneously evaluated for their phenotypic effects. A highly saturated genome of fully informative markers is also highly amenable to integrated QTL analysis over multiple pedigree-connected families of variable size with the FlexQTL™ software (www.flexqtl.nl), which supports the Pedigree-Based Analysis

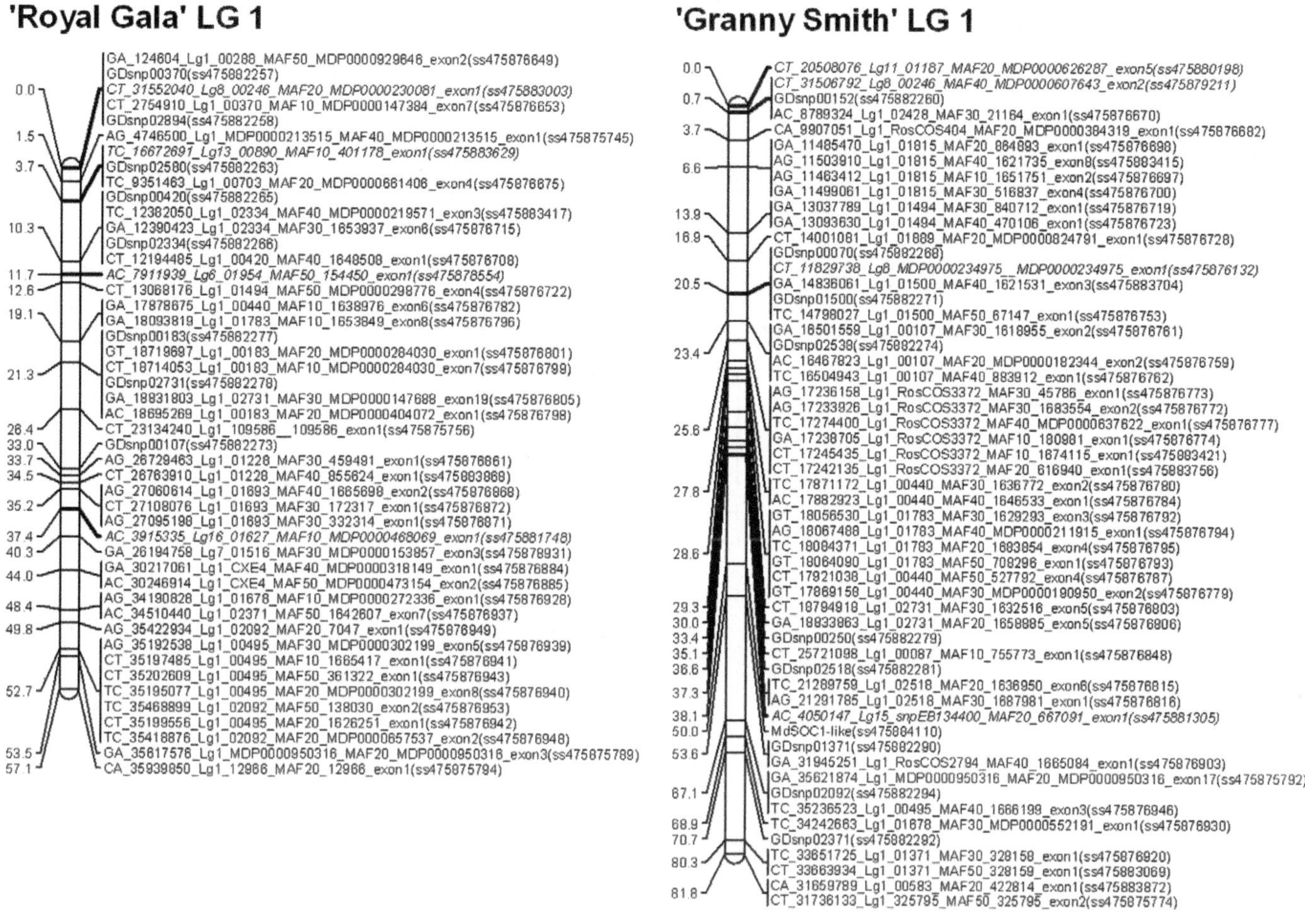

Figure 4. Genetic maps of linkage group (LG) 1 constructed using the 'Royal Gala'×'Granny Smith' segregating population. Marker names indicate the type of SNP, physical location on the 'Golden Delicious' genome assembly (in base pair), cluster marker, minor allelic frequency (MAF) in the 27 re-sequenced accessions, gene model, position in the predicted gene model, and NCBI dbSNP accession number (in parentheses).

approach [32]. We chose GDsnps as focal points for clusters because of their established genetic resolution of 1 cM. RosCOS loci were used because their genetic locations were known and they provide an efficient orthologous marker system for the assessment of comparative genome synteny across Rosaceae family genera [24]. Finally, candidate genes were included on the array because of their putative enhanced capacity for association of detected genetic variation with horticultural trait variation.

Our principal source of SNPs was from whole-genome resequencing of apple accessions for which pedigree connections were considered, in order to represent founder contributions to breeding germplasm efficiently and thereby increase the probability of detecting informative haplotypes segregating in breeding programs. Previously available SNP markers for apple were derived from the 'Golden Delicious' genome sequence [11] and while 'Golden Delicious' is a common founder in pedigrees of cultivated apple, it nevertheless accounts for only a fraction of potentially unique sources of genetic variation among modern apple cultivars [7]. Most accessions in the SNP detection germplasm set were included because they represent the most common founders in the ancestry of cultivated apple. The recent-generation cultivars 'CrimsonCrisp' and Co-op 15 have missing links in their ancestry and were included in the sequencing panel to help to capture informative haplotypes in current breeding programs.

To obtain sufficient sequencing depth for each accession, we chose the NGS technique provided by Illumina GA II instrumentation. Because of its high information output and relative affordability, such high-throughput sequencing by synthesis has revolutionized agricultural and horticultural genomics, enabling sequencing of complete genomes of plant species including several crops of the Rosaceae family: apple, peach, and strawberry. Further to these physical mapping successes, sequencing by synthesis has revealed detailed information on sequence polymorphism within species of interest [9,33,34,35,36].

The sequencing of 'Cox's Orange Pippin' at two locations enabled us to compare SNP detection in independent samples. We believe that the high proportion of SNPs with non-matching genotypes between the 'Cox's Orange Pippin' samples was due to low sequence coverage providing insufficient reads at a locus to support detection of both alleles at heterozygous SNP loci. Two strategies can be used for SNP detection using NGS. Our strategy was to sequence a wide range of cultivated apple accessions at relatively low coverage to capture as many breeding-relevant SNPs as possible. An alternative approach would have been to re-sequence fewer accessions at higher coverage to enable prediction of SNP genotypes more correctly for each accession. However, this latter strategy would have reduced opportunities to identify rare haplotypes carried by some breeding lineages. The sequences of 27 apple accessions generated in this study are an important resource

SNP marker name	Position in genome (Mb)	'Royal Gala'		x 'Braeburn'		-> 'Scifresh'		'(Royal) Gala'		x 'Splendour'		-> 'Sciros'	
		H3	*H1*	*H2*	*H4*	*H1*	*H2*	*H3*	*H1*	*H3*	*H2*	*H1*	*H3*
ss475876655		B	B	A	B	B	A	B	B	B	A	B	B
ss475876657	5.2	B	B	B	B	B	B	B	B	B	B	B	B
ss475876658		A	A	A	A	A	A	A	A	A	A	A	A
ss475876660		B	A	A	A	A	A	B	A	B	A	A	B
ss475882259 (GDsnp01826)		B	B	B	B	B	B	B	B	B	B	B	B
		H1	*H1*	*H2*	*H2*	*H1*	*H2*	*H1*	*H1*	*H2*	*H1*	*H1*	*H2*
ss475876662		B	B	B	B	B	B	B	B	B	B	B	B
ss475876664	5.8	A	A	A	A	A	A	A	A	A	A	A	A
ss475876666		A	A	A	A	A	A	A	A	A	A	A	A
ss475882260 (GDsnp00152)		A	A	B	B	A	B	A	A	B	A	A	B
ss475875746	6.1	B	A	B	B	A	B	B	A	B	A	A	B
ss475875748	7.9	A	A	A	B	A	A	A	A	A	A	A	A
		H1	*H1*	*H2*	*H2*	*H1*	*H2*	*H1*	*H1*	*H1*	*H2*	*H1*	*H1*
ss475882261 (GDsnp02428)		B	B	B	B	B	B	B	B	B	B	B	B
ss475876669	8.7	A	A	A	A	A	A	A	A	A	A	A	A
ss475876670		A	A	B	B	A	B	A	A	A	B	A	A
		H2	*H1*	*H1*	*H1*	*H1*	*H1*	*H2*	*H1*	*H1*	*H2*	*H1*	*H1*
ss475882262 (GDsnp00703)		A	A	A	A	A	A	A	A	A	A	A	A
ss475876675	9.3	B	A	A	A	A	A	B	A	A	B	A	A
		H3	*H1*	*H2*	*H2*	*H1*	*H2*	*H3*	*H1*	*H1*	*H4*	*H1*	*H1*
ss475876678		A	A	B	B	A	B	A	A	A	B	A	A
ss475876679	9.4	A	A	B	B	A	B	A	A	A	B	A	A
ss475882263 (GDsnp02580)		B	A	A	A	A	A	B	A	A	B	A	A
ss475876680		B	B	A	A	B	A	B	B	B	A	B	B
									X				
		H4	*H1*	*H2*	*H3*	*H1*	*H2*	*H4*	*H1*	*H2*	*H4*	*H4*	*H2*
ss475876681		B	A	B	B	A	B	B	A	B	B	B	B
ss475876682	9.9	B	B	B	B	B	B	B	B	B	B	B	B
ss475876684		B	B	A	A	B	A	B	B	A	B	B	A
ss475876685		B	B	A	B	B	A	B	B	A	B	B	A
ss475876687		B	B	A	A	B	A	B	B	A	B	B	A

Figure 5. Example of the usefulness of the International RosBREED SNP Consortium (IRSC) apple 8K SNP array v1 for developing haplotypes. SNP markers were the same as represented in the example of Figure 3. Haplotypes inferred using FlexQTL™ for each cluster of SNPs are numbered from *H1* to *H4*). Haplotypes inherited by the progeny from the parents are boxed and colored coded for each parent. A putative recombination is indicated by a cross for the '(Royal) Gala'×'Splendour'–>'Sciros' trio.

for apple geneticists, providing a comprehensive inventory of point mutations potentially underlying genetic variation for important horticultural traits beyond their use here in development of the 8K SNP array. The strategy of re-sequencing higher numbers of accessions at low coverage proved successful, as shown by evaluation of both the GoldenGate® and Infinium® arrays. While the *in silico* genotypes of the 27 accessions based on whole-genome sequencing are not always correct, most of the predicted SNPs are real and were converted into SNP markers in the final Infinium® II array, to provide more than 5500 polymorphic markers in a diverse breeding germplasm set.

To enhance the capacity to identify and characterize genetic control regions for horticultural traits – a common intended use of the array – hundreds of candidate genes were built into the array design. Many of these candidate genes have the advantage of being well studied by researchers [25,37,38,39,40,41,42], thus providing loci of known or readily interpretable genetic variation of immediate interest to genome scan users and a relatively fast path to practical application. Similarly, the choice of exonic SNPs for the array biases the genome scan to the preferred target of the gene space of apple genomes. Although the bias toward exonic SNPs was initially only a guide based on performance in the

GoldenGate® array, the sheer number of SNPs available from the detection step allowed the final design to feature only those SNPs putatively located in exons. However, it is unlikely that the SNP array will span causative mutations for all characters, as such mutations are often located in non-coding regulatory regions, such as the mutation demonstrated to control red pigmentation of the apple fruit flesh and foliage, which is located in the promoter region of *MYB10* [39].

SNP filtering and validation

A set of 7867 apple SNPs corresponding to 1355 clusters was employed to construct the IRSC apple 8K SNP array v1. These SNPs were chosen from more than 2 million SNPs detected across the apple genome, utilizing 67 Gb of NGS data from 27 apple accessions. The number of SNPs detected was more than 250 times as many as needed for array construction, allowing very fastidious filtering. Firstly, some SNP types were filtered out because of the choice of the technology for the final array, with C/G and A/T transversions being rejected as they require two Infinium® II probes for detection. Further decisions on the filtering criteria used to develop the final SNP array were made on the basis of empirically determined thresholds obtained from a GoldenGate® validation assay of a representative subset of 144 SNPs, in order to optimize the success rate for screening apple breeding germplasm. The success rate in the validation assay varied according to the type of SNPs and their parameters.

The main reason for choosing exonic SNPs for construction of the Infinium® II array was the observation that randomly distributed SNPs tended to perform more poorly in the GoldenGate® than SNPs located within coding regions. This imbalance in performance might be explained by a lower number of nucleotide polymorphisms in exonic regions, increasing stringency of hybridization between the array features and target genomic DNA. The presence of undetected SNPs in the sequences flanking genes has been observed to reduce both the SNP success rate [43,44] and the rate of SNP conversion [12]. This finding represents a major concern for those developing high-throughput SNP screening tools in highly heterozygous genomes, such as the forest tree species genomes [12,43]. Based on recent reports, apple is among the most genetically polymorphic agricultural species analyzed to date. For example, [5] estimated a frequency of 1 SNP per 149 bp based on public apple ESTs, while [7] identified an average of 1 SNP per 455 bp in a single cultivar and 1 SNP per 52 bp across germplasm. This can be compared with figures ranging from 1 SNP every 61 bp in maize [45], 1 SNP/117 bp in grapevine (based on the 'Pinot noir' genome sequence; [46]) or 1 SNP/64–104 bp (based on multi-locus cultivar analysis [46,47], to 1 SNP/5,700 bp in japonica rice cultivars [48]. To address this very specific issue, SNPs were discarded from consideration during the filtering step, whenever another SNP was detected within 50 bases. Nevertheless, because of the low sequencing coverage for each of the 27 accessions, it is probable that many flanking SNPs have not been detected.

MAF influenced the success of SNPs in the GoldenGate® assay, where SNPs with low MAF based on the NGS detection were less likely to be polymorphic than SNPs with higher MAF. Our strategy of choosing SNPs exhibiting a range of MAFs within clusters evenly spaced across the apple genome will help to capture available haplotypes with high-MAF SNPs (i.e., MAF>0.4), improving detection of heterozygosity for any assayed germplasm individual, while lower-MAF SNPs (MAF <0.2), tend to come from specific founders and improve identification of rare haplotypes.

IRSC apple 8K SNP array v1 evaluation

Based on information derived from a consensus apple genetic map of 1343 cM (M. Troggio, unpublished), the IRSC apple 8K SNP array v1 corresponds to an average density of one focal point every 1 cM. Overall, the IRSC apple 8K SNP array v1 was successful, as 5554 markers were polymorphic in the examined germplasm set, with 4368 SNPs having a MAF>0.05, and 3371 having a high call rate, reliability score, and MAF. While this result indicates that at least 3371 markers are likely to provide a genotype when the array is used for screening any germplasm set, any given genetic mapping experiment can expect to obtain ~4000 polymorphic bi-allelic markers or ~1000 fully informative haplotypes. The number of SNPs in the array that can successfully provide a genotype might be improved by enhancing DNA quality of the template used for the experiment. In our array evaluation, the set of accessions tested was from a breeding program where the number of individuals was maximized at the expense of the DNA quality.

We have demonstrated that the array is effective for use in genetic mapping, providing 994 focal points in a 'Royal Gala'×'Granny Smith' segregating population, a sufficient number for the construction of a highly saturated genetic map. This resolution will be also sufficient to implement genomic selection in apple [29]. However, the resolution will not be high enough for association mapping studies using unrelated germplasm, because a recent study indicates that linkage disequilibrium among apple cultivars decays faster than this, based on calculations in a apple germplasm collection [7].

Conclusion

The International RosBREED SNP Consortium apple 8K SNP array v1 has been developed for public use by apple geneticists worldwide. The design and evaluation of the array has indicated that it will be effective for a wide range of germplasm and applications such as high-resolution genetic mapping, QTL detection and characterization, marker-assisted introgression, and genomic selection.

Genomic resources

All SNPs detected, the SNPs chosen for the IRSC apple Infinium® II 8K array, and the GenomeStudio cluster file developed are deposited in the Genome Database for Rosaceae (www.rosaceae.org). SNPs are available in dbSNP (http://www.ncbi.nlm.nih.gov/projects/SNP/) under accessions ss475875741 to ss475892397.

Supporting Information

Table S1 The 160 apple accessions used for the GoldenGate® SNP validation assay.

Table S2 List of 7867 apple SNPs on the Apple Infinium® II array v1. The NCBI dbSNP accession, location on the 'Golden Delicious' genome assembly [11], and source of SNP are indicated.

Acknowledgments

We thank all international researchers who participated in the Rosaceae SNP Summit, Stellenbosch, South Africa, November 11–12, 2010, for their contribution to international needs assessment and design of SNP filtering criteria. We thank Dr J. Norelli from USDA for providing sequencing data from *Malus sieversii* PI613981. DC thanks Dianne Hyndman, Paul Fisher,

Rosemary Rickman and Matt Bixley (AgResearch Invermay, New Zealand) for providing the Illumina genotyping service and Susan Thomson for useful comments on the manuscript.

Author Contributions

Conceived and designed the experiments: DC MT MWD CL RPH RV DM JDR AI EVDW SEG NB CP. Performed the experiments: DC BG SV NB. Analyzed the data: DC RNC MT MWD CL SK AC TM LW EVDW CP. Contributed reagents/materials/analysis tools: RPH RV JDR AI SEG NB. Wrote the paper: DC MT NB CP.

References

1. Peace CP, Norelli JL (2009) Genomics approaches to crop improvement in Rosaceae. In: Folta K, Gardiner S, eds. Genetics and Genomics of Rosaceae. New York: Springer. pp 19–53.
2. Liebhard R, Gianfranceschi L, Koller B, Ryder CD, Tarchini R, et al. (2002) Development and characterisation of 140 new microsatellites in apple (*Malus×domestica* Borkh.). Molecular Breeding 10: 217–241.
3. Silfverberg-Dilworth E, Matasci CL, Van de Weg WE, Van Kaauwen MPW, Walser M, et al. (2006) Microsatellite markers spanning the apple (*Malus×domestica* Borkh.) genome. Tree Genetics & Genomes 2: 202–224.
4. Wang A, Aldwinckle H, Forsline P, Main D, Fazio G, et al. (2011) EST contig-based SSR linkage maps for *Malus×domestica* cv Royal Gala and an apple scab resistant accession of *M. sieversii*, the progenitor species of domestic apple. Molecular Breeding;doi: 10.1007/s11032-011-9554-1.
5. Chagné D, Gasic K, Crowhurst RN, Han Y, Bassett HC, et al. (2008) Development of a set of SNP markers present in expressed genes of the apple. Genomics 92: 353–358.
6. Han Y, Chagné D, Gasic K, Rikkerink EHA, Beever JE, et al. (2009) BAC-end sequence-based SNPs and Bin mapping for rapid integration of physical and genetic maps in apple. Genomics 93: 282–288.
7. Micheletti D, Troggio M, Zharkikh A, Costa F, Malnoy M, et al. (2011) Genetic diversity of the genus *Malus* and implications for linkage mapping with SNPs. Tree Genetics & Genomes 7: 857–868.
8. Wittwer CT, Reed GH, Gundry CN, Vandersteen JG, Pryor RJ (2003) High-resolution genotyping by amplicon melting analysis using LCGreen. Clinical Chemistry 49: 853–860.
9. Huang X, Wei X, Sang T, Zhao Q, Feng Q, et al. (2010) Genome-wide association studies of 14 agronomic traits in rice landraces. Nature Genetics 42: 961–967.
10. Myles S, Chia J-M, Hurwitz B, Simon C, Zhong GY, et al. (2010) Rapid genomic characterization of the genus *Vitis*. PloS One 5: e8219.
11. Velasco R, Zharkikh A, Affourtit J, Dhingra A, Cestaro A, et al. (2010) The genome of the domesticated apple (*Malus×domestica* Borkh.). Nature Genetics 42: 833–839.
12. Grattapaglia D, Silva OB, Kirst M, de Lima BM, Faria DA, et al. (2011) High-throughput SNP genotyping in the highly heterozygous genome of *Eucalyptus*: assay success, polymorphism and transferability across species. BMC Plant Biology 11: 65.
13. Shen G-Q, Abdullah KG, Wang QK (2009) The TaqMan method for SNP genotyping. Methods Mol Biol, 578: 293–306.
14. Iezzoni A, Weebadde C, Luby J, Yue CY, WegEvd, et al. (2010) RosBREED: Enabling marker-assisted breeding in Rosaceae. Acta Horticulturae. pp 389–394.
15. Li RQ, Yu C, Li YR, Lam TW, Yiu SM, et al. (2009) SOAP2: an improved ultrafast tool for short read alignment. Bioinformatics 25: 1966–1967.
16. Wang J, Li R, Li Y, Fang X, Feng B, et al. (2008) Genome resequencing and identification of variations by Illumina Genome Analyzer Reads. Protocol Exchange.
17. Newcomb RD, Crowhurst RN, Gleave AP, Rikkerink EHA, Allan AC, et al. (2006) Analyses of expressed sequence tags from apple. Plant Physiology 141: 147–166.
18. Wu TD, Watanabe CK (2005) GMAP: a genomic mapping and alignment program for mRNA and EST sequences. Bioinformatics 21: 1859–1875.
19. Maliepaard C, Alston FH, van Arkel G, Brown LM, Chevreau E, et al. (1998) Aligning male and female linkage maps of apple (*Malus pumila* Mill.) using multi-allelic markers. Theoretical and Applied Genetics 97: 60–73.
20. Dyk MMv, Rees DJG, van Dyk MM (2009) Bin mapping of EST-SSRs in apple (*Malus×domestica* Borkh.). Acta Horticulturae. pp 681–688.
21. Celton JM, Tustin DS, Chagne D, Gardiner SE (2009) Construction of a dense genetic linkage map for apple rootstocks using SSRs developed from *Malus* ESTs and *Pyrus* genomic sequences. Tree Genetics & Genomes 5: 93–107.
22. Kenis K, Keulemans J, Davey MW (2008) Identification and stability of QTLs for fruit quality traits in apple. Tree Genetics & Genomes 4: 647–661.
23. Cabrera A, Kozik A, Howad W, Arus P, Iezzoni AF, et al. (2009) Development and bin mapping of a Rosaceae Conserved Ortholog Set (COS) of markers. BMC Genomics 10: 562.
24. Illa E, Sargent DJ, Girona EL, Bushakra J, Cestaro A, et al. (2011) Comparative analysis of rosaceous genomes and the reconstruction of a putative ancestral genome for the family. BMC Evolutionary Biology 11: 9.
25. Giovannoni J (2001) Molecular biology of fruit maturation and ripening. Annual Review of Plant Physiology and Plant Molecular Biology 52: 725–749.
26. Mouradov A, Cremer F, Coupland G (2002) Control of flowering time: Interacting pathways as a basis for diversity. Plant Cell 14: S111–S130.
27. Wang Y, Li J (2008) Molecular basis of plant architecture. Annual Review of Plant Biology 59: 253–279.
28. Rowan DD, Hunt MB, Dimouro A, Alspach PA, Weskett R, et al. (2009) Profiling fruit volatiles in the progeny of a 'Royal Gala'×'Granny Smith' apple (*Malus×domestica*) cross. Journal of Agricultural and Food Chemistry 57: 7953–7961.
29. Kumar S, Bink M, Volz R, Bus V, Chagné D (2011) Towards genomic selection in apple (*Malus×domestica* Borkh.) breeding programmes: Prospects, challenges and strategies. Tree Genetics & Genomes;doi: 10.1007/s11295-011-0425-z.
30. Guitton B, Kelner J, Velasco R, Gardiner S, Chagné D, et al. (2011) Genetic control of biennial bearing in apple. Journal of Experimental Botany 63: 131–149.
31. Bushakra J, Sargent D, Cabrera A, Crowhurst R, Lopez Girona E, et al. (2011) Assessing genome synteny between *Malus* and *Fragaria* using Rosaceae conserved orthologous set (RosCOS) markers. Tree Genetics & Genomes;doi:10.1007/s11295-011-0450-y.
32. Van de Weg WE, Voorrips RE, Finkers R, Kodde LP, Jansen J, et al. (2004) Pedigree genotyping: a new pedigree-based approach of QTL identification and allele mining. Acta Horticulturae. pp 45–50.
33. Lam HM, Xu X, Liu X, Chen WB, Yang GH, et al. (2010) Resequencing of 31 wild and cultivated soybean genomes identifies patterns of genetic diversity and selection. Nature Genetics 42: 1053–1059.
34. Myles S, Boyko AR, Owens CL, Brown PJ, Grassi F, et al. (2011) Genetic structure and domestication history of the grape. Proceedings of the National Academy of Sciences of the United States of America 108: 3530–3535.
35. Nelson JC, Wang S, Wu Y, Li X, Antony G, et al. (2011) Single-nucleotide polymorphism discovery by high-throughput sequencing in sorghum. BMC Genomics 12: 352.
36. Nordborg M, Weigel D (2008) Next-generation genetics in plants. Nature 456: 720–723.
37. Costa F, Stella S, Van de Weg WE, Guerra W, Cecchinel M, et al. (2005) Role of the genes *Md-ACO1* and *Md-ACS1* in ethylene production and shelf life of apple (*Malus domestica* Borkh). Euphytica 141: 181–190.
38. Costa F, Van de Weg WE, Stella S, Dondini L, Pratesi D, et al. (2008) Map position and functional allelic diversity of *Md-Exp7*, a new putative expansin gene associated with fruit softening in apple (*Malus×domestica* Borkh.) and pear (*Pyrus communis*). Tree Genetics & Genomes 4: 575–586.
39. Espley RV, Brendolise C, Chagne D, Kutty-Amma S, Green S, et al. (2009) Multiple repeats of a promoter segment causes transcription factor autoregulation in red apples. Plant Cell 21: 168–183.
40. Espley RV, Hellens RP, Putterill J, Stevenson DE, Kutty-Amma S, et al. (2007) Red colouration in apple fruit is due to the activity of the MYB transcription factor, *MdMYB10*. Plant Journal 49: 414–427.
41. Gao ZS, van de Weg WE, Schaart JG, Schouten HJ, Tran DH, et al. (2005) Genomic cloning and linkage mapping of the *Mal d 1* (PR-10) gene family in apple (*Malus domestica*). Theoretical and Applied Genetics 111: 171–183.
42. Wang A, Yamakake J, Kudo H, Wakasa Y, Hatsuyama Y, et al. (2009) Null mutation of the *MdACS3* gene, coding for a ripening-specific 1-Aminocyclopropane-1-Carboxylate Synthase, leads to long shelf life in apple fruit. Plant Physiology 151: 391–399.
43. Eckert AJ, Pande B, Ersoz ES, Wright MH, Rashbrook VK, et al. (2009) High-throughput genotyping and mapping of single nucleotide polymorphisms in loblolly pine (*Pinus taeda* L.). Tree Genetics & Genomes 5: 225–234.
44. Pindo M, Vezzulli S, Coppola G, Cartwright DA, Zharkikh A, et al. (2008) SNP high-throughput screening in grapevine using the SNPlexTM genotyping system. BMC Plant Biology 8: 12.
45. Krzywinski M, Schein J, Birol I, Connors J, Gascoyne R, et al. (2009) Circos: An information aesthetic for comparative genomics. Genome Research 19: 1639–1645.
46. Lijavetzky D, Antonio Cabezas J, Ibanez A, Rodriguez V, Martinez-Zapater JM (2007) High throughput SNP discovery and genotyping in grapevine (*Vitis vinifera* L.) by combining a re-sequencing approach and SNPlex technology. BMC Genomics 8: 424.
47. Vezzulli S, Micheletti D, Riaz S, Pindo M, Viola R, et al. (2008) A SNP transferability survey within the genus *Vitis*. BMC Plant Biology 8: 128.
48. Yamamoto T, Nagasaki H, Yonemaru J-i, Ebana K, Nakajima M, et al. (2010) Fine definition of the pedigree haplotypes of closely related rice cultivars by means of genome-wide discovery of single-nucleotide polymorphisms. BMC Genomics 11: 267.

2

Comparative Analysis of Zinc Finger Proteins Involved in Plant Disease Resistance

Santosh Kumar Gupta[1,2], Amit Kumar Rai[1,2], Shamsher Singh Kanwar[2], Tilak R. Sharma[1]*

1 National Research Centre on Plant Biotechnology, Indian Agricultural Research Institute, New Delhi, India, **2** Department of Biotechnology, Himachal Pradesh University, Summer-Hill, Shimla, India

Abstract

A meta-analysis was performed to understand the role of zinc finger domains in proteins of resistance (*R*) genes cloned from different crops. We analyzed protein sequences of seventy *R* genes of various crops in which twenty six proteins were found to have zinc finger domains along with nucleotide binding sites - leucine rice repeats (NBS-LRR) domains. We identified thirty four zinc finger domains in the R proteins of nine crops and were grouped into 19 types of zinc fingers. The size of individual zinc finger domain within the *R* genes varied from 11 to 84 amino acids, whereas the size of proteins containing these domains varied from 263 to 1305 amino acids. The biophysical analysis revealed that molecular weight of *Pi54* zinc finger was lowest whereas the highest one was found in rice *Pib* zinc finger named as Transposes Transcription Factor (TTF). The instability ($R^2 = 0.95$) and the aliphatic ($R^2 = 0.94$) indices profile of zinc finger domains follows the polynomial distribution pattern. The pairwise identity analysis showed that the Lin11, Isl-1 & Mec-3 (LIM) zinc finger domain of rice blast resistance protein pi21 have 12.3% similarity with the nuclear transcription factor, X-box binding-like 1 (NFX) type zinc finger domain of Pi54 protein. For the first time, we reported that Pi54 (Pi-k^h-Tetep), a rice blast resistance (R) protein have a small zinc finger domain of NFX type located on the C-terminal in between NBS and LRR domains of the R-protein. Compositional analysis depicted by the helical wheel diagram revealed the presence of a hydrophobic region within this domain which might help in exposing the LRR region for a possible *R-Avr* interaction. This domain is unique among all other cloned plant disease resistance genes and might play an important role in broad-spectrum nature of rice blast resistance gene *Pi54*.

Editor: Tianzhen Zhang, Nanjing Agricultural University, China

Funding: Financial assistance received from the Department of Biotechnology (F.No. BT/AB/FG-2(PH-II)/2009-3A), Government of India, National Agricultural Innovation Project-Indian Council of Agricultural Research (C4/C1071) by TRS is gratefully acknowledged. The funders had no role in study design, data collection and analysis, decision to publish, or preparation of the manuscript.

Competing Interests: The authors have declared that no competing interests exist.

* E-mail: trsharma@nrcpb.org

Introduction

The zinc finger proteins are a super family of proteins involved in numerous activities of plant growth and development and are also known to regulate resistance mechanism for various biotic and abiotic stresses [1,2]. Any small, functional, freely folded domain in which coordination of one or more zinc ions required to stabilize its structure is known as zinc finger [3]. These domains are actively required to regulate various metabolic processes and stress conditions in plants [4,5,6]. The Zinc finger domains are widely dispersed in eukaryotic genomes [7,4,8,9] and are actively involved in sequence specific binding to DNA/RNA and contribute in protein-protein recognitions [10]. The presence of zinc finger DNA binding domain in nucleotide binding sites-leucine rich repeats (NBS-LRR) class of proteins determines the regulatory function of this protein in stress conditions. These domains are basically transcription factor in origin which makes the protein as regulator. Zinc finger binds to DNA through the interaction of amino acids at the periphery of the zinc finger with base pairs at the center of the DNA double helix [11]. It is a compact protein domain, and its small size allows it to have close relation with DNA base pairs. Zinc fingers basically bind with nucleic acids for their function or participate in transcriptional or translational regulation processes [12]. These are classified into nine types according to their structural and functional variation. These are C2H2, C8, C6, C3HC4, C2HC, C2HC5, C4, C4HC3 and CCCH (C and H represent cysteine and histidine, respectively) [13,14,15,16,17]. The presence of zinc finger domain has been reported in many disease resistance genes cloned from various crops. These domains are LSD1 (C2C2), LOL1 (C2C2), Zat 12 (C2H2), Zat 7 (C2H2) & AtNFX1 (NF-X1) of Arabidopsis [18], StZFP1 (C2H2) of potato [8], and OsLSD1 (C2C2), OsLOL1 (C2C2), OsRING-1 (RING H2, RING HC), OsRFP1, OsDOS (CCCH), OsZFP (C2H2) & SRZ1(C2C2) of rice [19,20]. These reports indicate that zinc finger motifs have important role in imparting host- plant resistance.

The NBS-LRRs are most prevalent class of R- proteins. These are basically three types such as TIR-NBS-LRR (TNL) and CC-NBS-LRR (CNL) and mixed type having either Toll/interleukin-1 receptor-like domain (TIR) or Coiled coil (CC) domains or both in one protein [21,22]. Proteins fused with these zinc finger domains make them as zinc finger proteins. Presence of these domains is required for the function of individual proteins under stress conditions. These domains are able to regulate the proteins like switches in rice and poplar [23,24]. Sometimes, R proteins carry extensions at carboxy-terminus (C- terminal) with a typical WRKY DNA-binding domain and between linkers of NBS and LRR domains, respectively [25,26].

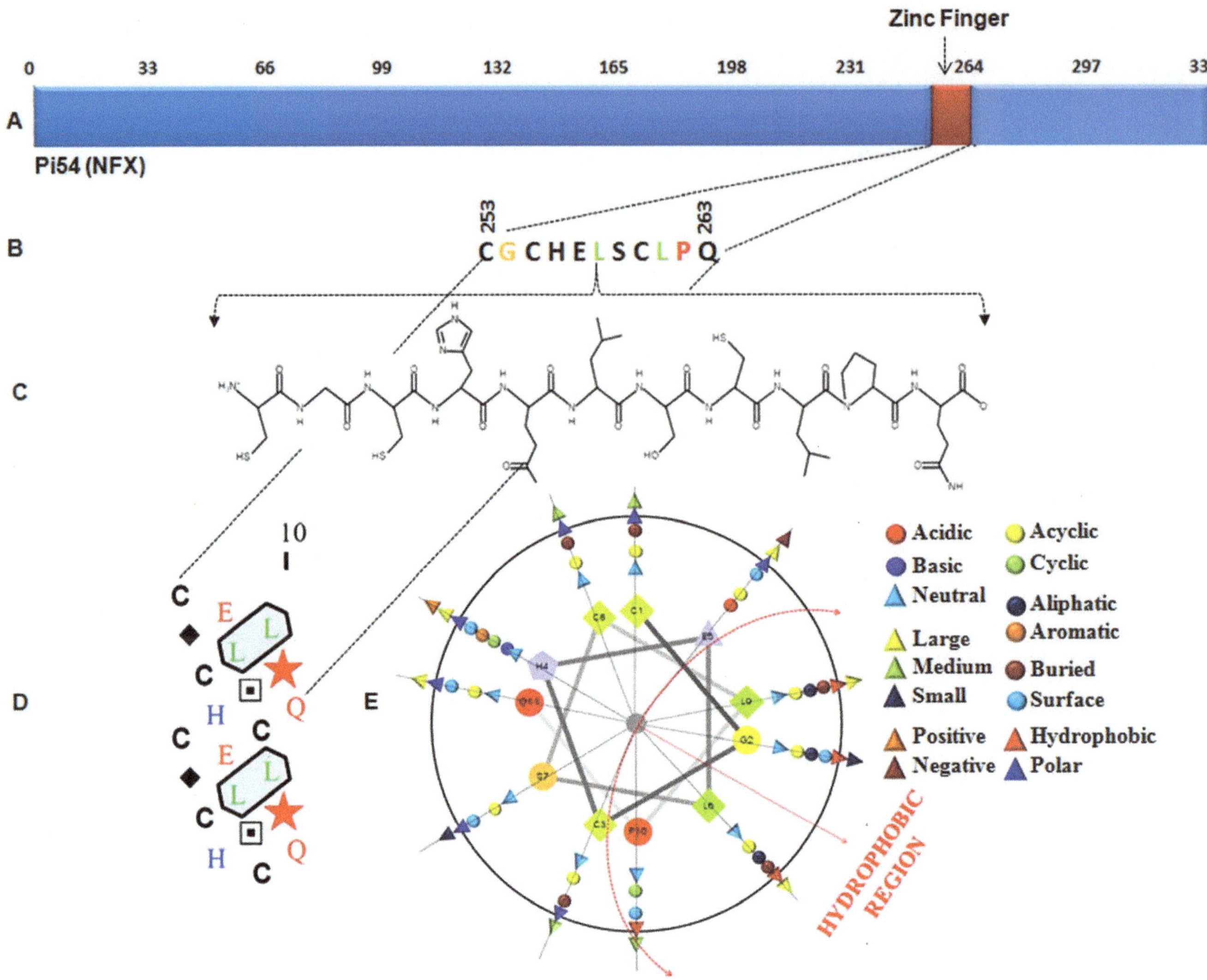

Figure 1. The structure of *Pi54* zinc finger domain. (A) Positional analysis of the domain showed that this domain is C- terminal in nature. The type of this zinc finger domain is NFX. (B) The amino acids, numbers and their positions in this domain. (C) Chemical structure of individual amino acids. (D) Secondary structure of zinc finger domain (E) Helical wheel diagram of *Pi54* zinc finger domain. The helical wheel is a plot of the amino acid residues around a potentially helical segment. The graphical representation showed the clustering of polar and/or non-polar residues toward one face of a helix.

Rice blast caused by *Magnaporthe oryzae* is one of the most important biotic stresses of rice resulting into huge yield loss every year [27]. The disease can be effectively managed by the resistance gene deployment. We have earlier identified and cloned a new rice blast resistance gene *Pi-k*^{h} from *indica* rice cultivar, Tetep showing resistance to different strains of *M. oryzae* [28,29]. The gene was only the third one to be cloned in the series of cloned rice blast resistance genes, after the cloning of *Pi-b* [30], and *Pi-ta* genes [31]. The gene was renamed as *Pi54*, after it was fine mapped at a slightly different location from the *Pi-k* cluster of the genes [32]. Functional validation of the gene has established that it confers a stable and high-level of resistance against geographically diverse strains of *M. oryzae* in India [33] and USA [34]. Expression analysis of the gene has revealed that it is induced by pathogen challenge. In turn, the gene was found to induce the synthesis of callose (β-1, 3 glucan) in response to pathogen challenge, suggesting its requirement in the initiation of a defense response cascade in the blast resistant plants [33]. Transcriptional and biochemical analysis revealed that rice transgenic lines containing *Pi54* single functional blast resistant gene show activation of a complex defense mechanism after pathogen inoculation [35]. The cloning of orthologue of *Pi54* gene has also been achieved from wild species of rice *O. rhizomatis* [36]. It has also been reported that *Pi54* protein contains an NBS-LRR domain in addition to a small zinc finger domain [37,26].

The *R* genes are also categorized into separate categories on the basis of status and position of zinc finger domains. These domains are present either at N– terminal or C– terminal of the proteins encoded by *R* genes and along with NBS-LRR domain, play a crucial role in regulating expression of the genes involved in plant resistance [38,39]. Many defense proteins of Arabidopsis and rice containing zinc finger domain have been shown to regulate programmed cell death (PCD). Despite having proven role in stress management in plants, the presence and involvement of zinc finger motifs along with NBS-LRR has not been analyzed in relation to plant disease resistance. The structural analysis of zinc finger domain present within these proteins is important for understanding the role of small protein domains that have diverse functions [13,40].

The objectives of present study were (i) to analyze the presence of zinc finger domains in all the cloned plant disease resistance genes, (ii) to determine the probable structure of zinc finger domain of blast resistance protein Pi54 (iii) computational analysis of biophysical properties of zinc finger domains in the proteins of cloned *R* genes and (iv) comparative analysis of zinc finger proteins in relation to Pi54 protein.

Results and Discussion

Identification and amino acid composition analysis of zinc finger domains

The amino acid sequence of protein encoded by rice blast resistance gene *Pi54* was downloaded and analyzed using various bioinformatics tools along with a careful manual inspection. A small (11AA) zinc finger motif of nuclear transcription factor, X-box binding-like 1 (NFX) type was identified between the positions 253–263 amino acids in this protein. This domain is C– terminal in nature and integrated within LRR. Earlier *Pi-k*^{h} (*Pi54*) gene responsible for the expression of this protein was found to be pathogen inducible in nature [35,33,26]. Amino acid composition analysis revealed that there are eight amino acids present within this domain with varying properties and mole percentage (Figure 1; Table 1). The amino acids in Pi54_ZnF domain were characterized on the basis of their side chains. These are negatively charged acidic (Glu E), non-polar aliphatic (G, L), polar uncharged (C, Q, S) and positively charged basic (H). The protein characteristics are believed to be related with the composition of amino acids and some of these structural factors were obtained due to the exchange of some amino acids [41].

Non-Polar amino acids. The nonpolar amino acids were characterized for having non polar atoms (only carbon and hydrogen) in their side chains. They include Glycine (Gly, G), Ala (Alanine, A), Val (Valine, V), Leu (Leucine, L), Ile (Isoleucine, I), Pro (Proline, P), and Met (Methionine, M). Presence of such residues makes domains more hydrophobic in nature. The hydrophobic amino acid residues can increase the rigidity and hydrophobicity of proteins [42]. Among the hydrophobic residues, Leu belongs to the aliphatic amino acids (Table 1). It has been found that aliphatic amino acids would contribute to the hydrophobic interaction and required to maintain the conformational stability in the inner part of the protein [43]. We identified two Leu residues in small zinc finger domain of Pi54 protein (Figure 1). More number of Leu residues result in higher average hydropathy and aliphatic index [44]. Besides, in Pi54 protein–Gly residues were also identified. Gly is known as the residue responsible to maintain or generate cavity in the inner part of protein structure [41]. These residues make domains more flexible for better folding in different way.

Table 1. Amino acid compositional analysis of Pi54 Zinc finger domain.

S.No.	Amino Acid	Number	Mol%	R Group
1	Glu E	1	9.09	Negatively Charged , Acidic
2	Gly G	1	9.09	Non Polar Aliphatic
3	Leu L	2	18.18	Non Polar Aliphatic
4	Cys C	3	27.27	Polar Uncharged
5	Gln Q	1	9.09	Polar Uncharged
6	Pro P	1	9.09	Non-polar
7	Ser S	1	9.09	Polar Uncharged
8	His H	1	9.09	Positively Charged , Basic

One Pro residue was also identified in the zinc finger domain of Pi54 protein. This residue contains a non-polar, uncharged R group (Figure 1). The Pro residue can only adopt a few configurations due to their pyrolidine ring and has the lowest conformational entropy. It thus restricts the configurations allowed for the preceding residue. Because of the presence of this residue rigidity and conformation in protein structure have been reported [45]. The Pro residue has been used to increase the protein stability in the several mutational studies and hence an increase in the stability of such domains of disease resistant proteins might be important for rice plant to resist more against *M. oryzae* [46].

Polar, uncharged amino acids. The polar nature of the side chain means that these amino acids are ready to interact with water (hydrophilic) and can form hydrogen bonds [42]. There are four numbers of polar uncharged amino acids (Cysteine C, Glutamine Q, Serine S and sometimes Histidine H with pK of 6.5) in small zinc finger domain of the Pi54 protein (Figure 1). These are the amino acids which possess oxygen, sulfur and/or nitrogen in the side chain and hence polar, but cannot have their side chain ionized and thus do not carry an overall charge. The amino acid Glutamine is known as thermolabile amino acids due to its tendency to undergo deamination at high temperature [41]. This indicates that presence of such residue makes the zinc finger domain of Pi54 more stable in various stress conditions; hence might help in maintaining the ability of rice plants to resist against the incoming *M. oryzae* pathogen under varied climatic conditions. The amino acid residue Serine is known as the best residues for interacting with the water molecules surrounding protein due to its hydrophilic nature [47]. The water molecules that are ready for interaction with these residues for hydrogen bond formation have been reported to release at high temperature. Hence, the protein structure around water-binding site changed to unstable which might increase the instability of proteins [48].

The side group of another polar, uncharged residue Cysteine present in zinc finger domain of Pi54 protein contains a sulfur atom (Figure 1). The sulfur group in Cysteine comes at the end of the hydrocarbon chain and therefore, has the potential to be more reactive. Cysteine is also known as thermolabile amino acids because it undergoes oxidation at high temperature like Methionine [49]. There are three Cysteine residues in Pi54 zinc finger domain. The proteins of maize, rice, and tomato contain 1.62%–1.69% cysteine whereas the yeast contained least cysteine (1.21%). More number of cysteine residues indicates existence of short-range intra-polypeptide chains interactions which plays important role in evolution [50]. Among the prokaryotes, Cyanobacteria, *E. coli*, *Psuedomonas aeruginosa*, and *Rhodobactor sphaeroides* contained 1.03%–1.13% cysteine. The cysteine contents of proteins of different species may be as low as 0.4%–0.5% in Archea. The extreme halophil *Haloarcula marismortui* and the thermophil *Thermus aquaticus* were reported to have the lowest 0.49% and 0.41%, respectively cysteine among the investigated species. The C-$(X)_2$-C motifs present at position 253–260 of Pi54 protein are important domains of metal binding proteins [51].

Polar, charged amino acids. The negatively charged, acidic R group containing Glutamic acid and positively charged, basic amino acid, Histidine were also been identified in zinc finger domain of the Pi54 protein (Figure 1). The polar side chains of these residues can also carry a positive charge or negative charge and are therefore highly hydrophilic in nature. The charged amino acids would contribute to the electrostatic interaction, which is an important force for maintaining conformational stability in the outer part of the proteins [52,53]. The conformational stability to expose the LRR domain for interaction with Avr proteins of the pathogen *M. oryzae* is necessary for *R* gene interactions [54]. Since the zinc finger domain of Pi54 protein is C– terminal in nature and integrated with LRR, it might be playing an important role in protein-protein interaction in rice-*M. oryzae* pathosystem. The charged amino acid residues are less labile and also retain the hydrogen bonding capacity. The charged residues may be involved in location of ion pairs within molecular structures which also appears to be important in determining protein stability [55]. The presence of these residues at the protein surface, increase ion interactions and enhances occurrence of salt bridges and ion pairs [56,55]. These amino acids are important for the flexibility of the proteins [6].

The Helical wheel diagram gives a view of a helix from a protein sequence looking down the axis of the helix. It is useful for highlighting the amphipathicity and other properties of residues around a helix. The hydrophobic region of Pi54-ZnF domain which constitutes of two Leucine residues, one Proline and one Glycine has been represented by a helical wheel diagram (Figure 1). The helical wheel is a plot of the amino acid residues around a potentially helical segment. The method was developed to find helices with a hydrophobic face buried away from a polar solvent, with the graphical representation showing the clustering of polar and/or non- polar residues toward one face of a helix. Earlier, the helical wheel facilitated the identification of potential helical segments in protein sequences [57]. Besides, its applicability has also been expanded to designer proteins such as leucine zipper proteins [58] and studying trans-membrane proteins like G-protein- coupled receptors [59].

Comparative analysis of proteins of *R* genes cloned from different crops

We analyzed protein sequences of 70 *R* genes cloned from different crops to delineate their zinc finger domain (Table S1 and Supplementary References, 1–69). The cloned disease resistance genes of the thirteen crops (Arabidopsis; Supplementary References (1–16), Barley (17–22), Pepper (23), Potato (24–27), Rice (28–43), Sunflower (44), Tomato (45–58), Tobacco (59), Wheat (60–63), Flax (64–66), Maize (67), Lettuce (68) and Beet (69)) were analyzed for identification and characterization of zinc finger domains in their respective proteins. The types of the proteins encoded by these *R* genes were varied in nature. The 'SMART' analysis revealed that twenty six genes contain zinc finger domains of different types, varying sizes and at different positions in their respective proteins. We found twenty six *R* genes from nine crops having nineteen types of zinc finger (Table 2). The *R* genes of rice represent 31% of all the zinc finger containing cloned *R* genes followed by tomato which has 23% proteins containing zinc finger domains. The *R* genes of wheat, sunflower, flax and tobacco were found to have zinc finger domains and constitute 4% of all the cloned *R* genes. Barley and potato represents 11% while Arabidopsis showed 8% share in this study of cloned *R* genes (Figure 2; Table S2). All zinc finger containing rice blast resistance genes encoded the NBS-LRR type of proteins except *Pi36* and *Pi5-1* which contains the CC-NBS-LRR type of proteins. The sunflower gene *pl8* was also found to contain zinc finger domain but is a CC-NBS-LRR type of protein. Tobacco *R* gene '*N*' encoded the TIR-NBS-LRR type of protein also have the zinc finger domains. The six *R* genes of Tomato have zinc finger

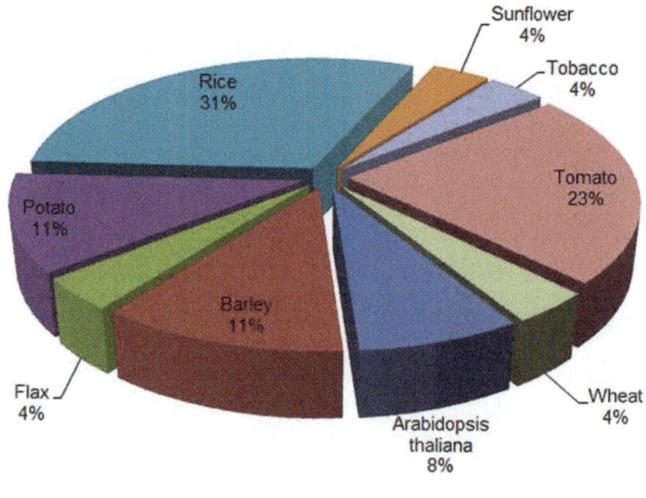

Figure 2. The distribution of zinc finger domains across different crops. The cloned *R* genes of various crops were found to have zinc finger domain in their proteins. The number of zinc finger domains represented by cloned *R* genes of each crop is shown in pie chart.

domain in their NBS-LRR, LZ-NBS-LRR and LRR types of the proteins (Table 2).

Analysis of zinc finger domain in R proteins

The zinc finger domains encoded by these proteins were used for further analysis. *R* genes of nine crops were further distributed into five families of eudicots and monocots representing their zinc finger domains (Table S2). Thirty four zinc finger sequences were found in twenty six *R* genes (Table S3). The domain architecture of these proteins depicts different types, size, position and number of zinc finger domains (Figure 3; Table 2). The blast resistance gene *Pi54* contain NFX type zinc finger domain in their protein. This domain is smallest and distinct from all others, though the smallest protein was encoded by *Pi21* of rice (Figure 3). The NFX1-type zinc finger proteins are a group of the human NFX1 transcription factors. This protein was identified as a protein that represses class II MHC (major histocompatibility complex) gene [60]. The NFX1-type zinc fingers containing proteins are found in protists, fungi, animals and plants. The NFX type zinc finger domain containing proteins are known to be involved in growth and survival of plants by managing reactive oxygen species (ROS), salicylic acid (SA), and also in biotic stress and abscisic acid (ABA) responses [61]. The only plant homologue AtNFXL1 of the NFX1 gene has been experimentally confirmed and found that it plays a crucial role in different stress responses [62,63,64]. In the present analysis, five genes were found to have two zinc finger domains whereas three zinc finger domains were obtained in one gene only. Twenty genes were found to contain only one zinc finger domains in their respective proteins (Table 2).

Characteristic features of zinc finger domains in R-proteins

Zinc finger domain was deduced in a total of eight blast resistance genes of rice (Table 2). These genes were found to have ten different types of zinc finger domains in their respective proteins. The R- gene *Pib* encoded NBS-LRR protein along with two zinc finger domains U1 and TTF. Eight types of zinc fingers such as U1, TTF, UBP, CHCC, NFX, PMZ, LIM and TAZ were found in the rice R proteins (Table 2). The zinc finger type UBP

Table 2. Summary of various Zinc finger domains present in cloned *R* genes.

S No.	Crop name	Name of *R* gene	Protein length (aa)	Zinc finger type	Size of ZnF domain	Position of ZnF protein	Number of ZnF domains	Accession number	Disease resistance type	References
1	Arabidopsis	*SSI4*	1055	DBF	52	910–961	1	AAN86124.1	TIR-NBS-LRR	[92]
2	Arabidopsis	*RCY1*	361	RAD18	22	48–69	2	NP_001077963.1	RLKs	[93]
				PMZ	28	129–156				
3	Barley	*Mla1*	958	BED	49	386–434	1	AAG37354.1	CC-NBS-LRR	[94]
4	Barley	*Mla12*	961	BED	43	893–935	1	AAO43441.1	CC-NBS-LRR	[95]
5	Barley	*Rpg1*	837	TTF	84	145–228	2	AAM76922.1	RLKs	[96]
				C3H1	21	543–563				
6	Flax	*M*	1305	Rad18	16	290–305	1	AAB47618.1	NBS-LRR	[97]
7	Potato	*Rx2*	912	BED	47	346–392	1	CAB55838.1	NBS-LRR	[98]
8	Potato	*Gpa2Rx1*	937	BED	47	346–392	1	CAB50786.1	NBS-LRR	[99]
9	Potato	*Gro 1.4*	1136	CHCC	41	113–153	2	AAP44390.1	TIR-NBS-LRR	[100]
				CDGSH	31	688–718				
10	Rice	*Pib*	1251	U1	35	546–580	2	BAA76282.2	NBS-LRR	[30]
				TTF	83	60–142				
11	Rice	*Pi-ta*	928	UBP	43	537–579	1	AAK00132.1	NBS-LRR	[31]
12	Rice	*Pi36*	1056	CHCC	40	845–884	1	ABI64281.1	CC-NBS-LRR	[101]
13	Rice	*Pi54*	330	NFX	11	253–263	1	AAY33493.1	NBS-LRR	[26]
14	Rice	*Piz-t*	1033	FMZ	20	645–664	1	ABC73398.1	NBS-LRR	[102]
15	Rice	*Pi21*	263	LIM	39	11–49	1	BAG72124.1	NBS-LRR	[76]
16	Rice	*Pi5-1*	1025	TAZ	67	749–815	1	ACJ54697.1	CC-NBS-LRR	[78]
17	Rice	*Pi-2*	974	UBP	45	74–118	1	ABC94597	NBS-LRR	[102]
18	Sunflower	*pl8*	1279	CHCC	45	292–236	3	AAT08955.1	CC-NBS-LRR	[103]
				GATA	43	754–796				
				C2C2	31	1083–1113				
19	Tobacco	*N*	1128	C2H2	30	568–597	1	BAD12594	TIR-NBS-LRR	[104]
20	Tomato	*Mi-1*	1257	UBR1	56	1064–1119	1	AAC97933	NBS-LRR	[105]
21	Tomato	*Sw5-e*	1241	UBP	37	42–78	1	AAG31017	LZ-NBS-LRR	[106]
22	Tomato	*Cf-9*	863	PMZ	20	160–179	1	AAA65235.1	LRR	[107]
23	Tomato	*I2C*	373	U1	27	127–153	1	AAB63276	NBS-LRR	[15]
24	Tomato	*Hero*	1283	ZZ	44	171–214	2	CAD29729	NBS-LRR	[108]
25	Tomato	*Cf-4*	806	ZNF_C4	39	46–84	1	CAA05268.1	LRR	[109]
26	Wheat	*Lr10*	636	C2C2	38	32–69	2	AAC49629.1	RLKs	[110]
				ZZ	39	238–276				

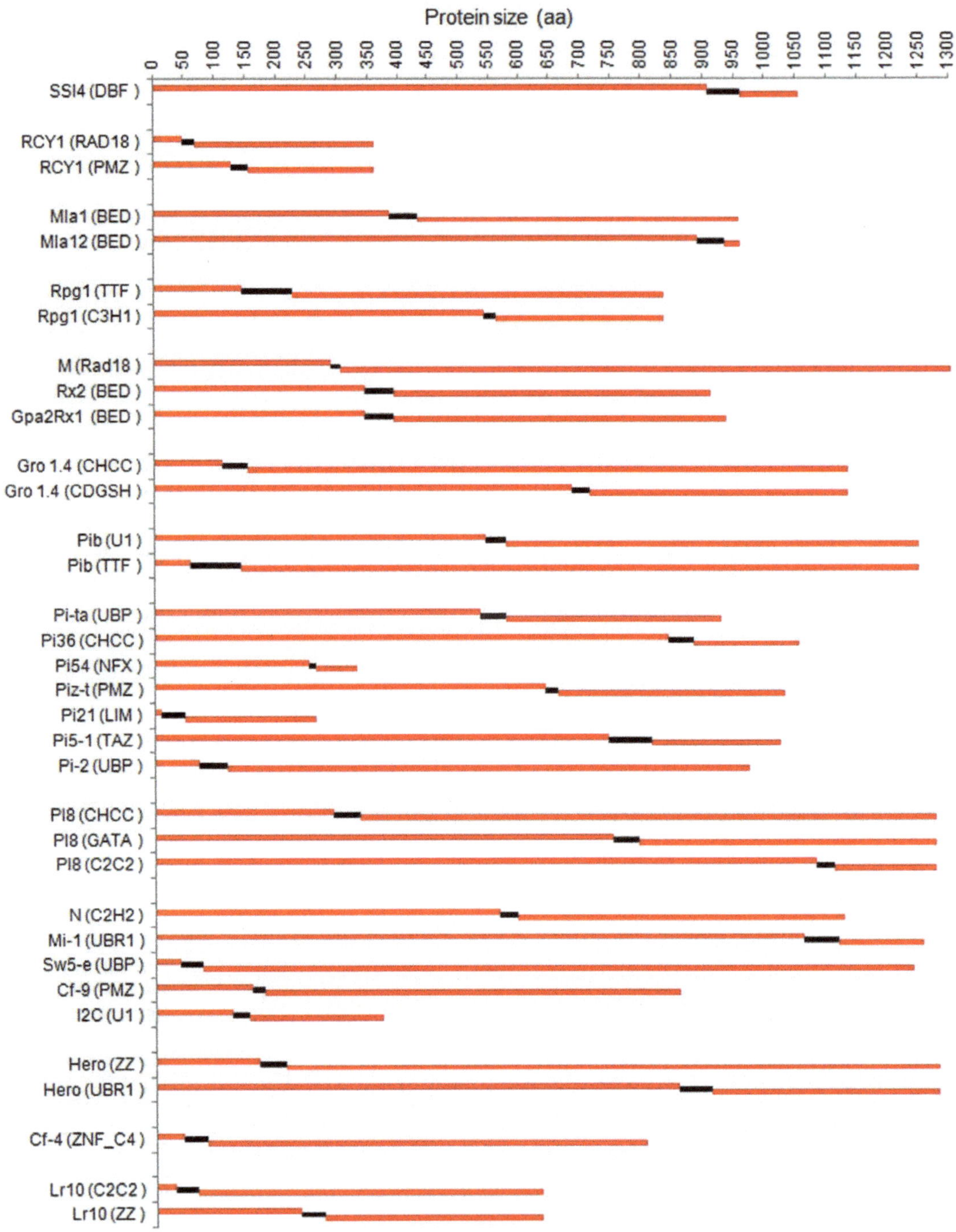

Figure 3. The architecture of zinc finger domains containing R proteins. The scale (0–1300) represents the size of proteins and their domains. The position and number of zinc finger domains in each *R* genes are also represented in the scale. The black broken lines indicate the zinc finger domains in individual R-proteins.

was deduced in blast resistance genes *Pi-ta* and *Pi-2*. The size of zinc finger domains varies from 11 to 84 amino acid residues. The NFX, LIM and TAZ types were only found in the *R* genes cloned from rice. The U1-ZnF of *Pib* was reported in *I2C* gene of tomato whereas TTF was found in barley Rpg1 protein. The PMZ of rice deduced in Piz-t protein having length (20 AA) similar to PMZ of Tomato R protein Cf-9. This zinc finger domain is also present in RCY1 protein of barley. The LIM zinc finger deduced in Pi21 protein of rice is smallest protein amongst all the proteins of cloned genes in plants, whereas the zinc finger domain deduced in Pi54 protein was found to be the smallest amongst all plant proteins cloned till date. The *Pib* gene encoded protein of rice was found to have largest zinc finger domain TTF (83AA) among all the cloned *R* genes of rice though TTF of barley *R* gene *Rpg1* is largest (84 AA) among all the crops (Table 2, S3).

Among other crops, tomato contains six types of zinc finger domains like UBR1, UBP, PMZ, U1, ZZ and C4 in their disease resistance genes. The smallest zinc finger domain of tomato is

PMZ (20AA) which was deduced in Cf-9 protein and is similar to the rice PMZ domain of blast resistance gene *Piz-t*. The UBR1 zinc finger of tomato protein Mi-1 is largest amongst all other cloned resistance proteins of tomato (Table 2, S3). Similarly, cloned R proteins of potato were found to have three types of zinc finger domains such as BED, CHCC and CDGSH. Of these, the zinc finger domain BED has also been deduced in *Mla1*& *Mla12* genes of barley.

Biophysical characterization of zinc finger domains of R-proteins

The Protparam analysis revealed that these R-proteins have varying numbers and types of amino acids (Table S1, S3). This analysis also includes the molecular weight, theoretical PI, instability index, aliphatic index and hydropathicity index of all zinc finger domains present in *R* genes (Table S1, S4). The molecular weight of *Pi54* zinc finger NFX is 1189.3 Daltons. It was lowest among all, whereas the highest one was 9905 Daltons in *Pib* zinc finger TTF (Table S4). The isoelectric point (pI) of zinc fingers of cloned R- genes ranged from 4.0 (UBP of *Sw5-e*) to 10.66 (PMZ of *RCY1*). The isoelectric point (pI) is the pH at which the surface of protein is covered with charge but net charge of protein is zero. The pI of Pi54 zinc finger domain was found to be 5.24 (Table S4) which indicates that this domain is slightly acidic in nature. The calculated isoelectric point (pI) will be useful because at this pI, solubility is least and mobility in an electro focusing system is zero [65].

The instability status of zinc finger proteins

The instability index value for the *R* gene zinc finger domains ranged from 6.37 to 120.89. The instability index of *Pi54* zinc finger domain was 120.89. This value indicates its highly unstable nature among all the cloned *R* genes. The zinc finger domain is arranged in their increasing order of stability profile which follows the polynomial trends ($R^2 = 0.95$) (Figure 4; Table S4). The instability index affords an estimate of the stability of protein in a test tube. This method assigns a weight value of instability, which can be used to compute an instability index. A protein whose instability index smaller than 40 is predicted as stable where as a value above 40 predicts that the protein may be unstable [66].

The aliphatic index of zinc finger domains of R proteins

The aliphatic index is defined as the relative volume of a protein occupied by aliphatic side chains. The aliphatic index of the zinc finger sequences ranged from 30.0 to 122.58. The very high aliphatic index of zinc finger sequences indicates that these domains are stable at wide temperature range [67]. The aliphatic index of *Pi54* zinc finger domain ZnF_NFX was recorded as 70.91. This value indicates that the NFX domain of Pi54 is thermo-stable as well as flexible in nature. The high aliphatic index 122.58 was recorded for CDGSH zinc finger domain of *Gro 1-4* gene of potato. The lower thermal stability of ZZ zinc finger domain of *Lr10* gene of wheat indicates that this domain is very flexible than others. The aliphatic index profile follows the polynomial distribution pattern ($R^2 = 0.94$) (Figure 5; Table S4). The aliphatic index (AI) occupied by aliphatic side chains (A, V, I and L) is regarded as a positive factor which is defined as the relative volume of a protein for the increase of thermal stability of globular proteins [68,69]. The very high aliphatic index of zinc finger sequences indicates that these domains are stable at wide temperature range [67].

Analysis of Grand Average of Hydropathy of zinc finger domains of R-proteins

The Grand Average of Hydropathy (GRAVY) indices of zinc finger domain of cloned *R* genes ranged from −1.072 to 0.421 (Figure 6; Table S4). GRAVY values determined to provide a view of the hydrophobicity of the whole protein [70]. The GRAVY values usually vary in the range of ±2. Positive scores indicate hydrophobicity and negative scores indicate hydrophilicity. This low range of GRAVY value indicates the possibility of better interaction with water [67,71]. The GRAVY value for *Pi54* ZnF domain was calculated as 0.191 (Figure 6) which indicates that this domain is hydrophobic in nature. This nature of the domain of R proteins makes them available for better interaction with their Avr counterparts as reported in case of Pto protein of tomato [72]. GRAVY values calculated for this domain of all the protein also follows the linear distribution pattern (($R^2 = 0.95$). The majority of zinc finger domains were hydrophilic in nature and some were found to hydrophobic.

Phylogenetic relationship among zinc finger motifs

To establish a relationship between thirty four taxa of zinc finger domains, a maximum parsimony tree was constructed to identify the significant correlation among the different and highly divergent zinc fingers present in *R* genes. Fourteen possible groups of taxa having sequence similarities between each other were obtained (Figure 7). Further, percentage identity plot was calculated between each taxa which results a significant association with maximum parsimony inference between T1 & T2 (40%) and T16 & T17 (80.8%). The taxa T16 & T17 belong to potato *R* genes *Rx2* and *Gpa2/Rx1*. Both these genes contained BED zinc finger domains. The size of zinc finger domains in both the proteins is similar (47AA) but the size of both the NBS-LRR proteins is 912 and 937 AA, respectively (Table 2). This zinc finger domain was first reported in 2000 after two drosophila proteins named BEAF and DREF were identified [23]. These domains are probably involved in regulatory function of transcriptional control in plant disease resistance proteins [73]. Similarly the T1 & T2 taxa were represented by rice (*Piz-t*) and tomato (*Cf-9*). These are having sequence similarities between PMZ zinc fingers. The PMZ zinc finger size is 20 AA in both the proteins though the size of the protein in both the genes was 1033 AA (NBS-LRR) and 863 AA (LRR), respectively (Figure 7; Table 2). The PMZ (Plant Mutator Transposase) zinc finger is basically a transcription factor that is induced during the senescence and pathogen infection. These domains are present in AN1 like protein families. The PMZ domain containing proteins are induced by the abscisic acid and chitin stimuli [74,75]. There are four groups of taxa having more than 10% parsimony inference between each other. These are T9 & T10 (10.2%), T20 & T21 (11.3%), T22 & T23 (10.8%) and T30 & T31 (11.6%). However, eight groups of taxa having less than 10% sequence similarities between each other were also obtained. The parsimony inference was not observed between zinc finger domain of *Pi54* and other genes. These findings indicate that the zinc finger domain of Pi54 protein is distinct amongst all the analyzed highly divergent zinc finger sequences of the proteins of plant disease resistance genes.

Identity plot of zinc fingers of cloned R- genes in relation to *Pi54* gene

This analysis encompasses the identity between the zinc finger domains of blast resistance gene *Pi54* and the genes cloned from other crops (Figure 8). The blast resistance genes cloned from rice were found to be closer to *Pi54*. The rice blast resistance protein

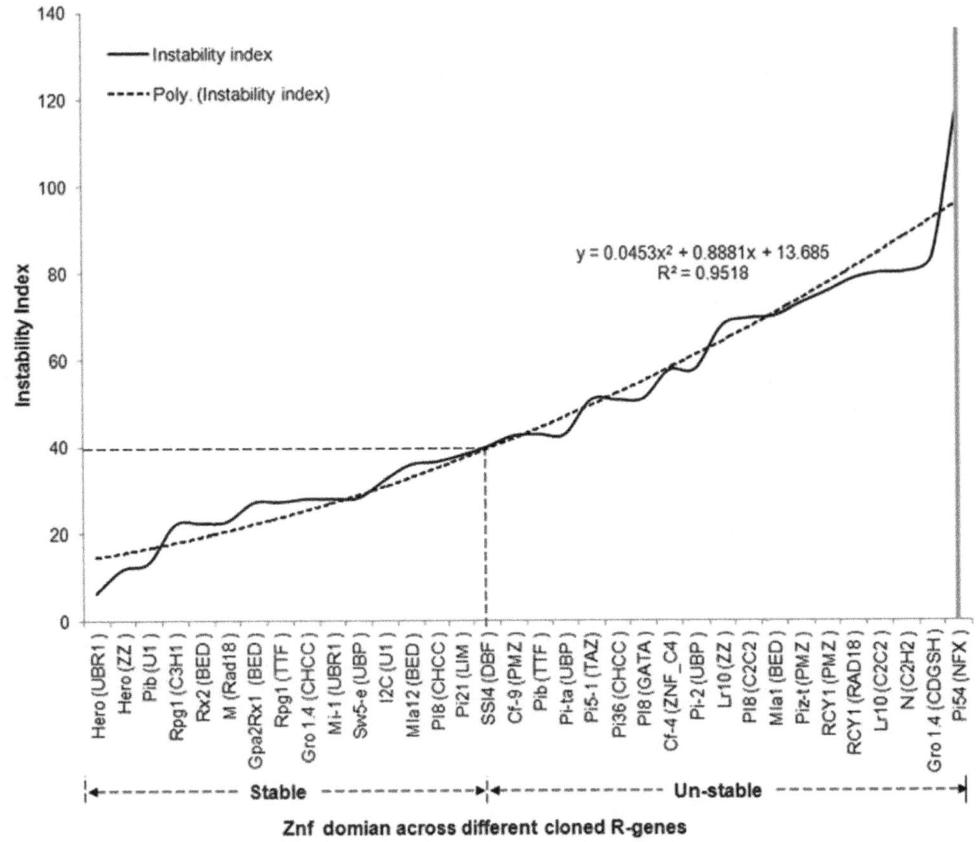

Figure 4. Instability index profile of identified zinc finger domain across cloned *R* gene. The data of 26 zinc finger *R* genes with their zinc finger types were included in this analysis.

pi21 is the smallest protein amongst all the zinc finger proteins which contains the LIM zinc finger domain [76]. This domain showed 12.3% sequence similarity (AA sequences) with zinc finger domain of Pi54 protein. The LIM domain is also a new functional motif deduced in *pi21* gene of rice and it contains a cysteine-rich motif of CX_2 - $CX_{17-19}HX_2CX_2CX_2CX_{16-20}CX_{2-3}C$. It was reported that LIM- containing proteins have been implicated in the transcriptional regulation of cell-differentiation and growth regulation and serve as the site for protein-protein interaction [17]. The similarity in amino acid sequences of zinc finger domains between Pi54 and pi21 supports our hypothesis that the LRR integrated NFX zinc finger of Pi54 might be involved in protein-protein interaction. We obtained 2.9% similarity between TAZ zinc finger domain of Pi5-1 and NFX domain of Pi54 proteins. The transcriptional adapter zinc binding (TAZ) domains are important sites for protein-protein interactions [77]. The similarity obtained between TAZ zinc finger domains of *Pi5-1* also supports our hypothesis. It was also reported that both the genes *Pi5-1* [78] and *Pi54* [26,33,35] expressed constitutively at a basal level in both transgenic as well as susceptible native lines up to 48 hours post inoculation and induced by *M. oryzae* infection in later hours. The pathogen inducible nature of both the genes also supports this analysis that similar amino acids in these proteins might play some important role in *R-Avr* interactions. The TTF zinc finger domain of Pib protein showed only 1.2% similarity with Pi54. The UBP zinc finger of Pi-z, PMZ zinc finger domain of Piz-t, UBP zinc finger of Pita, U1 zinc finger of Pib and CHCC of Pi36 protein were found to have no similarity with Pi54 zinc finger domain. The Lr10 disease resistance protein of wheat was found to have two zinc finger domains namely C2C2 and ZZ and have 10.5% and 7.6% similarity with zinc finger domain of Pi54 protein, respectively. The C2C2 zinc finger domain is reported as novel zinc finger in many disease resistance proteins of various crops like *Arabidopsis* (*LSD1& LOL1*) and its homologs in rice *OsLSD1* and *OsLOL1* that negatively regulates programmed cell death (PCD) and plant defense signaling pathways [19,30]. The similarity found between pairwise sequence alignment of zinc finger domains C2C2 and NFX showed that both are actively involved in negative regulation of up or down stream defense events. It has already been reported that NFX domains actively worked as negative regulators of various regulatory mechanisms which improves the physiological status of plants and supports growth and survival under various stress conditions. The expression of such type of zinc finger containing rice blast resistance gene *Pi54* was found under biotic stress conditions [64,79,35,33,26]. The ZZ zinc finger domain is a type of protein domain that was named because of its ability to bind two zinc ions [80]. The ZZ zinc finger domains containing proteins are found to be involved in protein-protein interactions and generation of hypersensitive responses [81]. The similarity between amino acid sequences of ZZ and NFX domains further supports the results of Pi54 zinc finger analysis. Tobacco *R* gene *N* having C2H2 zinc finger domain showed 6.6% similarity with NFX of *Pi54*. The C2H2 zinc finger proteins were mainly related to the plant development regulation or involved in various types of stress responses [82]. The one of the largest transcription factor families in eukaryotes are constituted by Cys2/His2-type zinc finger proteins [83]. Many stress-responsive C2H2-type zinc finger proteins have been identified in various plant species.

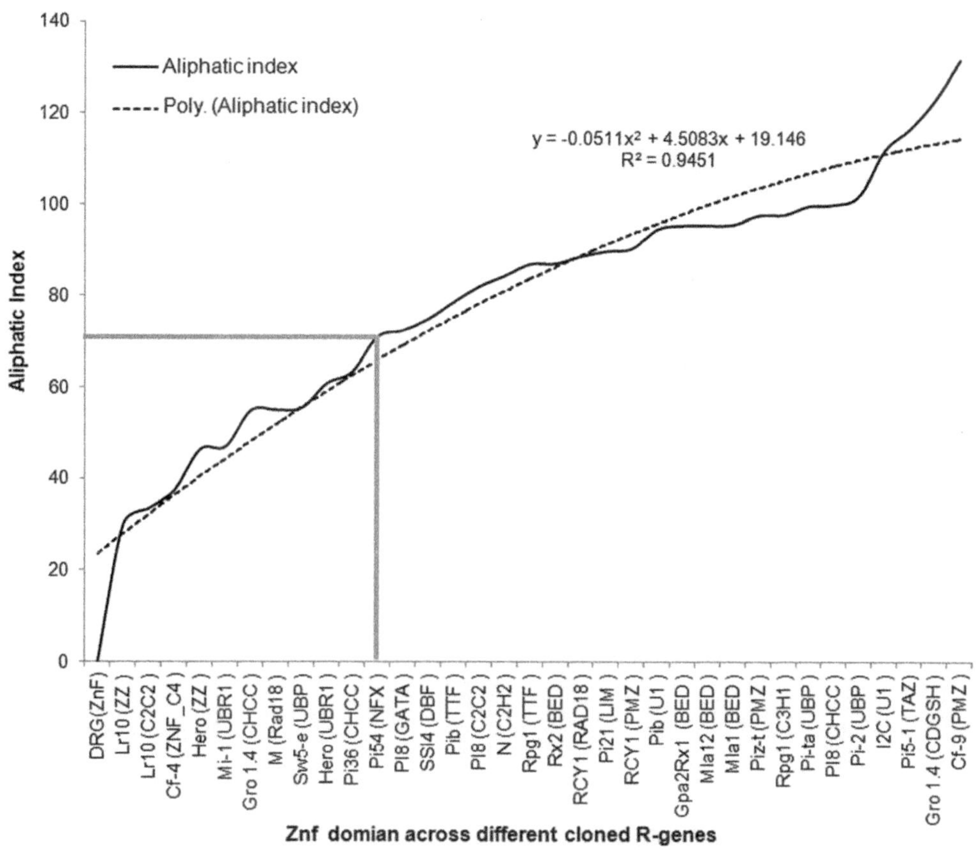

Figure 5. Aliphatic index profile of identified zinc finger domain across cloned *R* gene. The data of 26 zinc finger *R* genes with their zinc finger types were included in this analysis.

Several studies have reported that C2H2-type zinc finger proteins are responsible for both the activation of some stress-related genes and enhanced tolerance to salt, dehydration, and/or cold stresses [82]. The C2H2 zinc finger containing proteins are also reported for their ROS scavenging nature and enhanced expression of defense response genes in plants [84]. Hence, C2H2 zinc finger of N protein having similarity with Pi54 zinc finger also supports our results.

The *R* gene of sunflower *pl8* was found to be having three zinc finger domains viz C2C2, CHCC and GATA. Two zinc finger domains C2C2 and CHCC showed 6.4 and 4.4% similarities with NFX domain of Pi54 protein, respectively. These domains in pl8 protein were also present at -C terminal like that of the zinc finger domain of Pi54 protein. The CHCC zinc finger domains of pl8 protein are the variants of the classical zinc fingers like C2C2 and C2H2 [85]. The formation of CHCC is due to replacement of one cysteine residue by histidine residue. These are short zinc-finger domains conserved from fungi to humans with a consensus sequence of Cx8Hx14Cx2C. Both the domains, NFX and CHCC are short in length, have three cysteine and one histidine residues and structural fold comprising a β-hairpin followed by a short α-helix that adopt two different conformations. This makes the structure of such type of zinc fingers highly divergent from other eukaryotic zinc fingers and these motifs are reported to have a group of DNA binding proteins from *Archea* [86]. However, the third zinc finger domain GATA of this gene showed no similarity with NFX-*Pi54*. There were no significant similarity between Pi54 and other zinc finger domains of barley proteins. The zinc finger domains of Tomato disease resistance proteins, viz., Hero, I2C, Sw5-e and Mi-1 have 5.6% and 3.4%, 2.7% and 1.7% similarity with Pi54 protein, respectively. The zinc finger domains of Arabidopsis and Potato R proteins showed 5.7 and 2.1% sequence similarities with NFX zinc finger domain of Pi54 protein, respectively. However, no similarity was found between Rad18 zinc finger domain of Flax M protein and NFX of Pi54 (Figure 8).

An identity matrix shows the proportion of identical residues between all of the sequences in the alignment as they are originally aligned. A total of 561 combinations of pairwise sequence identity were generated with the given data sets, out of which 90.55% of the combinations shows less than 10% sequence identity (which represents a significant statistical support of zinc finger sequence divergence) whereas 8.73% of the combinations shows an identity between 10–25% and the highest pairwise sequence identity (80.8%) exist between Gpa2Rx1_BED-Potato & Rx2_BED-Potato followed by Cf-9_PMZ-Tomato & Piz-t_PMZ-Rice (40.0%), Mla1_BED-Barley & Rx2_BED-Potato (38.7%) and Mla1_BED-Barley & Gpa2Rx1_BED-Potato (36.7%).

Pairwise identity matrix created from the zinc finger proteins, zinc finger domains and 11AA trimmed sequence aligned with Pi54_NFX-Rice is given in Figure 9. The analysis showed that Pi21 protein has 0.5% identity with Pi54. In case of zinc finger domain, LIM zinc finger of *pi21*gene of rice showed maximum identity (10%) with the zinc finger domain (NFX) of Pi54 protein (Figure 9). The 15 zinc finger domain sequences analyzed in the present study did not show any identity with zinc finger domain of Pi54 protein (Figure 9). A total of fifteen zinc finger domains of proteins (Pib, Mla-1, GPa2Rx1, Rx2, Sw5-e, Pi5-1, Gro1.4, I2C, RCY1, pl8 & N) were found with increasing identity with zinc

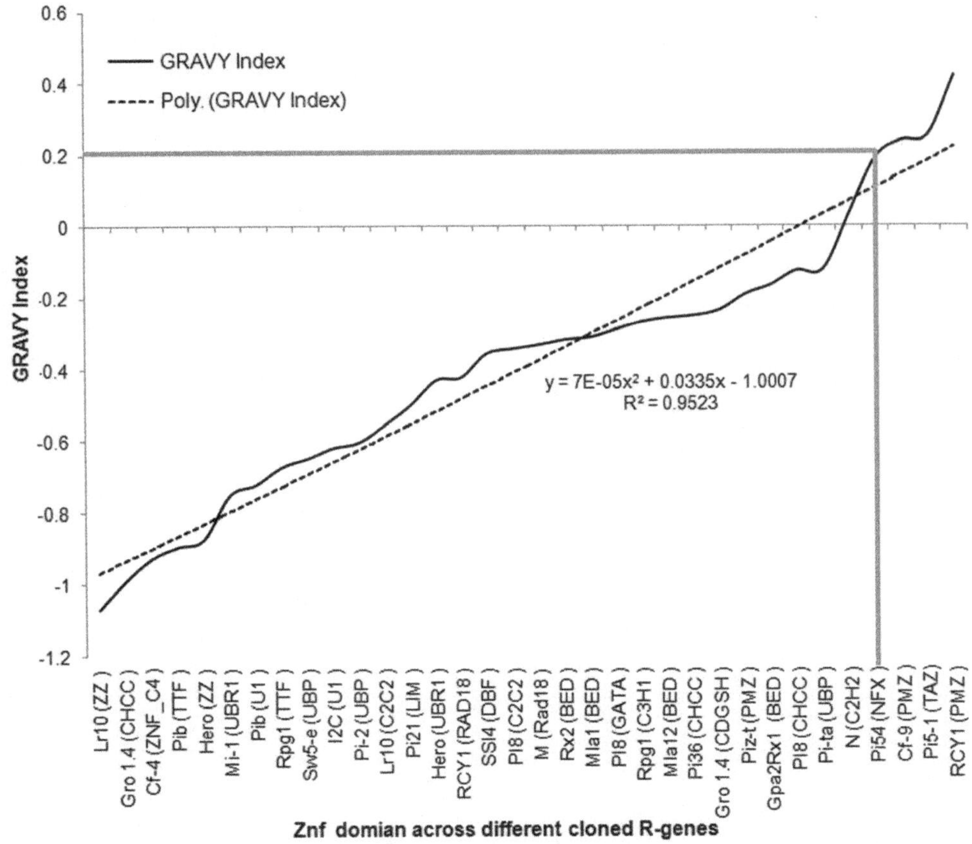

Figure 6. Grand Average of hydropathicity index profile of identified zinc finger domain across cloned *R* gene. The data of 26 zinc finger *R* genes with their zinc finger types were included in this analysis.

finger domain of Pi54 blast resistance protein. Three zinc finger domains of wheat (Lr10_ZZ-wheat, Lr10_C2C2-wheat), and rice (pi21_LIM-rice) were found to be very close to NFX zinc finger domain of Pi54 protein. These domains are known to be actively involved in protein-protein interaction and as a potential regulator of various regulatory mechanisms, besides helping to maintain the physiological status of these proteins in various abiotic and biotic stresses [17,64,79]. After alignment of the eleven amino acids of the Pi54 zinc finger with other zinc finger domains as trimmed sequence, we found that LIM zinc finger domain of rice blast resistance pi21 protein showed maximum similarity (more than 40%) with zinc finger domain of rice blast resistance protein Pi54. The TTF zinc finger of Pib, UBR1 of Mla1, BED of the GPa2Rx1 and UBP of Sw5-e showed 10% identity to the Pi54 zinc finger while TAZ zinc finger of Pi5-1 showed 20% identity. The CHCC zinc finger of pI8, UBR1 of Hero, DBF of SSI4 and BED zinc finger of Mla1 showed 30% identity. The C4 zinc finger of Cf-4, and TTF of Rpg1 protein showed no (0%) identity with the Pi54 zinc finger (Figure 9).

Conclusions

This study presents an analysis of plant disease resistance protein sequences in which zinc finger domains were found to be present along with other previously described domains like NBS and LRRs. The *Pi54* gene conferring durable resistance to blast disease in rice encodes a NBS-LRR protein along with a typical zinc finger domain. The zinc finger domain of Pi54 protein is NFX type and located on the C-terminal in between NBS and LRR domains of the R-protein. Compositional analysis depicted by the helical wheel diagram revealed the presence of a hydrophobic region within this domain which might help in exposing the LRR region for a possible *R-Avr* interaction. We also found the presence of different types of Zinc finger domains in rice blast resistance and in other plant disease resistance proteins. Maximum numbers of zinc finger domains were found in the proteins of disease resistance genes cloned from rice and tomato. Many disease resistance genes like *pI8* (Sunflower), *Pib* (Rice), *Lr10* (Wheat), *Gro1-4* (Potato), *RCY1* (Arabidopsis) and *Rpg1* (Barley) contains multiple number of zinc finger domains. The Pi54 protein contains the smallest zinc finger domain, despite the fact that smallest protein among these plant disease resistance proteins is pi21, another blast resistance gene. Instability, aliphatic and hydropathicity profile of these zinc finger domains gave a representation of the biochemical features of these proteins. We identified thirty four zinc finger domains in twenty six plant disease resistance proteins. These were found to be of nineteen different types of zinc fingers belonging to nine crops of five different families. Resistance proteins are known to play a crucial role in pathogen resistance by utilizing NBS and LRR domains for receiving signals from the pathogen. However, the presence of zinc finger domains, in combination with NBS-LRR domains in resistance proteins may reflect a major role of these domains in host- pathogen interaction.

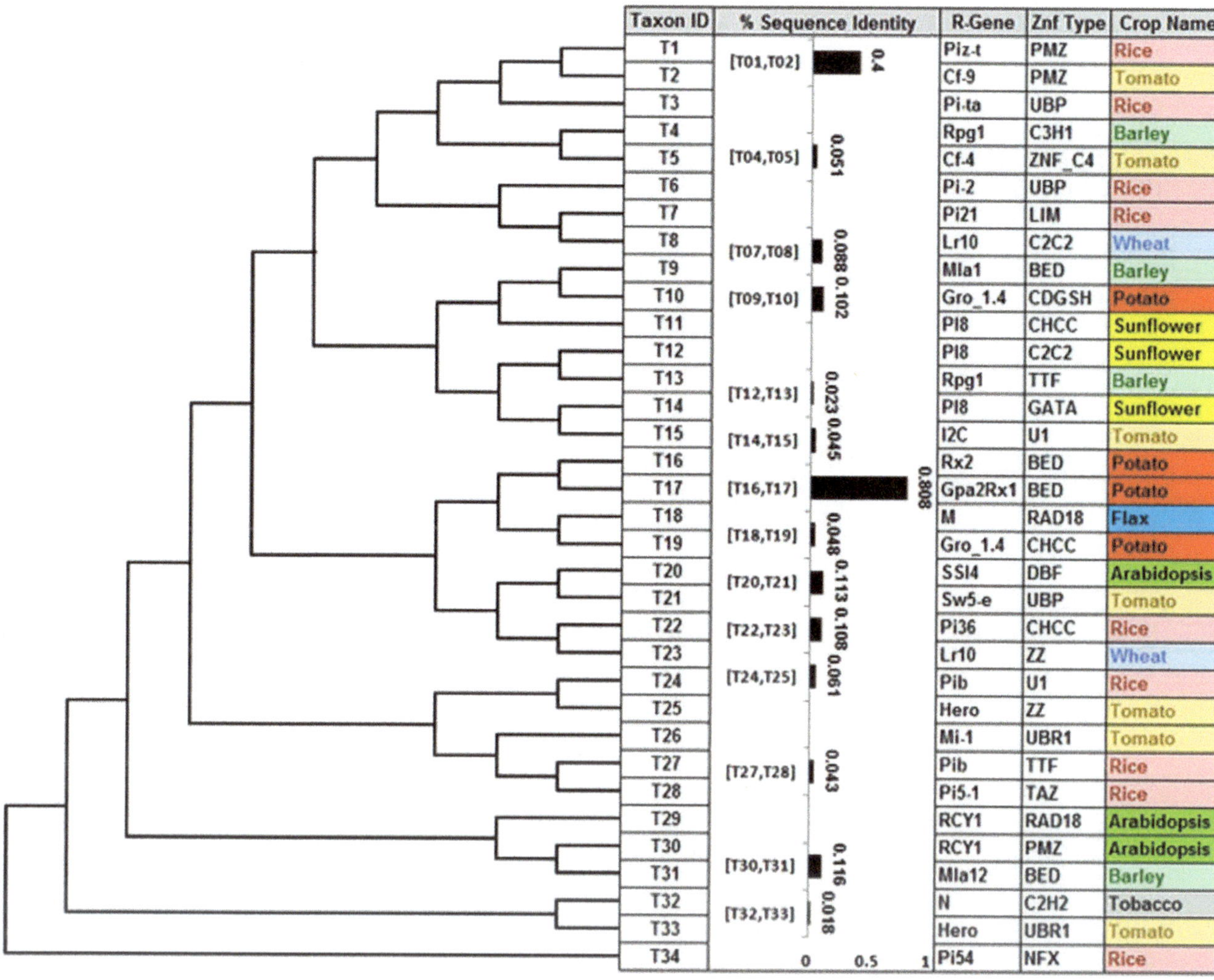

Figure 7. Maximum parsimony tree of 34 *R* genes containing zinc finger proteins. The bootstrap consensus tree (1000 replicates) is taken to represent the relationship between the taxa. Branches corresponding to partitions reproduced in less than 50% bootstrap replicates are collapsed. The percentage of replicate trees in which the associated taxa clustered together in the bootstrap test (1000 replicates) is shown next to the branches. The name of *R* gene, zinc finger types and crops are given on the termini of branches.

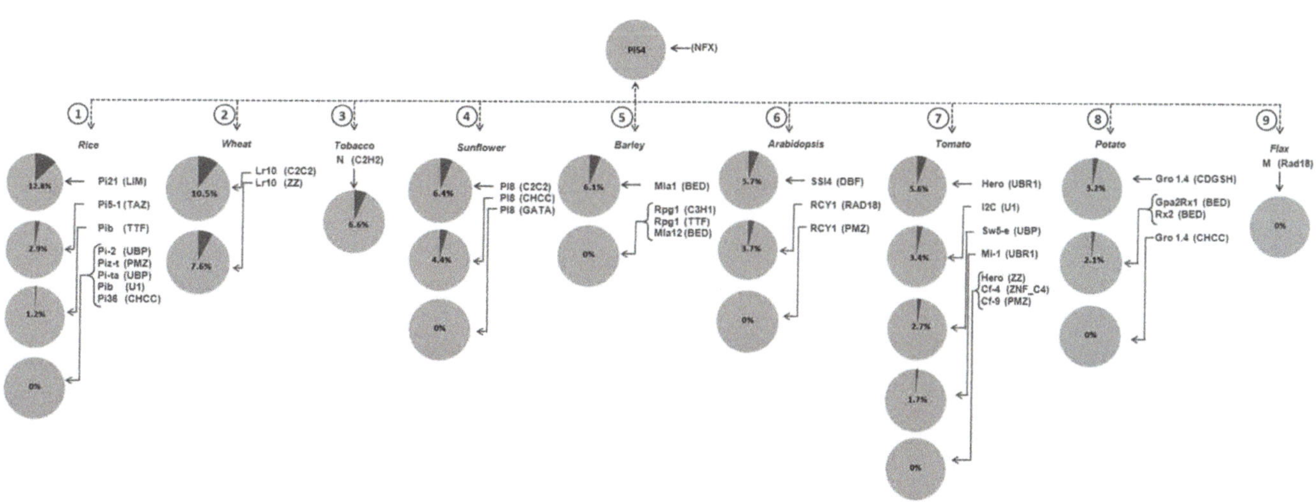

Figure 8. Identity plot of zinc finger domains. Identity plot of zinc fingers represents the identity analysis between amino acid sequences of Pi54 zinc finger domain and zinc finger domains present in cloned *R* genes of various crops.

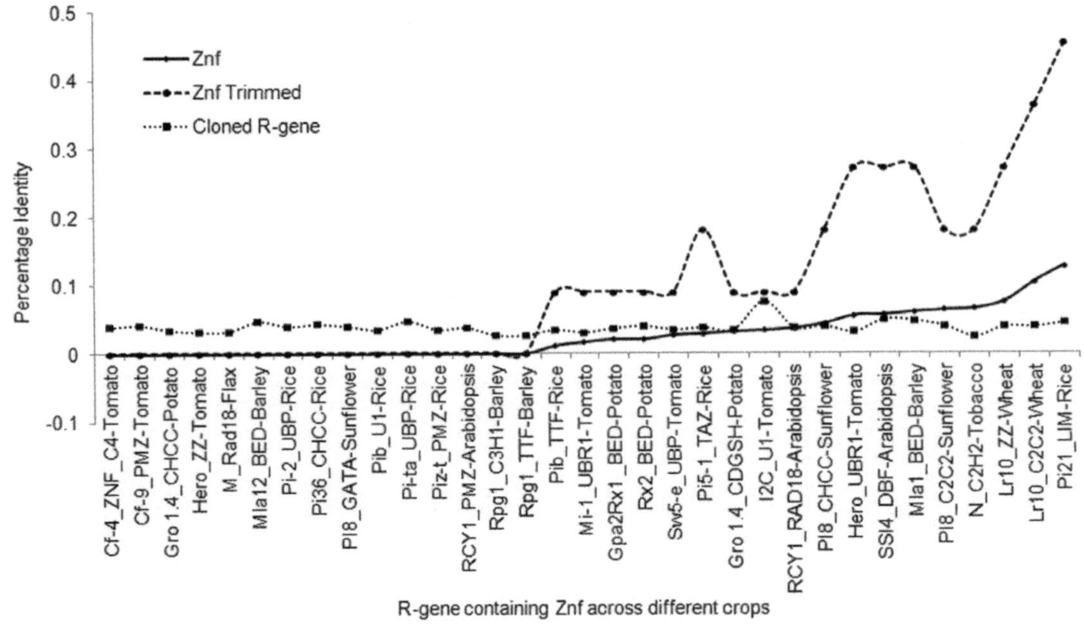

Figure 9. Pair-wise percentage identity profile. The Graphical representation showed the pair wise percentage identity profile of zinc finger domains of cloned *R* genes with respect to NFX domain of *Pi54* gene. This analysis includes the comparison between the complete protein sequences of zinc finger containing R-proteins, Zinc finger domains of these proteins and 11 amino acid trimmed sequences showed similarity with *Pi54* zinc finger domains. The X-axis of the graph represents the zinc finger containing cloned *R* genes across different crops whereas Y-axis represent the percentage identity profile between zinc fingers with respect to Pi54.

Materials and Methods

In- silico analysis of zinc finger motif

The *in-silico* examination was performed to deduce the zinc finger domain in cloned *R* genes. The protein sequences of more than 70 genes present in National Centre for Biotechnology Information (NCBI) database (www.ncbi.nlm.nih.gov) were downloaded and analyzed (Table S1). The Expasy proteomic tool SMART (Simple Modular Architecture Research Tool) (www.smart.embl-heidelberg.de) was used for the identification of zinc finger motif [87,88]. The identified zinc finger domain sequences were used for multiple sequence alignment using Clustal X version 1.83 (www.clustal.org) and Bio-edit 2.0 (http://www.mbio.ncsu.edu/bioedit) using default parameters [89,90].

Biophysical characterization

The structural and functional prediction of *Pi-k^h^* (*Pi54*) and other ZnF domains were studied using Expasy proteomic tool Protparam (http://us.expasy.org/tools/protparam.html) [70]. The physico-chemical parameters like Molecular weight, theoretical pI, instability index [66], aliphatic index [91] and grand average of hydropathicity (GRAVY) [70] were computed using Expasy's ProtParam Proteomics server. The Grand Average hydropathy (GRAVY) value for a peptide or protein is calculated as the sum of hydropathy values of all the amino acids, divided by the number of residues in the sequence.

Phylogenetic analyses

The predicted zinc finger domains of R-proteins were further undertaken for phylogenetic analysis. The aligned sequences were inspected and adjusted manually to minimize the number of gaps and insertions. The manual adjustments were based on the sequence similarities. The phylogenetic tree was constructed according to the Neighbor-Joining method and visualized by MEGA program version 4.0 [56]. To validate the reproducibility of the branching pattern, bootstrap analysis (1000 replicates) and distance analysis were performed.

Comparative pairwise identity profiling of *Pi54* zinc finger domain across different crops

For the pairwise identity between the zinc finger domain of blast resistance gene Pi54 and the genes cloned from other crops, 2-D pairwise identity data matrix was generated using Bio-Edit version 5.0.6 [89]. Further, 2-D pairwise identity data matrix was resolved with respect to Pi54 to delineate its comparative status in relation to other zinc finger domains across different crops.

Supporting Information

Supplementary References List of References for Supplementary files.

Table S1 The cloned plant disease resistance genes and their specific features.

Table S2 Distribution of Zinc finger R-proteins across different crops.

Table S3 Details of different Zinc finger domains across various crops cloned *R* genes.

Table S4 Biophysical parameters of identified Zinc finger domains across various cloned *R* genes.

Acknowledgments

Authors are thankful to Mr. Niraj Kumar of Xplorigen Technologies, New Delhi for his invaluable contribution in statistical data analyses, making graphics and Mr. Deepak Kumar Gupta for his help in domain architecture analysis of zinc finger proteins.

Author Contributions

Conceived and designed the experiments: TRS. Analyzed the data: SKG AKR. Wrote the paper: SKG AKR TRS. Provided input for data analysis: SSK.

References

1. Feurtado JA, Huang D, Wicki-Stordeur L, Hemstock LE, Potentier MS, et al. (2011) The Arabidopsis C2H2 zinc finger Indeterminate Domain1/Enhydrous promotes the transition to germination by regulating light and hormonal signaling during seed maturation. The Plant Cell 23: 1772–1794.
2. Giri J, Vij S, Dansana PK, Tyagi AK (2011) Rice A20/AN1 zinc-finger containing stress-associated proteins (SAP1/11) and a receptor-like cytoplasmic kinase (OsRLCK253) interact via A20 zinc-finger and confer abiotic stress tolerance in transgenic Arabidopsis plants. New Phytol 191: 721–732.
3. Laity JH, Lee BM, Wright PE (2001) Zinc finger proteins: New insights into structural and functional diversity. Curr Opin Struct Biol 11: 39–46.
4. Gourcilleau D, Lenne C, Armenise C, Moulia B, Julien J, et al. (2011) Phylogenetic study of plant Q-type C2H2 zinc finger proteins and expression analysis of poplar genes in response to osmotic, cold and mechanical stresses. DNA Research 18: 77–92.
5. Kodaira K, Qin F, Tran LP, Maruyama K, Kidokoro S, et al. (2011) Arabidopsis Cys2/His2 zinc-finger proteins AZF1 and AZF2 negatively regulate abscisic acid-repressive and auxin-inducible genes under abiotic stress conditions. Plant Physiol 157: 742–756.
6. Parthasarathy S, Murthy MR (2000) Protein thermal stability: Insights from atomic displacement parameters (B values). Protein Engg 13: 9–13.
7. Boocock GRB, Marit MR, Rommens JM (2006) Phylogeny, sequence conservation, and functional complementation of the SBDS protein family. Genomics 87: 758–771.
8. Emerson RO, Thomas JH (2009) Adaptive evolution in zinc finger transcription factors. PLoS Genet 5: e1000325.
9. Englbrecht CC, Schoof H, Böhm S (2004) Conservation, diversification and expansion of C2H2 zinc finger proteins in the *Arabidopsis thaliana* genome. BMC Genomics 5: 39.
10. Mackay JP, Crossley M (1998) Zinc fingers are sticking together. Trends Biochem Sci 23: 1–4.
11. Klug A, Rhodes D (1987) 'Zinc fingers': a novel protein motif for nucleic acid recognition. Trends Biochem Sci 12: 464–469.
12. Klug A (2005) Towards therapeutic applications of engineered zinc finger proteins. FEBS Lett 579: 892–894.
13. Berg JM, Shi Y (1996) The galvanization of biology: A growing appreciation for the roles of zinc. Science 271: 1081–1085.
14. Jenkins TH, Li J, Scutt CP, Gilmartin PM (2004) Analysis of members of the *Silene latifolia* Cys2/His2 zinc-finger transcription factor family during dioecious flower development and in a novel stamen-defective mutant ssf1. Planta 220: 559–571.
15. Ori N, Eshed Y, Paran I, Presting G, Aviv D, et al. (1997) The *I2C* family from the wilt disease resistance locus I2 belongs to the nucleotide binding, leucine-rich repeat superfamily of plant resistance genes. Plant Cell 9: 521–532.
16. Schumann U, Prestele J, O'Geen H, Brueggeman R, Wanner G, et al. (2007) Requirement of the C3HC4 zinc RING finger of the Arabidopsis *PEX10* for photorespiration and leaf peroxisome contact with chloroplasts. Proc Natl Acad Sci U S A 104: 1069–1074.
17. Takatsuji H (1999) Zinc-finger proteins: The classical zinc finger emerges in contemporary plant science. Plant Mol Biol 39: 1073–1078.
18. Ciftci-Yilmaz S, Mittler R (2008) The zinc finger network of plants. Cell Mol Life Sci 65: 1150–1160.
19. Dietrich RA, Richberg MH, Schmidt R, Dean C, Dangl JL (1997) A novel zinc finger protein is encoded by the Arabidopsis *LSD1* gene and functions as a negative regulator of plant cell death. Cell 88: 685–694.
20. Epple P, Mack AA, Morris VRF, Dangl JL (2003) Antagonistic control of oxidative stress-induced cell death in arabidopsis by two related, plant-specific zinc finger proteins. Proc Natl Acad Sci U S A 100: 6831–6836.
21. Meyers BC, Chin DB, Shen KA, Sivaramakrishnan S, Lavelle DO, et al. (1998) The major resistance gene cluster in lettuce is highly duplicated and spans several megabases. Plant Cell 10: 1817–1832.
22. Głowacki S, Macioszek VK, Kononowicz AK (2011) R proteins as fundamentals of plant innate immunity. Cell Mol Biol Lett 16: 1–24.
23. Aravind L (2000) The BED finger, a novel DNA-binding domain in chromatin-boundary-element-binding proteins and transposases. Trends Biochem Sci 25: 421–423.
24. Tuskan GA, DiFazio S, Jansson S, Bohlmann J, Grigoriev I, et al. (2006) The genome of black cottonwood, *Populus trichocarpa* (torr. & gray). Science 313: 1596–1604.
25. Deslandes L, Olivier J, Peeters N, Feng DX, Khounlotham M, et al. (2003) Physical interaction between RRS1-R, a protein conferring resistance to bacterial wilt, and PopP2, a type III effector targeted to the plant nucleus. Proc Natl Acad Sci U S A 100: 8024–8029.
26. Sharma TR, Madhav MS, Singh BK, Shanker P, Jana TK, et al. (2005) High-resolution mapping, cloning and molecular characterization of the *Pi-k^h* gene of rice, which confers resistance to *Magnaporthe grisea*. Mol Genet Genomics 274: 569–578.
27. Sharma TR, Rai AK, Gupta SK, Vijyan J, Devanna BN, et al. (2012) Rice Blast Management through Host Resistance: Retrospect and Prospects. Agric Res 1: 37–52.
28. Sharma TR, Shanker P, Singh BK, Jana TK, Madhav MS, et al. (2005) Molecular mapping of rice blast resistance gene *Pi-k^h* in rice variety Tetep. J Plant Biochem Biotechnol 14: 127–133.
29. Sonah H, Deshmukh RK, Singh VP, Gupta DK, Singh NK, et al. (2011) Genomic resources in horticultural crops: Status, utility and challenges. Biotechnol Adv 29: 199–209.
30. Wang ZX, Yano M, Yamanouchi U, Iwamoto M, Monna L, et al. (1999) The *Pib* gene for rice blast resistance belongs to the nucleotide binding and leucine-rich repeat class of plant disease resistance genes. Plant J 19: 55–64.
31. Bryan GT, Wu K, Farrall L, Jia Y, Hershey HP, et al. (2000) A single amino acid difference distinguishes resistant and susceptible alleles of the rice blast resistance gene *Pi-ta*. Plant Cell 12: 2033–2046.
32. Sharma TR, Rai AK, Gupta SK, Singh NK (2010) Broad-spectrum Blast Resistance Gene *Pi-k^h* Cloned from Rice Line Tetep Designated as *Pi54*. J Plant Biochem Biotechnol 19: 87–89.
33. Rai AK, Kumar SP, Gupta SK, Gautam N, Singh NK, et al. (2011) Functional complementation of rice blast resistance gene *Pi-k^h* (*Pi54*) conferring resistance to diverse strains of *Magnaporthe oryzae*. J Plant Biochem Biotechnol 20: 55–65.
34. Costanzo S, Jia Y (2010) Sequence variation at the rice blast resistance gene *Pi-km* locus: Implications for the development of allele specific markers. Plant Science 178: 523–530.
35. Gupta SK, Rai AK, Kanwar SS, Chand D, Singh NK, et al. (2012) The single functional blast resistance gene *Pi54* activates a complex defence mechanism in rice. J Exp Bot 63: 757–772.
36. Das A, Soubam D, Singh PK, Thakur S, Singh NK, et al. (2012) A novel blast resistance gene, *Pi54rh* cloned from wild species of rice, *Oryza rhizomatis* confers broad spectrum resistance to *Magnaporthe oryzae*. Funct Integr Genomics 12: 215–228.
37. Madhav MS, Plaha P, Singh NK, Sharma TR (2009) Molecular Characterization of a Genomic Fragment Containing *Pi-k^h* Gene from the Genomic Library of *indica* Rice Line Tetep. J Phytopathol 157: 322–324.
38. Eulgem T, Somssich IE (2007) Networks of WRKY transcription factors in defense signaling. Curr Opin Plant Biol 10: 366–371.
39. Ulker B, Somssich IE (2004) WRKY transcription factors: From DNA binding towards biological function. Curr Opin Plant Biol 7: 491–498.
40. Klug A, Schwabe JW (1995) Protein motifs 5. zinc fingers. FASEB J 9: 597–604.
41. Zhou X, Wang Y, Pan Y, Li W (2007) Differences in amino acids composition and coupling patterns between mesophilic and thermophilic proteins. Amino Acids 34: 25–33.
42. Yu H, Li J, Zhang D, Yang Y, Jiang W, et al. (2008) Improving the thermostability of N-carbamyl-d-amino acid amidohydrolase by error-prone PCR. Appl Microbiol Biotechnol 82: 279–285.
43. Pack SP, Yoo YJ (2004) Protein thermostability: Structure-based difference of amino acid between thermophilic and mesophilic proteins. J Biotechnol 111: 269–277.
44. Lawrence GJ, Finnegan EJ, Ayliffe MA, Ellis JG (1995) The *L6* gene for flax rust resistance is related to the Arabidopsis bacterial resistance gene *RPS2* and the tobacco viral resistance gene *N*. Plant Cell 7: 1195–1206.
45. Watanabe K, Iwabuchi K, Sun J, Tsuji Y, Tani T, et al. (2009) RAD18 promotes DNA double-strand break repair during G1 phase through chromatin retention of 53BP1. Nucleic Acids Res 37: 2176–2193.
46. Van den Burg B, Vriend G, Veltman OR, Venema G, Eijsink VGH (1998) Engineering an enzyme to resist boiling. Proc Natl Acad Sci U SA 95: 2056–2060.
47. Mattos C (2002) Protein-water interactions in a dynamic world. Trends Biochem Sci 27: 203–208.
48. Xu J, Baase WA, Quillin ML, Baldwin EP, Matthews BW (2001) Structural and thermodynamic analysis of the binding of solvent at internal sites in T4 lysozyme. Protein Science 10: 1067–1078.
49. Russell RJ, Hough DW, Danson MJ, Taylor GL (1994) The crystal structure of citrate synthase from the *Thermophilic archaeon*, *Thermoplasma acidophilum*. Structure 2: 1157–1167.
50. Cammack R, Johnson CE, Hall DO, Rao KK (1971) The state of the iron atoms in hydroxylase iron-sulphur proteins. Biochem J 125: 18–19.
51. Miseta A, Csutora P (2000) Relationship between the occurrence of cysteine in proteins and the complexity of organisms. Mol Biol Evol 17: 1232–1239.
52. Dill KA (1990) Dominant forces in protein folding. Biochemistry 29: 7133–7155.

53. Kumar S, Nei M, Dudley J, Tamura K (2008) MEGA: biologist-centric software for evolutionary analysis of DNA and protein sequences. Brief Bioinformatics 9: 299–306.
54. Flor HH (1971) Current status of the gene-for- gene concept. Annu Rev Phytopathol 9: 275–296.
55. Xiao L, Honig B (1999) Electrostatic contributions to the stability of hyperthermophilic proteins. J Mol Biol 289: 1435–1444.
56. Tamura K, Dudley J, Nei M, Kumar S (2007) MEGA4: Molecular evolutionary genetics analysis (MEGA) software version 4.0. Mol Biol Evol 24: 1596–1599.
57. Schiffer M, Edmundson AB (1967) Use of helical wheels to represent the structures of proteins and to identify segments with helical potential. Biophys J 7: 121–135.
58. Kamtekar S, Schiffer JM, Xiong H, Babik JM, Hecht MH (1993) Protein design by binary patterning of polar and nonpolar amino acids. Science 262: 1680–1685.
59. Baldwin JM (1993) The probable arrangement of the helices in G protein-coupled receptors. EMBO J 12: 1693–1703.
60. Song Z, Krishna S, Thanos D, Strominger JL, Ono SJ (1994) A novel cysteine-rich sequence-specific DNA-binding protein interacts with the conserved X-box motif of the human major histocompatibility complex class II genes via a repeated cys-his domain and functions as a transcriptional repressor. J Exp Med 180: 1763–1774.
61. Müssig C, Schröder F, Usadel B, Lisso J (2010) Structure and putative function of NFX1-like proteins in plants. Plant Biology 12: 381–394.
62. Asano T, Yasuda M, Nakashita H, Kimura M, Yamaguchil K, et al. (2008) The AtNFXLl gene functions as a signaling component of the type A trichothecene-dependent response. Plant Signal Behav 3: 991–992.
63. Larkindale J, Vierling E (2008) Core genome responses involved in acclimation to high temperature. Plant Physiol 146: 748–761.
64. Lisso J, Altmann T, Müssig C (2006) The *AtNFXL1* gene encodes a NF-X1 type zinc finger protein required for growth under salt stress. FEBS Lett 580: 4851–4856.
65. Arora N, Banerjee AK, Mutyala S, Murty US (2009) Comparative characterization of commercially important xylanase enzymes. Bioinformation 3: 446–453.
66. Guruprasad K, Reddy BV, Pandit MW (1990) Correlation between stability of a protein and its dipeptide composition: A novel approach for predicting in vivo stability of a protein from its primary sequence. Protein Engg 4: 155–161.
67. Roy S, Maheshwari N, Chauhan R, Sen NK, Sharma A (2011) Structure prediction and functional characterization of secondary metabolite proteins of Ocimum. Bioinformation 6: 315–319.
68. Liao H, Yeh W, Chiang D, Jernigan RL, Lustig B (2005) Protein sequence entropy is closely related to packing density and hydrophobicity. Protein Eng Des Sel 18: 59–64.
69. Ponnuswamy PK, Muthusamy R, Manavalan P (1982) Amino acid composition and thermal stability of proteins. Int J Biol Macromol 4: 186–190.
70. Kyte J, Doolittle RF (1982) A simple method for displaying the hydropathic character of a protein. J Mol Biol 157: 105–132.
71. Sahay A, Shakya M (2010) *In-silico* analysis and homology modelling of antioxidant proteins of spinach. J Proteomics Bioinform 3: 148–154.
72. Xing W, Zou Y, Liu Q, Liu J, Luo X, et al. (2007) The structural basis for activation of plant immunity by bacterial effector protein AvrPto. Nature 449: 243–247.
73. Hacquard S, Petre B, Frey P, Hecker A, Rouhier N, et al. (2011) The poplar-poplar rust interaction: Insights from genomics and transcriptomics. Mol Plant Microbe Interact 25: 279–293.
74. Breeze E, Harrison E, McHattie S, Hughes L, Hickman R, et al. (2011) High-resolution temporal profiling of transcripts during Arabidopsis leaf senescence reveals a distinct chronology of processes and regulation. The Plant Cell 23: 873–894.
75. Barth C, Moeder W, Klessig DF, Conklin PL (2004) The timing of senescence and response to pathogens is altered in the ascorbate-deficient Arabidopsis mutant vitamin C-1. Plant Physiol 134: 1784–1792.
76. Fukuoka S, Saka N, Koga H, Ono K, Shimizu T, et al. (2009) Loss of function of a proline-containing protein confers durable disease resistance in rice. Science 325: 998–1001.
77. De Guzman RN, Wojciak JM, Martinez-Yamout MA, Dyson HJ, Wright PE (2005) CBP/p300 TAZ1 domain forms a structured scaffold for ligand binding. Biochemistry 44: 490–497.
78. Lee S, Song M, Seo Y, Kim H, Ko S, et al. (2009) Rice *Pi5*-mediated resistance to *Magnaporthe oryzae* requires the presence of two coiled-coil-nucleotide-binding-leucine-rich repeat genes. Genetics 181: 1627–1638.
79. Krishna SS, Majumdar I, Grishin NV (2003) Structural classification of zinc fingers: Survey and summary. Nucleic Acids Res 31: 532–550.
80. Ponting CP, Blake DJ, Davies KE, Kendrick-Jones J, Winder SJ (1996) ZZ and TAZ: New putative zinc fingers in dystrophin and other proteins. Trends Biochem Sci 21: 11–13.
81. Kang X, Chong J, Ni M (2005) Hypersensitive to red and blue 1, a ZZ-type zinc finger protein, regulates phytochrome B–Mediated red and cryptochrome-mediated blue light responses. The Plant Cell 17: 822–835.
82. Kim S, Park J, Park S, Mitchell TK, Lee Y (2010) Identification and analysis of in planta expressed genes of *Magnaporthe oryzae*. BMC Genomics 11: 104–104.
83. Kubo K, Sakamoto A, Kobayashi A, Rybka Z, Kanno Y, et al. (1998) Cys2/His2 zinc-finger protein family of petunia: Evolution and general mechanism of target-sequence recognition. Nucleic Acids Res 26: 608–615.
84. Sun S, Guo S, Yang X, Bao Y, Tang H, et al. (2010) Functional analysis of a novel Cys2/His2-type zinc finger protein involved in salt tolerance in rice. J Exp Bot 61: 2807–2818.
85. Matthews JM, Sunde M (2002) Zinc fingers–folds for many occasions. IUBMB Life 54: 351–355.
86. Cornilescu CC, Porter FW, Zhao KQ, Palmenberg AC, Markley JL (2008) NMR structure of the mengovirus leader protein zinc-finger domain. FEBS Lett 582: 896–900.
87. Letunic I, Doerks T, Bork P (2012) SMART 7: recent updates to the protein domain annotation resource. Nucleic Acids Res 40: 302–305.
88. Schultz J, Milpetz F, Bork P, Ponting CP (1998) SMART, a simple modular architecture research tool: Identification of signaling domains, Proc Natl Acad Sci U S A 95: 5857–5864.
89. Hall TA (1999) BioEdit: A user-friendly biological sequence alignment editor and analysis program for windows 95/98/NT. Nucleic Acids Symp Ser 41: 95–98.
90. Thompson JD, Gibson TJ, Plewniak F, Jeanmougin F, Higgins DG (1997) The CLUSTAL_X windows interface: Flexible strategies for multiple sequence alignment aided by quality analysis tools. Nucleic Acids Res 25: 4876–4882.
91. Ikai A (1980) Thermostability and aliphatic index of globular proteins. J Biochem 88: 1895–1898.
92. Shirano Y, Kachroo P, Shah J, Klessig DF (2002) A gain-of-function mutation in an Arabidopsis Toll Interleukin1 receptor-nucleotide binding site-leucine-rich repeat type *R* gene triggers defense responses and results in enhanced disease resistance. Plant Cell 14: 3149–3162.
93. Takahashi H, Miller J, Nozaki Y, Takeda M, Shah J, et al. (2002) *RCY1*, an Arabidopsis thaliana *RPP8/HRT* family resistance gene, conferring resistance to cucumber mosaic virus requires salicylic acid, ethylene and a novel signal transduction mechanism. Plant J 32: 655–667.
94. Zhou F, Kurth J, Wei F, Elliott C, Valè G, et al. (2001) Cell-Autonomous Expression of Barley *Mla1* Confers Race-Specific Resistance to the Powdery Mildew Fungus Via a *Rar1*-Independent Signaling Pathway. The Plant Cell 13: 337–350.
95. Shen Q, Zhou F, Bieri S, Haizel T, Shirasu K, et al. (2003) Recognition specificity and *RAR1/SGT1* dependence in barley *Mla* disease resistance genes to the powdery mildew fungus. Plant Cell 15: 732–744.
96. Brueggeman R, Rostoks N, Kudrna D, Kilian A, Han F, et al. (2002) The barley stem rust-resistance gene *Rpg1* is a novel disease-resistance gene with homology to receptor kinases. Proc Natl Acad Sci U S A 99: 9328–9333.
97. Anderson PA, Lawrence GJ, Morrish BC, Ayliffe MA, Finnegan EJ, et al. (1997) Inactivation of the flax rust resistance gene *M* associated with loss of a repeated unit within the leucine-rich repeat coding region. The Plant Cell 9: 641–651.
98. Bendahmane A, Querci M, Kanyuka K, Baulcombe DC (2000) Agrobacterium transient expression system as a tool for the isolation of disease resistance genes: Application to the *Rx2* locus in potato. Plant J 21: 73–81.
99. van der Biezen EA, Freddie CT, Kahn K, Parker JE, Jones JDG (2002) Arabidopsis *RPP4* is a member of the *RPP5* multigene family of TIR-NB-LRR genes and confer downy mildew resistance through multiple signalling components. Plant J 29: 439–451.
100. Paal J, Henselewski H, Muth J, Meksem K, Menéndez CM, et al. (2004) Molecular cloning of the potato *Gro1-4* gene conferring resistance to pathotype Ro1 of the root cyst nematode *Globodera rostochiensis*, based on a candidate gene approach. Plant J 38: 285–297.
101. Liu X, Lin F, Wang L, Pan Q (2007) The *in-silico* map-based cloning of *Pi36*, a rice coiled-coil nucleotide-binding site leucine-rich repeat gene that confers race-specific resistance to the blast fungus. Genetics 176: 2541–2549.
102. Zhou B, Qu S, Liu G, Dolan M, Sakai H, et al. (2006) The eight amino-acid differences within three leucine-rich repeats between *Pi2* and *Piz-t* resistance proteins determine the resistance specificity to *Magnaporthe grisea*. Mol Plant Microbe Interact 19: 1216–1228.
103. Radwan O, Mouzeyar S, Nicolas P, Bouzidi MF (2005) Induction of a sunflower CC-NBS-LRR resistance gene analogue during incompatible interaction with *Plasmopara halstedii*. J Exp Bot 56: 567–575.
104. Konagaya K, Matsushita Y, Kasahara M, Nyunoya H (2004) Members of 14-3-3 protein isoforms interacting with the resistance gene product *N* and the elicitor of tobacco mosaic virus. J Gen Plant Pathol 70: 221–231.
105. Vos P, Simons G, Jesse T, Wijbrandi J, Heinen L, et al. (1998) The tomato *Mi-1* gene confers resistance to both root-knot nematodes and potato aphids. Nature Biotechnol 16: 1365–1369.
106. Brommonschenkel SH, Frary A, Frary A, Tanksley SD (2000) The broad-spectrum tospovirus resistance gene *Sw5-e* of tomato is a homolog of the root-knot nematode resistance gene *Mi*. Mol Plant-Microbe Interact 13: 1130–1138.
107. Jones DA, Thomas CM, Hammond-Kosack KE, Balint-Kurti PJ, Jones JD (1994) Isolation of the tomato *Cf-9* gene for resistance to *Cladosporium fulvum* by transposon tagging. Science 266: 789–793.
108. Ernst K, Kumar A, Kriseleit D, Kloos D, Phillips MS, et al. (2002) The broad-spectrum potato cyst nematode resistance gene (*hero*) from tomato is the only member of a large gene family of NBS-LRR genes with an unusual amino acid repeat in the LRR region. Plant J 31: 127–136.

109. Parniske M, Hammond-Kosack KE, Golstein C, Thomas CM, Jones DA, et al. (1997) Novel disease resistance specificities result from sequence exchange between tandemly repeated genes at the *Cf-4/9* locus of tomato. Cell 91: 821–832.

110. Feuillet C, Schachermayr G, Keller B (1997) Molecular cloning of a new receptor-like kinase gene encoded at the *Lr10* disease resistance locus of wheat. Plant J 11: 45–52.

Heterotic Trait Locus (HTL) Mapping Identifies Intra-Locus Interactions That Underlie Reproductive Hybrid Vigor in *Sorghum bicolor*

Imri Ben-Israel[1], Benjamin Kilian[2], Habte Nida[1], Eyal Fridman[1]*

1 The Robert H. Smith Institute of Plant Sciences and Genetics in Agriculture, Faculty of Agriculture, Food and Environment, The Hebrew University of Jerusalem, Israel, **2** Institute of Plant Genetics and Crop Plant Research (IPK), Gatersleben, Germany

Abstract

Identifying intra-locus interactions underlying heterotic variation among whole-genome hybrids is a key to understanding mechanisms of heterosis and exploiting it for crop and livestock improvement. In this study, we present the development and first use of the heterotic trait locus (HTL) mapping approach to associate specific intra-locus interactions with an overdominant heterotic mode of inheritance in a diallel population using *Sorghum bicolor* as the model. This method combines the advantages of ample genetic diversity and the possibility of studying non-additive inheritance. Furthermore, this design enables dissecting the latter to identify specific intra-locus interactions. We identified three HTLs (3.5% of loci tested) with synergistic intra-locus effects on overdominant grain yield heterosis in 2 years of field trials. These loci account for 19.0% of the heterotic variation, including a significant interaction found between two of them. Moreover, analysis of one of these loci (*hDPW4.1*) in a consecutive F2 population confirmed a significant 21% increase in grain yield of heterozygous vs. homozygous plants in this locus. Notably, two of the three HTLs for grain yield are in synteny with previously reported overdominant quantitative trait loci for grain yield in maize. A mechanism for the reproductive heterosis found in this study is suggested, in which grain yield increase is achieved by releasing the compensatory tradeoffs between biomass and reproductive output, and between seed number and weight. These results highlight the power of analyzing a diverse set of inbreds and their hybrids for unraveling hitherto unknown allelic interactions mediating heterosis.

Editor: Jinfa Zhang, New Mexico State University, United States of America

Funding: This research was supported by the Hebrew University Innovative grant no. 0346591. E.F. was supported in part by the Birk Foundation. I.B.-I. was supported by a BARD graduate grant no. GS-16-2011. The funders had no role in study design, data collection and analysis, decision to publish, or preparation of the manuscript.

Competing Interests: The authors have declared that no competing interests exist.

* E-mail: fridmane@agri.huji.ac.il

Introduction

Associating of causal polymorphism with complex phenotypes can provide a better understanding of the mechanisms underlying developmental and biochemical constraints, thus enabling accelerated crop improvement. Advances in genotyping technologies and computational approaches have reshaped the way genetic analysis is conducted and have allowed capturing a substantial proportion of the global genetic diversity for genome-wide association studies [1]. While different "association mapping" approaches in plants are focused mainly on additive variation between inbred lines, much less attention has been paid to developing experimental and computational approaches for identifying and deciphering the possible role of intra-locus interactions that contribute to the phenotypic landscape of the hybrids. The underlying genetics of heterosis has long been debated–ever since it was first observed and documented by Charles Darwin [2] and later studied experimentally with natural and artificial populations of various organisms. Although it is agreed that increased homozygosity often lowers fitness-related characters (survival, growth rate and fertility), at the heart of the debate is the extent to which this can be attributed to increased homozygosity for partially recessive detrimental mutations (dominance), rather than changes in homozygosity for alleles at loci with heterozygote advantage (overdominance model [3]). This debate also holds for agricultural yield: is the vigor of the hybrids the outcome of many dominant loci with intermediate effects working in a multiplicative manner on different yield-associated traits, or are those overdominant loci key regulators in several pathways throughout plant development?

Two recent reports illuminate this debate, although the final conclusion of both may still be linked to the type of plant system and phenotype analyzed, as well as to the genetic background (whole genome hybrids compared to isogenic homozygous background). Riedelsheimer *et al.* [4] crossed 285 diverse Dent inbred lines from worldwide sources with two testers and predicted their combining abilities for seven biomass- and bioenergy-related traits using 56,110 sngle-nucleotide polymorphisms and 130 metabolites. Then they performed genome-wide association scans using a Q+K model [5] and found no strong association signals, even though population size, heritabilities of the traits, extent of linkage disequilibrium and marker density were sufficiently high to detect large quantitative trait loci (QTLs). Based on these results, it was concluded that the genetic architecture that underlies heterosis is close to an infinitesimal model. On the other hand, Krieger *et al.* [6]

searched for genes that cause heterosis in tomato (*Solanum lycopersicum*) by crossing 33 diverse fertile mutants with the matching non-mutagenized parent known as 'M82' to create isogenic mutant heterozygotes. Comparison of their yield traits identified six mutant heterozygotes that showed yield heterosis, with the individual effects ranging from 36% to 88%.

To date, no individual overdominant locus has been isolated by unbiased QTL mapping and recent attempts to zoom in on genomic loci associated with heterosis have mostly made use of biparental populations. This is performed by estimating the heterosis phenotype through test-crossing introgression lines or recombinant inbred lines to the recurrent parent [7,8] and linking the heterotic mode of inheritance with marker-defined genetic intervals. Additional approaches have included the use of immortalized F2 populations [9] and the fine-mapping of heterotic QTLs in nearly isogenic lines [10]. Lariepe *et al.* [11] recently reviewed the use of biparental populations for dissecting heterosis and extending the use of the North Carolina III (NCIII) design with markers [12] to develop and implement a multiparental-connected model. This allowed studying heterosis in families derived from both related and unrelated parents, and comparing not only contrasts between homozygous and heterozygous genotypes, but also contrasts between heterozygous genotypes at each locus [11]. One approach for studying heterosis, devised very early in plant and animal breeding, was generation of a cross matrix (diallel) between founder lines (FLs) followed by phenotypic analysis of these inbred lines and their hybrids. Kearsey [13] was the first to compare the merits of five designs–full diallel, half diallel, partial diallel, NCI and NCII. These studies, which were conducted before the molecular concepts of genetics had been formulated, were concerned with dissecting the heterotic variation into its components (additive, dominant and epistatic), as well as with determining general and specific combining abilities between the different founder inbreds. With the advent of genetic marker technologies, it became feasible to estimate the relationship between overall genetic distance and the magnitude of heterosis calculated by the general combining ability of the parents [14]. Cho *et al.* [15] connected genetic heterogeneity and heterosis at each locus by calculating the differences in the levels of best parent heterosis (BPH) between different genotypes. The ANOVA comparison between BPH values of homozygotes and heterozygotes at each locus concluded that more than 30% of the markers were associated with a heterotic mode of inheritance. Note that this analysis did not distinguish between different allelic combinations among the hybrids and treated the heterozygotes in each locus as a single group.

Previous studies have reported a high level of heterosis in *Sorghum bicolor*, the fifth most important crop in the world (www.FAO.org), which exhibits a consistent yield increase in hybrids vs. varieties (up to more than double) under a wide range of growing and management conditions [16,17]. Interestingly, these early studies led to the identification of one of the few cases of whole-plant heterosis resulting from the heterozygous condition of a single mutation that affects duration of growth [18]. However, since the underlying gene was not identified, the possibility that this non-additive mode of inheritance is due to pseudo-overdominance [7] of more than one gene still remains open. Heterotic groups are not as clearly defined in sorghum as in maize, and studies using molecular markers have shown that prediction of heterosis is enhanced by using particular linkage groups in models attempting to associate genetic distance and hybrid-group performance [19]. Although these observations lend some support to the attractive possibility of identifying single genes underlying heterosis in this major crop plant, the heterosis map of sorghum, which would allow alignment with existing maps in maize [9,11,12,20], rice [21] and other plants, is still lacking.

Here we present the first use of heterotic trait locus (HTL) mapping to identify intra-locus interactions mediating an over- or underdominant mode of inheritance. The wide diallel population of *S. bicolor* is derived from 19 founder lines (FLs) selected from a wide gene pool. This population was phenotyped for 2 consecutive years and statistical and computational tools were integrated to test for association between specific intra-locus interactions and overdominant mode of inheritance. We discuss possible mechanisms underlying heterosis for grain yield in hybrids based on our experimental data and demonstrate that the results of this fast-track mapping are validated by the overdominant mode of inheritance of one HTL in the consecutive F2 population. Finally, we propose the incorporation of individual resequenced genomes of only diallel FLs to zoom in on hitherto unknown loci underlying heterosis, thereby paving the way for an improved understanding of molecular mechanisms underlying this elusive and important phenomenon.

Materials and Methods

Sampling, Genotyping and Genetic Analysis of Wide Collection of *S. bicolor*

This study was initiated by obtaining 173 available *S. bicolor* ssp. *bicolor* accessions from the Israeli Plant Gene Bank (IGB; http://igb.agri.gov.il/main/index.pl; Table S1). In addition, the USDA cultivated sorghum collection in the Germplasm Resources Information Network database (GRIN; http://www.ars-grin.gov) was mined and an additional 100 accessions were added. All plants were grown in pots in a greenhouse for DNA extraction and seed collection. DNA was extracted as follows: young leaf samples were ground in tubes using a TissueLyser (Qiagen, Hilden, Germany); 0.7 mL nuclei lysis solution (0.2 M Tris pH 7.5, 0.05 M EDTA, 0.2% w/v CTAB and 2 M NaCl) was added to each sample before incubating at 65°C for 1 h, and then 800 μL chloroform:isoamyl alcohol (24:1, v/v) was added. Samples were shaken for 15 min and centrifuged for 15 min at 10,000 *g*; the supernatant was transferred to a new tube, and 2.5 μL RNase (Qiagen) was added before incubating at 37°C for 15 min. DNA was precipitated by centrifugation with 500 μL ice-cold isopropanol and the pellet was washed with 0.2 M sodium acetate (pH 5.2) in 76% ethanol at room temperature for 5 min before centrifugation for 5 min at 10,000 *g*. This was followed by an additional wash of the pellet with 10 mM ammonium acetate in 76% ethanol at room temperature for 5 min and centrifugation for 5 min at 10,000 *g*. DNA was dried and dissolved in 30 μL DDW.

Microsatellite [simple sequence repeat (SSR)] genotyping was conducted using multiplexes of three markers with primers labeled with FAM, HEX or TAMARA (Table S2). PCR products were separated and analyzed using the MegaBACE Genetic Profiler and Fragment Profiler software tools (Amersham Biosciences, München, Germany). The raw data were entered into the database and examined for typing errors, false-positive alleles and data authenticity.

Analysis of genetic diversity and allele frequency, phylogenetic tree construction and genetic distances were calculated with PowerMarker version 3.25 [22] as described previously [23]. FLs for the diallel were selected using the "line selection" function in the design tool, in the analysis option, using "allele number" as the major criterion for selection.

Construction of Diallel and Plant Phenotype

A crossing scheme was designed to allow continuous and synchronous flowering of the FLs. To perform crosses between all FLs, the panicles of the female plant were emasculated by hot-water treatment (45°C for 10 min) with dehiscence control using a plastic bag (Stephens and Quiniby, 1933 in [24]). Non-emasculated flowering panicles were used for manual pollination 1 to 3 days after the hot-water treatment. Prior to the field trials, the hybrid seeds were sown in trays and DNA was extracted from leaves of 2- to 3-week-old seedlings in 96-well plates using the high-throughput method described by Xin *et al.* [25]. This DNA was subjected to high resolution melting (HRM) genotyping on a LightCycler (Roche, Basel, Switzerland) using SSR markers Xtxp57 or Xtxp321 (Table S2). Inbred and artificial hybrid DNA (1:1 mix of inbreds) was used as a control in each 96-well plate, and HRM conditions were as described previously [26] for validation of hybrids. Hybrids and FLs were transplanted into the experimental farm at the Robert H. Smith Faculty of Agriculture, Food and Environment (Rehovot, Israel) as single plants (15 cm between plants, 80 cm between rows, which is comparable for sorghum density at commercial field [27]) in a complete-randomized block design. In 2010, two FLs and their derived hybrids were excluded from the analysis due severe chemical damage of these FLs during germination in trays. A total of 123 hybrids (out of 136 possible) and 27 reciprocal crosses were analyzed. In 2011, a total of 157 hybrids (out of 171 possible) and 15 reciprocal crosses were analyzed (Figure S1A and S1B). Trials included 7 or 4 replicates per hybrid and 14 or 10 replicates per parent line, respectively.

Number of days to flowering was scored for each plant when half of the panicle was at anthesis. Traits evaluated at harvest were: plant height (from the soil to the base of the panicle); stem diameter (of the lower third of the stem); leaf weight, and stem weight. Panicles were oven-dried for 2 days at 65°C and dry panicle weight (DPW) was measured. Next, the panicle was dissected following Brown *et al.* [28] to determine primary, secondary and tertiary branching number, as well as rachis length, whorl number and primary branch length. Seed dry weight (SDW) was obtained by weighing a total of 50 grains, then seed number (SN) was calculated as SN = (DPW/SDW) x 50.

Calculation of Overdominant Heterosis (ODH) Value

The ODH parameter of the r progeny of a cross FLx X FLy was determined as follows ($P2_{xy}$ and $P1_{xy}$ are the high and low mean values of the two parents, respectively):

(1) If the phenotypic value of replicate r of the cross, $F1_{xyr}$, is greater than or equal to $P2_{xy}$, then the ODH parameter of replicate r, ODH_{xyr}, is calculated using the algebraic expression: $ODH_{xyr} = (F1_{xyr} - P2_{xy})/P2_{xy}$.

(2) If the phenotypic value of the replicate r, $F1_{xyr}$, is less than or equal to $P1_{xy}$, then ODH_{xyr} is calculated using the algebraic expression: $ODH_{xyr} = (F1_{xyr} - P1_{xy})/P1_{xy}$.

(3) If the phenotypic value of the replicate r, $F1_{xyr}$, is between $P1_{xy}$ and $P2_{xy}$, then $ODH_{xyr} = 0$.

The mean ODH_{xy} for the cross was then calculated for all replicates of the cross using the algebraic expression $\overline{ODH_{XY}} = \frac{\sum_r ODH_{xyr}}{R}$, where R is the total number of replicates of the cross.

Correlation Patterns of Traits and of Heterotic Values

JMP software (SAS Institute, Cary, NC) was used to perform multiple correlation analyses using the "stepwise" option under the "fit model" function. A forward direction was taken to select factors for the final model that correlate between traits or heterosis levels (this test was conducted using the hybrid means).

HTL Mapping

Data derived from each of the 2 years were analyzed separately. In the first stage, a general linear regression was implemented using TASSEL software [29] to assess the effect of the genetic heterogeneity of each marker on the ODH in the diallel (Figure S2A and S2B). The model was: $Y_{ij} = \mu + \alpha_i + e_{ij}$, where μ is the mean ODH of the diallel population, α_i is the effect of the genotypic group and e_{ij} is the random error, i.e. the variation between the ODH means of the different crosses of same genotypic group. The non-distribution-dependent, experiment-wise error level of $\alpha = 0.05$ was computed based on 1000 permutations [29,30].

The significant loci were then analyzed in the second step: A non-linear Kolmogorov-Smirnov pair-wise test (R environment http://www.r-project.org/) was used to compare the ODH distribution of the different hetero-genotypic groups to the homozygous group in these loci (Figure S2C and S2D): $D_{n,n'} = \sup_x [F_{1,n}(x) - F_{2,n'}(x)]$, where $\sup_x$ is the supermum of the set of distances, and $F_{1,n}$ and $F_{2,n'}$ are the empirical ODH distribution function of a specific hetero-genotypic group and the homozygous allelic combinations, respectively. The null hypothesis is rejected at level α if $n'\sqrt{\frac{nn'}{n+n'}} > ka$.

HTL Mapping with Consideration of the FLs Genomic Structure

Estimation of cluster likelihood was calculated based on five independent runs (STRUCTURE software [31]) for a variable number of clusters, from K = 2 to K = 10 (length of burning period: 20,000 and number of MCMC repeats: 20000). K = 4 was chosen due to the low variation of probability values and repetitive clustering.

For each hybrid, the combination of FLs clustering was determined and this new variable was considered a co-factor in the analysis. The GLM model was $Y_{ij} = \mu + \alpha_i + \beta_k + e_{ij}$, where μ is the mean ODH of the diallel population, α_i is the effect of the genotypic group, β_k is the effect of the clustering assignment of the FLs and e_{ij} is the random error. For the Kolmogorov-Smirnov test, the ODH values of the hybrids were adjusted, considering the effect of the clustering assignment of the FLs. The effect of each combination was measured as the deviation of this combination from the weighted average of ODH values.

Three-way and Epistatic Relation Analyses

In each HTL, allelic combinations were assembled into two genetic states: P (positive intra-locus interaction) or N (neutral, other allelic combinations). JMP software (SAS Institute, Cary, NC) was used to perform two- and three-way ANOVA using the "Standard Least Squares" option under the "fit model" function with the EMS method. The two-way model was $Y_{ijk} = \mu + \alpha_i + \gamma_j + \alpha\gamma_{ij} + \beta_k + e_{ijk}$, where μ is the mean ODH value of the diallel population, α_i and γ_j are the main fixed effects of the HTLs, $\alpha\gamma_{ij}$ is the genetic interaction effect, β_k is the random effect of the year, and e_{ijk} is the random error.

The three-way model was: $Y_{ijrk} = \mu + \alpha_i + \gamma_j + \theta_r + \alpha\gamma_{ij} + \beta_k + e_{ijrk}$, where μ is the mean ODH value of the diallel population, α_i, γ_j and θ_r are the main fixed effects of the HTLs (*hDPW1.1*, *hDPW1.2* and *hDPW4.1*, respectively), $\alpha\gamma_{ij}$ is the effect of interaction between HTLs *hDPW1.1* and *hDPW1.2*; β_k is the random effect of the year, and e_{ijrk} is the random error.

Analysis of F2 Population

A single validated hybrid between SB018 and SB153 that carries the 154 and 162 alleles of the Dsenhabm39 SSR marker was selfed to obtain the F2 population. One hundred plants were transplanted into the field during the 2011 trial and phenotyped with the diallel's plants of this year. Linkage between the Dsenhabm39 locus and the different traits was tested by ANOVA under the "fit Y by X" function, and the "Compare with the best" option under the Compare means.

Analysis of Sorghum-maize Synteny

The 'block view' tool of Symap (http://www.symapdb.org/) was used to align the sorghum HTLs to the syntenic regions in the maize genome which were then assigned to the corresponding genomic bins (http://www.maizegdb.org/). Maize overdominant QTLs found on these bins were assembled to the synteny map by the physical location of the associated markers (http://www.maizegdb.org/).

Results

Genetic Analysis of the Wide Collection of *Sorghum bicolor* ssp. *bicolor*

This study was initiated by obtaining 273 *S. bicolor* accessions, including BTx623 which was used previously for whole-genome sequencing [32]. These accessions originated from different geographical regions worldwide but there were no details of their breeding history (Figure 1A and Table S1). To assess the diversity in this collection, the plants were first genotyped with a set of 50 microsatellite (SSR) markers spanning telomeric and centromeric regions (2 and 1 for each region, respectively [33,34]; Table S2). This set included 23 EST-derived markers (eSSR) and an additional 27 markers from non-coding genomic regions (gSSR). A subset of 19 FLs was chosen with the goal of harvesting maximum genetic diversity by a reasonable number of homozygous lines, while balancing for equal frequency of alleles in each locus. Such equal frequency allows statistical comparison within the derived diallel mapping population of both heterozygous and homozygous groups in each locus (see discussion). Figure 1B depicts a phylogenetic tree of the original wide collection, including indications for the external nodes of the selected FLs. These FLs represented 58% or 44% of the original allelic diversity based on eSSRs and gSSRs, respectively. On average, based on these marker types, there are 5.7 or 7.5 alleles per locus, respectively. Overall, in each of these loci, there are at least two alleles shared by a minimum of three FLs.

Next, the genetic analysis of the 19 FLs was supplemented with an additional 35 SSR markers and both Bayesian and genetic distance approaches were taken to further infer the genetic relationship among these inbred lines (see Materials and Methods). Both approaches indicated identical clustering of the FLs into four major genomic groups which were named A, B, C and D (green, blue, red and yellow, respectively; Figure 2A and 2B). Both the fixation index (Fst) values and branching architecture of the phylogenetic tree (Figure 2A) indicated different levels of similarity within each group. Excluding group C, in which lines were relatively clustered (Fst = 0.435), the lines in the other groups were relatively dispersed (Fst = 0.27, 0.27 and 0.21 for groups A, B and D, respectively).

Analysis of Phenotypic and Heterotic Variation

The FLs selected for the heterosis analysis were intercrossed to achieve a diallel sufficient for two replicated field experiments in 2010 and 2011. The experiments were set up to achieve a reasonable compromise between gaining phenotypic and heterosis values from a large number of hybrids and growing plants under conditions resembling common practice for sorghum under dense planting conditions ([27]; see Materials and Methods). In the first diallel (2010), plants were measured for dry panicle weight (DPW) and days to flowering; in 2011, the phenotyping was extended to include vegetative traits: leaf weight, stem weight, stem diameter and plant height (see Materials and Methods). The weighed panicles were further dissected to obtain levels of primary, secondary and tertiary branching (PBN, SBN and TBN, respectively), as well as rachis length, whorl number and primary branch length following Brown *et al.* [28]. Seed number was estimated based on DPW and seed dry weight (SDW) (see Materials and Methods), due to the high correlation ($r = 0.88$, $P < 0.001$) found between DPW and total seed number within a subset of 60 different FLs and hybrids.

Different estimates of heterosis were computed for all traits: (1) mid parent heterosis, (2) BPH [35] and (3) ODH. The ODH parameter is a measure of the extent to which the phenotype of the progeny of a cross deviates from the phenotypic bounds of its parents, thus emphasizing overdominant or underdominant mode of inheritance (see Materials and Methods). Comparison of the traits' ODH distribution showed significantly higher values for the reproductive trait DPW (median 0.51, quartile 0.216) than for any of the vegetative traits (Figure 3). Within the vegetative traits, leaf weight (median 0.27, quartile 0.05) and stem weight (median 0.19, quartile 0.035) showed significantly higher values than stem diameter (median 0.093, quartile 0) or plant height (median 0.068, quartile 0.01). Notably, a comparison of the reciprocal hybrids showed that there is one or no hybrid per trait which can be considered significantly different ($P \leq 0.01$). According to these results, the imprinting effect on heterosis was considered insignificant in our mapping populations.

Grain yield components such seed number and weight are known to be negatively correlated [36]. To determine whether these kinds of relations are also maintained at the heterosis level, the correlation pattern between the individual yield assemblers and DPW was compared to that between their heterotic modes of inheritance. At first, multiple regression analysis based on trait values found significant correlations between the three branching levels (PBN, SBN and TBN), SDW and total grain yield (DPW): the dominant effect was that of SDW (7.36, 0.04; effect and P value respectively), and a comparison of the effects of the different branching levels on DPW showed a stronger effect of SBN (4.78, <0.0001) relative to PBN (0.61, <0.001) and TBN (0.12, 0.014). Overall, this four-factor model explained 68% of the variation in grain yield (DPW). Next, a similar analysis between the heterosis values (BPH) of these traits showed that only SBN and PBN heterosis values are significantly and positively correlated with that of DPW (1.37, <0.0001; 0.396, 0.019, respectively), and together these two explained 35% of the DPW heterosis.

A similar comparison of the relationships within the grain yield components, both on the trait per se and the heterosis values, showed significant differences in the direction of the correlations between the two levels. Multiple regression analysis between the panicle architecture traits and seed number showed that SBN and PBN explain 65% of the variation in seed number across all hybrids in the 2011 experiment, with SBN showing an approx. eightfold stronger effect (275.5, <0.0001) than PBN (34.7, <0.0001). There was no significant correlation between TBN and seed number. A correlation test between trait values of SDW and these significant contributors for seed number indicated negative correlations for both SBN and PBN to SDW ($r = -0.25$ and -0.33; $P = 0.0001$ and 0.001, respectively; Figure 4A and 4B).

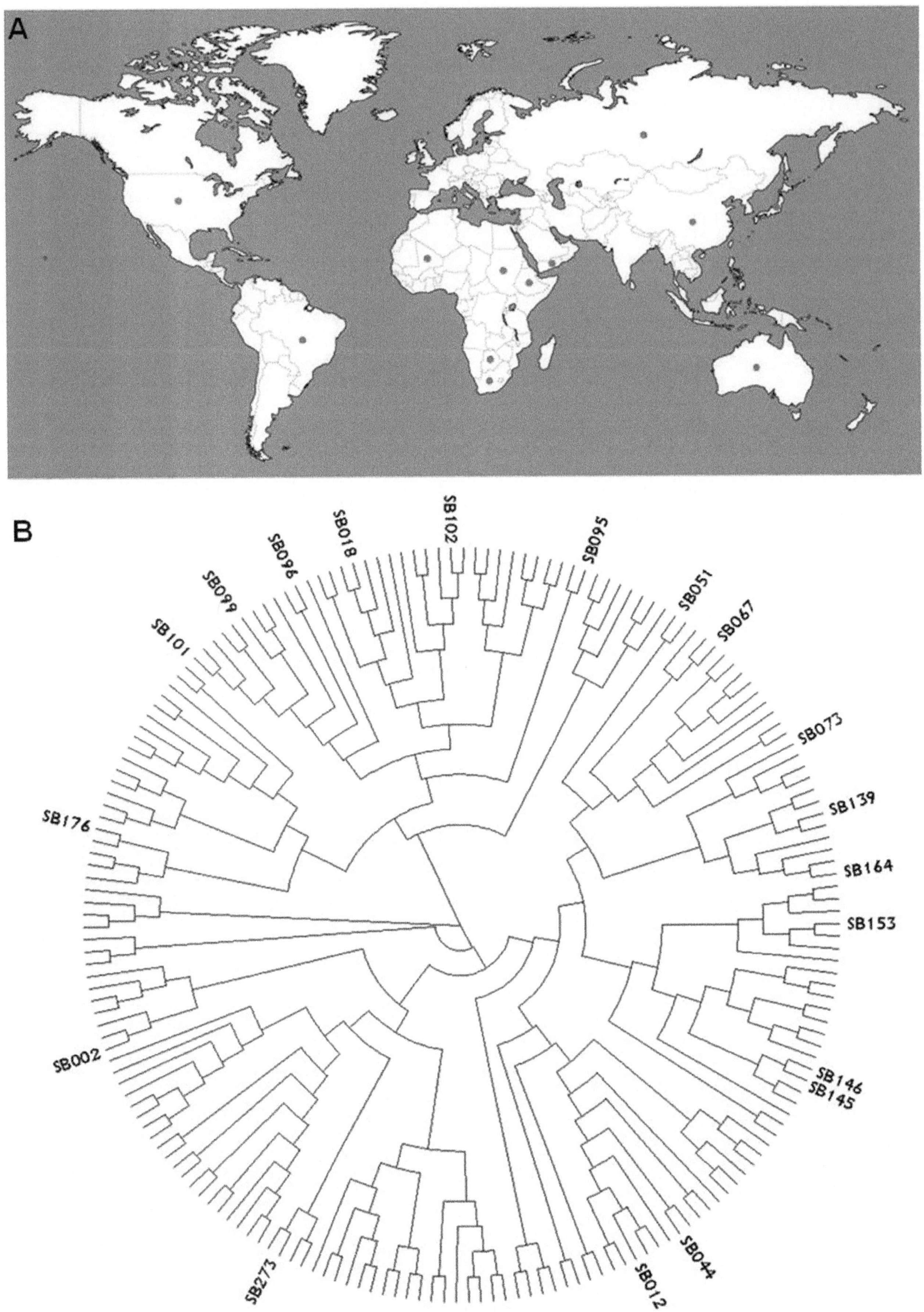
A
B
SB102
SB018
SB096
SB099
SB101
SB095
SB051
SB067
SB073
SB139
SB164
SB153
SB176
SB002
SB146
SB145
SB044
SB012
SB273

Figure 1. The *Sorghum bicolor* lines used for heterosis mapping. A. Origin of the wide collection of lines includes accessions collected worldwide (see Table S1 for details). **B.** Phylogenetic tree of the wide *Sorghum bicolor* ssp. bicolor collection. The external nodes and coding of founder lines (FLs) are indicated.

On the other hand, analysis of the heterosis values showed no correlation between those of SBN and SDW (Figure 4C) and notably, a positive correlation between those of PBN and SDW ($r = 0.24$, $P = 0.003$, Figure 4D). A similar opposite pattern between traits' relationships and their associated modes of inheritance was found with vegetative weight and SDW. While a significant negative correlation was found between the traits' values ($r = -0.39$, $P < 0.0001$, Figure 4E), their heterosis values were positively correlated ($r = 0.2$, $P = 0.02$; Figure 4F).

Genomic Scan for HTLs

HTL mapping combines multiple alleles in the analysis and focuses on associations between specific allelic combinations and overdominant mode of inheritance. Therefore, the use of ODH rather than the classical mid parent heterosis or BPH [35] is not meant to describe all of the cryptic heterotic variation between two or more inbred parents but rather extract major loci which exhibit their overdominant effects over a wide genetic background. Optimal implementation of HTL mapping should include ultra-high-resolution genotyping of the inbred FLs, i.e. genotype by sequencing [37]. At this stage, we present the analysis using relatively low-resolution genotyping (a total of 85 genetic markers across 10 chromosomes; Table S2).

The nature of the mapping population and the homozygosity of the selected parents allowing to project the allelic state in each locus from the inbred parents onto their hybrids (Figure 5A and 5B). In each locus, the hybrids are sorted to the different genotypic groups: the different heterozygous combinations are treated as separate hetero-genotypic groups, and the homo-genotypic group includes hybrids that are homozygous for the different alleles (In this study, due to the relatively high number of rare alleles in each locus, which is characteristic of SSR markers, the homozygotes in each locus were grouped together; Figure 5C). Next, the mean ODH values of each hybrid were considered (Figure 5B) and the distributions of these values were compared between genotypes to associate specific allelic interactions with overdominance (see Materials and Methods). A locus was considered a HTL if at least one heterozygous combination showed significantly advantageous heterosis values as compared to the homozygous combinations in both years (Figure 5C). Overall, this analysis revealed three significant HTLs for grain yield (DPW) residing on chromosomes 1 and 4 (namely *hDPW1.1, 1.2,* and *4.1*; Figure S3). Overall, these loci represent 3.5% of the loci tested.

A structured population is a known limitation when performing association mapping [5]. We therefore examined whether the clustering pattern of the FLs bias the mapping results due to co-variation of heterosis levels and FL assignment to genomic groups. Sorting the FLs into four groups ([31], $K = 4$, Figure 2) allowed determining the genomic combinations for each hybrid. When this new variable was considered a co-factor in each of the two steps of the mapping procedure (see Materials and Methods), the results remained the same. All three, and only those three previously identified HTLs, were consistently significant over 2 years.

Analysis of HTLs for Grain Yield in Consecutive F2 Generation

Perhaps the major concern with the HTL mapping approach is possible confounding effects, i.e. the association of certain local intra-locus interactions to the heterotic mode of inheritance where in fact this association is due to either a linked or non-linked locus found in linkage disequilibrium. This scenario is best illustrated by the example of loci in which the FLs are sorted in an identical segregation. This will lead to false associations between certain heterogeneities and the heterotic mode of inheritance where in fact, only one of these loci carries the functional causal polymorphism. We therefore tested the validity of the causal relationship identified between one of the mapped HTLs and the overdominant mode of inheritance in consecutive generations. One of the hybrids participating in the 2010 diallel (cross between SB018 and SB153, Figure 2) was caged to obtain F2 progeny. These plants were planted in the 2011 trial for genetic analysis of *hDPW4.1* (Figure 6A), which showed a significant association between heterozygosity for the 154/162 allelic combination and increased ODH values for DPW in the diallel in both years (Figure 6B). All F2 progeny were genotyped with the Dsenhabm39 marker and analyzed for their DPW. A comparison of the grain yield values of the three genotypic groups showed that the mean DPW of plants carrying the two specified alleles of *hDPW4.1* (154/162) was significantly higher than those of both homozygous plants (154/154 and 162/162, respectively; $P = 0.048$ and $P = 0.037$; Hsu's multiple comparisons with the best test), with a mean effect of 21% (Figure 6C). These results thus support the overdominant mode of inheritance associated with this allelic combination, as suggested by the original HTL mapping (Figures 6A, and S3).

Contribution of HTLs to Heterotic Variation and Their Epistatic Interactions

We next investigated the contribution of each HTL to the overall heterotic variation found in the diallel population, as well as the possible epistatic interactions between the three loci. Due to the multiple-parent population structure and the nature of the SSR markers, a direct two-way analysis is in fact unattainable. For this reason, in each of the identified HTLs the interacting allelic combinations were considered to be one genetic group (P; positive), and all other combinations were treated as a single neutral non-interacting combination (N). *hDPW4.1* showed the highest contribution to heterosis (12%; $P < 0.0001$) as a single factor, and *hDPW1.1* and *hDPW1.2* contributed 7% and 4% ($P < 0.0001$ and $P < 0.0026$), respectively, to this mode of inheritance. Next, two-way interactions were analyzed to test for epistatic relationships between the different loci. Of the three possible interactions, only that between the two HTLs on chromosome 1 was significant ($P = 0.014$; Figure 7). In addition, to estimate the cumulative contribution of all HTLs to the overall heterotic variation, a three-way linear regression was performed (Figure 7B and 7C). Overall, this final model explains 19.0% of the heterotic variation found in this population (Figure 7B and 7C).

Discussion

HTL Mapping and its Implications for Rapid Identification of Overdominant Loci

The search for the genetic architecture of quantitative and complex phenotypes in plants has experienced a quantum leap in the depth of genomic architecture descriptions [38], as well as in the speed of unraveling causative polymorphisms. This is mainly due to the adoption of population-based association approaches [39,40] and application of higher throughput resequencing and

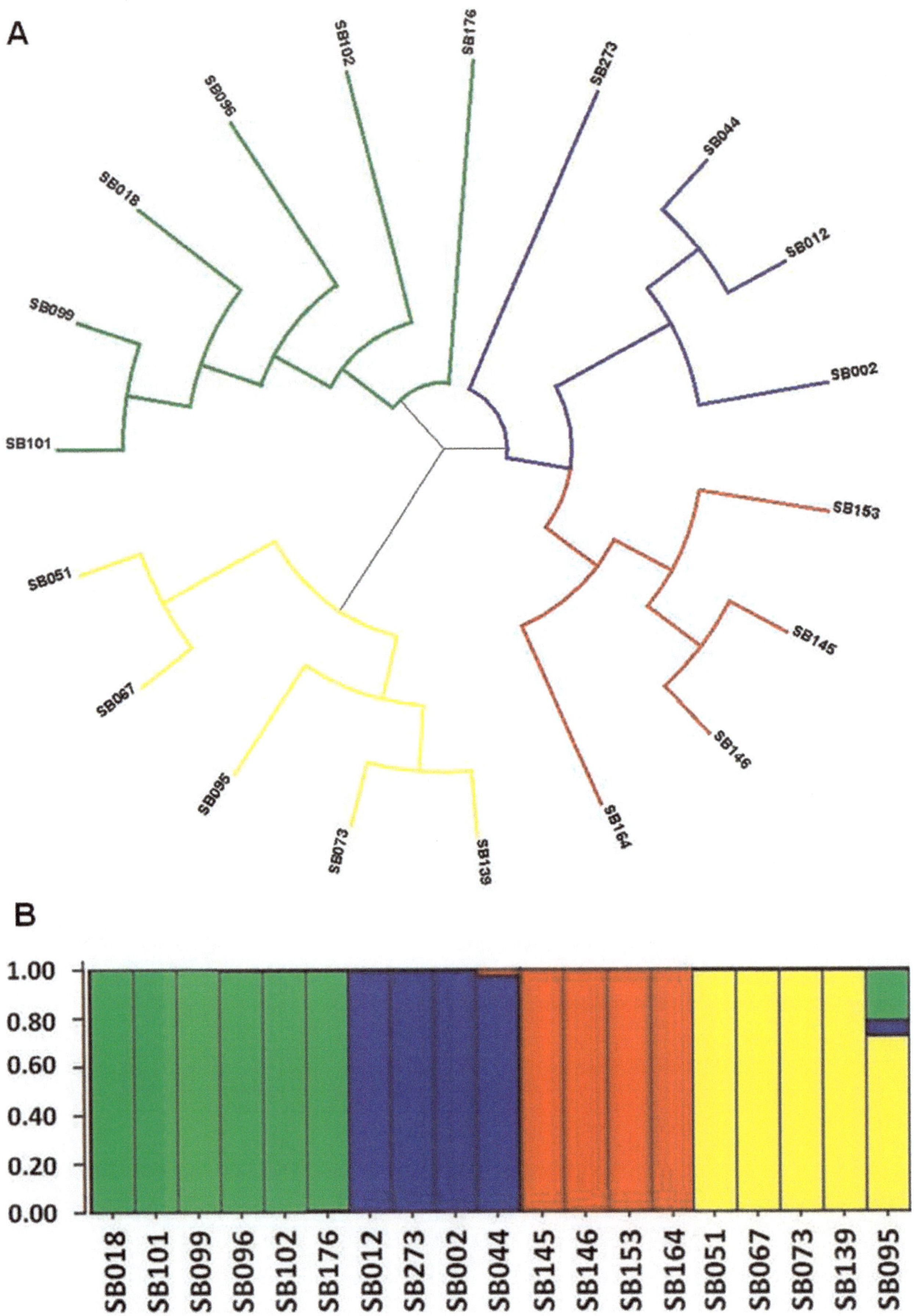
A
SB176
SB102
SB096
SB018
SB099
SB101
SB273
SB044
SB012
SB002
SB153
SB145
SB146
SB164
SB139
SB073
SB095
SB067
SB051
B
1.00
0.80
0.60
0.40
0.20
0.00
SB018
SB101
SB099
SB096
SB102
SB176
SB012
SB273
SB002
SB044
SB145
SB146
SB153
SB164
SB051
SB067
SB073
SB139
SB095

Figure 2. Genetic analysis of the diallel founder lines. A. Clustering analysis of the 19 *Sorghum bicolor* inbreds based on unrooted neighbor joining tree. Color coding representing the four identified clusters. **B.** Model-based ancestry for each founder line with enforcement of the cluster number (K) to 4 (see Materials and Methods). Distruct plot is shown with color coding representing the four clusters of the STRUCTURE analysis and the name of each founder line is depicted below.

marker genotyping [41]. In this study, we address the challenge of developing new genetic mapping strategies to direct and optimize the discovery of genetic variation [1] underlying heterosis, which is a most important component in modern breeding.

Although the diallel population does not offer the advantage of permanent populations, such as the ability to propagate it and phenotype each genotype under different environments, there are several attributes which make it an attractive alternative for unbiased discovery of novel allelic variation in plant genetic resources. It is important to note that the identification of hitherto unknown heterosis loci in this study was achieved with no prior knowledge of the level of heterosis or its genetic basis. While this may limit the validation of our results, no comparable heterosis map for sorghum exists; however, we do show that for *hDPW4.1*, the overdominant mode of inheritance is maintained in a consecutive F2 population (Figure 6). Although mapping heterosis loci might be simpler using diallelic populations such as F2 crosses, advanced inter-crosses and recombinant inbred lines descended from two parents, such an approach requires a priori knowledge of the mode of inheritance of their hybrid for the different traits analyzed. The multi-parent and multi-allelic system characteristic of the mapping population, in fact, increases the likelihood of detecting allelic combinations with synergistic effects on yield, or on any other trait. Nevertheless, this relatively rapid mapping procedure, including the harvesting of allelic variation into a single population of hybrids, raises both operative and statistical challenges. Given that the power to compare the heterotic values of any allelic combination at each locus relies on a minimal genotypic frequency in the hybrids (Figure S4), the choice of FLs from a wide collection should attempt to provide a balanced representation for the different alleles. Such a representation could be achieved by simulating the different FL combinations to achieve an optimal allelic combination in most of the loci tested. This might be achieved, as in this study, by relatively low-resolution genotyping. In practice, such selection should be well thought out to include the possibility of obtaining all hybrids required for the HTL mapping in a reasonable time. Therefore, in this study, relevant characteristics for sorghum such as day-length sensitivity were also considered (data not shown).

The basic premise underlying the HTL mapping approach is that intra-locus interactions play a significant role in the manifestation of heterosis. The genetic analysis of each locus within a diallel population is highly subjected to epistatic interactions due to the genome-wide intersection of many genetic backgrounds and, although each locus is analyzed separately (single-point analysis), the HTL mapping results reflect the influence of inter-allelic interactions as well. Despite the fact that this genetic "background noise" may increase intra-genotypic-group variation (Figure S2), thereby reducing the power of discrimination between the phenotypic distributions of the different groups, we hypothesize that the major loci will still emerge. Moreover, in our opinion, such noise is biologically relevant due to the significant role of these interactions in manifesting complex phenotypes [26,42].

This study differs from the analysis of heterosis in a diallel population reported previously by Cho *et al.* [15] in that (1) the different hetero-genotypic groups were separated and the allelic combination mediating the heterosis in each locus was specified, (2) the heterosis was analyzed with values that emphasize overdominant mode of inheritance (ODH), and (3) a less permissive permutation approach was used to set the genome-wide type I error. This probably explains the smaller proportion of markers associated with ODH in this study (Figure S3), as compared to over 30% positive associations in Cho *et al.*'s [15] analysis. Notably, the analysis of sorghum heterosis in this study, although separating the hetero-genotypic groups, treats the homozygous hybrids in each locus as a single reference group. This originates from the hypothesis that homo-allelic combinations

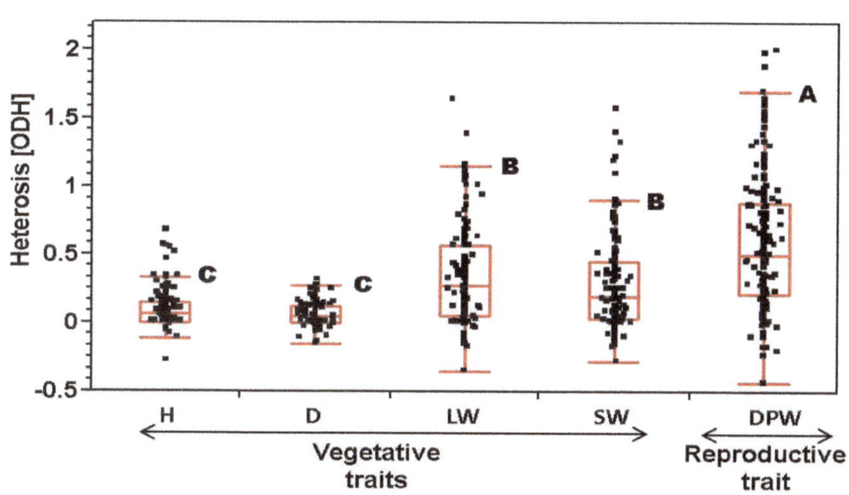

Figure 3. Overdominant heterosis (ODH) in the diallel. ODH distribution of vegetative (height, H; diameter, D; leaf weight, LW; stem weight, SW) and reproductive (dry panicle weight, DPW) traits. Different letters denote significant difference between ODH distributions (Kruskal-Wallis test, $P<0.0001$). Quantile boxes show the range between the 25th and 75th percentiles, including the 50th percentile indicated in between. The bottom and upper outer lines depict the 10th and 90th percentiles, respectively.

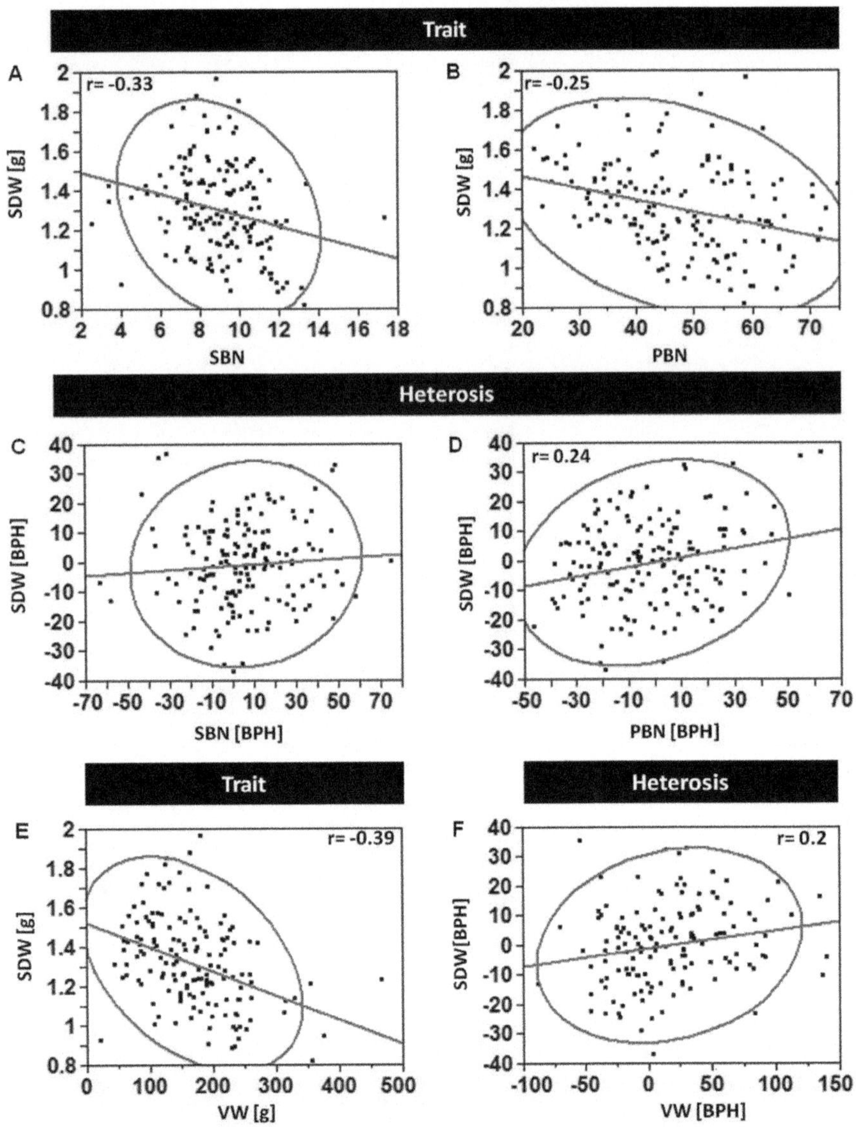

Figure 4. Hybrid reproductive superiority is induced by release of tradeoff relationship. Correlations between components of seed number (secondary branching number, SBN; primary branching number, PBN) and seed dry weight (SDW). Analyses of trait values show negative correlations (**A, B**) while analysis of heterotic values (best parent heterosis; BPH) show either no (**C**) or positive (**D**) correlation. **E–F.** Correlations between vegetative weight (VW) and seed dry weight (SDW). Analysis of trait values (**E**) shows negative correlation whereas that of heterosis values (**F**) shows positive correlation.

will not result in synergistic or antagonistic interactions. It is also possible to further separate the homo-genotypic group and compare each hetero-genotype to hybrids carrying each of the two interacting alleles in a homozygous state (such an approach is currently being tested in another study with *Saccharomyces cerevisiae* hybrids; Ben-Israel and Fridman, unpublished results).

It should be noted that as a methodology, the HTL mapping approach can also be used to map the opposite trend in hybrids, i.e. to scan the genome for intra-locus interactions associated with hybrid inferiority. Whereas in our study, the general positive ODH values for all of the traits (Figure 3) directed HTL mapping for overdominant loci, recent studies have shown that such interactions at a single gene play key roles in the evolution of plant populations and mediate an underdominant mode of inheritance [43]. It would therefore be beneficial to implement the HTL mapping approach in studies aimed at identifying 'genes of speciation' in models such as *Arabidopsis thaliana*, in order to test the prevalence of additional loci governing speciation.

Significance of Intra-locus Interactions in Heterosis

Despite the fact that the HTL mapping was performed with a relatively small number of genetic markers, it allowed us to apply a novel mapping approach to study the relative contribution of intra-locus interactions to the heterotic variation in the gene pool. Overall, at this resolution, we identified three HTLs which were reproducibly associated with grain yield heterosis over 2 years, and additional loci which were not considered because the effect was only found in one growing season (data not shown). A comparison of our results to other studies shows that the number of loci with reproducible and significant overdominant effects on grain yield is comparable to that found in rice, for example. Although QTL studies show that heterosis is controlled by a large number of loci,

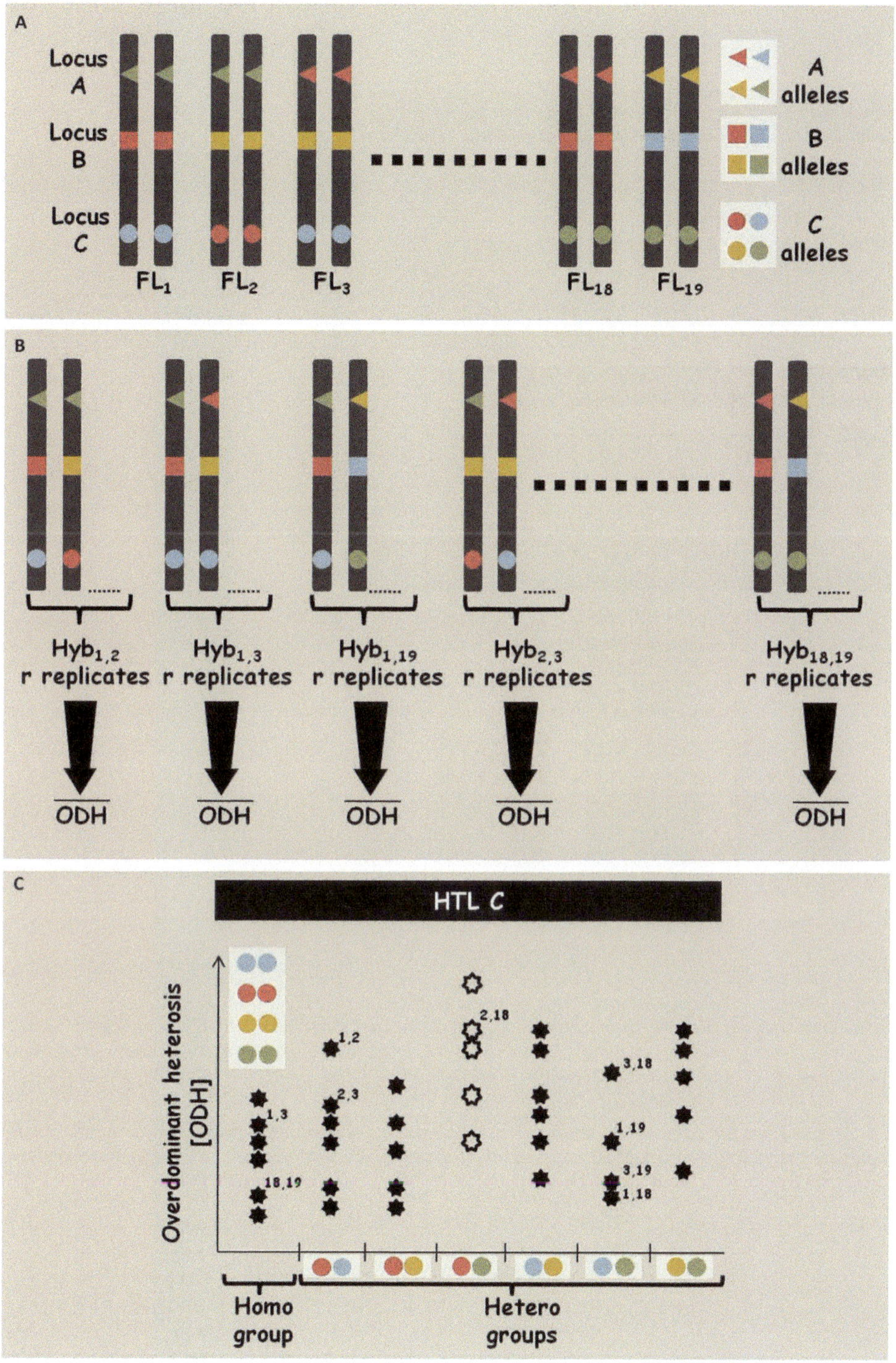

Figure 5. Illustration of the HTL mapping. A. Genotyping of the selected FLs is represented by three loci with different shapes (A, B and C), with each harboring 4 different alleles among the FLs of the diallel. **B.** Projection of the FLs genotype to the derived hybrids and calculation of the mean heterosis values (ODH) for the r replicates of an hybrid. **C.** Statistical analysis to identify specific hetero combinations with advantageous ODH values (the purple/green hetero-group in this illustration).

and includes the involvement of complex epistasis, the number of true overdominant loci for grain yield with reproducible effects in more than 1 year sum to less than 4% of the markers used [44,45]. Based on the analysis in our study, it can be concluded that the portion of heterosis explained by the overdominant model [7] is significant, contributing approximately 20% to the overall heterotic variation. This value is either an underestimate of the effects of intra-locus interactions due to the low resolution of the mapping, or it might suggest that the dominance genetic model is highly prevalent. In addition, the ability to identify significant HTLs in this study implies that, as in standard models for quantitative variation, there may be a small number of major loci

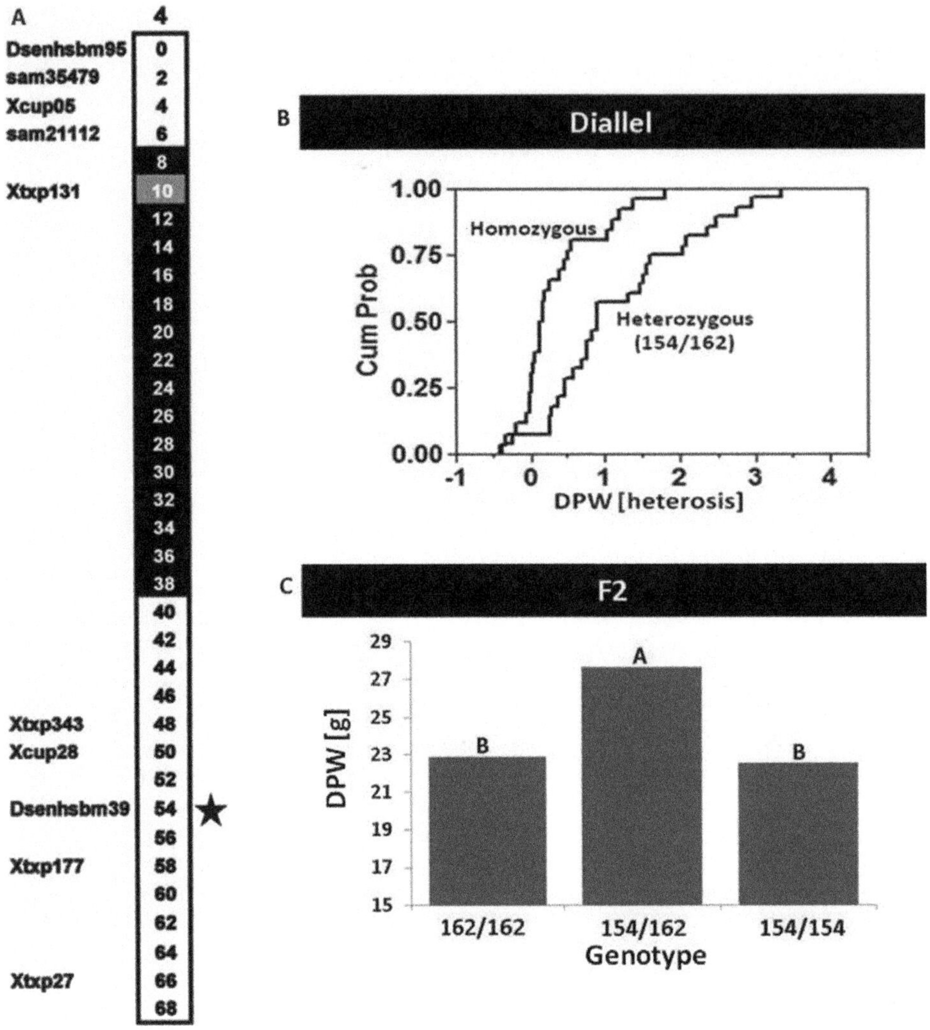

Figure 6. The *hDPW4.1* grain yield heterotic trait locus (HTL). A. Chromosomal location of the Dsenhabm39 SSR marker is indicated by star on the physical map of chromosome 4. Black and white coloring indicate pericentric-heterochromatic and telomeric-euchromatic chromosomal regions, respectively. Gray indicates markers within pericentric-heterochromatic chromosomal regions. **B.** Cumulative distribution function plot showing the ODH values of the significant hetero-genotypic (154/164) and homo-genotypic (H:H) groups for the same marker in the diallel (year 2011). **C.** Linkage analysis of the *hDPW4.1* locus with dry panicle weight (DPW) in the F2 population. Different letters above bars denote significant difference ($P<0.05$; Hsu's MCB test) between mean values.

underlying heterosis (the overdominant ones), and a large number of loci which make a small contribution following the infinitesimal genetic model [46].

hDPW4.1: HTL for Grain Yield in *S. Bicolor*

This study provides the identification of grain yield heterotic loci in a major crop plant, through genetic association over a wide genetic background, all in a relatively short time. Starting from a wide collection (Figure 1 and Table S1), the HTL mapping approach pinpointed three significant loci associated with heterosis. Furthermore, the superiority of heterozygosity for *hDPW4.1* in an F2 population (Figure 6), despite possible masking effects of inbreeding and epistatic interactions from the rest of the genome, strongly support this association obtained in the diallel. Further phenotypic and genetic analyses are required to determine the nature of this locus, which increases grain yield by more than 20%. This will include looking for pleiotropic effects which would support a multiplicative model [47], and perhaps the more challenging task of identifying and defining the causal polymorphisms underlying these effects. This will require fine mapping of this locus, either by screening for recombination events in large F2 populations, followed by phenotypes of derived F3 progeny, or alternatively, selecting for relevant recombined haplotypes in recombinant inbred populations. In sorghum, as in maize, these populations have become a key component in dissecting additive variation using the nested-association mapping (NAM) approach [48,49]. Therefore, once the ultrahigh genotype of these FLs will be determined using same approaches it will be imperative to compare the haplotype structure between these to the sorghum NAM founder lines. This may assist in zooming in on these HTLs at high resolution and determining the genetic factors that mediate these major heterotic effects.

S. Bicolor: an old-new Model for Studying Heterosis

There are several genetic and developmental attributes of *S. bicolor* which position this crop plant as an ideal model system for in-depth dissection of heterosis, and for projecting the outcome of these studies on its relatives in grasses. Genetic analysis of the wide

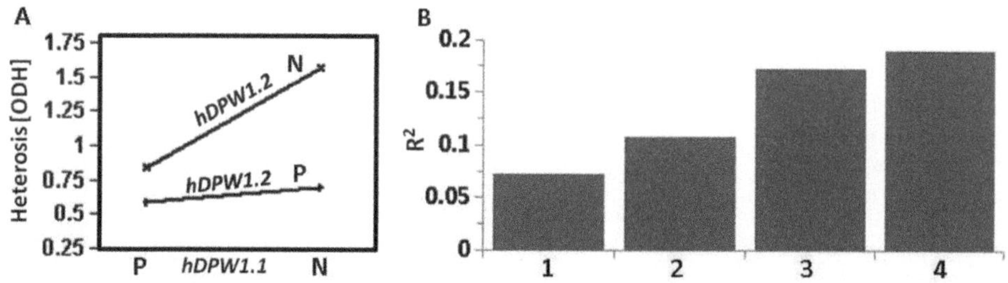

C

Model	Random factor	Fixed Factor				R^2
1	Year	*hDPW1.1*				0.07
2	Year	*hDPW1.1*	*hDPW1.2*			0.11
3	Year	*hDPW1.1*	*hDPW1.2*	*hDPW4.1*		0.17
4	Year	*hDPW1.1*	*hDPW1.2*	*hDPW4.1*	*hDPW1.1 X hDPW1.2*	0.19

Figure 7. Contribution of heterotic trait loci and their epistatic interactions to heterotic variation in a diallel population. A. Heterosis least squared means plot. The x axis represents the genotypes of *hDPW1.1*. The lines represent different genotypes of *hDPW1.2*. N, neutral non-interacting genotypic allelic combinations; P, positive interacting combination. **B.** Accumulated variation explained by the model (R^2) with each additional factor. **C.** The factors included in each of the models.

core collection of *S. bicolor* showed two main genetic attributes. First, the amount of allelic richness found in this material far exceeds that found in other cultivated crops; it is, in fact, comparable to that of the wild ancestors of other major crops. For example, recent analysis of a large core collection of both cultivated and wild barley accessions (*Hordeum vulgare* and *H. spontaneum*, respectively–The Barley1K) with a similar genetic marker system showed the significant genetic bottleneck experienced by this crop during domestication, including reduction in allelic richness from 13.4 to 6 alleles per locus [23,50]. In the cultivated *S. bicolor*, on the other hand, the allelic richness–which is a prerequisite for testing multiple allelic interactions in a population structure such as the one presented in this study–is in fact comparable to that of *H. spontaneum*: more than 13 alleles per locus in the wide collection and 7 alleles per locus in the FLs of the diallel. In addition, while there is ample genetic heterogeneity in the wide collection, each of the accessions is, by itself, highly homozygous, reflecting the partial allogamous mode of reproduction in this plant [17]. These two attributes are critical in the implementation of HTL mapping since they provide the opportunity to extract a rich allelic and phenotypic repertoire (Figure 1) while retaining the ability to project the genotype of the hybrids from their fully homozygous parents (Figure 5).

Notably, comparison of the sorghum DPW HTL mapping results with previous QTL studies for grain yield in maize [9,12,20] shows synteny of two sorghum HTLs with maize bins that were reported to harbor overdominant QTLs (Figure S5). On average, the syntenic blocks on sorghum corresponded to a three-fold larger block in maize and in addition, *hDPW4.1* was syntenic with two heterotic maize QTLs for yield that reside on bins 2.04 and 10.04 (Figure S5). These observations reflect the evolutionary history of maize and sorghum with approximately 3× genome expansion in maize since its divergence from sorghum 12 MYA. They lend further support to the usefulness of the latter for associating causative polymorphism with complex traits due to reduced genomic and possibly functional redundancy between paralogs.

Possible Mechanisms Governing Grain Yield Heterosis in Sorghum

The wide diallel used in this study creates a unique phenotypic and genetic framework for a systematic analysis of the transition from inbreds to hybrids, and of how hybrid superiority over inbreds is orchestrated via the different developmental pathways in the plant. Smith and Fretwell [51] predicted that once the resources available for reproduction are fixed, a tradeoff between seed size and number is inevitable: any increase in the size of individual offspring must be compensated for by a reduction in offspring number. Indeed, negative correlations were found in this study between grain yield components across the hybrid populations and these results are in agreement with previous studies [37]. This was observed for reproductive assemblers–seed number and seed weight (Figure 4A and 4B), and for productive competitors–vegetative weight and seed weight accumulation (Figure 4E). To answer the question of whether hybrids can surpass their inbred parents for two traits under such a tradeoff relationship, correlation analyses using the heterosis values of both traits were conducted. There was a lack of correlation between heterosis values of SBN and SDW (Figure 4C), as compared to positive correlations between those of PBN and SDW (Figure 4D) and between vegetative and reproductive heterosis (Figure 4F). These differences between trait values and their associated heterotic mode of inheritance suggest that introduction of new allelic combinations within hybrids eliminated the limiting intra-plant compensatory tradeoff. The new genomic pattern in the hybrids can be interpreted as an environment with better growth potential for the plant compared with their inbred parents [52]. Developmental and biochemical reasons for the hybrid's ability to increase both vegetative and reproductive output compare to its inbred parents indicate that either the efficiency of the whole system is much higher, or rate-limiting factors are modulated due to hybridization.

More specifically, the mechanism underlying such relaxation of metabolic flow or increase in growth potential may originate from changes in the activities of gene products with critical spatial and temporal distribution, at the cellular level, at early developmental stages or both. Such a sequential developmental pathway underlying naturally occurring variation, though not modulated by inter-allelic interactions, was exemplified in the previously characterized tomato QTLs *fw2.2* and *Brix9-2-5*. In those studies, it was found that changes in a single gene (*ORFx* or the apoplastic invertase *LIN5*) occurring early in ovary development lead to increases in cell number or total sugar yield in the mature fruit [53,54]. It would therefore be interesting to determine the causative polymorphism underlying the overdominant effects of the HTLs and (1) test whether these genes share similar early effects on plant development that lead sequentially to increased growth and if so, (2) whether overdominant QTLs have unique characteristics relative to additive QTLs (see above), and eventually (3) determine the molecular and biochemical bases for the differences between homo-alleles and hetero-alleles, the stoichiometry of their products, and how this relates to modulation of the reproductive output.

Supporting Information

Figure S1 Scheme of the two mapping populations from A. 2010 and B. 2011 experiments. Dark and light purple blocks show crosses for which accessions indicated on the left or bottom, respectively, were used as females. Blue blocks show hybrids for which two reciprocal crosses were analyzed. Brown indicates hybrids whose two reciprocal crosses were field-trialed, but there were not enough replicates (n <3 for each) for statistical comparison between them. White blocks indicate missing hybrids.

Figure S2 Heterotic trait locus (HTL) mapping. Two-step genomic scan as performed with overdominant heterosis (ODH) values derived from the 2011 field experiment, shown for representative markers. **A.** Dsenhsbm99 showing similar ODH distributions for the different genotypic groups, i.e. did not pass the first mapping step (GLM). The line across each diamond and the vertical span represent the group mean and the 95% confidence interval for each group, respectively. **B.** Xcup64 showing significant difference (GLM, perm. $P = 0.002$) between ODH values of the different genotypic groups, albeit with no advantage for specific hetero-genotypic group as compared to the homo-genotypic group (H:H). This marker passed the first step and failed in the second (Kolmogorov-Smirnov). **C.** Dsenhsbm40 that passed the first step (GLM, perm. $P = 0.005$) and the second step with a significant advantage only for the hetero-genotypic groups 154:156 and 154:164 (in gray) vs. the homo-genotypic group (H:H; Kolmogorov-Smirnov test, $P = 0.0008$, 0.0165, respectively). **D.** Cumulative distribution function plot showing the ODH values of the significant hetero-genotypic (154:156, 154:164) and homo-genotypic groups (H:H) for the same marker (Dsenhsbm40). DPW, dry panicle weight.

Figure S3 Chromosomal physical map (in Mbp) of markers used in this study and the identified heterotic trait loci (HTLs) for grain yield (dry panicle weight; DPW). Position is only shown for HTLs that were mapped over 2 years–each HTL is marked with a black diamond. Black and white coloring indicate pericentric-heterochromatic and telomeric-euchromatic chromosomal regions, respectively. Gray indicates markers within pericentric-heterochromatic chromosomal regions.

Figure S4 Illustration of the allelic distribution in a diallel. A. Allelic distribution in founder lines (FLs) including the common A1 and A2 alleles shared by 4 and 3 parents, respectively, and the rare A3 allele found in FL18 and FL19. A rare allele is defined as one carried by less than three FLs. The genotype of each hybrid is projected from the two homozygous parents. **B.** Six homozygous hybrids derived from the four FLs that share the A1 allele (homo-genotypic group A1/A1). **C.** Eight heterozygous hybrids derived from four FLs that share the A1 allele and two FLs that share the A3 allele (hetero-genotypic group A1/A3). **D.** A rare homo-genotypic group (A3/A3) including a single hybrid derived from a cross between FLs sharing the rare A3 allele. This group cannot be compared to the corresponded hetero-genotypic groups (any groups that harbor one copy of A3).

Figure S5 Comparative heterosis mapping of sorghum and maize. For each sorghum heterotic trait locus (HTL; white diamonds), genomic intervals which are syntenic with the maize genome are shown. The locations of the maize overdominant QTL for yield-associated traits (kernel number, KN; grain yield, GY) on the maize bin map are drawn based on a combination of data from Tang *et al.* [9], Cockerham and Zeng) [12] and Frascaroli *et al.* [20].

Table S1 List of the accessions included in the wide Sorghum bicolor sps. bicolor collectionTable S2: SSR markers used in this study.

Table S2 SSR markers used in this study.

Acknowledgments

We wish to thank Jianming Yu, Chengsong Zhu (Department of Agronomy, Kansas State University, Manhattan, KS), Yuval Eshed (Weizmann Institute of Science, Rehovot), and Noam Shental (Open University, Ra'anana) for helpful discussion on the genetic analysis. The technical assistance of Leah Rosenthal, Heelee Bonfil, Shimrit Mekonen, Michal Gadamo, Shir-El Azaria and Shula Blum (all from The Robert H. Smith Institute of Plant Sciences and Genetics in Agriculture, The Hebrew University of Jerusalem, Rehovot) is highly appreciated. We also thank Kathrin Baake, Corinna Trautewig and Kerstin Wolf for their technical assistance in the microsatellite genotyping, as well as Raj K. Pasam for his support in the SSR analysis (all at IPK, Gatersleben).

Author Contributions

Conceived and designed the experiments: IBI BK HN EF. Performed the experiments: IBI BK HN. Analyzed the data: IBI BK HN EF. Contributed reagents/materials/analysis tools: IBI BK EF. Wrote the paper: IBI EF. Designed the software used in analysis: IBI EF. Performed and analyzed the microsatellite genotyping: IBI BK.

References

1. Morrell PL, Buckler ES, Ross-Ibarra J (2011) Crop genomics: advances and applications. Nature Reviews Genetics 13: 85–96.
2. Darwin CR (1876) The effects of cross- and self-fertilization in the vegetable kingdom. London: John Murray.
3. Charlesworth D, Willis JH (2009) Fundamental concepts in genetics The genetics of inbreeding depression. Nature Reviews Genetics 10: 783–796.
4. Riedelsheimer C, Czedik-Eysenberg A, Grieder C, Lisec J, Technow F, et al. (2012) Genomic and metabolic prediction of complex heterotic traits in hybrid maize. Nat Genet 44: 217–220.
5. Yu J, Pressoir G, Briggs WH, Vroh Bi I, Yamasaki M, et al. (2006) A unified mixed-model method for association mapping that accounts for multiple levels of relatedness. Nat Genet 38: 203–208.

6. Krieger U, Lippman ZB, Zamir D (2010) The flowering gene SINGLE FLOWER TRUSS drives heterosis for yield in tomato. Nat Genet 42: 459–463.
7. Lippman ZB, Zamir D (2007) Heterosis: revisiting the magic. Trends Genet 23: 60–66.
8. Meyer RC, Kusterer B, Lisec J, Steinfath M, Becher M, et al. (2010) QTL analysis of early stage heterosis for biomass in Arabidopsis. Theoretical and Applied Genetics 120: 227–237.
9. Tang J, Yan J, Ma X, Teng W, Wu W, et al. (2010) Dissection of the genetic basis of heterosis in an elite maize hybrid by QTL mapping in an immortalized F2 population. Theoretical and Applied Genetics 120: 333–340.
10. Pea G, Paulstephenraj P, Cane MA, Sardaro ML, Landi P, et al. (2009) Recombinant near-isogenic lines: a resource for the mendelization of heterotic QTL in maize. Mol Genet Genomics 281: 447–457.
11. Lariepe A, Mangin B, Jasson S, Combes V, Dumas F, et al. (2012) The genetic basis of heterosis: multiparental qtl mapping reveals contrasted levels of apparent overdominance among traits of agronomical interest in maize (*Zea mays* L.). Genetics 190: 795–811.
12. Cockerham CC, Zeng ZB (1996) Design III with marker loci. Genetics 143: 1437–1456.
13. Kearsey MJ (1965) Biometrical analysis of a random mating population - a comparison of five experimental designs. Heredity 20: 205–235.
14. Melchinger AE, Lee M, Lamkey KR, Hallauer AR, Woodman WL (1990) Genetic diversity for restriction-fragment-length-polymorphisms and heterosis for 2 diallel sets of maize inbreds. Theoretical and Applied Genetics 80: 488–496.
15. Cho YI, Park CW, Kwon SW, Chin JH, Ji HS, et al. (2004) Key DNA markers for predicting heterosis in F-1 hybrids of japonica rice. Breeding Science 54: 389–397.
16. Quinby JR, Kramer NW, Stephens JC, Lahir KA, Karper RE.(1958) Grain sorghum production in Texas; College Station, Tex., USA.
17. Doggett H (1988) Sorghum. Harlow, Essex, England New York: Longman Scientific & Technical; Wiley. xviii, 512 p.
18. Quinby JR, Karper RE (1972) Heterosis in sorghum resulting from heterozygous condition of a single gene that affect duration of growth. American Journal of Botany 33: 716–721.
19. Jordan DR, Tao Y, Godwin ID, Henzell RG, Cooper M, et al. (2003) Prediction of hybrid performance in grain sorghum using RFLP markers. Theoretical and Applied Genetics 106: 559–567.
20. Frascaroli E, Cane MA, Landi P, Pea G, Gianfranceschi L, et al. (2007) Classical genetic and quantitative trait loci analyses of heterosis in a maize hybrid between two elite inbred lines. Genetics 176: 625–644.
21. Li L, Lu K, Chen Z, Mu T, Hu Z, et al. (2008) Dominance, overdominance and epistasis condition the heterosis in two heterotic rice hybrids. Genetics 180: 1725–1742.
22. Liu KJ, Muse SV (2005) PowerMarker: an integrated analysis environment for genetic marker analysis. Bioinformatics 21: 2128–2129.
23. Hubner S, Hoffken M, Oren E, Haseneyer G, Stein N, et al. (2009) Strong correlation of wild barley (*Hordeum spontaneum*) population structure with temperature and precipitation variation. Mol Ecol 18: 1523–1536.
24. Schertz KF, Clark LE (1967) Controlling dehiscence with plastic bags for hand crosses in Sorghum. Crop Science 7: 540–542.
25. Xin ZG, Velten JP, Oliver MJ, Burke JJ (2003) High-throughput DNA extraction method suitable for PCR. Biotechniques 34: 820–825.
26. Ben-Israel I, Yu G, Austin MB, Bhuiyan N, Auldridge M, et al. (2009) Multiple biochemical and morphological factors underlie the production of methylketones in tomato trichomes. Plant Physiol 151: 1952–1964.
27. Tesso T, Tirfessa A, Mohammed H (2011) Association between morphological traits and yield components in the durra sorghums of Ethiopia. Hereditas 148: 98–109.
28. Brown PJ, Klein PE, Bortiri E, Acharya CB, Rooney WL, et al. (2006) Inheritance of inflorescence architecture in sorghum. Theoretical Applied Genetics 113: 931–942.
29. Bradbury PJ, Zhang Z, Kroon DE, Casstevens TM, Ramdoss Y, et al. (2007) TASSEL: software for association mapping of complex traits in diverse samples. Bioinformatics 23: 2633–2635.
30. Churchill GA, Doerge RW (1994) Empirical threshold values for quantitative trait mapping. Genetics 138: 963–971.
31. Pritchard JK, Stephens M, Rosenberg NA, Donnelly P (2000) Association mapping in structured populations. American Journal of Human Genetics 67: 170–181.
32. Paterson AH, Bowers JE, Bruggmann R, Dubchak I, Grimwood J, et al. (2009) The *Sorghum bicolor* genome and the diversification of grasses. Nature 457: 551–556.
33. Bhattramakki D, Dong JM, Chhabra AK, Hart GE (2000) An integrated SSR and RFLP linkage map of *Sorghum bicolor* (L.) Moench. Genome 43: 988–1002.
34. Srinivas G, Satish K, Madhusudhana R, Seetharama N (2009) Exploration and mapping of microsatellite markers from subtracted drought stress ESTs in *Sorghum bicolor* (L.) Moench. Theoretical and Applied Genetics 118: 703–717.
35. Falconer DS, Mackay TFC (1996) Introduction to quantitative genetics. Essex, England: Longman. xiii, 464 p.
36. Kambal AE, Webster OJ (1966) Manifestations of hybrid vigor in grain sorghum and relations among components of yield weight per bushel and height. Crop Science 6: 513–515.
37. Huang X, Feng Q, Qian Q, Zhao Q, Wang L, et al. (2009) High-throughput genotyping by whole-genome resequencing. Genome Res 19: 1068–1076.
38. Buckler ES, Holland JB, Bradbury PJ, Acharya CB, Brown PJ, et al. (2009) The genetic architecture of maize flowering time. Science 325: 714–718.
39. Hamblin MT, Buckler ES, Jannink JL (2011) Population genetics of genomics-based crop improvement methods. Trends Genet 27: 98–106.
40. Huang X, Wei X, Sang T, Zhao Q, Feng Q, et al. (2010) Genome-wide association studies of 14 agronomic traits in rice landraces. Nat Genet 42: 961–967.
41. Yu J (2009) The potential of ultrahigh throughput genomic technologies in crop improvement. Plant Genome 2: 2.
42. Phillips PC (2008) Epistasis–the essential role of gene interactions in the structure and evolution of genetic systems. Nature Reviews Genetics 9: 855–867.
43. Smith LM, Bomblies K, Weigel D (2011) Complex evolutionary events at a tandem cluster of *Arabidopsis thaliana* genes resulting in a single-locus genetic incompatibility. PLoS Genet 7: e1002164.
44. Hua JP, Xing YZ, Xu CG, Sun XL, Yu SB, et al. (2002) Genetic dissection of an elite rice hybrid revealed that heterozygotes are not always advantageous for performance. Genetics 162: 1885–1895.
45. Luo LJ, Li ZK, Mei HW, Shu QY, Tabien R, et al. (2001) Overdominant epistatic loci are the primary genetic basis of inbreeding depression and heterosis in rice. II. Grain yield components. Genetics 158: 1755–1771.
46. Clark SA, Hickey JM, van der Werf JH (2011) Different models of genetic variation and their effect on genomic evaluation. Genet Sel Evol 43: 18.
47. Williams W (1959) Heterosis and the genetics of complex characters. Nature 184: 527–530.
48. Yu J, Holland JB, McMullen MD, Buckler ES (2008) Genetic design and statistical power of nested association mapping in maize. Genetics 178: 539–551.
49. Yu J, Hamblin MT, Tuinstra MR (2011) Association genetics strategies and resources. In: Paterson AH, editor. Genetics and Genomics of the Saccharinae: Springer Verlag.
50. Hubner S, Günther T, Flavell A, Fridman E, Graner A, et al. (2012) Islands and Streams: Clusters and gene flow in wild barley populations from the levant. Mol Ecol 21: 1115–1159.
51. Smith CC, Fretwell SD (1974) The optimal balance between size and number of offspring. The American Naturalist 108 : 499–506.
52. Adams MW (1967) Basis of yield component compensation in crop plants with special reference to field bean *Phaseolus vulgaris*. Crop Science 7: 505–510.
53. Fridman E, Carrari F, Liu YS, Fernie AR, Zamir D (2004) Zooming in on a quantitative trait for tomato yield using interspecific introgressions. Science 305: 1786–1789.
54. Frary A, Nesbitt TC, Grandillo S, Knaap E, Cong B, et al. (2000) fw2.2: a quantitative trait locus key to the evolution of tomato fruit size. Science 289: 85–88.

4

Genetic Relatedness of Low Solitary Nests of *Apis dorsata* from Marang, Terengganu, Malaysia

Najmeh Sahebzadeh[1*¶], Makhdzir Mardan[2¶], Abdul Manaf Ali[3], Soon Guan Tan[4], Nur Azura Adam[1], Wei Hong Lau[1]

1 Department of Plant Protection, Faculty of Agriculture. Universiti Putra Malaysia, Serdang, Selangor, Malaysia, 2 Department of Agriculture Technology, Faculty of Agriculture, Universiti Putra Malaysia, Serdang, Selangor, Malaysia, 3 Department of Biotechnology, Faculty of Agriculture and Biotechnology, Universiti Sultan Zainal Abidin, Kuala Terengganu, Terengganu, Malaysia, 4 Department of Cell and Molecular Biology, Faculty of Biotechnology and Biomolecular Sciences, Universiti Putra Malaysia, Serdang, Selangor, Malaysia

Abstract

Knowledge on the population of genetic structure and ecological behaviour of *Apis dorsata* from Peninsular Malaysia is needed for effective management and conservation of this species since unsustainable whole solitary low nest cutting for product harvesting is the current common practice here. The analysis of 15 single locus DNA microsatellite markers on samples from 20 solitary nests of *A. dorsata* showed that while these markers were polymorphic, high intracolonial relatedness existed. Furthermore, in general, slightly negative values of intercolony relatedness (R) among the nests of *A. dorsata* were found. However, positive values of mean intercolony relatedness were observed between 54 pairs of nests out of 190 possible combinations. The R values among nest pairs 3–4 and 3–5 was higher than 0.50 showing that their queens were half siblings, whereas nest pair 19–20 showed relatedness of 0.95 indicating that the same queen was sampled. The results that we obtained could not conclusively support the hypothesis of this study that the honey hunters in Marang district of Malaysia repeatedly harvest the same nest located at a different site and at a different time during the same honey harvesting season. However, our finding of an appreciable level of intercolonial relatedness between several pairs of nests in this pioneer study indicated that a comprehensive study with a larger sample size of solitary nests found throughout the region would be necessary to provide concrete proof for this novel idea.

Editor: Stephanie Ann White, UCLA, United States of America

Funding: University Putra Malaysia supported this research via FRGS grant (07-04-10-819FR). The funders had no role in study design, data collection and analysis, decision to publish, or preparation of the manuscript.

Competing Interests: The authors have declared that no competing interests exist.

* E-mail: najmeh.sahebzadeh@gmail.com

¶ These authors contributed equally to this work.

Introduction

The Indo-Malaya region is known as the centre of honeybee diversity, whereby seven out of the nine honeybee species of the world are sympatric and endemic to this region [1]. Among the *Apis* species, the nests of *A. dorsata* are known as the sources of wild honey in Malaysia [2]. Many colonies of the single-comb, open-nesting and sedentary of *A. dorsata* are found to nest either singly or low to the ground, or high in aggregates on tree limbs of tall bee trees along the coastal, submerged *Melaleuca* forest in Marang, Terengganu, Malaysia [3]. Of the two nest types, it is the low solitary nests that are usually harvested in Peninsular Malaysia and thus is the subject of this study.

Seasonal migration and aggregation found densely on a bee tree are known as unique characteristics in *A. dorsata*, which may differentiate this particular species from other *Apis* species [4]. The seasonal migration between alternative nesting sites is done to find available forage [5] and control the levels of *Tropilaelaps clarea* (parasitic mite) [6]. While the aggregation structure (up to 50 or 100 nests of *A. dorsata* on a bee tree high in the air [7], usually from 5 to 40 metres above the ground in the rainforest) is a unique feature of this honeybee species, abundance of low aggregations (2 to 10 nests on a tree) and solitary nests (only one or rarely two nests on a tree) are more common in this species [7], especially in the secondary forests in the district of Marang, Terengganu, Malaysia (Figure 1). Both low aggregations and low solitary nests are found on trees of less than 5 metres in height, which makes it easy for honey hunters to harvest these nests. The honey hunters in Marang district climb the tree and remove the whole comb of this type of *A. dorsata* nests easily during the days of harvesting season (Personal observation).

In Malaysia, the interaction between man, the forest and *A. dorsata* has been established for several hundreds of years [8]. Most honey hunters are found in the states of Kedah, Terengganu and Negeri Sembilan in the Peninsular Malaysia [8]. Collection of wild honey is lucrative and generates an income of about RM6,000 (USD2000) per harvesting season in about three months [8]. Because of the inherent organic properties, high medical and nutritional values of wild honey, its price is much higher than the commercial honey, which may be produced by *A. mellifera* and *A. cerana* [9]. Due to the large size of the *A. dorsata* nest, a considerable amount of wild honey (up to 45 kg) [10] may be stored by a nest. This amount of honey tempts local honey hunters to harvest the nests of *A. dorsata* during flowering season frequently. Therefore,

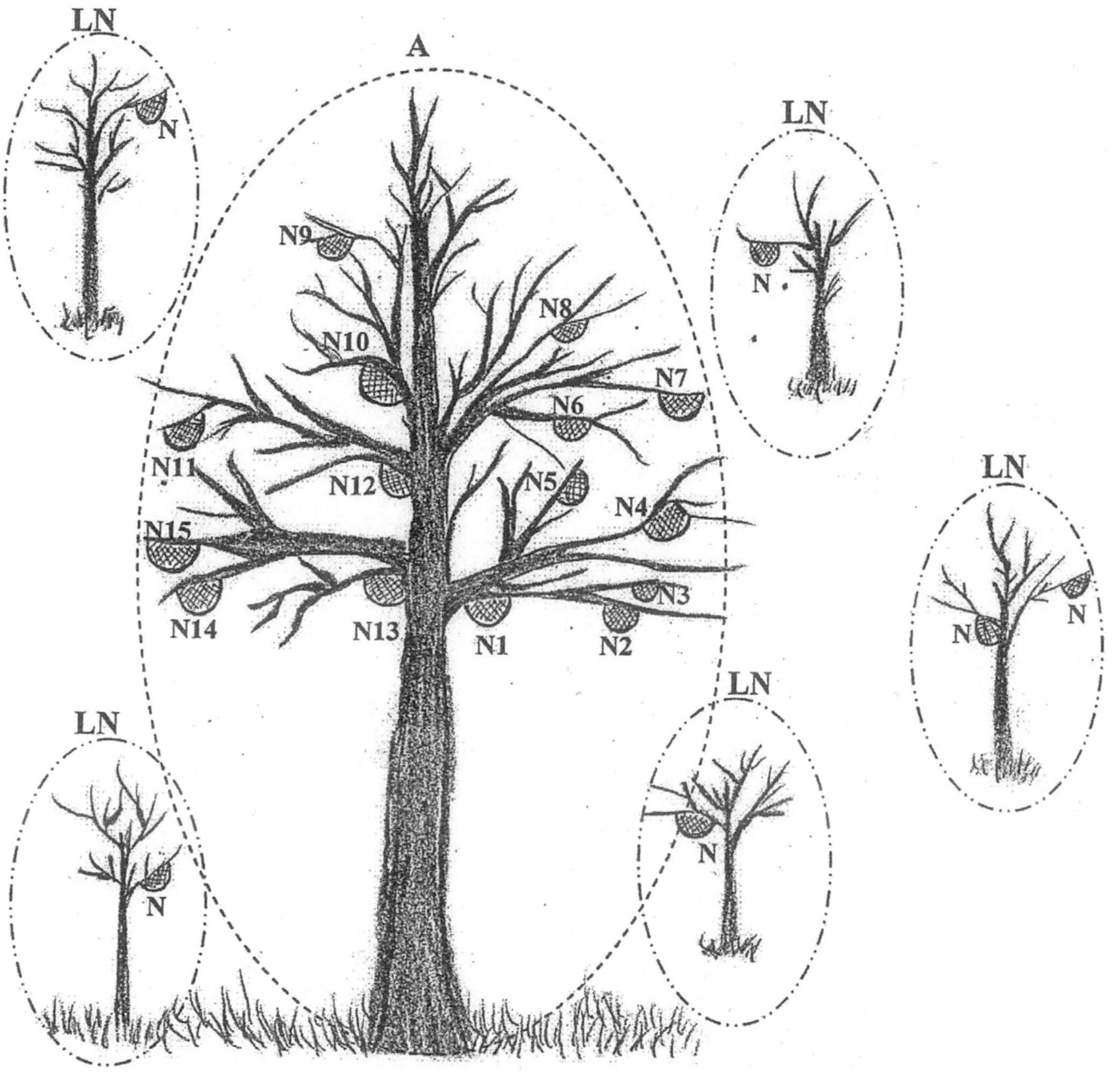

Figure 1. A diagram of *A. dorsata* aggregation (high nesting) and low solitary nesting (Marang, Terengganu, 2010). A: An aggregation on a bee tree of ~ 40 metres in height. **LN**: Low solitary nests on a tree of between 2–5 metres in height. **N**: Nest.

Malaysian honey hunters harvest the whole nest of *A. dorsata* by cutting and taking the different parts of a nest for selling. This method of nest harvesting is especially done on solitary nests, which are spread throughout the rainforest of Malaysia. This common harvesting method is deemed unsustainable as the bees may avoid this area as their nesting site in the future causing a decrease of the *A. dorsata* population here.

Honey hunters in Marang claim that they harvest about 600 or more nests of *A. dorsata* per harvesting season (June, July and August) in this area (Personal communication). This figure is seemingly huge, and based on the *A. dorsata* biology and plant flora, it is not possible to have this huge number of *A. dorsata* nests in this area. Furthermore, *A. dorsata* needs specific plant species for providing nectar, pollen and establishment of its nest. Saberioon *et al.* (2010) reported that in Marang, *A. dorsata* constructs its nest on the branches of available nesting support of *Acaia auriculiformis* and *Melaleuca cajuputi* trees which are the sources of nectar and pollen for producing strong flavour and weak density of wild honey in the forested areas of Marang district [11].

We hypothesize that the honey hunters in Marang harvest the same nests, in which the queen of each nest heads away from the original nest site together with her nest members after the first harvest to construct a new nest at another site in the same season and in the available nesting support in the vicinity of the same geographical location. Information on the genetic structure and relatedness among the solitary nests of *A. dorsata* could provide us with the estimation of honey harvesting dates from the new nests to help rectify the current unsustainable nest harvesting practices.

Paar *et al.* (2004) studied the genetic structure and relatedness among and within the *A. dorsata* aggregations in northeast India [5]. Their results revealed significant genetic differentiation among the aggregations. They also found positive relatedness among nests within aggregations which indicated that the colonies within aggregations were more highly related than random colonies. However, they did not study the low solitary nests it was necessary for us to do so since this nest type is mainly harvested in Malaysia.

There is currently no data on the genetic relatedness and population structure of *A. dorsata* solitary nests from Marang district (Terengganu, Malaysia). The knowledge on the genetic structure of *A. dorsata* is needed for effective management and conservation strategies of this species in Malaysia, especially to prevent over harvesting of honey. Overall, the objectives of this study were to estimate the level of intercolonial and intracolonial relatedness and to identify the population structures of Malaysian *A. dorsata* low solitary nests using single locus DNA microsatellites. According to Insuan *et al.* (2007), microsatellites are efficient markers to reveal the detailed genetic structure of *A. dorsata* population [12].

Results and Discussion

Microsatellite Amplification and Genetic Variation

All the 15 loci were successfully amplified with the expected sizes of PCR products (Table 1).The inferred genotypes at 15 loci of the queen of each solitary nest are presented in Table 2. The number of the effective alleles displayed per locus (allelic diversity) acts as a useful measure of genetic variability within a population [13]. A total of 134 alleles were found at the 15 microsatellite loci in the honeybees of Marang district population. The number of alleles per locus ranged from 6 alleles for locus Ap273 to 11 alleles for locus Ag005a. All loci were found to be polymorphic. H_o ranged from 0.30 to 0.60 across the 15 loci, with mean of 0.41±0.025 and mean of H_e 0.82±0.007 (Table 3).

Higher PIC values of a molecular marker indicate higher heterozygote frequency in a population, as well as more genetic information [14]. In this research, all loci showed high polymorphism. It was found that all PIC values were less than their related heterozygosity. The obtained results indicated that the loci with more alleles contained higher rates of heterozygosity and PIC values, but this was not absolute because of the effect of allele frequencies. For example, Ag005a had 11 alleles, while its related heteozygosity and PIC values were lower (H_e = 0.829, PIC = 0.646) than locus Ap049 with 10 alleles (H_e = 0.854, PIC = 0.718). Thus, it was concluded that at Ap049 locus, the rare alleles contributed little effect on heterozygosity. Furthermore, the heteozygosity at this locus exceeded the expected heterozygosity computed from the number of alleles as a result of mutation drift equilibrium [15]. The mean effective number of alleles (5.738±0.223) across the 15 microsatellite loci indicated that the sample size of this study was enough [16].

A polymorphic locus must have at least 0.10 of heterozygosity to reflect the variation of genetic structure [17]. The mean heterozygosity (0.822±0.007) across the 15 microsatellite loci was sufficient to show genetic variation in the studied population.

Genetic Structure

The deviations from HWE were assessed using the method of Guo and Thompson (1992) for each locus- nest combination using a Markov chain of 100 000 steps and 1000 dememorization steps [18]. For social insects, the colony structure may show significant deviations from HWE and linkage disequilibrium tests whereas no deviation exists in a population if the workers' genotypes are used [5]. So, in this research, these analyses were carried out using the queen's alleles only. The results showed that no nest had significant deviation from HWE (P>0.05). There was also no significant deviation from genotypic linkage disequilibrium (P>0. 34). The results also demonstrated that the low solitary nests which were sampled for this study were mating randomly.

Population Genetic Differentiation

Among closely related populations, genetic drift is assumed to be the main factor in genetic differentiation [19]. Hence, F_{ST} index is suggested to be used for estimating the differences between populations [20,21]. Theoretically, F_{ST} vales change between 0 and 1 [22]. All F_{ST} values between the solitary nests of *A. dorsata*

Table 1. Detailed information of microsatellite loci.

SSR	Name	Core sequence	Forward primer (5'→3')	Reverse primer(5'→3')	Expected size	Reference
SSR1	At003	$(CT)_6TT(CT)_{14}$	GATCATTTCTTTCATTCTTCTCTCTC	ATGCTCGACTATTCCGCG	199	[31]
SSR2	A088	$(GA)_{15}$ … $(GCTCG)_5$	CGAAGGTAAGGTAAATGGAAC	GGCGGTTAAAGTTCTGG	150	[31]
SSR3	AB124	$(CT)_8$ (CT)14$(GGCT)_8$	GCAACAGGTCGGGTTAGAG	CAGGATAGGGTAGGTAAGCAG	250	[31]
SSR4	AB024	$(CT)_{11}$	CACAAGTTCCAACAATGC	CACATTGAGGATGAGCG	100	[31]
SSR5	Ag005a	$(A)_8G(A)_6G(A)_5$	TGTTCCGGCAAGCTGAAG	GTGCTCCGCAACAACGTG	108	[31]
SSR6	Ap085	$(GA)_6(GA)_{11}$	GATCAAACACACAAACGAAAGC	ACCGGAAGCCTAATCAAGG	196	[31]
SSR7	Ap226	$(CT)_8$	AACGGTGTTCGCGAAACG	AGCCAACTCGTGCGGTCA	231	[31]
SSR8	A107	$(GCTC)_2(GCT)_2(CT)_{23}$	CCGT GGGAGG58T TTATTGTCG	CCTTCGTAACGGATGACACC	200	[32]
SSR9	Ap049	$(AGG)_7$	CCAATAGCGGCGAGTGTG	GGGCTTCGTACGTCCACC	142	[31]
SSR10	Ap243	$(TCC)_9$	AATGTCCGCGAGCATCTG	TGTTTACGAGAATTCGACGGG	260	[31]
SSR11	Ap273	$(CT)_8$	GATCTTGTGTTAAACAGCCG	GATCTCTGGCAGACGAAGAG	108	[31]
SSR12	Ad3	$(CT)_7$	CCGTAACTGGACTTCTTTCCCTCC	GACAATGGCGTACTTTGTGG	160	[5]
SSR13	Ap036	$(GA)_{21}$	CTACGCGCTTACAGGGCA	GCCGAAATTCAACGCTCA	159	[31]
SSR14	Ap256	$(GA)_{12}$ $AT(GA)_3$	CCAAGTCGCTTCATCATCGT	CCTAAGGTCTACACCCCCGT	162	[31]
SSR15	A14	$(CT)_{13}$…$(GGT)_9$	GTGTCGCAATCGACGTAACC	GTCGATTACCGATCGTGACG	206	[32]

Table 2. Inferred queen genotypes of each solitary nest at each locus (Marang, Terengganu, Malaysia, 2010).

Plot	Queen	SSR 1	SSR 2	SSR 3	SSR 4	SSR 5	SSR 6	SSR 7	SSR 8	SSR 9	SSR 10	SSR 11	SSR 12	SSR 13	SSR 14	SSR 15
A	Nest 1	218/218	150/150	250/250	92/92	110/110	200/200	229/229	204/204	150/150	260/260	110/110	167/167	160/160	206/206	230/230
A	Nest 2	218/220	150/150	250/250	92/92	112/112	200/200	229/229	198/198	142/142	254/260	110/120	157/157	168/168	214/214	214/230
A	Nest 3	218/218	150/150	254/254	100/100	112/112	198/198	233/233	198/198	150/150	254/260	120/120	157/157	160/160	214/214	214/214
A	Nest 4	218/220	150/150	254/254	100/100	112/112	198/198	229/233	204/204	152/152	254/254	120/120	167/167	160/160	214/214	230/230
A	Nest 5	218/218	150/150	254/254	100/100	110/110	200/200	233/233	198/198	150/150	254/260	120/120	157/157	168/168	206/206	230/230
A	Nest 6	218/218	154/154	254/254	100/100	110/110	198/198	233/233	204/204	150/150	260/260	110/110	157/157	160/160	206/214	214/214
A	Nest 7	200/200	154/154	244/244	104/104	100/110	210/210	229/243	206/216	150/160	254/270	114/114	169/175	168/180	200/210	2102/14
A	Nest 8	212/212	148/148	252/252	108/108	118/118	196/196	237/237	206/206	144/144	256/256	116/116	161/169	178/178	216/216	220/220
A	Nest 9	212/212	146/146	254/254	110/110	116/116	196/196	235/237	200/200	148/148	256/262	108/120	169/169	172/178	220/220	220/220
B	Nest 10	212/212	146/146	254/254	108/108	116/116	192/192	235/237	200/200	144/148	256/256	116/116	169/169	172/172	216/216	214/214
B	Nest 11	216/216	148/148	252/252	108/108	118/118	192/196	235/235	206/206	148/148	262/262	108/108	161/161	178/178	216/216	214/214
B	Nest 12	216/216	148/148	254/254	110/110	118/118	196/196	237/237	206/206	148/148	256/256	110/110	161/161	172/172	216/220	220/220
B	Nest 13	212/212	146/146	252/252	110/110	118/118	196/196	235/237	206/206	144/148	262/262	108/108	169/169	172/172	220/220	214/214
B	Nest 14	218/220	140/150	242/246	100/102	104/106	202/204	231/237	212/220	138/146	264/272	112/120	159/171	166/172	202/212	210/216
B	Nest 15	200/210	142/142	246/246	106/106	118/118	210/210	229/243	208/208	138/148	256/260	120/120	173/173	174/180	208/208	224/224
B	Nest 16	212/218	150/156	246/248	102/110	108/114	196/198	231/237	196/206	142/158	254/270	114/120	171/177	168/178	206/218	210/218
B	Nest 17	200/204	148/152	254/256	106/108	104/110	206/210	229/235	216/220	146/160	256/268	108/114	157/169	162/174	202/214	212/226
B	Nest 18	200/204	158/160	244/248	102/104	116/120	196/198	237/245	196/212	140/152	254/264	110/116	167/173	170/170	210/220	216/220
B	Nest 19	218/220	150/154	240/250	92/98	102/110	198/208	233/247	198/218	140/158	254/260	114/120	159/165	166/180	202/216	218/230
C	Nest 20	218/220	150/158	240/250	92/98	102/110	198/208	233/247	198/218	142/158	260/260	114/114	165/167	166/180	202/216	230/230

were significantly different from zero ($P<0.01$). A moderate level of genetic differentiation (F_{ST}) among the sampled solitary nests (mean of $F_{ST}=11.6\%$) was found. In addition, the AMOVA indicated that this percentage of differentiation was due to the differences among the solitary nests (11%), an appreciable amount of genetic variation among workers of all nests (59%), and within workers of each nest (30%), as shown in Figure 2. The low percentage of differentiation (11%) among the solitary nests could

Table 3. A summary of microsatellite genetic variation in the low solitary nests of *A. dorsata*.

Locus	N_a	N_e	H_o	H_e	PIC
At003	7	4.598	0.450	0.783	0.627
A088	10	5.096	0.300	0.804	0.656
AB124	9	4.819	0.300	0.793	0.640
AB024	8	6.504	0.300	0.846	0.621
Ag005a	11	5.839	0.350	0.829	0.646
Ap085	9	5.369	0.350	0.814	0.672
Ap226	8	5.442	0.600	0.816	0.691
A107	10	6.667	0.350	0.850	0.709
Ap049	10	6.838	0.500	0.854	0.718
Ap243	8	5.063	0.600	0.803	0.698
Ap273	6	4.469	0.400	0.776	0.648
Ad3	10	6.667	0.400	0.850	0.697
Ap036	9	6.838	0.400	0.854	0.707
Ap256	10	6.557	0.450	0.848	0.666
A14	9	5.298	0.400	0.811	0.688
Mean±SE	8.933±0.345	5.738±0.223	0.410±0.025	0.822±0.007	0.672±0.0067

N_a is the number of different alleles, while N_e is the number of effective alleles. H_o, H_e and PIC indicate the observed and expected heterozygosity and polymorphism information content, respectively.

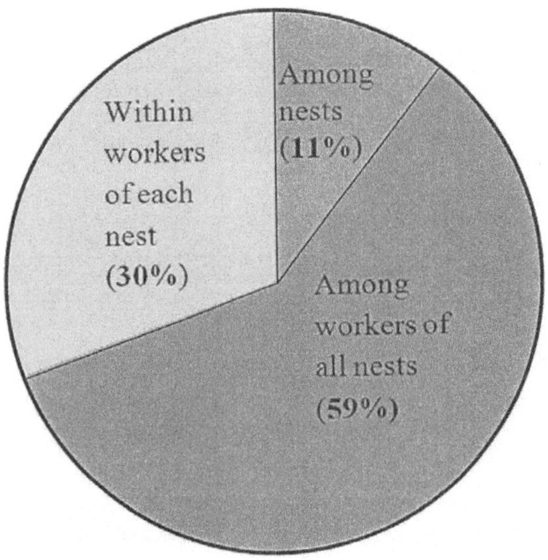

Figure 2. The hierarchical AMOVA based on microsatellite markers for showing the genetic differentiation of *A. dorsata* solitary nests.

support this research hypothesis, which states that the honey hunter may be harvesting the same nest at a different site of a same geographical area and at a different time. The genetic variation among *A. dorsata* workers of all nests and within workers of each nest were resulted as the reproductive system in honeybees and high levels of polyandry [23].

Genetic Relatedness among the Solitary Nests of *A. dorsata*

We followed Oldroyd *et al.*'s (1995) analytical approach to determine whether the spatial distributions of the solitary nests of *A. dorsata* were random or otherwise [16]. The Poisson method and negative binomial method were used. An organism under Poisson distribution is dispersed randomly, whereas an aggregated organism follows negative binomial distribution in the environment (Dale, 2000 cited in [16]). Both types of distribution were tested, because two types of nests (the solitary and aggregations) were found in the study site. This paper is focused on the results of our work on the more frequently harvested solitary nests. The observed and expected distributions of nests under these models showed that the observed distribution differed significantly ($P<0.01$) from the negative binomial distribution, indicating that the solitary nests were distributed randomly throughout the plots (Figure 3).

The degree of intercolonial genetic relatedness (R) among the solitary nests of *A. dorsata* revealed positive relatedness between some of the nest pairs based on the queen's heading of them. The mean relatedness coefficient among the solitary nests of *A. dorsata* was negative ($R = -0.053 \pm 0.016$). The 95% confidence interval was determined by bootstrap resamplings and limits of -0.092 and 0.094 were found. Under Kin selection hypothesis, the majority of relatedness coefficients among pairs of queens should be positive. However, positive values of mean intercolonial relatedness were observed only between 54 pairs of nests out of 190 possible combinations (Table 4). The R values among nest pairs 3–4 and 3–5 was higher than 0.50 showing that their queens were half sibling, whereas nest pair 19–20 showed a relatedness of 0.95 indicating that the same queen was sampled (Table 2 and Table 4).

Despite the tendency of *A. dorsata* to aggregate on a single tree which suggested that the colonies within the aggregations might show higher degree of relatedness than random colonies, Paar *et al.* (2004) found that the long distance migration between the original and alternative sites might minimize the genetic differentiation among the geographical locations as well as within the aggregations of this species [5]. Hence, the low genetic differentiation among and within the *A. dorsata* aggregations was due to gene exchange among them. The slightly negative values of intercolony relatedness within the aggregations of *A. dorsata* were found using microsatellites [5]. In their study, no pairs of colonies within the aggregations showed high relatedness. Only one pair of the queens carried at least one identical allele at every locus. These queens were considered to have formed mother and daughter colonies [5]. Paar *et al.* (2004) further explained that the limited number of bee workers per colony and the loci which they used affected their results.

Based on the objectives of the present study, the pair wise nests numbered 19 and 20 showed the highest intercolonial relatedness ($R = 0.95$) among all the pairs of nests (Table 4) while queen pair carried out two same identical alleles at 10 loci and at least one allele for the remaining five loci. On the contrary, the nest number 16 had the lowest degree of relatedness ($R = 0.016$) with nest numbers 19 and 20. Despite the distance between plot B and plot C, and the different date of sampling at these sites, 95% of genetic relatedness was found between one pair of nests (numbered 19 and 20). This high level of relatedness between this pair of nests raised the possibility that the honey hunters might have harvested the same nest after the nest shifted to a new site in the same season and at the same geographical area. Hence, it was concluded that the plundering or destruction of a particular *A. dorsata* nest for honey harvesting caused the queen to head the nest members away to establish a new nest at different site but within the same geographical area.

Along the degree of intercolonial relatedness, intracolonial pedigree relatedness (r) was calculated based on the effective number of drones which mated with the queen of each nest (Crozier, 1970 cited in [5]). In the present study, the effective number of drones was found for each nest and locus ranged between 7.89 ± 2.1 and 14.29 ± 3.07, followed by a mean average of intracolonial relatedness of 0.281 ± 0.019. Oldroyd *et al.* (1996) reported that the *A. dorsata* queen mates with an effective number of 20.0 ± 6.6 and a mean average intracolonial relatedness of 0.29 ± 0.007 [23]. They determined the (r) levels using three microsatellites and a large number of workers per nest (42–194). Despite the number of microsatellites and workers per nest used in our study and that of Oldroyd *et al.* (1996), a similar mean average of intracolonial relatedness was found. Our results approach was in line with the view of Takazaki and Nei (1996) [20] that using more microsatellite loci is better for obtaining genetic relationships when the closely related samples are being studied and the average heterozygosity is high.

Our finding on the nesting behaviour of the *A. dorsata* bees sampled along the coastal, submerged *Melaleuca* forest in Marang (Malaysia, Terengganu) should be taken into account by the Forestry Department of Peninsular Malaysia in their effort to formulated possible sustainable methods of nest product harvesting rather than the present practice of cutting off the entire nest.

Materials and Methods

Study Area

This study was conducted in Marang district, which is located in the state of Terengganu at the northern east of Peninsular

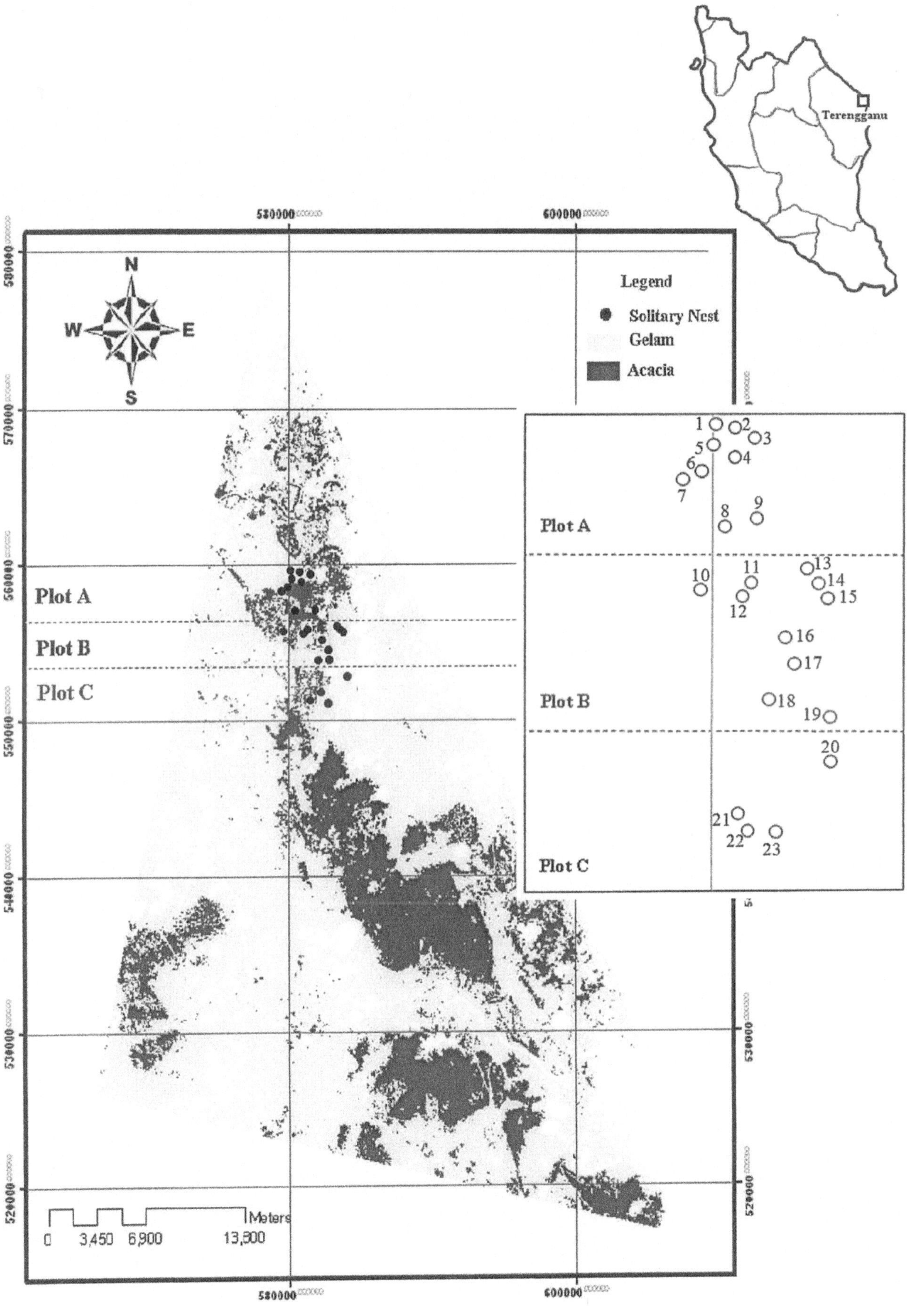
Terengganu
580000
600000
N
S
E
W
Legend
Solitary Nest
Gelam
Acacia
Plot A
Plot B
Plot C
1 2 3 4 5 6 7 8 9 10 11 12 13 14 15 16 17 18 19 20 21 22 23
Meters
0 3,450 6,900 13,800
580000
570000
560000
550000
540000
530000
520000

Figure 3. The spatial distribution of *A. dorsata* solitary nests from Marang, Terengganu, Malaysia (2010). Numbers 1 to 20 were the solitary nests, while numbers 21 to 23 were the aggregated nests.

Malaysia, between the upper left of 5°01′ N, 103°11′ E and the lower right of 4°50′ N, 103°24′ E. Mangrove, *Melaleuca*, *Acacia*, rubber and coconut trees are the dominant trees species in this area [11]. The locations of these trees are not in any way restricted information. Hence, no specific permits were required for this study and the authors also confirm that the nests sampled did not involve any endangered or protected species.

In this research, we followed Saberioon *et al.*'s (2010) prediction map for finding the solitary nests of *A. dorsata* in Marang district [11]. The samplings were done during the harvesting season of 2010. The map was prepared for the *A. dorsata* nesting sites [11], and the positions of the solitary nests were added to that map for the present study (Figure 3). The nests numbered 1 to 9, 10 to 19, and 20 to 23 were sampled during the last week of June, third and fourth weeks of July, and third week of August of 2010, respectively. The distribution data of the solitary nests is shown in and Table 5. The nests numbered 1 to 20 were solitary nests, while those numbered 21 to 23 were the aggregated nests on three different bee trees (i.e. up to 20 nests on a bee tree). Therefore, in this study, only the low solitary nests were analyzed to achieve the research objectives and kept the aggregated nests for the next research. Furthermore, one-kilometre distance was estimated as the forage area of each nest [11]. Then, at least one-kilometre distance was kept between the pair wise nests of each plot (Figure 3).

Sampling and DNA Extraction

Based on the effective number of mating in *A. dorsata* (20.0±6.6) [23], 20 bee workers per nest were collected. The entire nests were removed and the bee workers were taken from the honey part of each nest separately. The bee workers were preserved at 95% of ethanol and stored at −80°C for DNA extraction later.

The total genomic DNA was isolated from the thorax of each bee worker of each nest using the Wizard® Genomic DNA Purification Kit (Promega, USA). The protocol of DNA extraction followed the animal tissue protocol, which is available at http://www.promega.com/resources/protocols/. The concentrations of the extracted DNA stocks were measured using a spectrophotometer (Eppendorf Biophotometer Plus). The concentrations of DNA ranged from 200–370 ng. Then, the needed volume of DNA stocks was diluted to obtain 10 ng/ul concentration of working DNA. After testing for the purity and dilution, all the DNAs were labelled. So that each DNA sample can be traced to the worker and the nest from this was originated. The DNAs were kept at −20°C for PCR amplification later.

Analysis of Microsatellite Polymorphisms

Twenty two (22) heterospecific primer sets of DNA microsatellites (including Ad3, A007, AB024, A088, AB124, Ap036, Ap043, Ap049, Ap068, Ap085, Ap207, Ap226, Ap243, Ap273, Ap297, At003, Ag005a, A107, A76, B124 and A14) were screened and 15 polymorphic loci (Table 1) were selected for achieving the objectives of the research using the reaction conditions described in Paar *et al.* (2004). In the present research, however, the non-labelled primers were used. The loci were size fractionated on 8% of denaturating polyacrylamide gels, visualized by a modified silver staining protocol [24] and then their sizes were estimated by a

Table 4. Labelled pair wise relatedness matrix [30].

1	2	3	4	5	6	7	8	9	10	11	12	13	14	15	16	17	18	19	20	
0																				**1**
0.36	0																			**2**
0.09	0.34	0																		**3**
0.24	0.22	0.55	0																	**4**
0.36	0.35	0.53	0.28	0																**5**
0.37	0.02	0.58	0.28	0.40	0															**6**
−0.09	−0.10	−0.16	−0.19	−0.09	−0.01	0														**7**
−0.26	−0.28	−0.31	−0.29	−0.30	−0.29	−0.16	0													**8**
−0.28	−0.28	−0.19	−0.16	−0.18	−0.21	−0.20	0.19	0												**9**
−0.26	−0.24	−0.13	−0.20	−0.21	−0.11	−0.13	0.32	0.48	0											**10**
−0.25	−0.22	−0.21	−0.27	−0.29	−0.19	−0.12	0.44	0.06	0.15	0										**11**
−0.19	−0.25	−0.23	−0.21	−0.22	−0.12	−0.20	0.43	0.37	0.13	0.35	0									**12**
−0.27	−0.24	−0.23	−0.29	−0.31	−0.20	−0.09	0.31	0.53	0.28	0.41	0.31	0								**13**
−0.13	−0.11	−0.05	−0.02	−0.04	−0.16	−0.19	−0.19	−0.17	−0.16	−0.24	−0.14	−0.16	0							**14**
−0.13	−0.12	−0.15	−0.13	−0.14	−0.20	0.04	−0.08	−0.11	−0.15	−0.07	−0.04	−0.11	−0.05	0						**15**
−0.13	−0.03	−0.06	−0.04	0.02	−0.18	−0.04	0.02	−0.02	−0.22	−0.14	−0.06	−0.02	0.08	−0.14	0					**16**
−0.15	−0.10	−0.13	−0.14	−0.12	−0.08	0.14	−0.05	−0.06	0.02	−0.01	−0.09	−0.12	−0.10	0.03	−0.30	0				**17**
−0.13	−0.21	−0.21	−0.03	−0.26	−0.15	0.05	0.00	0.02	−0.09	−0.20	0.01	−0.10	−0.04	−0.11	−0.02	−0.19	0			**18**
0.13	0.21	0.10	0.09	0.18	−0.02	−0.13	−0.27	−0.33	−0.28	−0.26	−0.32	−0.35	−0.03	−0.16	0.01	−0.23	−0.21	0		**19**
0.13	0.21	0.10	0.09	0.18	−0.02	−0.13	−0.27	−0.33	−0.28	−0.26	−0.32	−0.35	−0.03	−0.16	0.01	−0.23	−0.21	0.950	0	**20**

The bold numbers (**1–20**) indicate *A. dorsata* solitary nests.

Table 5. The distribution data of low solitary nests of *Apis dorsata* in Marang district.

Nest	Coordinate X	Coordinate Y	Number of nest on tree
1	584212.23	551979.28	1
2	585361.07	551090.93	1
3	584449.37	551973.10	1
4	583015.41	556065.82	1
5	578453.71	558196.83	1
6	581476.68	560048.02	1
7	584303.09	555828.73	1
8	581075.12	557498.23	1
9	590707.35	544311.77	1
10	581651.14	557439.59	1
11	581108.79	556896.12	1
12	583409.25	554716.91	1
13	583095.50	556114.89	1
14	593358.41	538489.08	1
15	583132.63	556317.61	1
16	593108.59	537899.25	1
17	583905.88	556283.56	1
18	576381.47	560296.06	1
19	583083.15	556010.49	1
20	582379.28	552542.33	1

comparison to a 20-bp ladder marker (Fermentas). The fragments were scored using the UVIDoc version 99.02 software. The DNA fragment sizes as the base pairs were entered into the Excel sheets as co-dominant markers and were saved for data analysis.

Statistical Analysis

In social insects that follow the haplodiploidy sex determination system, the genetic structure may be determined using the genotypes of the queen and paternal genotypes. In this study, the workers' genotypes per nest of *A. dorsata* were converted to COLONY [25] software format. This software determines the inferred genotypes of the queen and paternal genotypes of each nest. The inferred genotypes were subjected to genetic analysis. The inferred queen's genotype of each nest was conducted to calculate the number of alleles per locus (N_a), effective number of alleles (N_e), observed and expected heterozygosity (H_o and H_e), polymorphism information content (PIC), and relatedness (R) among the queens which head the low solitary nests separately. The genetic differentiation measure (F_{ST}) was estimated using the paternal inferred genotypes. According to Paar *et al.* (2004), the paternal genotypes were diploidized as co-dominant marker. GENALEX version 6.41 [26] was used to estimate the following measures of genetic variation N_a, N_e, H_o, H_e, and R, and genetic differentiation structure (F_{ST}). Hierarchical genetic structuring based on the analysis of molecular variance (AMOVA) [27] and bootstrap resampling were executed by using GENALEX version 6.41 [26]. Meanwhile, PIC values were calculated using the microsatellite toolkit package [28]. Hardy Weinberg equilibrium (HWE) tests and genotypic linkage disequilibrium were performed using GENEPOP version 3.3 [29], whereas the intercolonial relatedness between the solitary nests of *A. dorsata* was computed based on the queen's inferred genotype using an algorithm as described by Queller and Goodnight (1989) [30].

Conclusion

This study was conducted to show the genetic relatedness among the solitary nests of *A. dorsata* in the Marang district, Malaysia. The results revealed the high intracolonial relatedness within the studied nests based on the data obtained from the use of 15 single locus DNA microsatellite markers. The presence of appreciable levels of intercolonial relatedness between several single pair of nests indicated the necessity for a comprehensive study involving a large sample size to confirm whether the honey hunters in this area of Malaysia harvest the same nest at a different sites and dates during a single harvesting season or not. Such research is needed to formulate effective management strategies to replace the current common unsustainable *A. dorsata* nest harvesting practice of whole solitary nest removal. Genetic studies on the aggregate nests should also be done in the future.

Acknowledgments

The authors would like to express their sincere appreciation and gratitude to University Sultan Zainal Abidin (UniSZA), Kuala Terengganu, Malaysia, for providing the facilities for this research. We are really grateful to the science officers and laboratory assistants of the Molecular Biology Laboratory, UniSZA, for their corporation. Our thanks also go to the honey hunters for their assistance in bee sampling. We would also like to convey our appreciation to Professor B. Oldroyd of the University of Sydney, Professor M. Nei of the Pennsylvania State University and Professor J. Wang of the Zoological Society of London for their valuable advice on the use of related software packages for the data analysis.

Author Contributions

Conceived and designed the experiments: NS MM. Performed the experiments: NS. Analyzed the data: NS. Contributed reagents/materials/analysis tools: NS MM AMA SGT NAA WHL. Wrote the paper: NS MM SGT.

References

1. Oldroyd BP, Wongsiri S (2006) Asian honey bees (biology, conservation, and human interactions). Harvard University Press. 340 p.
2. Mardan M (1989) Thermoregulation in the Asiatic giant honeybee *Apis dorsata* (Hymenoptera: Apidae). Ph.D. dissertation, University of Guelph, Ontario.
3. Mardan M (2007) More about Malaysia bees and bee trees. Bees for Development 83: 7–9.
4. Kastberger G, Sharma DK (2000) The predator-prey interaction between blue-bearded bee eaters (Nyctyornis athertoni Jardine and Selby 1830) and giant honeybees (*Apis dorsata* Fabricius 1798). Apidologie 31: 727–736.
5. Paar J, Oldroyd BP, Huettinger E, Kastberger G (2004) Genetic structure of an *Apis dorsata* population: the significance of migration and colony aggregation. Journal of Heredity 95: 119–126.
6. Rinderer T, Oldroyd B, Lekprayoon C, Wongsiri S, Boonthai C, et al. (1994) Extended survival of the parasitic honey bee mite *Tropilaelaps clareae* on adult workers of *Apis mellifera* and *Apis dorsata*. Journal of Apicultural Research 33: 171–173.
7. Oldroyd B, Osborne K, Mardan M (2000) Colony relatedness in aggregations of *Apis dorsata* Fabricius (Hymenoptera, Apidae). Insectes Sociaux 47: 94–95.
8. Hamid AH (1997) Honey or money: Changing nature of honey collectors-environment relationship in rural Malaysia. In: Proceedings of the International Conference on Tropical Bees: The environment Universiti Pertanian Malaysia (Mardan M, Sipat A, Yusoff KM, Kiew HMS, Abdullah MM, eds): 60–71.
9. Yong PL, Haji Othman MS (2007) Economic values of honey bees, Peninsular Malaysia. Forestry Department Peninsular Malaysia (FDPM). Mashka sdn. Bhd. 120.
10. Ruttner F (1988) Biogeography and taxonomy of honeybees. Berlin: Springer-Verlag.
11. Saberioon MM, Mardan M, Nordin L, Alias MS, Gholizadeh A (2010) Predict location(s) of *Apis dorsata* nesting sites using remote sensing and geographic

information system in Melaleuca forest. American Journal of Applied Sciences 7 (2): 252–259.
12. Insuan S, Deowanish S, Klinbunga S, Sittipraneed S, Sylvester HA, et al. (2007) Genetic differentiation of the giant honey bee (*Apis dorsata*) in Thailand analyzed by mitochondrial genes and microsatellites. Biochemical Genetics 45(3/4): 340–360.
13. Norris AT, Bradley DG, Cunningham EP (1999) Microsatellite genetic variation between and within farmed and wild Atlantic salmon (*Salmo salar*) populations. Aquaculture 180: 247–264.
14. Ting J, Ling Y, Wen-bin B, Guo-hong C (2008) Study on genetic diversity of *Apis cerana cerana* and *Apis mellifera ligustica* in China with microsatellite markers. Research Journal of. Animal Sciences 2(6): 178–182.
15. Cornuet JM, Luikart G (1997) Description and power analysis of two tests for detecting recent population bottlenecks from allele frequency data. Genetics 144: 2001–2014.
16. Oldroyd BP, Smolenski A, Lawler S, Estoup A, Crozier R (1995) Colony aggregations in *Apis mellifera*. L. Apidologie 26: 119–130.
17. Ott J (2001) Analysis of human genetic linkage (revised edition). Baltimore: Johns Hopkins University Press.
18. Guo SW, Thompson EA (1992) Performing the exact test of Hardy - Weinberg proportion for multiple alleles. Biometrics 48: 361–372.
19. Weir BS, Cockerham CC (1984) Estimating F-Statistics for the analysis of population structure. Evolution 38: 1358–1370.
20. Takezaki N, Nei M (1996) Genetic distances and reconstruction of phylogenetic trees from microsatellite DNA. Genetics 144: 389–399.
21. Weir BS (1990) Genetic data analysis. Sinauer Associates, Sunderland, USA. pp 222–289.
22. Hartl DL, Clark AG (2007) Principles of population genetics. Sinauer Associates Inc. Publishers. Sunderland, Massachussettes. 652 p.
23. Oldroyd BP, Smolenski AJ, Cornuet J, Wongsiri S, Estoup A, et al. (1996) Levels of polyandry and intracolonial genetic relationships in *Apis dorsata* (Hymenoptera: Apidae). Annals of the Entomological Society of America 89: 276–283.
24. Creste S, Tulmann NA, Figueira A (2001) Detection of single sequence repeat polymorphisms in denaturing polyacrylamide sequencing gels by silver staining. Plant Molecular Biology Reporter 19: 299–306.
25. Wang J (2008) COLONY. Version 2.0.1.1. Zoological Society of London website. Available: http://www.zsl.org/science/research-projects/software/colony,1154,AR.html. Accessed 2012 January 02.
26. Peakall R, Smouse PE (2010) GenAlEx 6.41: Genetic analysis in Excel. Population genetic software for teaching and research. The Australian National University website (Canberra, Australia). Available: http://www.anu.edu.au/BoZo/GenAlEx/. Accessed 2012 January 02.
27. Michalakis Y, Excoffier L (1996) A generic estimation of population subdivision using distances between alleles with special reference for microsatellite loci. Genetics 142: 1061–1064.
28. Park S (2008) Excel Microsatellite Toolkit. Version 3.1.1. Animal Genomics Lab website, (University College, Dublin, Ireland). Available: http://animalgenomics.ucd.ie/sdepark/ms-toolkit/. Accessed 2012 January 02.
29. Raymond M, Rousset F (1995) GENEPOP (version 3.3): population genetics software for exact tests and ecumenicism. Journal of Heredity 86: 248–249.
30. Queller D, Goodnight KF (1989) Estimating relatedness using genetic markers. Evolution 43: 258–275.
31. Solignac M, Vautrin D, Loseau A, Mougel F, Baudry E, et al. (2003) Five hundred and fifty microsatellite markers for the study of the honeybee (*Apis mellifera* L.) genome. Molecular Ecology Notes 3: 307–311.
32. Estoup A, Garnery L, Solignac M, Cornuet JM (1995) Microsatellite variation in honeybee (*Apis mellifera* L.) populations: hierarchical genetic structure and test of the infinite allele and stepwise mutation models. Genetics 140: 679–695.

Limited Fitness Advantages of Crop-Weed Hybrid Progeny Containing Insect-Resistant Transgenes (*Bt/CpTI*) in Transgenic Rice Field

Xiao Yang[1], Feng Wang[2], Jun Su[2], Bao-Rong Lu[1]*

1 Ministry of Education Key Laboratory for Biodiversity Science and Ecological Engineering, Department of Ecology and Evolutionary Biology, Fudan University, Shanghai, China, **2** Fujian Province Key Laboratory of Genetic Engineering for Agriculture, Fujian Academy of Agricultural Sciences, Fuzhou, China

Abstract

Background: The spread of insect-resistance transgenes from genetically engineered (GE) rice to its coexisting weedy rice (*O. sativa f. spontanea*) populations *via* gene flow creates a major concern for commercial GE rice cultivation. Transgene flow to weedy rice seems unavoidable. Therefore, characterization of potential fitness effect brought by the transgenes is essential to assess environmental consequences caused by crop-weed transgene flow.

Methodology/Principal Findings: Field performance of fitness-related traits was assessed in advanced hybrid progeny of F_4 generation derived from a cross between an insect-resistant transgenic (*Bt/CpTI*) rice line and a weedy strain. The performance of transgene-positive hybrid progeny was compared with the transgene-negative progeny and weedy parent in pure and mixed planting of transgenic and nontransgenic plants under environmental conditions with natural *vs.* low insect pressure. Results showed that under natural insect pressure the insect-resistant transgenes could effectively suppress target insects and bring significantly increased fitness to transgenic plants in pure planting, compared with nontransgenic plants (including weedy parent). In contrast, no significant differences in fitness were detected under low insect pressure. However, such increase in fitness was not detected in the mixed planting of transgenic and nontransgenic plants due to significantly reduced insect pressure.

Conclusions/Significance: Insect-resistance transgenes may have limited fitness advantages to hybrid progeny resulted from crop-weed transgene flow owning to the significantly reduced ambient target insect pressure when an insect-resistant GE crop is grown. Given that the extensive cultivation of an insect-resistant GE crop will ultimately reduce the target insect pressure, the rapid spread of insect-resistance transgenes in weedy populations in commercial GE crop fields may be not likely to happen.

Editor: Guy Smagghe, Ghent University, Belgium

Funding: This research was supported by the Natural Science Foundation of China (Grant no. 30730066 and 30871503), and National Program of Development of Transgenic New Species of China 2011ZX08011-006. The funders had no role in study design, data collection and analysis, decision to publish, or preparation of the manuscript.

Competing Interests: The authors have declared that no competing interests exist.

* E-mail: brlu@fudan.edu.cn

Introduction

The commercial application of genetically engineered (GE) crops in agricultural production has aroused great biosafety concerns worldwide. The potential environmental impacts caused by the cultivation of the GE crops are the most debated issues [1–3]. Transgene flow from a GE crop into populations of wild or weedy relatives and its potential ecological risks is considered as a key environmental problem [4–6]. (Trans) gene flow from a crop to its wild relatives has been widely documented in the last decades [7–9]. However, our knowledge on the role of introgressed transgenes that confer novel traits with a strong selective advantage in changing the evolutionary process of wild or weedy populations is still limited. Therefore, assessing potential environmental risks caused by the extensive cultivation of GE crops prior to their commercialization becomes a common practice [1,5,8]. The study of potential ecological consequences created by transgene flow to wild relatives particularly the coexisting and conspecific weeds will provide solid bases for environmental risk assessment [10–13]. The fate of weedy populations that acquired transgenes through gene flow is largely different, depending on the fitness effect of the introgressed transgenes under given environmental conditions [5,14]. If the introgressed transgenes can increase fitness, the transgenes will enhance the competitiveness and invasiveness of the weedy populations, leading to the rapid spread of the transgenes in the weedy populations, and *vice versa.* Thus, estimating fitness effect of transgenes on weedy populations is essential.

Rice is an important world crop providing staple food for nearly one half of the global population [15]. Research and development of GE rice has been extensively practiced in China, and consequently a many GE rice lines with novel traits such as insect-resistance, herbicide-tolerance, and improved grain quality have been developed [5,11,16]. In 2009, Chinese authority granted biosafety certificates for two insect-resistant rice lines containing a *Bt* transgene (*Bacillus thuringiensis*) [17], meaning that GE rice may enter commercial production in near future. Transgene spread from insect-resistant GE rice to its coexisting weedy rice populations through gene flow becomes a major environmental biosafety concern. Weedy rice (also known as red rice, *O. sativa f. spontanea*) is a noxious weed that causes significant losses of rice yield and quality worldwide [18,19]. The introgression of transgenes with selective advantages may largely enhance the spread of the weedy rice, causing more serious weed problems. Weedy rice is an annual weed conspecific to cultivated rice [20], and transgene flow from cultivated rice to weedy rice cannot be avoided owing to their similar flowering phenology [21]. It is therefore valid to determine the fitness effect of insect-resistance transgenes on weedy rice populations under different insect pressure for risk assessment.

Fitness of an insect-resistance transgene is largely associated with the environment in which insect pressure can largely be variable [22–25]. The relationship between fitness effect brought by an insect-resistance transgene and the ambient insect pressure has not been well described. A number of studies of crop-weed hybrids showed variable fitness effect brought by insect-resistant transgenes under different insect occurrences [25,26]. It is apparent based on these studies that the enhanced fitness benefit of insect-resistant transgenes was always associated with the high target herbivore pressure, whereas under low herbivore pressure the benefit brought by the insect-resistant transgenes was reduced. Therefore, to analyzed fitness effect brought by an insect-resistance transgene under designed experimental conditions with different insect pressure should be the key to detect the fitness effect of such a transgene.

Studies have already revealed the effects of *Bt* transgene in crop-wild or crop-weed hybrid progeny under different environments [23,25,26–28]. Our previous studies of F_1–F_3 crop-weed hybrids and their derived lineages from insect-resistant rice revealed increased fecundity brought by *Bt/CpTI* transgenes when target insects were abundant [25,29]. However, the relationship between variation in different insect pressure and the corresponding fitness change was not well addressed by a properly designed experiment. In this study, we intended to address the following questions: (i) Are the insect-resistant transgenes (*Bt/CpTI*) still effective on resisting rice target herbivores in the transgenic F_4 populations derived from a GE crop-weed hybrid under natural insect pressure? (ii) Does the transgenic F_4 population have significantly higher fitness than the nontransgenic F_4 population and weedy rice parent under natural insect pressure? (iii) Does the mixed planting of transgenic plants significantly reduce the ambient insect pressure of the nontransgenic plants, and consequently, reduce the potential fitness benefit brought by insect-resistance transgenes in the experimental fields? The answer of above questions is essential for assessing the long-term ecological impacts caused by insect-resistant transgene flow from to its weedy rice populations.

Methods

Experimental Materials and their Sources

Our experiment included two sets of F_4 populations, with or without transgenes derived from a crop-weed hybrid, and its weedy rice parent. The F_4 populations were generated by continued self-pollination from the F_1 hybrids of an insect-resistant GE rice line crossed with a weedy rice strain (used as the pollen recipient) [25,29]. The GE rice line produced by Fujian Academy of Agricultural Sciences in Fuzhou of China had the *Bt/CpTI* (*Bacillus thuringiensis* (*cryIAc*) and cowpea trypsin inhibitor) transgenes in a double insertion, tightly linked with the selectable marker gene *hyg* (hygromycin resistance), transformed by an agrobacterium method [30]. This GE rice line was produced to control lepidopteran pests such as rice stem borers (*Scirpophaga incertulas*, *Chilo suppressalis*, and *Sesamia inferens*) and rice leaf-folders (*Cnaphalocrocis medinalis*) [24]. The weedy rice strain (coded as W) was donated by Dr. H. S. Suh of Yeungnam University from S Korea [29]. The identification and separation of the transgene homozygous population (coded as TP) from non-transgene homozygous population (TN) were made from screening the seedlings derived from seeds of F_2 plants using the hygromycin-B water solution treatment [25]. The selected transgene-positive and transgene-negative F_3 populations were self-pollinated for one more generation. Therefore, plants of both transgene-positive and transgene-negative F_4 populations were homozygous at the transgenic locus.

Experimental Design

The field experiment was conducted in 2010 in the environment with a natural insect occurrence at the designated Biosafety Assessment Center in Fuzhou, Fujian Province, China. The experimental plots were allocated in two separate blocks (natural insect *vs.* low insect pressure) isolated from each other for >100 m. The block with natural insect pressure was free of any insecticide treatment, whereas the block with low insect pressure was achieved by spraying different insecticides (Methamidophos, Folimat, Buprofezin, and Monosultap) every 7–10 days starting at the tillering stage. The main comparison was made among the weedy parent, transgenic and nontransgenic hybrid lineages under the different insect pressure. Therefore, under each of the insect-pressure environmental conditions (natural *vs.* low), the following treatments were designated: (1) pure planting of weedy rice parent, transgene-positive, and transgene-negative F_4 populations; (2) mixed planting between transgene-positive and -negative F_4 populations, or transgene-positive F_4 population and weedy rice parent, or transgenic-negative F_4 population and weedy rice parent. Consequently, there were a total of twelve treatments included in the experiment.

For each treatment, four replicates (plots) were included. In the pure planting design, each plot included 64 plants in an 8×8 grid with 20 cm spacing between plants. In the mixed planting design, each plot included 32 plants of either type of the experimental materials grown alternatingly in an 8×8 grid with 15 cm spacing between plants. The field layout of plots was arranged in a complete randomized design in the two blocks. Seeds of all the plant materials were germinated in 50×100 cm pots filled with paddy soil from fields and placed in a green house. Seedlings were transferred to a nursery bed 20 days after seed germination, and then transplanted into field plots ~40 days after seed germination. To avoid accidental seed shattering, panicles of all plants were enclosed by nylon mesh-bags 10–15 days after the flowering.

Data Collection and Analysis

For estimating insect damage and fitness-related traits, all plants in a plot were characterized at various vegetative and reproductive stages, except for those in the border rows that were excluded to avoid the edge effect. As a result, a total 36 plants from a pure-planting plot and 18 plants for each of the two types of populations from a mixed-planting plot were characterized. Blasted tillers (by rice stem borers) and folded leaves (by leaf-folder) at the beginning of flowering stage were measured to estimate the insect damage. An insect index was used to estimate the insect pressure that was calculated as the sum of percentage of both blasted tiller and

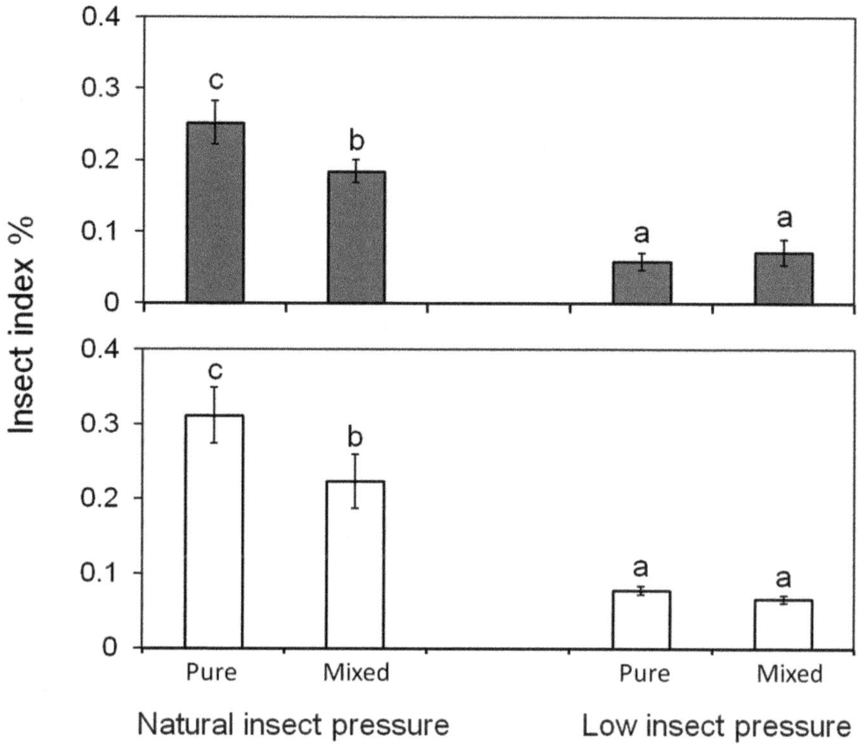

Figure 1. Insect index (%) calculated based on the ratio of blasted tillers and folded leaves on weedy parent (above) and transgene-negative hybrid progeny (below) in different cultivation modes (pure *vs.* mixed) under natural or low insect pressure. The comparison was made among pure and mixed planting under natural insect and low insect pressure. Different letters above the columns indicate significant differences according to Duncan's multiple range tests after Bonferroni correction. "Pure" indicates pure planting, "Mixed" indicates mixed planting. Bars represent standard error. Levels of significance: $P<0.05$.

folded leaves in the plots of weedy parent and nontransgenic F_4 populations. The following fitness-related traits were included for measurement: plant height, number of tillers, number of panicles, and number of filled seeds per plant, seed set, and 1000-seed weight at various growth stages.

One way ANOVA (Duncan's multiple range test) was used to examine significant differences for insect index obtained from weedy rice and transgene-negative plants in pure and mixed planting under natural and low insect pressure, and to test significant differences in insect damage and fitness-related traits among weedy parent, transgene-positive, and transgene-negative populations in pure planting, followed by the Bonferroni correction. The paired *t*-test was used to examine significant differences between transgene-positive and transgene-negative, transgene-positive and weedy parental, or transgene-negative and weedy parental populations in mixed planting. All statistical analyses were performed using the software package IBM SPSS Statistics ver. 19.0 for Windows (SPSS Inc., IBM Company Chicago, IL, USA, 2010).

Results

Insect Pressure Under the Two Environmental Conditions

The insect index calculated from the insect damage of blasted tillers and folded-leaves demonstrated a significant difference between the two blocks with natural insect pressure and low insect pressure (Fig. 1; Table 1 and 2). As estimated from the weedy parental and transgene-negative F_4 populations in pure planting, the insect index was significantly higher under the natural insect pressure (25.3% and 31.5%) than that the under low insect pressure (5.8% and 7.8%; $P<0.001$). The significant differences of insect pressure allowed us to determine the net benefit caused by the transgenes when all the plants were exposed to herbivores, and the net cost caused by the transgenes when herbivores were absent or reduced to a low level.

Noticeably, the insect index was significantly lower in the mixed planting of transgene-positive populations than in the pure planting under natural insect pressure (Fig. 1), indicating that the mixed planting with transgene-positive plants will generally reduce the ambient insect pressure. However, such reduction of insect damage was not observed in the mixed planting of transgene-positive plants under low insect pressure.

Performance of F_4 and Weedy Parental Populations in Pure Planting

In general, transgene-positive F_4 populations with insect-resistance transgenes showed significantly less insect damage than transgene-negative F_4 populations and weedy parents, under natural insect pressure (Table 1). Transgene-positive F_4 populations had a better performance for most of the fitness-related traits than transgene-negative F_4 populations and weedy parents, under natural insect pressure (Table 1). For example, transgene-positive plants produced a significantly higher number of panicles (25% increase) and ratio of seed set (17% increase) than their transgene-negative plants (Table 1). Particularly, the number of seeds produced by transgene-positive plants was significantly higher than both transgene-negative plants (44% increase) and weedy parents (31% increase) (Fig. 2). Under low insect pressure, nearly all the fitness-related traits did not show significant differences

Table 1. Average values and standard errors (±) of insect damage and fitness-related traits in weedy rice parent, transgene-positive, and transgene-negative crop-weed hybrid progeny (F_4) in pure planting under natural (above) vs. low (below) insect pressure.

Trait	Plant material		
	Weedy parent	Transgene-positive	Transgene-negative
Natural insect			
Blasted tillers %	13.4±1.1 b	7.0±1.3 a	15.6±2.0 b
Folded leaves %	11.8±2.1b	1.5±0.3 a	15.6±3.1 b
Plant height (cm)	91.9±4.0 a	111.9±4.2 b	112.3±2.6 b
No. of tillers	17.3±1.1 a	18.2±0.3 a	16.0±1.2 a
No. of panicles	10.4±1.0 a	15.0±0.5 b	12.0±0.9 a
Seed set (%)	50.0±1.5 b	47.8±2.5 b	40.8±0.9 a
1000-seed weight (g)	20.3±0.2 a	21.4±0.3 b	20.7±0.2 ab
Low insect			
Blasted tillers %	4.8±1.2 a	5.0±0.8 a	6.5±0.7 a
Folded leaves %	1.0±0.0 b	0.2±0.2 a	1.3±0.2 b
Plant height (cm)	107.3±1.7 a	111.0±2.5 a	120.5±2.7 b
No. of tillers	13.9±0.6 a	13.3±0.4 a	13.0±0.9 a
No. of panicles	13.6±0.7 b	10.6±0.7 a	11.0±0.4 a
Seed set (%)	51.3±1.3 ab	52.2±1.4 b	48.0±0.8 a
1000-seed weight (g)	19.7±0.2 a	22.1±0.2 b	21.6±0.3 b

Different letters following the average values in the same rows indicate significant differences according to Duncan's multiple range tests after Bonferroni correction (N = 4 plots).

Table 2. Average values and standard errors (±) of insect damage and fitness-related traits in weedy rice parent (W), transgene-positive (TP), and transgene-negative (TN) crop-weed hybrid progeny (F_4) in mixed planting under natural (above) vs. low (below) insect pressure.

Trait	Plant material					
	TP *vs.* TN		TP *vs.* W		TN *vs.* W	
Natural insect						
Blasted tillers %	11.4±2.0	17.3±0.9*	9.6±2.1	13.0±2.3	12.9±1.4	18.1±3.0
Folded leaves %	2.9±1.7	5.0±3.0	4.2±1.3	5.4±1.2	13.6±2.7	10.3±3.7
Plant height (cm)	101.4±2.6	103.4±4.8	96.9±2.9	93.1±3.0	107.9±6.4	99.4±6.7
No. of tillers	11.0±1.0	9.9±0.8	13.2±1.2	12.0±1.0	11.8±0.5	9.8±1.0
No. of panicles	8.9±0.4	8.6±0.8	9.7±1.0	8.3±0.4	8.9±0.7	8.2±0.9
Seed set (%)	42.5±4.2	37.1±0.6	43.5±1.7	49.3±2.6	37.0±2.0	47.8±3.8*
1000-seed weight (g)	20.4±0.6	20.5±0.2	20.6±0.1	19.6±0.3*	21.1±0.6	19.1±0.3*
Low insect						
Blasted tillers %	5.4±0.3	5.8±0.4	5.3±1.5	6.5±1.9	5.7±2.2	8.2±1.7
Folded leaves %	0.4±0.3	0.8±0.3	0.4±0.1	0.7±0.2	2.3±0.4	0.7±0.3*
Plant height (cm)	112.3±5.0	116.7±3.4	104.5±4.7	98.5±3.4	111.4±2.8	104.1±1.4
No. of tillers	7.6±0.1	8.2±0.3	8.5±0.5	7.8±0.5	8.8±0.4	8.3±0.5
No. of panicles	6.6±0.4	7.2±0.5	6.8±0.4	6.2±0.3*	7.3±0.6	7.2±0.3
Seed set (%)	53.9±0.9	49.0±3.2	49.9±5.7	61.9±2.4	51.6±2.6	55.7±2.9
1000-seed weight (g)	22.3±0.8	21.9±0.5	21.9±0.7	19.3±0.4*	21.7±0.2	18.8±0.4**

Level of significance was calculated based on the paired *t*-test between TP and TN, TP and W, or TN and W (N = 4 plots).
*$P<0.05$,
**$P<0.01$.

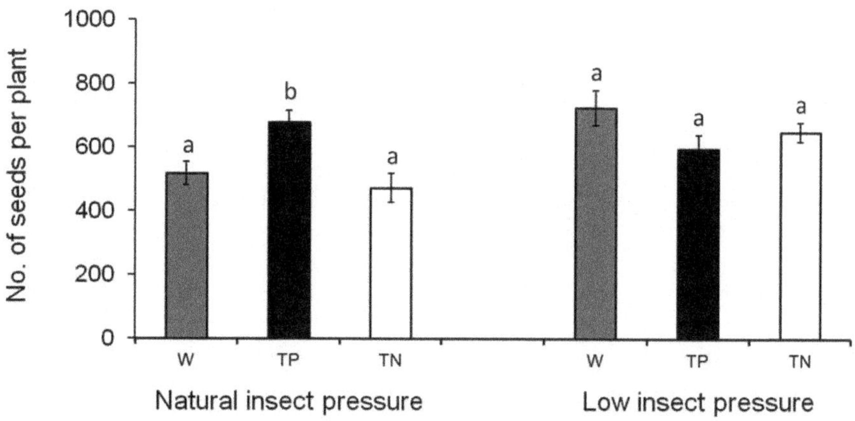

Figure 2. The average number of filled seeds per plant produced by weedy rice parent (W), transgene-positive (TP), and transgene-negative (TN) hybrid progeny under natural vs. low insect pressure in pure cultivation. The comparison was made among weedy rice parent, transgene-positive, and transgene-negative hybrid progeny. Different letters above the columns indicate significant differences according to Duncan's multiple range tests after Bonferroni correction. Bars represent standard error. Levels of significance: $P<0.05$.

between transgene-positive and transgene-negative populations (Table 1; Fig. 2) in both pure and mixed planting, suggesting no obvious fitness cost from the transgenes.

Noticeably, transgene-negative F_4 populations showed significantly higher values of plant height and 1000-seed weight than their weedy rice parental populations under both natural and low insect pressures (Table 1). This indicated that crop-weed hybridization may bring benefit through the enhanced performance of some traits of hybrids and hybrid lineages in later generations.

Performance of F_4 and Weedy Parental Populations in Mixed Planting

In contrast to the pure planting, transgene-positive and transgene-positive F_4 populations showed apparently less pronounced differences in insect damage in mixed planting than in pure planting under natural insect pressure (Table 2). The same trend was observed between transgene-positive F_4 and weedy parental populations (Table 2) in mixed planting. As a result, transgene-positive F_4 populations did not show significant differences for nearly all the fitness-related traits, compared with transgene-negative F_4 and weedy parental populations under natural insect pressure (Table 2). These results indicated that the mixed-planting plots with transgene-positive plants would result in a reduced insect damage to transgene-negative and weedy rice plants, and consequently reduce the differences in fitness-related traits between transgene-positive and transgene-negative plants.

Similarly, under low insect pressure, nearly all the fitness-related traits did not show significant differences among transgene-positive F_4, transgene-negative F_4 and weedy parental populations (Table 2; Fig. 3). However, it is worthy of notice that the 1000-seed weight of transgene-positive and transgene-negative F_4 plants was significantly higher than that of weedy parental plants, suggesting again the benefit of hybridization brought to hybrid progeny compared with their weedy parents in terms of seed weight, regardless of the presence of transgenes.

Discussion

The key point of assessing potential ecological risks caused by an insect-resistant transgene introgressed to weedy populations is to analyze the increased fitness of transgenic plants in comparison with their nontransgenic counterparts derived from the same crop-weed hybridization under controlled environmental conditions [11,22,24,25,31]. Our data from this study indicated significant differences in insect pressure (as measured by the insect index, Fig. 1) between the two rice planting environments with natural *vs.* low insect pressure, achieved by artificial spray of insecticides. The insect pressure in both pure and mixed planting plots of weedy rice parents and transgene-negative F_4 populations was significantly higher under the natural-insect condition than that under the controlled low-insect condition. These results indicate the effectiveness of insect control in our experiment. In addition, as indicated by the insect damage on weedy parents and transgene-negative plants, significantly higher insect index was also detected in pure planting plots than in the mixed-planting plots of nontransgenic plants (including parents) with transgenic plants under natural insect. In contrast, the mixed-planting plots with transgenic plants did not show such significant differences in insect index under low insect pressure. All together, these results demonstrate that the differential environment allowed us to analyze the fitness effect of insect-resistance transgenes in weedy rice plants against their nontransgenic counterparts, and that the mixed planting of transgenic plants in a field plot can significantly reduce the ambient insect pressure.

In pure planting, transgene-positive crop-weed F_4 plants showed significantly greater values of fecundity-related traits, such as the number of panicles and well-filled seeds and the ratio of seed set than their transgene-negative plants under natural insect pressure. Particularly, the number of panicles and filled seeds of transgene-positive F_4 plants showed significantly higher values than the weedy rice parents. However, such significant differences between the transgenic plants and nontransgenic plants were not detected in these traits under the low insect pressure. The increased fecundity of transgenic crop-weed F_4 plants is most likely brought by the introgression of insect-resistance transgenes that significantly reduced insect attack to the plants. The increased fecundity in the transgenic F_4 plants demonstrates that the introgressed insect-resistant transgenes (*Bt/CpTI*) are still effective on controlling rice target herbivores in the advanced generation of crop-weed hybrid populations. This result is consistent with our previous study in which the F_2–F_3 lineages derived from the crop-weed hybrids

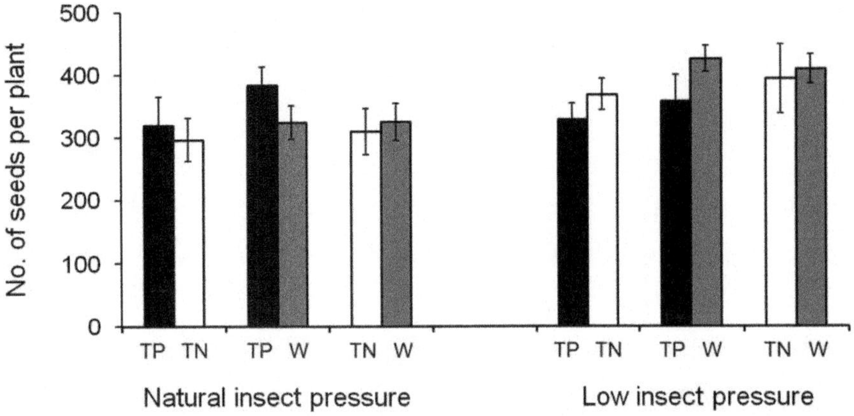

Figure 3. The average number of filled seeds per plant produced by weedy rice parent, transgene-positive (W), and transgene-negative (TP) hybrid progeny under natural vs. low insect pressure in mixed cultivation. The comparison was made between transgene-positive and transgene-negative, transgene-positive and weedy rice parent, or transgene-negative and weedy rice parent, based on the paired *t*-tests. Bars represent standard error. Levels of significance: $P<0.05$.

with the same transgenic event showed effective control of insect damages to the hybrid lineage compared with both transgene-negative hybrid lineages and the weedy parents, under a high level of insect pressure [25,29]. Many previous studies on crop-weed and crop-wild hybrid lineages also demonstrated that insect-resistance transgenes can dramatically reduce the target herbivores and increase the fecundity in hybrids and their advanced generation of hybrid populations [23,25,29]. These findings are consistent with our results in this study. It seems possible to make a conclusion from the field experimental data of above studies including ours that the introgressed insect-resistance transgenes will maintain their strong ability to control herbivores and increase fecundity of transgenic populations derived from crop-weed or crop-wild transgene introgression. The transgenic plants may have a strong ability to compete with their nontransgenic counterparts including nontransgenic hybrid and parental populations [23,25,28].

However, it is necessary to point out that in the mixed-planting plots the fecundity traits (particularly seed production) of the transgene-positive plants did not show significantly increased values, compared to the transgene-negative plants and weedy rice parents under natural insect pressure. This result suggests apparent losses of fitness benefit that should have been brought by the insect-resistance transgenes in mixed planting under natural insect pressure. The similar phenomenon was also observed in a number of studies in which the fitness benefit and cost of insect-resistant GE rice lines and their nonGE parental lines was analyzed under natural and low insect pressure, respectively [22,24]. Associated with the data of insect index from this study where significantly lower insect pressure was recorded in the mixed planting plots than in the pure planting plots under the same environmental condition of natural insect (Fig. 1), we think that the loss of fitness benefit to insect-resistant transgenic crop-weed hybrid populations in mixed planting is most likely owing to the considerably reduced insect pressure. In other words, the generally low ambient insect pressure in the experimental field caused by the mixture of insect-resistant transgenic plants will significantly reduce the potential fitness advantages that should have been brought by insect-resistance transgenes. As a result, the expected long-term persistence and rapid spread of insect-resistance transgenes in weedy rice populations caused considerably increased fitness advantages following transgene flow from a GE rice variety may not happen in the realistic situation if such fitness advantages are extremely limited. Under the actual situation where an insect-resistant GE rice variety is cultivated, the ambient insect pressure in an extensive field area should be reduced to a much greater extent than that in our experimental plots. The spread of transgenes in weedy rice populations might be considerably limited owning to the negligible fitness increase in insect-resistance GE rice fields where the target insect pressure is significantly reduced, although the issue of transgene flow from an insect-resistant GE rice variety to its coexisting weedy rice populations should not be neglected.

Our finding has its important implications for the risk assessment of transgene flow from insect-resistant GE crop to its wild relatives, and to the conspecific weedy populations in particular. Given a determined frequency of transgene flow, the magnitude of potential environmental risks should largely depend on the fitness effect of a transgene. It is generally recognized that fitness effect of an insect-resistance transgene is determined by the ambient insect pressure in the environment where wild/weedy populations occur. Therefore, the assessment of environmental risks caused by transgene flow from an insect-resistant GE crop to wild populations should first consider the pressure of target insects in the concerned environment. This principle may also applied to the risk assessment of transgene flow to wild populations, considering that insect pressure in natural habitats is significantly lower than in agriculture habitats [32]. For the crop conspecific weedy populations co-occurring with a crop (and GE crop), crop-to-weed transgene flow cannot be avoided. The expected environmental impact from transgene flow from insect-resistant GE crop could be large because hypothetically insect-resistance transgenes will bring fitness benefit to the weedy populations. However, the fitness advantages might be limited due to the fact that weedy plants will be surrounded by insect-resistant plants in a GE crop field, reducing the ambient pressure of target herbivores significantly. Consequently, the spread of the transgenes in weedy populations would be limited. In reality, the extensive commercial cultivation of an insect-resistant GE crop will largely reduce target herbivores in a GE deployed area [33]. Under such a circumstance, the environmental impacts caused by the crop-to-weed transgene flow from an insect-resistant GE crop will also be limited.

Author Contributions

Conceived and designed the experiments: BRL XY. Performed the experiments: XY JS. Analyzed the data: BRL XY FW. Contributed reagents/materials/analysis tools: FW JS. Wrote the paper: XY BRL.

References

1. Andow DA, Zwahlen C (2006) Assessing environmental risks of transgenic plants. Ecol Lett 9: 196–214.
2. Conner AJ, Glare TR, Nap JP (2003) The release of genetically modified crops into the environment - Part II. Overview of ecological risk assessment. Plant J 33: 19–46.
3. Snow AA, Palma PM (1997) Commercialization of transgenic plants: Potential ecological risks. Bioscience 47: 86–96.
4. Haygood R, Ives AR, Andow DA (2003) Consequences of recurrent gene flow from crops to wild relatives. P Roy Soc B-Biol Sci 270: 1879–1886.
5. Lu BR, Yang C (2009) Gene flow from genetically modified rice to its wild relatives: Assessing potential ecological consequences. Biotechnol Adv 27: 1083–1091.
6. Stewart CN, Halfhill MD, Warwick SI (2003) Transgene introgression from genetically modified crops to their wild relatives. Nat Rev Genet 4: 806–817.
7. Ellstrand NC, Prentice HC, and Hancock JF (1999) Gene flow and introgression from domesticated plants into their wild relatives. Annu Rev Ecol Evol S 30: 539–563.
8. Ellstrand NC (2003) Current knowledge of gene flow in plants: implications for transgene flow. Philos T R Soc B 358: 1163–1170.
9. Raybould AF, Gray AJ (1993) Genetically modified crops and hybridization with wild relatives: A UK Perspective. J Appl Ecol 30: 199–219.
10. Chen LJ, Lee DS, Song ZP, Suh HS, Lu BR (2004) Gene flow from cultivated rice (*Oryza sativa*) to its weedy and wild relatives. Ann Bot 93: 67–73.
11. Lu B-R, Snow AA (2005) Gene flow from genetically modified rice and its environmental consequences. BioScience 559: 669–678.
12. Gealy DR, Mitten DH, Rutger JN (2003) Gene flow between red rice (*Oryza sativa*) and herbicide-resistant rice (*O. sativa*): implications for weed management. Weed Technol 17: 627–45.
13. Warwick SI, Légère A, Simard MJ, James T (2008) Do escaped transgenes persist in nature? The case of an herbicide resistance transgene in a weedy *Brassica rapa* population. Mol Ecol 17: 1387–1395.
14. Hails RS, Morley K (2005) Genes invading new populations: a risk assessment perspective. Trend Ecol Evol 20: 245–252.
15. FAO (2004) The state of food and agriculture 2003–2004. Agricultural biotechnology: meeting the needs of the poor? http://www.fao.org/docrep/006/Y5160e/Y5160e00.HTM (Accessed on April 22, 2012).
16. Huang JK, Hu RF, Rozelle S, Pray C (2005) Insect-Resistant GM Rice in Farmers' Fields: Assessing Productivity and Health Effects in China. Science 308: 688–690.
17. MOA (Ministry of Agriculture, China) (2009) The second batch of genetically modified organisms that are granted biosafety certificates in 2009. (in Chinese) http://www.stee.agri.gov.cn/biosafety/spxx/P020091127591594596689.pdf. (Accessed on April 22, 2012).
18. Delouche JC, Burgos NR, Gealy DR, Zorilla-San MG, Labrada R, et al (2007) Weedy rices: origin, biology, ecology and control. FAO of the United Nations, Rome. 144. http://www.fao.org/docrep/010/a1023e/a1023e00.htm (Accessed on April 22, 2012).
19. Londo JP, Schaal BA (2007) Origins and population genetics of weedy red rice in the USA. Mol Ecol 16: 4523–4535.
20. Harlan JR, De Wet J (1971) Toward a rational classification of cultivated plants. Taxon 20: 509–517.
21. Cohen MB, Arpaia S, Lan LP, Chau LM, Snow AA (2008) Shared flowering phenology, insect pests, and pathogens among wild, weedy, and cultivated rice in the Mekong Delta, Vietnam: implications for transgenic rice. Environ Biosafety Res 7: 73–85.
22. Chen LY, Snow AA, Wang F, Lu BR (2006) Effects of insect-resistance transgenes on fecundity in rice (*Oryza sativa*, Poaceae): A test for underlying costs. Am J Bot 93: 94–101.
23. Snow AA, Pilson D, Rieseberg LH, Paulsen MJ, Pleskac N, et al (2003) A Bt transgene reduces herbivory and enhances fecundity in wild sunflowers. Ecol Appl 13: 279–286.
24. Xia H, Chen LY, Wang F, Lu BR (2010) Yield benefit and underlying cost of insect-resistance transgenic rice: Implication in breeding and deploying transgenic crops. Field Crop Res 118: 215–220.
25. Yang X, Xia H, Wang W, Wang F, Su J, et al (2011) Transgenes for insect resistance reduce herbivory and enhance fecundity in advanced generations of crop-weed hybrids of rice. Evol Appl 4: 672–684.
26. Halfhill MD, Sutherland JP, Moon HS, Poppy GM, Warwick SI, et al (2005) Growth, productivity, and competitiveness of introgressed weedy *Brassica rapa* hybrids selected for the presence of *Bt cry1Ac* and *gfp* transgenes. Mol Ecol 14: 3177–3189.
27. Mason P, Braun L, Warwick SI, Zhu B, Stewart CN (2003). Transgenic *Bt*-producing *Brassica napus*: *Plutella xylostella* selection pressure and fitness of weedy relatives. Environ Biosafety Res 2: 263–276.
28. Vacher C, Weis AE, Hermann D, Kossler T, Young C, et al (2004) Impact of ecological factors on the initial invasion of Bt transgenes into wild populations of birdseed rape (*Brassica rapa*). Theor Appl Genet 109: 806–814.
29. Cao QJ, Xia H, Yang X, Lu BR (2009) Performance of hybrids between weedy rice and insect-resistant transgenic rice under field experiments: implication for environmental biosafety assessment. J Integr Plant Biol 51: 1138–1148.
30. Rong J, Song ZP, Su J, Xia H, Lu BR, et al (2005) Low frequency of transgene flow from *Bt/CpTI* rice to its nontransgenic counterparts planted at close spacing. New Phytol 168: 559–566.
31. Jenczewski E, Ronfort J, Chèvre AM (2002). Crop-to-wild gene flow, introgression and possible fitness effects of transgenes. Environ Biosaf Reserv 2: 9–24.
32. Chen YH, Welter SC (2002) Abundance of a native moth *Homoeosoma electellum* (Lepidoptera: Pyralidae) and activity of indigenous parasitoids in native and agricultural sunflower habitats. Environ Entomol 31: 626–636.
33. Wu KM, Lu YH, Feng HQ, Jiang YY, Zhao JZ (2008) Suppression of cotton bollworm in multiple crops in China in areas with Bt toxin-containing cotton. Science 321: 1676–1678.

6

Evaluation of Candidate Reference Genes for Gene Expression Normalization in *Brassica juncea* Using Real Time Quantitative RT-PCR

Ruby Chandna, Rehna Augustine, Naveen C. Bisht*

National Institute of Plant Genome Research, Aruna Asaf Ali Marg, New Delhi, India

Abstract

The real time quantitative reverse transcription PCR (qRT-PCR) is becoming increasingly important to gain insight into function of genes. Given the increased sensitivity, ease and reproducibility of qRT-PCR, the requirement of suitable reference genes for normalization has become important and stringent. It is now known that the expression of internal control genes in living organism vary considerably during developmental stages and under different experimental conditions. For economically important Brassica crops, only a couple of reference genes are reported till date. In this study, expression stability of 12 candidate reference genes including *ACT2, ELFA, GAPDH, TUA, UBQ9* (traditional housekeeping genes), *ACP, CAC, SNF, TIPS-41, TMD, TSB* and *ZNF* (new candidate reference genes), in a diverse set of 49 tissue samples representing different developmental stages, stress and hormone treated conditions and cultivars of *Brassica juncea* has been validated. For the normalization of vegetative stages the *ELFA, ACT2, CAC* and *TIPS-41* combination would be appropriate whereas *TIPS-41* along with *CAC* would be suitable for normalization of reproductive stages. A combination of *GAPDH, TUA, TIPS-41* and *CAC* were identified as the most suitable reference genes for total developmental stages. In various stress and hormone treated samples, *UBQ9* and *TIPS-41* had the most stable expression. Across five cultivars of *B. juncea,* the expression of *CAC* and *TIPS-41* did not vary significantly and were identified as the most stably expressed reference genes. This study provides comprehensive information that the new reference genes selected herein performed better than the traditional housekeeping genes. The selection of most suitable reference genes depends on the experimental conditions, and is tissue and cultivar-specific. Further, to attain accuracy in the results more than one reference genes are necessary for normalization.

Editor: Christian Schönbach, Kyushu Institute of Technology, Japan

Funding: The study was financially supported by the core grant from National Institute of Plant Genome Research (NIPGR), New Delhi, India. RC was supported by the Research Associate fellowship from NIPGR. RA was funded by a fellowship from Council of Scientific and Industrial Research (CSIR), India. The funders had no role in study design, data collection and analysis, decision to publish, or preparation of the manuscript.

Competing Interests: The authors have declared that no competing interests exist.

* E-mail: ncbisht@nipgr.res.in

Introduction

Gene expression analysis is extremely important in many fields of biological research. Understanding the expression pattern of genes provides a useful mean of studying the complex regulatory networks occurring in living organism. Among the widely used methods to measure the levels of gene expression, real time quantitative reverse transcription PCR (qRT-PCR) represents a suitable technology [1]. Being an efficient, sensitive, and reliable method, qRT-PCR provides a rapid mean towards simultaneous measurement of gene expression across different samples [2]. Since this platform is relatively simple coupled with a high level of sensitivity, qRT-PCR is rapidly being adopted as a standard method for performing in-depth expression analysis of number of target genes. For accurate and reliable analysis of target gene expression, normalization of qRT-PCR data with suitable internal reference gene(s) is required [3]. Normalization is essential to correct the non-specific variations arising because of the difference in amount of template used and its quality that can affect the efficiency of the qRT-PCR reactions [4]. Normalization also allows the direct comparison of normalised transcript expression levels between samples [5]. An ideal reference gene should express at constant level in all tissues and at all developmental stages, regardless of the experimental conditions or treatments [6,7].

Commonly used reference genes are mostly cellular maintenance genes (also known as housekeeping genes), which are involved in basic and ubiquitous cellular processes such as components of the cytoskeleton, glycolytic pathway, protein folding, protein degradation, synthesis of ribosome subunits. Most frequently used housekeeping genes including β-actin (*ACT*), α-tubulin (*TUA*), ubiquitin (*UBQ*), glyceraldehde-3-phosphate dehydrogense (*GAPDH*), *18S* or *26S* ribosomal RNA and elongation factors (*EF*) have been validated as suitable internal control genes in many plants [5,8–11]. These genes were assumed to be expressed constitutively and also have constant expression levels between different samples. However, there are evidences that transcripts levels of housekeeping genes vary considerably across the developmental stages and under variable conditions [6]. This variability in expression across experimental samples may be because these housekeeping genes not only participate in basic cell metabolism but also in other cellular processes [12]. Therefore, selecting multiple stably expressed reference genes, other than the commonly used housekeeping genes need to be considered for accurate normalization of gene expression studies [13].

Recognising the importance of validation of reference gene(s) for normalization of qRT-PCR data initiated the development of number of software packages such as geNorm [6,13] and NormFinder [4] Usage of these statistical algorithms have greatly simplified the selection of appropriate reference genes by calculating the expression stability and determining the optimal number of candidate reference genes required for normalization under specific conditions in various organisms, including plants [14,15]. A number of attempts for reference gene validation have been reported in plants such as rice [16], chickpea [10], potato [8], soybean [9,17], tomato [18], chrysanthemum [11], grape [19], cabbage [20], wheat [21], *Brassica napus* [22] and poplar [23].

The *Brassica* species have diverse characteristics and are of great agronomic importance as vegetables, condiments, fodder, and oil crops. *Brassica* crops are globally the third most important sources of vegetable oil after soybean and groundnut [24]. *Brassica juncea* (brown or Indian mustard) is an important oilseed crop cultivated mainly in Indian sub-continent besides some parts of east Europe, Africa, Canada and China. Only a limited number of gene expression studies have been carried out in *B. juncea*, wherein *ACT* and *18S* are the commonly used reference genes [25–29].

Comparison of several candidate reference genes in Brassica crops, particularly in *B. juncea*, is not yet reported, thereby limiting our knowledge about the choice of best reference gene which could be used for normalization of gene expression across developmental stages and variable growth conditions. In present study, we have compared the performance of 12 candidate reference genes (consisting of five commonly used housekeeping genes of plants, and seven new candidate reference genes selected from *B. rapa* and *B. napus* microarray platforms) in 49 diverse samples of *B. juncea*, broadly categorized into five distinct experimental sets. Our results reveal that new reference genes are more stably expressed than the traditionally used housekeeping genes across all the five experimental sets. Further, combination of most stable reference genes provides a more accurate and reliable mean of normalization during qRT-PCR analysis.

Results

Selection of Candidate Reference Genes and Primer Design

A total of 12 candidate reference genes, including five traditional housekeeping genes namely, actin2 (*ACT2*), elongation factor 1B (*ELFA*), glyceraldehyde-3-phosphate- dehydrogenase (*GADPH*), α-tubulin (*TUA*) and ubiquitin 9 (*UBQ9*) and seven new reference genes, acyl carrier proteins (*ACP*), clathrin adaptor complex (*CAC*), sucrose non fermenting-1 protein kinase (*SNF*), tonoplastic intrinsic proteins-41 (*TIPS-41*), trans membrane proteins (*TMD*), tryptophan synthase-β (*TSB*) and zinc finger protein (*ZNF*) were used in this study. The new candidate reference genes were selected on the basis of their stable expression profiles across developmental stages and during abiotic stress conditions, as determined using microarray data of *B. rapa* (www.brassica.info/resource/trancriptomics.php) [30] and *B. napus* (www.rapeseed.plantsignal.cn) [31,32] and analysed using *A. thaliana* gene expression tool (www.jsp.weigelworld.org) [33] (File S1, S2 and S3).

Since, sequence information of *B. juncea* is very limited, we used publicly available gene sequences from Arabidopsis and related Brassica species to design the gene specific primers. The Brassica specific expressed sequence tags (EST) and genome survey sequences (GSS) of these candidate reference genes were largely obtained from the Brassica genome gateway (http://brassica.bbsrc.ac.uk) [34] and the recently available *B. rapa* genome portal (http://brassicadb.org/brad) [35], by providing Arabidopsis complementary DNA sequence (CDS) as a query. To ensure gene amplification in *B. juncea*, the Arabidopsis and Brassica sequences of each candidate gene were aligned together and the primers were designed from the consensus regions of the aligned sequences, preferably spanning an intron (File S4).

Verification of Primer Specificity and PCR Efficiency Analysis

In order to determine specificity of primers designed in the current study, agarose gel electrophoresis and melting curve analyses were performed following the qRT-PCR experiment on seedling stage of *B. juncea* L. cv. Varuna. All the primer pairs amplified single PCR product of expected size (File S5) and the specificity of amplicon was confirmed by the presence of single peak during melt curve and sequencing analysis (File S6 and S7). A standard curve was generated using 10-fold serial dilutions of cDNA to calculate the gene specific PCR efficiency. The slopes of the standard curves were used to calculate the correlation coefficient (R^2) and PCR efficiency (File S8). The linear R^2 for all the primers ranged between 0.994–0.999 over 1000 fold of cDNA dilution. Further, PCR efficiencies of primers ranged from 94%–106% (Table 1).

Expression Profiling of *B. juncea* Reference Genes

A real-time qRT-PCR assay, based on SYBR Green detection, was designed for transcript profiling of the 12 candidate reference genes (*ACP, ACT2, CAC, ELFA, GAPDH, SNF, TIPS-41, TMD, TSB, TUA, UBQ9* and *ZNF*) in 49 diverse samples of *B. juncea* (Table 1). In order to minimize the variability associated with qRT-PCR analysis, all RNA samples were adjusted to same concentration and quality pass prior to their conversion into cDNA (File S9). The expression level of the candidate genes obtained during qRT-PCR experiments are presented as threshold cycle (Ct) values. The 12 reference genes used in the current study showed relatively wide ranges of Ct values across 49 samples of *B. juncea*. The mean Ct values of reference genes ranged from 22.02–29.36 (Figure 1). *GAPDH* was the most abundant reference gene of the set (mean Ct = 22.02), whereas *TMD* was the least abundant reference gene (mean Ct = 29.36). Interestingly, three new candidate reference genes *CAC, TIPS-41* and *SNF* showed least expression variation (Co-variance, CV of 4.34, 4.96 and 4.98, respectively); while the commonly used housekeeping genes *GAPDH* and *ELFA* (CV of 17.77 and 14.57, respectively) had the most variable expression profile across all the 49 experimental samples (Figure 1).

In order to perform in-depth expression analysis of candidate reference genes, the 49 diverse samples were categorized under five experimental sets comprising of defined development or condition-specific samples of *B. juncea* (Table 2). The first experimental set comprised of six vegetative stages (cotyledon, seedling, young leaf, mature leaf, stem and roots), whereas the second set included six reproductive stages (flower bud, flower, pod 5 days post anthesis (dpa), pod 10 dpa, pod 15 dpa and pod 30 dpa). In the third experimental set, all the 12 developmental stages of *B. juncea* (both vegetative and reproductive stages as mentioned above) were analysed together. The fourth set consisted of 12 samples treated with different abiotic stress conditions and hormones. The fifth experimental set included the vegetative stages of Indian and east-European *B. juncea* cultivars (Pusa Bold, Kranti, Donskaja, Early Heera-2, and Zem 84500).

Table 1. Surveyed references genes with their amplification and expression characteristics in *B. juncea*.

Internal Reference	Primers (F/R)	Amplicon length (bp)	T_m (°C)	*PCR efficiency (%)*	*Regression coefficient (R^2)*	*Covariance (%)*
ACP	5'- GTTGATTATGGGAAGAAGTCTAAGCT	110	82.98	99.6	0.998	8.34
	5'- TTGTAAGGCTCTACAACAGCAGTA					
ACT2	5'-TGGGTTTGCTGGTGACGAT	290	83.72	104.3	0.994	8.07
	5'- TGCCTAGGACGACCAACAATACT					
CAC	5'- CAATCGATTGCTTGGTTTGG	110	78.33	94.5	0.990	4.34
	5'- CAAGTCCAAGATTTCTTCTCTCC					
ELFA	5'- CCAAGAATGGGCTTTATGC	130	81.18	99.0	0.998	14.57
	5'- GTGATAGAGTGTCCAACAAGGTAAGTA					
GAPDH	5'- TCAGTTGTTGACCT CACGGTT	100	81.48	102.3	0.999	17.77
	5'- CTGTCACCAACGAAGTCAGT					
SNF	5'- CAAAGTCAACTGTTGGTACTCCTG	150	79.98	103.1	0.985	4.98
	5'- ATGGATATGCATCAACCAACAT					
TIPS-41	5'- TGAAGAGCAGATTGATTTGGCT	100	76.99	103.4	0.999	4.96
	5'- ACACTCCATTGTCAGCCAGTT					
TMD	5'- ACTCAATTCTATCTCCGCCTCT	85	81.03	97.4	0.999	7.53
	5'- AACAGAGCTCCGCATATTCC					
TSB	5'- AAGTACGTCCCCGAAACTCTAATG	145	82.23	106.3	0.995	7.23
	5'- CTGCGAAGTAGAGAGGACTTTC					
TUA	5'- GCTGGGTCACTCCAGATTTTG	80	77.58	99.7	0.998	6.26
	5'- CCATCGCCTTGTCTGCAAG					
UBQ9	5'- GAAGACATGTTCCATTGGCA	160	80.73	99.5	0.997	12.27
	5'- ACACCTTAGTCCTAAAAGCCACCT					
ZNF	5'- ATTTTCAGGCGGTTTATGGC	150	82.07	99.8	0.974	11.14
	5'- CTCTTGCTTTCTTCTTGGCGT					
#BjDREB2	5'-TGTATGAAAGGCAAAGGAGGA	130	85.0	96.56	0.985	–
	5'-GAAAGTACCAAGCCAAAGCCT					

#- used for normalization validation during abiotic stress conditions.

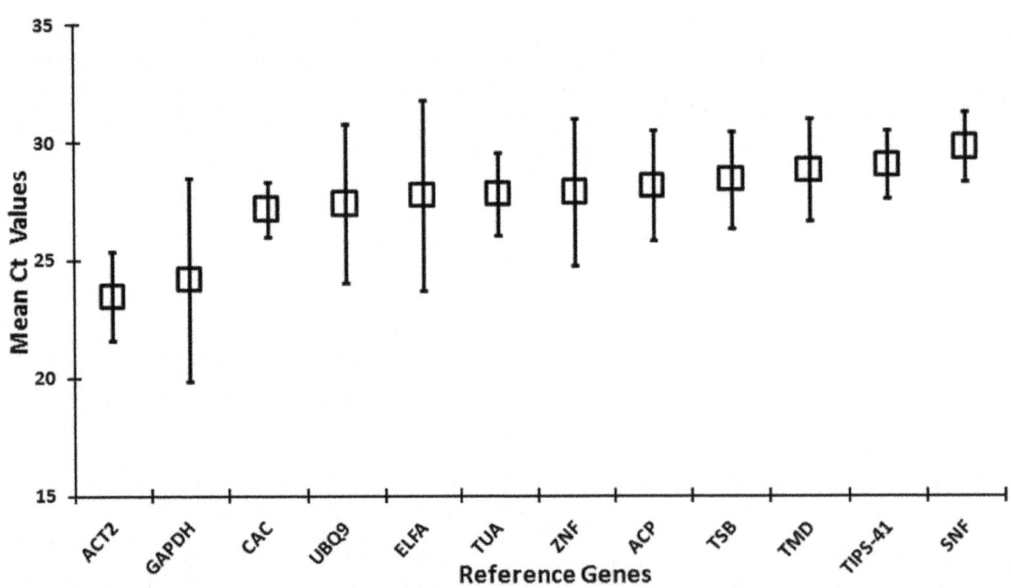

Figure 1. Expression levels of 12 candidate reference genes across all the five experimental sets of *B. juncea*. The boxes represent mean Ct values and bars correspond to the standard deviation. The Ct values were calculated on 1:50 diluted cDNA samples.

Table 2. Expression levels of 12 reference control genes across five experimental sets of *B. juncea* using geNorm.

Genes	VegetativeStages Mean Ct ± SD*	Reproductive Stages Mean Ct ± SD	Total Development stages Mean Ct ± SD	Treatments Mean Ct ± SD	Cultivars Mean Ct ± SD
ACP	27.6±1.4	25.8±0.8	26.7±1.5	26.7±1.1	29.5±2.3
ACT2	22.4±0.6	23.9±2.7	23.1±2.1	23.4±0.6	23.5±2.2
CAC	27.3±0.9	26.0±0.6	26.7±0.9	26.9±0.7	27.4±1.4
ELFA	29.3±0.6	29.6±2.3	29.4±1.6	29.7±1.1	25.9±4.8
GAPDH	21.4±1.6	19.8±0.6	20.6±1.4	21.0±1.3	27.3±3.9
SNF	28.5±0.9	28.9±1.4	28.7±1.1	27.7±0.7	29.4±1.5
TIPS-41	28.7±0.9	27.6±0.8	28.2±0.9	28.2±0.6	29.6±1.6
TMD	29.3±0.5	30.1±1.8	29.7±1.3	29.1±0.7	28.6±2.8
TSB	28.0±1.3	27.2±2.8	27.6±2.1	28.7±0.8	28.6±2.4
TUA	28.1±1.6	28.4±0.7	28.3±1.2	27.6±0.9	27.9±2.1
UBQ9	26.7±3.9	28.9±1.9	27.8±3.1	25.6±0.5	29.3±2.1
ZNF	24.9±3.3	22.4±1.1	23.7±2.7	27.9±0.6	29.8±1.8

*- mean of Ct values from all analyzed samples in individual experimental sets along with the standard deviations (SD) observed.

It was observed that the tissue type and experimental conditions does affect the expression of reference genes. In vegetative stages, *TMD*, *ELFA* and *ACT2* showed the least expression variation whereas a large variation in the expression levels of *UBQ9* and *ZNF* was observed (Table 2). In reproductive stages *TSB*, *ACT2* and *ELFA* showed large variation in their expression profiles, whereas *CAC* and *GAPDH* were the least variable transcripts. Interestingly, across total developmental stages, the two new reference genes *TIPS-41* and *CAC* emerged as most stable transcripts showing least variation and *UBQ9* and *ZNF* were the most variable transcripts (Table 2). In contrast, the expression of *UBQ9* was least variable in stress and hormone treated samples, whereas expression of *GAPDH* varied widely in various treated samples (Table 2). A significant variation in the expression of *ELFA* and *GAPDH* was also observed across the cultivars, while *CAC* and *TIPS-41* showed least expression variation (Table 2). In general, expression of all the 12 reference genes selected herein showed a large variation across the five cultivars of *B. juncea*. This probably reflects the presence of allelic variation in these genes across the *B. juncea* cultivars. Thus, transcript levels of none of the candidate reference genes remain constant throughout the developmental stages, stress and hormone treatments and across the *B. juncea* cultivars.

Gene Expression Stability Analysis and Ranking of *B. juncea* Reference Genes

The expression stability of 12 candidate reference genes, across five experimental sets, was measured and ranked using two different programs namely geNorm [6] and NormFinder [4]. Based on geNorm analysis, *ELFA*, *ACT2*, *CAC*, and *TIPS-41* (in order) were identified as the most stable genes across the vegetative stages whereas *TIPS-41*, *CAC*, *GAPDH* and *TUA* remained the most stable genes across the reproductive stages (Figure 2). When all the 12 samples were analysed together, *GAPDH*, *TUA*, *TIPS-41* and *CAC* were the most stable genes, while *UBQ9* and *ZNF* were least stable transcripts, in order. It is interesting to note that most stable housekeeping genes of vegetative stage (*ELFA* and *ACT2*), were not among the most stable reference genes of the total developmental set in *B. juncea*. Further, in response to various stress and hormone treatments, *UBQ9*, *TIPS-41*, *ZNF* and *CAC* were the most stable genes whereas across the *B. juncea* cultivars, *TIPS-41*, *CAC*, *ZNF* and *TUA* were identified as the most stable genes, in order. The commonly used housekeeping gene *GAPDH* was, however, found to be among the least stable transcript in treated samples as well as across the cultivars of *B. juncea* (Figure 2).

The NormFinder also ranked *ELFA*, *TIPS-41*, *CAC* and *ACT2* as the most stable reference genes in vegetative stages (in order), whereas in reproductive stages *TIPS-41*, *CAC*, *ZNF* and *TMD* were identified as the most stable reference genes (Table 3). Across total developmental stages *GAPDH*, *TIPS-41*, *ELFA* and *TUA* were the most stable genes while *ZNF* and *UBQ9* were the least stable genes (Table 3). Both, *CAC* and *TIPS-41* also emerged as most stable reference genes in stress and hormone treated samples and also across the *B. juncea* cultivars. Thus using both programs, similar genes were identified as stable reference genes across all the five experimental sets although the gene expression stability rankings for some reference genes were slightly altered.

Optimal Number of Reference Gene for Normalization Across the Experimental Sets

The geNorm software was further used to calculate the optimal number of reference genes necessary for normalization across different sets of experiment. The pairwise variation (Vn/Vn+1) between sequential normalization factors, NFn and NFn+1 was used to determine the number of genes required for reliable normalization [6]. As shown in Figure 3, differences in the expression stability values of the candidate reference genes were less marked in reproductive stages, than in other series (Figure 3). The V2/3 value for reproductive stage was 0.108 (geNorm V <0.15 when comparing a normalization factor based on the two or three most stable targets), so *CAC* together with *TIPS-41* would be sufficient for normalization purpose in reproductive stages (Figure 3). In vegetative stages, V4/5 value was 0.132 thereby suggesting that the optimal number of reference targets would be four, namely *ELFA*, *ACT2*, *CAC* and *TIPS-41*. However on analysing all the developmental stages together, four reference genes namely *GAPDH*, *TUA*, *TIPS-41* and *CAC* should be considered (V4/5 value $=0.137$). In treated samples, the pair of *TIPS-41* and *UBQ9* produced a V2/3 value of 0.12, therefore these two candidate reference genes can be used for normalization

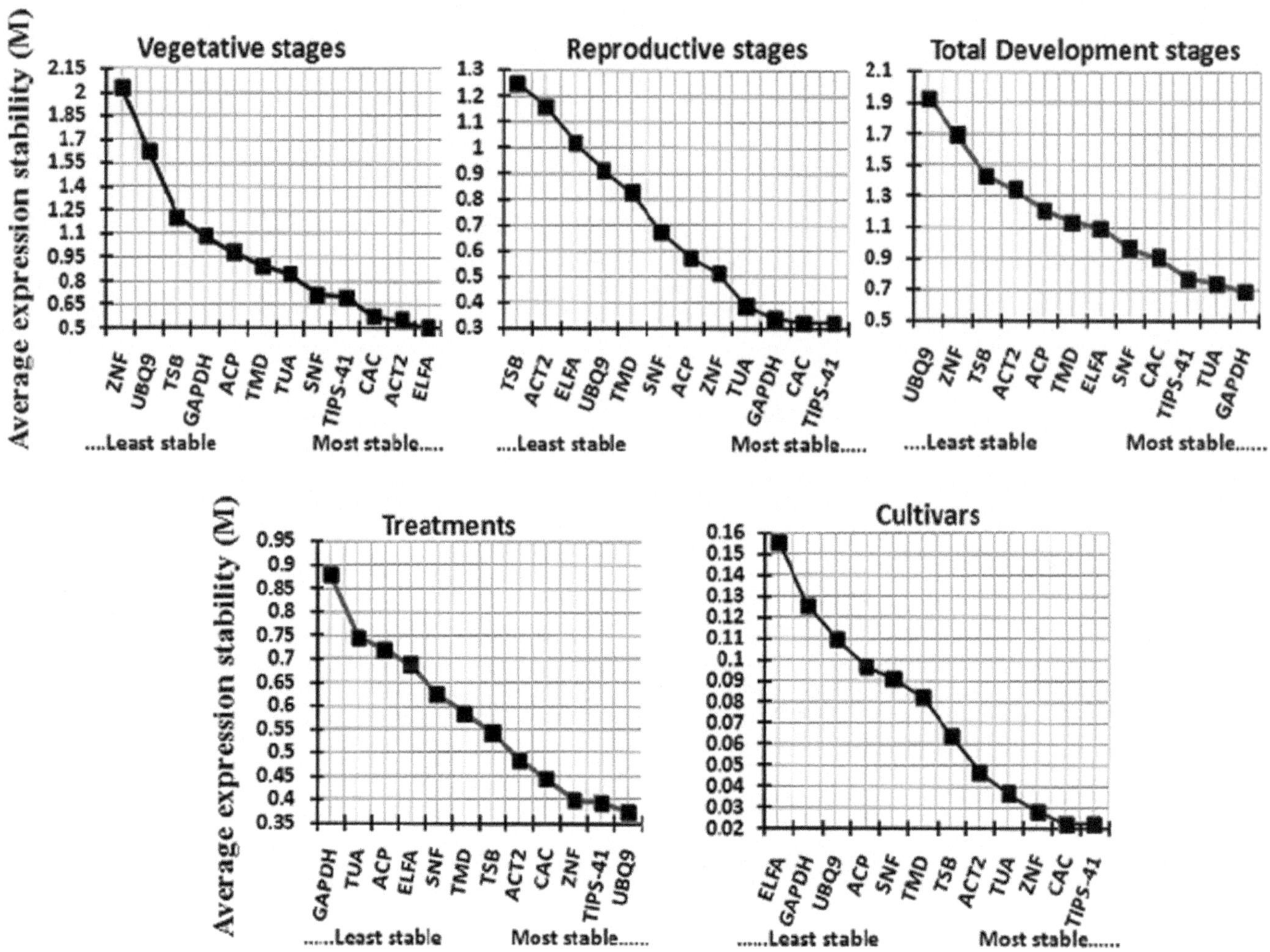

Figure 2. Gene expression stability of the 12 candidate genes of *B. juncea* as predicted by geNorm. Mean expression stability (M) following stepwise exclusion of the least stable gene across all the samples within an experimental set. The least stable genes are on the left, and the most stable on the right.

Table 3. Expression stability of the 12 reference control genes of *B. juncea* as calculated by Normfinder.

Rank	Vegetative stages		Reproductive stages		Total Development stages		Treatments		Cultivars	
	Gene	Stability	Gene	Stability	Gene	Stability	Gene	Stability	Gene	Stability
1	*ELFA*	0.223	*TIPS-41*	0.443	*GAPDH*	0.526	*TIPS-41*	0.304	*CAC*	0.179
2	*TIPS-41*	0.286	*CAC*	0.542	*TIPS-41*	0.540	*CAC*	0.361	*TIPS-41*	0.253
3	*CAC*	0.335	*ZNF*	0.581	*ELFA*	0.688	*UBQ9*	0.407	*ZNF*	0.278
4	*ACT2*	0.505	*TMD*	0.608	*TUA*	0.820	*SNF*	0.444	*TUA*	0.341
5	*ACP*	0.784	*UBQ9*	0.671	*CAC*	0.954	*ZNF*	0.461	*ACT2*	0.410
6	*SNF*	0.786	*TUA*	0.773	*TMD*	0.989	*TSB*	0.513	*TSB*	1.241
7	*TMD*	0.835	*SNF*	0.971	*SNF*	1.025	*ACT2*	0.556	*TMD*	1.248
8	*GAPDH*	1.045	*GAPDH*	1.011	*ACP*	1.050	*ELFA*	0.658	*UBQ9*	1.554
9	*TUA*	1.120	*ELFA*	1.031	*TSB*	1.191	*TMD*	0.661	*SNF*	2.120
10	*TSB*	1.534	*ACP*	1.265	*ACT2*	1.239	*ACP*	0.690	*ACP*	3.134
11	*UBQ9*	3.572	*ACT2*	1.553	*UBQ9*	1.820	*TUA*	0.729	*GAPDH*	3.961
12	*ZNF*	4.133	*TSB*	1.568	*ZNF*	2.055	*GAPDH*	1.492	*ELFA*	4.724

for a wide range of stressed and hormone treated tissue samples in *B. juncea*. Across cultivars, *TIPS-41* and *CAC* could be the choice of optimal reference genes (V2/3 value of 0.124). Thus across all the five experimental sets of *B. juncea*, the two new candidate reference genes namely *TIPS-41* and *CAC* were identified as the most suitable reference genes for normalization in gene expression studies.

Reference Gene Validation

To validate the selection of candidate reference genes for normalization, we analyzed the normalized fold expression of *BjDREB2* (a dehydration responsive element binding proteins-2 homolog of *B. juncea*). DREB2 is a transcription factor that imparts stress endurance to plants and plays a crucial role in providing tolerance to heat, dehydration, wounding and salt stresses. *UBQ9*, *TIPS-41* and *ZNF* reference genes identified as the most stable genes using geNorm analysis among all stress treated samples, were further tested for normalization. The transcript abundance of *BjDREB2* increased in dehydration stressed sample when normalized using all the three genes independently, although at different level (Figure 4). The expression pattern of *BjDREB2* transcript was higher in heat stress in case of *UBQ9*, whereas the increased expression pattern was observed in dehydration samples using either *ZNF* or *TIPS-41* as the reference gene. However, when the expression of *BjDREB2* was normalized using a combination of *UBQ9* and *TIPS-41*, identified by geNorm as most stable reference genes, the fold expression of *BjDREB2* was highest in dehydration stressed sample. This data clearly suggests that the use of more than one reference genes for normalization provides more accurate representation of target gene expression tested across the variable experimental conditions. As expected, the *BjDREB2* expression remains unaltered in hormone treated samples (data not shown).

Discussion

Quantitative real time PCR has become a powerful technique for gene expression studies, because of its high throughput, sensitivity and accuracy [2]. The choice of stably expressed reference genes for normalization is the paramount to accurate interpretation of the results. The normalization takes care of the variation introduced by the quantity and quality of input RNA, its cDNA conversion and the various steps involved in the qRT-PCR assay. Normalization studies with multiple reference genes validated for their expression stability is required for reliable gene expression results, as no single reference gene in plants has been shown to have a stable expression during variable experimental conditions including different samples/treatments. The usage of statistical algorithms, such as geNorm and NormFinder, has greatly simplified the selection and validation of reference genes by calculating the expression stability of reference genes as well as determining the number of reference genes required for accurate normalization across the experimental conditions tested [4,6].

This study describes a comprehensive analysis on the validation of 12 candidate reference genes (including five commonly used housekeeping genes of plants and seven new candidate reference genes) in 49 diverse samples of *B. juncea*, divided broadly into five experimental sets. Our analysis based on geNorm and NormFinder algorithms indicated that the choice of reference genes for normalization should be experiment and stage-specific. For example, across total developmental stages of *B. juncea* (including both vegetative and reproductive stages) four genes namely *GAPDH*, *TUA*, *TIPS-41* and *CAC* are ideal for normalization factor. However, when the total developmental stages were subdivided into two distinct experimental sets viz., the vegetative and reproductive stages, different sets of candidate reference genes appeared to be the best for normalization in each stage (Figure 2; Table 3). For vegetative stages, the four reference genes viz.,

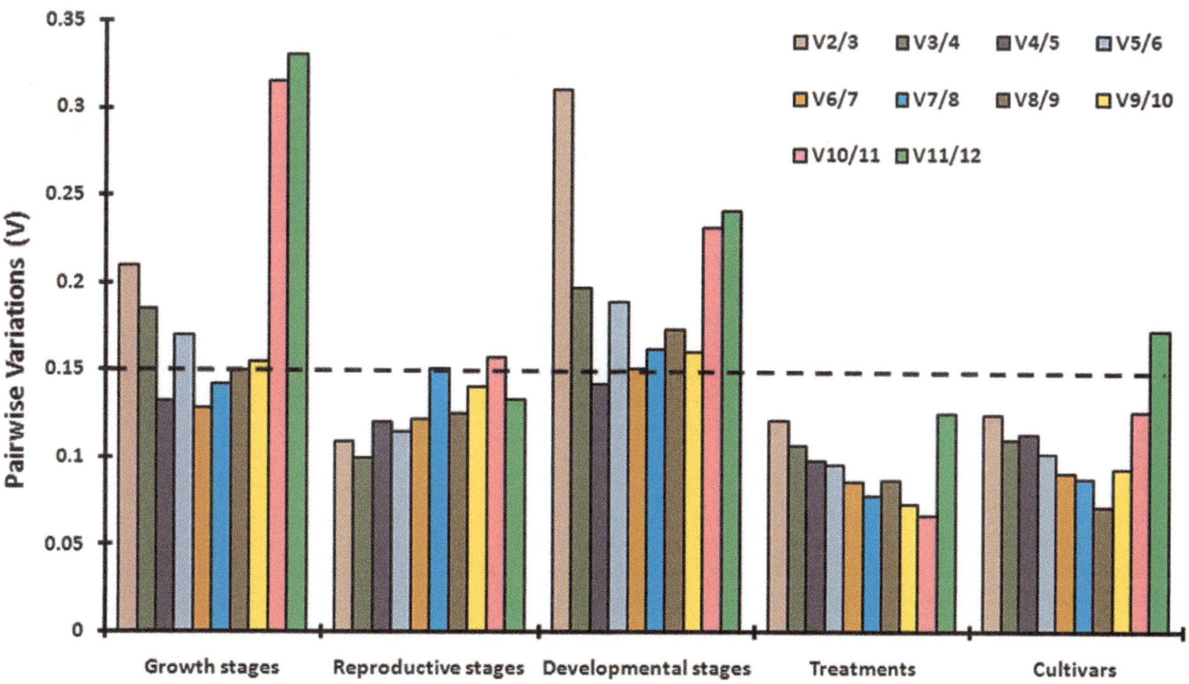

Figure 3. The optimal number of reference genes required for effective normalization in each experimental sets of *B. juncea*. The pairwise variation (Vn/Vn+1) was analyzed between normalization factors NFn and NFn+1 by geNorm program to determined the optimal number of reference genes.

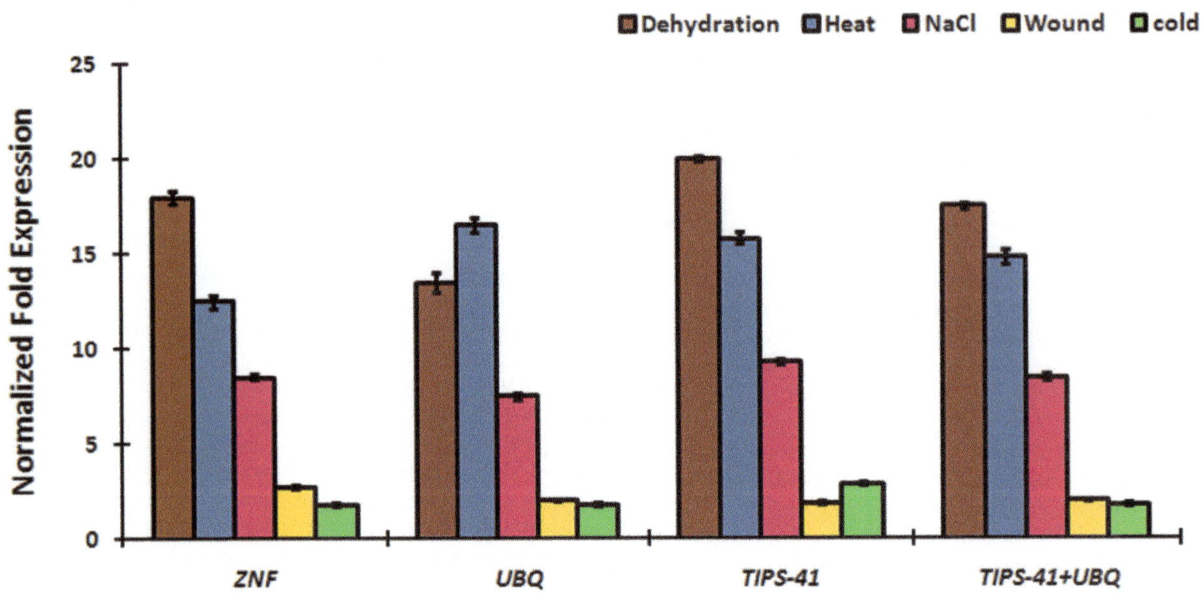

Figure 4. Relative quantification of *BjDREB-2* expression using identified stable reference genes and their combination for normalization under various stress conditions.

ELFA, *ACT2*, *CAC* and *TIPS-41* are ideal for calculating normalization factor, however for reproductive stages *TIPS-41* and *CAC* are enough for calculating the normalization factor. Since seed development is an economically important and distinct phase in Brassica crops, identification of suitable reference genes, particularly in reproductive stages will greatly assist in performing in-depth expression analysis of seed-specific genes in the said crop plants.

Some of the novel candidate reference genes selected in the current study performed better than the traditional housekeeping genes under each experimental set. *TIPS-41* (Tonoplast intrinsic proteins) and *CAC* (Clathrin adaptor complex), were identified amongst the list of most stable reference genes in all the five experimental sets of *B. juncea* tested in this study. *TIPS-41* (or *TIP41*) is one of the top ranked reference genes identified in the both vegetative and reproductive stages of *B. juncea* and *B. napus*, probably reflecting its similar and stable transcription regulation across the Brassica crops [22]. Previous studies in selection of reference genes during tomato development, *Fagopyrum esculentum* developmental stages and soybean under varied light regimes [9,18,36] also identified *TIPS-41* as the most stable reference gene for the vegetative sample. Similarly, *CAC* has also proved to be the best candidate for normalization in banana fruit, tomato, coffee, buckwheat and *Cucurbita pepo* studies [18,37–39].

In current study, we tested the expression stability of commonly used housekeeping genes like *GAPDH*, *UBQ*, *ACT*, *ELFA* and *TUA* that have been previously described as "candidate controls" in various plant studies. Some of these housekeeping genes qualify among the best reference genes under specific experimental sets of *B. juncea*, however none of them was found to be the suitable reference gene across all the five experimental sets of *B. juncea*. For example, *UBQ9* was found to be the most appropriate gene in stress and hormone treated samples while expression stability of *UBQ9* ranked late in different sets of samples including total developmental stages (Figure 2; Table 3). Our findings are in accordance with the previous studies, wherein *UBQ9* and its putative homologues were the least stably expressed genes during total developmental stages in rice, soybean, grapeberry, zucchini and chicory [5,16–17,19,39] Expression of homologues of *UBQ* was also found to be stable when evaluated in chrysanthemum and banana under various stresses [11,38].

Housekeeping genes like *GAPDH* and *ELFA* were also not found to be expressed stably across the diverse experimental sets of *B. juncea*, tested in the current study. The expression of *GAPDH* and *ELFA* was only found to be stable in total development and vegetative stages, respectively (Figure 3; Table 3). Earlier studies have shown that *GAPDH* has constant expression in total developmental stages of chickpea and sugarcane, while it was observed as the least stable transcript in wheat developmental series [10,21,40]. Similarly, *ELFA* was also reported as one of the best reference control gene across vegetative stages of rice, perennial ryegrass, chickpea, *Linum usitatissimum*, *Brachypodium distachyon*, Arabidopsis and chicory [5,10,16,23,41–43]. However, there are evidences that the expression profile of *ELFA* was not as consistent as that of other reference genes in soybean, *Salvia miltiorrhiza* and tomato [17,18,44]. The poor performance of *ACT2* across total development, reproductive stages, treatments and cultivars of *B. juncea* was surprising since this gene has been used as a reference control in earlier gene expression studies [27–28]. However, *ACT2* can be used in combination of other selected genes for normalization in vegetative stages. Recently, several studies have also shown that the use of *ACT* for normalization is not reliable in rice [16], potato [8], Arabidopsis [23] and peach [45].

The varied expression profiles of commonly used housekeeping genes may be because they are reported to be involved in many other cellular processes besides their basic cellular metabolic functions. For example, *GAPDH* not only acts as a component of the glycolytic pathway, but it is also involved in other processes such as cell proliferation [46]. Similarly, ubiquitin which primarily participates in proteolytic degradation, also has non-proteolytic functions [47]. The varied expression of actin may be due to its participation in cytoplasmic streaming, cell division and the distribution of the plasma membrane proteins other than being a major component of eukaryotic cytoplasmic microfilaments [48].

The newly selected reference genes of *B. juncea*, are superior to traditional ones in terms of their expression stability. The microarray expression data of *B. rapa* and *B. napus* also showed better expression stability of new reference genes identified, compared to the traditional ones (File S1 and S2). Further, the findings of this study are also in good accordance with the Arabidopsis microarray expression data of these 12 candidate reference genes available at AtGenExpress Visualization Tool portal [3], which compiles the expression profiles of five datasets of Arabidopsis (developmental, hormones, abiotic stress, light and pathogen; File S3). Finally, using the gene expression profiling of *BjDREB2* in abiotic stressed samples of *B. juncea*, we found that normalization involving the combination of more than one stable reference genes resulted in improved accuracy (Figure 4).

Conclusion

In the present study, we evaluated the expression stability of 12 candidate reference genes across large number of *B. juncea* samples in an effort to identify a set of stable reference gene(s) for normalization during gene expression studies. Analysis of expression stability using geNorm and NormFinder revealed that the expression of *TIPS-41* and *CAC* are most stable across variable experimental tissues. In addition, data analysis using geNorm suggested that three housekeeping genes (*ELFA*, *UBQ9* and *GAPDH*) can be used in combination with *CAC* and/or *TIPS-41* to calculate the normalization factor based on multiple reference genes. Although no candidate reference gene was constantly superior to the others, our data suggest that the novel genes performed better that commonly used housekeeping genes of *B. juncea*. We conclude that the results outline in the present study will facilitate sensitive and accurate quantification of gene expression in *B. juncea* which could also be extrapolated to related Brassica crops. Further, in the absence of enriched genome and transcriptome information from *B. juncea* and its diploid progenitor parents, the current study will greatly assist the Brassica research community to select a set of novel reference genes which could potentially be used for large arrays of experimental conditions and more-importantly in a cross-species manner.

Methods

Plant Materials

Mustard (*Brassica juncea* L. cv. Varuna) was used for the experiments. A total of 12 tissues including six vegetative stages (cotyledons, seedlings, young leaf (20 days post sowing), mature leaf (40 dps), root and stem) and six reproductive stages (bud, flower, pod 5 dpa (days post anthesis), pod 10 dpa, pod 15 dpa and pod 30 dpa) were collected from the plant growing in field condition.

Five *B. juncea* cultivars were also included in the study: Early Heera2 (EH2), Pusa Bold (PB), Kranti (KR), Donskaja (DK) and Zem 84500 (ZM). Five vegetative stages (cotyledons, seedlings, leaf, stem and roots) were harvested from these cultivars.

Stress treatments: For stress treatments, 7-day old seedlings were used. Seeds were grown in half strength Murashige-Skoog (MS) media till five days. Elicitors were added after adapting 6-day old seedlings in sucrose free liquid medium for 24 h in dark. Thereafter, seedlings were transferred to beakers containing MS along with sodium chloride (NaCl, 300 mM), abscisic acid (ABA, 100 μM), methyl jasmonate (MeJa, 200 μM), salicylic acid (SA, 200 μM), glucose (3%), indole-3-acetic acid (IAA, 100 μM) and 1-aminocyclopropane-1-carboxylate deaminase (ACC,100 μM), and incubated for 6 h.

For drought treatment seedlings were air dried for 6 h duration; for cold and heat shock treatments the seedlings were kept at 4±1°C and 42±1°C respectively, for 6 h. For wound treatment seedlings were wounded with blunt forceps and collected after 10 min. The mock treated seedling for same interval served as control.

Totally, the experimental samples comprised of 12 developmental stages, 12 exposed to various stress treatments and 25 samples of vegetative stages involving five different cultivars, thereby consisting a total of 49 different tissues.

RNA Isolation, Quality Control and cDNA Synthesis

RNA was extracted from all the plant tissues using the Spectrum Plant Total RNA Kit (Sigma Life Sciences, USA) according to manufacturer's instructions. The quantity and quality of RNA sample was checked using Nano spectrophotometer (ND-1000 Thermo scientific); and all RNA samples were adjusted to the same concentration. RNA quality was further assessed using the Agilent-2100 Bioanalyzer and RNA 6000 Nano chips (Agilent Technologies, Singapore). RNA samples with 260/280 ratio from 1.9 to 2.1, 260/230 ratio from 2.0–2.5 and RIN (RNA integrity number) more than 7, were used for further analysis (File S9). The integrity of RNA samples were also checked by agarose gel electrophoresis. For each method the measurement was done in duplicates.

First strand cDNA was synthesized by reverse transcribing 2 μg of total RNA with high-capacity cDNA Reverse Transcription kit (Applied Biosystems, USA) in a 20 μl reaction using mixture of random primers and oligo-dT's in 1:1 ratio according to manufacturer's instructions. cDNA was diluted 50 times for the use of real-time qRT-PCR reaction. All cDNA were stored at −20°C until PCR.

Selection of Reference Genes and Primer Designing

The five traditional housekeeping genes often used as references control in plants were selected: *ACT* (Actin2, At3G46520), *UBC9* (Ubiquitin9, At4G27960), *ELFA* (Elongation factor 1B, At1G09640), *TUA* (α-Tubulin, At5G19770), *GAPDH* (Glyceraldehyde-3-phospho dehydrogense, At3G04120).

In addition, seven new reference control genes were selected using the available *B. rapa* www.brassica.info/resource/trancriptomics.php [30] and *B. napus* www.rapeseed.plantsignal.cn [31,32] microarray experiments, covering various abiotic stresses and developmental stages. The ratio of expression levels between control and treatment experiments for each gene that was within the limit of two-fold were selected as putative candidate reference genes. Based on the stable expression profiles during abiotic stress and development stages in *B. rapa* microarray transcriptome data, sequences of four potential new reference genes: *ACP* (Acyl carrier proteins, Arabidopsis ortholog At1G54630), *SNF* (Sucrose non fermenting-1 protein kinase, At3G50500), *TMD* (Leucine rich trans membrane domain proteins, At5G22600), *TIPS-41* (Tonoplastic intrinsic proteins, At4G34270) were selected (File S1). The three remaining genes, *CAC* (Clathrin adaptor complex, At5G46630), *TSB* (Tryptophan synthase-β, At4G27070) and *ZNF* (Zinc finger protein, At1G01930) were selected on the basis of non-significant change in their expression profiles during seed development in *B. napus* transcriptome analysis (File S2).

The full length complementary DNA (CDS) of the Arabidopsis genes were used to query homologous *Brassica* sequences. Expressed sequence tags (ESTs) and Genome survey sequences (GSS) were obtained from the publically available platform at NCBI, Brassica genome gateway http://brassica.bbrc.ac.uk [34] and recently available *B. rapa* genome sequence portal http://www.brassica-rapa.org [49]. The primer for *DREB2* (dehydration responsive binding element) were designed using CDS of Arabidopsis *DREB2* gene and its *B. rapa* gene orthologous (Bra005852 and Bra009112; available on Brassica database http://brassicadb.org [50]. MegAlign module of DNASTAR was used to align the sequenced Brassica EST's/GSS's with the known Arabidopsis CDS. The qRT-PCR primer was designed from the consensus sequences preferably spanning the intron(s) (File S4). The details of primer sequences are given in Table 1.

For all genes, primer pairs were designed using the online available Sigma DNA Calculator with the following parameters: optimal length 20–25 nucleotides, melting temperature 60–65°C, GC content <50%, product size range 80–200 base pairs, minimum or no self complementarities at 3′ end, absence for the hairpin structures and self-dimers. In order to confirm the sequences of the amplicons, PCR was performed on cDNA for all designed primer pairs. The products were analyzed on 2% agarose gel and sequenced (File S7). A series of 10 fold of three dilutions of cDNA (10–1,000 fold dilution), were made to determine the gene specific PCR amplification efficiency for each primer pair in qRT-PCR experiments. Based on the Ct values for all dilution points in a series, a standard curve was generated using linear regression and the slope. The qbase http://medgen.ugent.be/̃jvdesomp/genorm [51] calculated the gene specific PCR amplification efficiency of the primer using the following equation: Efficiency % $= 10^{(-1/\text{slope})} \times 100\%$.

Mean of expression levels for all the 12 genes studied in 49 tissue samples were calculated. Standard deviation was calculated using Microsoft Excel and Co-variance was calculated as Standard Deviation/Mean ×100.

Real-time qRT-PCR Assays

Real-Time PCR was performed in an optical 96-well plate with an 7900 HT real time PCR machine (Applied Biosystems) and universal cycling conditions (95°C for 10 min, 40 cycles of 15 s at 95°C and 60°C for 60 s) in final volume of 20 µl. Reactions contained SYBR Green Master Mix (Kapa Biosystems), 10 pM of a gene specific forward and reverse primers and 2 µl of the diluted cDNA. A no template control (NTC) was also included in each run for each gene; in this study 2 µl RNase free water was used. Each experiment was conducted in three technical replicates with at least two biological replicates for each tissue. To check for the specificity of PCR amplification dissociation curve was generated. The Ct values were automatically determined for each reaction using SDS version 2.3 and RQ manager version 1.2 (Applied Biosciences) software with default parameters.

Statistical Analysis

Following PCR data collection, two publically available software tools, called geNorm v3.2 [5] and NormFinder v0.953 [3] were used to rank the expression stability of reference genes across all the five experimental sets. The procedures outlined in the user's manuals, http://medgen.ugent.be/̃jvdesomp/genorm [51] and http://www.mdl.dk/publicationsnormfinder.htm [52], were followed to calculate instability values of gene expression.

Briefly, the geNorm program is based on pairwise comparisons and stepwise exclusion of candidate genes according to their expression stability measures (M) values. In general, lower the M value, higher the gene expression stability. geNORM recommends M<1.5 to identify sets of reference genes with stable expression. Further, the pairwise variation (Vn/Vn+1) between sequential normalization factors, NFn and NFn+1 was used to determine the number of genes required for reliable normalization. A threshold value of 0.15 was reported by Vandesompele et al. [6] for normalization of expression stability. It has been suggested that below this threshold values there is no need of an additional internal control gene.

The NormFinder reference tool was also applied to rank the candidate reference genes expression stability for all the samples with no subgroup determination. NormFinder used ANOVA based model to estimate intra- and inter-group variation, and combines these estimates to provide a direct measure of the variations in expression for each gene [4]. It ranks the genes according to their stability under given set of experimental conditions. Genes with lower average expression stability values are more stable.

Supporting Information

File S1 Gene expression profile of the genes considered in present study, as observed in *B. rapa* microarray database (http://www.brassica-rapa.org).

File S2 Expression profile of the selected genes as observed in previous study by Niu et al., 2009 [31] on *B. napus* seed development and fatty acid metabolism (www.rapeseed.plantsignal.cn).

File S3 Arabidopsis microarray expression data of development stages showing the expression profile of 12 candidate reference genes considered in this study (available at jsp.weigelworld.org).

File S4 An example of DNA sequence alignment of Arabidopsis CDS and its homologous Brassica EST and GSS. The nucleotide bases highlighted in grey are the conserved sequences for designing primers for *TIPS*-41.

File S5 The confirmation of expected amplicon size of the primer pairs.

File S6 Melt curves of the candidate reference genes under different experimental conditions.

File S7 Sequencing data of the PCR amplicons.

File S8 Amplification efficiencies of the primers designed in the current study.

File S9 The integrity values of RNA samples determined using Bioanalyzer.

Acknowledgments

The central instrumentation facility at National Institute of Plant Genome Research (NIPGR) is acknowledged. We thank Dr. Manoj Majee for his critical comments on the manuscript.

Author Contributions

Conceived and designed the experiments: RC NCB. Performed the experiments: RC RA NCB. Analyzed the data: RC NCB. Contributed reagents/materials/analysis tools: NCB. Wrote the paper: RC NCB.

References

1. Bustin SA, Benes V, Nolan T, Pfaffl MW (2005) Quantitative real-time RT-PCR – a perspective. J Mol Endocrinol 34: 597–601.
2. Bustin SA, Benes V, Garson JA, Hellemans J, Huggett J, et al. (2009) The MIQE guidelines: minimum information for publication of quantitative real-time PCR experiments. Clin Chem 55: 611–622.
3. Czechowski T, Stitt M, Altmann T, Udvardi MK, Scheible WR (2005) Genome wide identification and testing of superior reference genes for transcript normalization in Arabidopsis. Gen Anal 139: 5–17.
4. Andersen CL, Jensen JL, Orntoft TF (2004) Normalization of real-time quantitative reverse transcription-PCR data: a model-based variance estimation approach to identify genes suited for normalization, applied to bladder and colon cancer data sets. Cancer Res 64: 5245–5250.
5. Maroufi A, Bockstaele EV, Loose MD (2010) Validation of reference genes for gene expression analysis in chicory (*Cichorium intybus*) using quantitative real-time PCR. BMC Mol Biol 11: 15–27.
6. Vandesompele J, De Preter K, Pattyn F, Poppe B, Van Roy N, et al. (2002) Accurate normalization of real-time quantitative RT-PCR data by geometric averaging of multiple internal control genes. Genome Biol 3: 1–11.
7. Radonic A, Thulke S, Mackay IM, Landt O, Siegert W, et al. (2004) Guideline to reference gene selection for quantitative realtime PCR. Biochem Biophys Res Commun 313: 856–862.
8. Nicot N, Hausman JF, Hoffmann L, Evers D (2005) Housekeeping gene selection for real-time RT-PCR normalization in potato during biotic and abiotic stress. J Exp Bot 56: 2907–2914.
9. Hu R, Fan C, Li H, Zhang Q, Fu Y-F (2009) Evaluation of putative reference genes for gene expression normalization in soybean by quantitative real-time RT-PCR. BMC Mol Biol 10: 93–104.
10. Garg R, Sahoo A, Tyagi AK, Jain M (2010) Validation of internal control genes for quantitative gene expression studies in chickpea. Biochem Biophys Res Commun 396: 283–288.
11. Gu C, Chen S, Liu Z, Shan H, Luo H, et al. (2011) Reference gene selection for quantitative real-time PCR in Chrysanthemum subjected to biotic and abiotic stress. Mol Biotechnol 49: 192–197.
12. Guenin S, Mauriat M, Pelloux J, Van Wuytswinkel O, Bellini C, et al. (2009) Normalization of qRT-PCR data: the necessity of adopting a systematic, experimental conditions-specific, validation of references. J Exp Bot 60: 487–493.
13. Hellemans J, Mortier G, De Paepe A, Speleman F, Vandesompele J (2007) qBase relative quantification framework and software for management and automated analysis of real-time quantitative PCR data. Genome Biol 8: R23.
14. Brunner AM, Yakovlev IA, Strauss SH (2004) Validating internal controls for quantitative plant gene expression studies. BMC Plant Biol 13: 4–14.
15. Dheda K, Huggett JF, Chang JS, Kima LU, Bustin SA, et al. (2005) The implications of using an inappropriate reference gene for real-time reverse transcription PCR data normalization. Anal Biochem 37: 141–143.
16. Jain M, Nijhawan A, Tyagi AK, Khurana JP (2006) Validation of housekeeping genes as internal control for studying gene expression in rice by quantitative real-time PCR. Biochem Biophy Res Commun 345: 646–651.
17. Jian B, Liu B, Bi Y, Hou W, Wu C, et al. (2008) Validation of internal control for gene expression study in soybean by quantitative real-time PCR. BMC Mol Biol 9: 59–72.
18. Exposito-Rodriguez M, Borges A, Borges-Perez A, Perez J (2008) Selection of internal control genes for quantitative real-time RT-PCR studies during tomato development process. BMC Plant Biol 8: 131–142.
19. Reid KE, Olsson N, Schlosser J, Peng F, Lund ST (2006) An optimized grapevine RNA isolation procedure and statistical determination of reference genes for real-time RT-PCR during berry development. BMC Plant Biol 6: 27–37.
20. Qi J, Yu S, Zhang F, Shen X, Zhao X, et al. (2010) Reference Gene Selection for Real-Time Quantitative Polymerase Chain Reaction of mRNA Transcript Levels in Chinese Cabbage (*Brassica rapa* L. ssp. pekinensis). Plant Mol Biol Rep 28: 597–604.
21. Long X, Wang JR, Ouellet T, Rocheleau H, Wei Y, et al. (2010) Genome-wide identification and evaluation of novel internal control genes for Q-PCR based transcript normalization in wheat. Plant Mol Bio 74: 307–311.
22. Chen X, Truksa M, Shah S, Weselake RJ (2010) A survey of quantitative real-time polymerase chain reaction internal reference genes for expression studies in *Brassica napus*. Anal Biochem 405: 138–140.
23. Gutierrez L, Mauriat M, Guenin S, Pelloux J, Lefebvre JF, et al. (2008) The lack of a systematic validation of reference genes: a serious pitfall undervalued in reverse transcription polymerase chain reaction (RT-PCR) analysis in plants. Plant Biotechnol J 6: 609–618.
24. A.O F (2009) Food and Agriculture Organization of the United Nations.
25. Muthukumar B, Yakubov B, Salt DE (2007) Transcriptional activation and localization of expression of *Brassica juncea* putative metal transport protein *BjMTP1*. BMC Plant Bio 7: 32–44.
26. Jami SK, Clark GB, Turlapati SA, Handley C, Roux SJ, et al. (2008) Ectopic expression of an annexin from *Brassica juncea* confers tolerance to abiotic and biotic stress treatments in transgenic tobacco. Plant Physiol Biochem 46: 1019–1030.
27. Srivastava S, Srivastava AK, Suprasanna P, D'Souza SF (2009) Comparative biochemical and transcriptional profiling of two contrasting varieties of *Brassica juncea* L. in response to arsenic exposure reveals mechanisms of stress perception and tolerance. J Exp Bot 60: 3419–3431.
28. Xu J, Chai T, Zhang Y, Lang M, Han L (2009) The cation-efflux transporter BjCET2 mediates zinc and cadmium accumulation in *Brassica juncea* L. Leaves. Plant Cell Rep 28: 1235–1242.
29. Kaur P, Jost R, Sivasithamparam K, Barbetti MJ (2011) Proteome analysis of the *Albugo candida–Brassica juncea* pathosystem reveals that the timing of the expression of defence-related genes is a crucial determinant of pathogenesis. J Exp Bot 62: 1285–1298.
30. *Brassica rapa* transcriptome database. Available: http://www.brassica.info/resource/trancriptomics.php. Accessed 2011 Dec, 20.
31. Niu Y, Wu GZ, Lin WH, Shi QM, Xue LJ, et al. (2009) Global analysis of gene expression profiles in *Brassica napus* developing seeds reveals a conserved lipid metabolism regulation with *Arabidopsis thaliana*. Mol Plant 2: 1107–1122.
32. *Brassica napus* transcritome database. Available: http://www.rapeseed.plantsignal.cn. Accessed 2011 Dec, 20.
33. AtGenExpress visualization tool. Available: http://jsp.weigelworld.org Accessed 2011 Nov, 30.
34. Brassica genome gateway. Available: http://brassica.bbsrc.ac.uk. Accessed 2011 Nov, 28.
35. *B. rapa* genome portal. Available: http://brassicadb.org/brad. Accessed 2011 Dec, 5.
36. Demidenko NV, Logacheva MD, Penin AA (2011) Selection and validation of reference genes for quantitative real-time PCR in Buck wheat (*Fagopyrum esculentum*) based on transcriptome sequence data. PLoS One 6: e19434.
37. Cruz F, Kalaoun S, Nobile P, Colombo C, Almeida J, et al. (2009) Evaluation of coffee reference genes for relative expression studies by quantitative real time RT-PCR. Mol Breed 23: 607–616.
38. Chen L, Zhong H, Kuang J, Li J, Lu W, et al. (2011) Validation of reference genes for RT-qPCR studies of gene expression in banana fruit under different experimental conditions. Planta 234: 377–390.
39. Obrero A, Die JV, Roman B, Gomez P, Nadal S, et al. (2011) Selection of reference genes for gene expression studies in Zucchini (*Cucurbita pepo*) using qPCR. J Agric Food Chem 59: 5402–5411.
40. Iskandar HM, Simpson RS, Casu RE, Bonnet GD, Maclean DJ, et al. (2004) Comparison of reference genes for quantitative real-time polymerase chain reaction analysis of gene in sugarcane. Plant Mol Biol Rep 22: 325–337.
41. Hong SY, Seo PJ, Yang MS, Xiang F, Park CM (2008) Exploring valid reference genes for gene expression studies in *Brachypodium distacyon* by real-time PCR. BMC Plant Biol 8: 112–122.
42. Lee JM, Roche JR, Donaghy DJ, Thrush A, Sathish P (2010) Validation of reference genes for quantitative RT-PCR studies of gene expression in perennial ryegrass (*Lolium perenne* L.). BMC Mol Biol 11: 8–22.
43. Huis R, Hawkins S, Neutelings G (2010) Selection of reference genes for quantitative gene expression normalization in flax (*Linum usitatissimum* L.). BMC Plant Biol 10: 71–86.
44. Yang Y, Hou S, Cui G, Chen S, Wei J, et al. (2010) Characterization of reference genes for quantitative real-time PCR analysis in various tissue of *Salvia miltiorrhiza*. Mol Biol Rep 37: 507–513.
45. Tong Z, Gao Z, Wang F, Zhou J, Zhang Z (2009) Selection of reliable reference genes for gene expression studies in peach using realtime PCR. BMC Mol Biol 10: 71–82.
46. McNulty S, Toscano W (1995) Transcriptional regulation of glyceraldehyde-3-phosphate dehydrogenase by 2,3,7,8-tetrachlorodibenzo- p-dioxin. Biochem Biophys Res Commun 212: 165–171.
47. Hochstrasser M (2000) Evolution and function of ubiquitin-like protein-conjugation systems. Nature Cell Biology 2: E153–157.
48. Stürzenbaum SR, Kille P (2001) Control genes in quantitative molecular biological techniques: the variability of invariance. Comp Biochem Physiol Biochem Mol Biol 130: 281–289.
49. *B. rapa* genome sequence portal. Available: http://www.brassica-rapa.org/BrEMD/microarray_overview.jsp. Accessed 2011 Dec, 01.
50. Brassica database. Available: http://brassicadb.org/brad. Accessed 2011 Dec, 20.
51. geNorm software manual update. Available: http://medgen.ugent.be/jvdesomp/genorm/. Accessed 2011 Nov, 22.
52. Normfinder software manual updates. Available: http://www.mdl.dk/publicationsnormfinder.htm. Accessed 2011 Nov, 24.

7

Indica and *Japonica* Crosses Resulting in Linkage Block and Recombination Suppression on Rice Chromosome 12

Yulin Jia[1]*, Melissa H. Jia[1], Xueyan Wang[2], Guangjie Liu[3]

1 Dale Bumpers National Rice Research Center, Agricultural Research Service, United States Department of Agriculture, Stuttgart, Arkansas, United States of America, **2** Department of Biology, College of Life Sciences, China Jiliang University, Hangzhou, Zhejiang, China, **3** Rice Research and Extension Center, University of Arkansas, Stuttgart, Arkansas, United States of America

Abstract

Understanding linkage block size and molecular mechanisms of recombination suppression is important for plant breeding. Previously large linkage blocks ranging from 14 megabases to 27 megabases were observed around the rice blast resistance gene *Pi-ta* in rice cultivars and backcross progeny involving an *indica* and *japonica* cross. In the present study, the same linkage block was further examined in 456 random recombinant individuals of rice involving 5 crosses ranging from F_2 to F_{10} generation, with and without *Pi-ta* containing genomic *indica* regions with both *indica* and *japonica* germplasm. Simple sequence repeat markers spanning the entire chromosome 12 were used to detect recombination break points and to delimit physical size of linkage blocks. Large linkage blocks ranging from 4.1 megabases to 10 megabases were predicted from recombinant individuals involving genomic regions of *indica* and *japonica*. However, a significantly reduced block from less than 800 kb to 2.1megabases was identified from crosses of *indica* with *indica* rice regardless of the existence of *Pi-ta*. These findings suggest that crosses of *indica* and *japonica* rice have significant recombination suppression near the centromere on chromosome 12.

Editor: Jinfa Zhang, New Mexico State University, United States of America

Funding: Financial supports of National Natural Science Foundation of China (Grant No. 31000847) and of Science and Technology Department of Zhejiang Province (Grant No. 2011R10038) to XW are also acknowledged. The funders had no role in study design, data collection and analysis, decision to publish, or preparation of the manuscript.

Competing Interests: The authors have declared that no competing interests exist.

* E-mail: Yulin.jia@ars.usda.gov

Introduction

Recombination suppression is a complex biological phenomenon often resulting in large linkage blocks (LBs) where a set of genes is inherited together. Recombination suppression is known to be determined by the location of a genomic region, the method of selection, and genetic factors that determine recombination [1,2]. Although it is known that recombination suppression can easily occur in regions with more repetitive DNA sequences and/or near the centromere regions [3,4,5,6], the molecular and genetic bases of recombination suppression have not been clearly defined.

Rice (*Oryza sativa* L.) is one of the most important food crops in the world. Blast disease of rice caused by the filamentous fungal pathogen *Magnaporthe oryzae* has been a serious threat to rice production for years in most regions where rice is grown. Breeding for resistance to blast has been one of the most important objectives for rice breeders worldwide. Over 80 major and 35 quantitative *R* genes have been incorporated, some individually and/or in clusters, into diverse elite cultivars by classical plant breeding [7]. In practice, resistant resources for *japonica* rice growing areas have been commonly identified from subspecies *indica* whereas *japonica* resistant resources are useful for *indica* rice growing areas [8]. Crosses between *indica* and/or *japonica* rice have been routinely made for incorporating these blast *R* genes. The major *R* gene *Pi-ta* in rice was introduced from landrace *indica* varieties, Tetep and Tadukan, into a series of *japonica* cultivars in the Southern US and other regions of the world. A total of 11 newly developed rice cultivars, Katy, Drew, Madison, Kaybonnet, Banks, Ahrent, Spring, Cybonnet, Catahoula, CL111, and Templeton, were confirmed to contain *Pi-ta* [9]. *Pi-ta* encodes a cytoplasmic receptor with nucleotide binding domain and imperfect leucine rich repeats preventing infections by races of *M. oryzae* that contain *AVR-Pita* [10]. *AVR-Pita,* known to encode a metalloprotease, was predicted to be involved in pathogenicity [11]. Distribution, evolution and interaction of *Pi-ta* and *AVR-Pita1* have been investigated [12,13,14,15,16,17,18,19,20,21]. As a result, DNA markers have been identified from portions of the cloned gene for marker assisted selection [22,23]. In a previous study, large LBs around the *Pi-ta* locus were identified from backcrossed progeny and several rice cultivars [24]. It was unclear if the observed large LBs were due to multiple genes required for *Pi-ta* mediated resistance that is clustered within the same genomic region [25] and/or were due to recombination suppression because of incompatibility between *indica* and *japonica*.

The objective of the present study was to examine LBs at the *Pi-ta* locus using five random mapping populations involved in recombination events of *indica* with *japonica* and *indica* with *indica*.

Results

LB between a Cross of a *japonica* Line (*pi-ta/pi-ta*) with a *japonica* Cultivar Carrying a Portion of *indica* Genome (*Pi-ta/Pi-ta*)

To further investigate the recombination suppression observed at the *Pi-ta* locus in germplasm selected for blast resistance [24], we investigated a random RIL population consisting of 87 lines or 174 recombination events of a *japonica* breeding line RU9101001 (Early) with a *japonica* cultivar Katy with a large fragment of *Pi-ta* containing *indica* genome on chromosome 12. Markers RM27941, RM1337, RM27940 and RM7102 co-segregated with *Pi-ta*. SSR marker OSM89 physically located at 8.8 megabases (MB) was found at the left border at 2.2 cM from the *Pi-ta* gene; whereas, SSR marker RM511 physically located at 17.4 MB was found at the right border 6.4 cM from the *Pi-ta* gene. LB in this population was reduced to 8.6 MB (Figure 1A).

LBs between Crosses of Two *japonica* Cultivars (*pi-ta/pi-ta*) with Two *indica* Cultivars (*pi-ta/pi-ta*)

Analysis of 186 recombinant events from Lemont (*pi-ta/pi-ta*) and Jasmine 85 (*pi-ta/pi-ta*) RIL population identified SSR markers OSM91 and RM7102 that were found to co-segregate perfectly with *Pi-ta*. The physical distance between these two markers is 3.1 MB (Table 1). However, SSR marker RM3246, located at 9.1 MB, was identified at 0.4 cM from the *Pi-ta* gene that defines the left border; whereas SSR marker RM277 located at 18.3 MB is 5.5 cM from *Pi-ta* that defines the right border (Figure 1B). These two border markers indicated an LB of 9.2 MB.

A population derived from a cross of a different *pi-ta/pi-ta indica* cultivar with a different *pi-ta/pi-ta japonica* cultivar was studied to determine if LB can be reduced if a different *indica* genome on chromosome 12 is used. A total of 93 random doubled haploid lines or 186 recombination events of a tropical *japonica* cultivar Cocodrie (*pi-ta/pi-ta*) and an *indica* derived germplasm MCR 01-0277 (*pi-ta/pi-ta*) were analyzed. SSR marker RM7003 physically located at 6.8 MB was found at the left border of the *pi-ta* gene. RM3246, RM1047, OSM91, RM27940, RM1337, RM27973 and RM7102, the furthest marker whose physical location is 13.2 MB, co-segregated with *pi-ta*. This analysis identified a LB is greater than 6.4 MB (Figure 1C). Unfortunately, the exact LB block was undetermined because we failed to identify the right border in this population. This finding suggests that observed linkage block is largely due to genomic contents on chromosome 12.

LB between a Cross of a *japonica* Cultivar Carrying *Pi-ta indica* Region with an *indica* Cultivar with *Pi-ta* and a Defective Component

To investigate if LB can be reduced in germplasm containing *Pi-ta* from two different origins, a total of 94 individuals (188 recombination events) of an F_2 population derived from a cross of Katy (*Pi-ta/Pi-ta*) with an *indica* cultivar 172 from Sri Lanka (*Pi-ta/Pi-ta*) but missing additional critical components for blast resistance where both contain different *indica* origins at chromosome 12 were analyzed. SSR marker RM3246 whose physical location is 9.1 MB was identified at 1.2 cM from the *Pi-ta* gene. SSR marker RM27941 co-segregated with *Pi-ta*. The SSR marker RM1047 whose physical location is 11.2 MB was identified at 0.8 cM from the *Pi-ta* gene defining the right border. LB was further delimited within a 2.1 MB fragment (Figure 1D).

LB between a Cross of *japonica* Carrying an *indica* Genome at the *Pi-ta* Locus with an *indica* Cultivar (*pi-ta/pi-ta*)

We investigated if LB can be reduced when a *japonica* cultivar carrying *indica* genome on chromosome 12 containing *Pi-ta* was crossed with an *indica* cultivar from a different rice growing area lacking *Pi-ta*. A total of 89 individuals of an F_{10} RIL population of a cross of an *indica* Zhe733 (*Pi-ta/Pi-ta*) and tropical genetic stock Kaybonnet-*lpa* (*Pi-ta/Pi-ta*) with chromosome 12 identical to that in Katy was analyzed (Table 1, Figure 1). SSR marker OSM91 (10.1 MB) was identified at 1.2 cM from the *Pi-ta* gene and RM27946 (10.9 MB) was identified 0.1 cM at the right border of the *Pi-ta* gene. This analysis further determined that LB was independent to *Pi-ta*, and was within 0.8 MB (Figure 1E).

Discussion

Investigation of the molecular and genetic basis of LB and recombination suppression can benefit plant breeding and genetic studies. In a previous study, we demonstrated a large linkage block from 43 BC_5F_2 progeny and several elite rice cultivars around the rice blast resistance gene *Pi-ta* [24]. It was proposed that selection for genes required for blast resistance could be one of the major reasons for the observed linkage blocks. In the present study, we examined if the observed linkage block could also be due to the

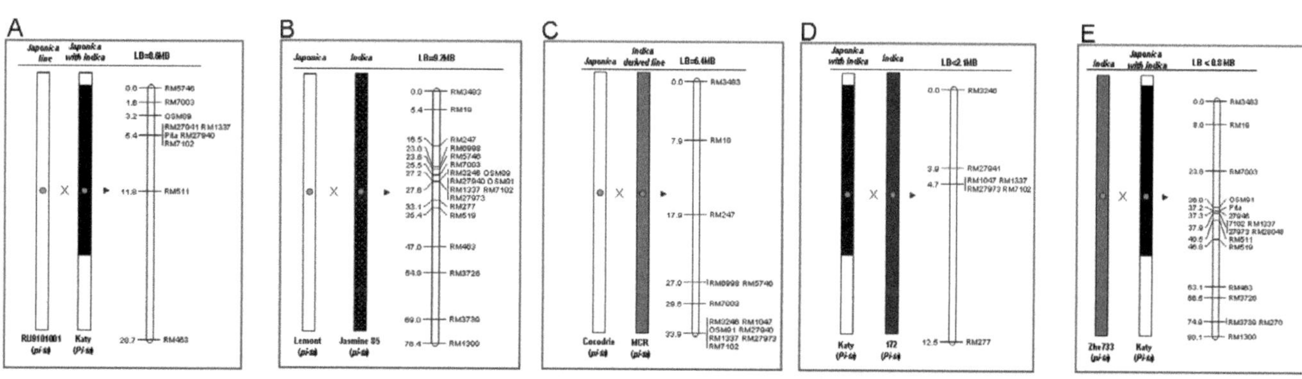

Figure 1. Genetic maps showing the size of linkage block among different crosses. A: Lemont and Jasmine 85; B: Early and Katy; C. Cocodrie and MCR 01-0277; D. Katy and Amane (172); E: Zhe733 and Katy. Graphic presentation of chromosome 12 with centromere, and difference of genotypes were indicated by different color shading (Left). Genetic distance in CentiMorgan defined by SSR marker and physical size of the linkage block defined by the two closest SSR markers was shown as LB = (Right) were shown.

Table 1. Physical distances of SSR markers co-segregated with *Pi-ta* and with both borders of the recombination break points in mapping populations.

	Left border		Co-segregated		Right border	
	Marker	Physical location (MB)[1]	Marker	Physical distance between two markers (MB)[2]	Marker	Physical location
Lemont X Jasmine 85	RM3246	9.1	OSM91-RM7102	3.1	RM277	18.3
RU9101001 X Katy	OSM89	8.8	*Pi-ta*-RM7102	2.6	RM511	17.4
Zhe733 X Katy	OSM91	10.1	*Pi-ta*	0	RM 27946	10.9
Cocodrie X MCR	RM7003	6.8	RM3246-RM7102	4.1	None	
Katy X 172	RM3246	9.1	RM27941	0	RM1047	11.2

[1]Physical location of each marker was obtained from www.gramene.org.
[2]MB denotes megabases.

incompatibility between *indica* and *japonica* in crosses without selection pressure for blast resistance. A total of 5 random recombinant populations consisting of 456 individuals with different resistant and susceptible *Pi-ta* haplotypes were examined using 40 SSR markers spanning the centromeric region of chromosome 12 (Figure 2). Remarkably, differential recombinant events between individuals derived from japonica and indica, indica and indica crosses were detected, LB was delimited within 800 KB between the *indica* cultivar Zhe733 without the resistant *Pi-ta* allele (Figure 1E), and a tropical *japonica* cultivar Katy carrying resistant *indica Pi-ta* allele from an *indica* landrace variety Tetep; whereas a large 9.2 MB linkage block was observed between a *japonica* lacking *Pi-ta* and *indica* containing *Pi-ta* and further reduced to be 6.4 MB with *japonica* and *indica* derived lines lacking the resistant *Pi-ta* allele. Taken together, this study suggests that the existence of a large linkage block at the centromeric region at the *Pi-ta* locus is mostly due to incompatibility between *indica* and *japonica*. To our knowledge this is the first demonstration of this magnitude of lack of recombination between *indica* and *japonica* rice crosses involving chromosome 12.

We knew that an individual of an F2 or DH population has less recombination than an individual of a RIL population. In the present study, three RIL populations were used to investigate the magnitude of LBs. Of the three RIL populations studied two were indica by japonica, and one was indica by indica to ensure similar recombination rates. It was found that indica by indica RIL population had significantly higher recombination rates than indica by japonica crosses. We might have observed an even smaller LB in the Katy, a japonica cultivar with large indica genomic region by an indica cultivar "172" mapping population if it had been advanced to RIL generation. With less recombination events we still detected 2.1 MB LB that is smaller than other indica and japonica crosses and larger than other indica and indica cross. In the present study, the numbers of individuals ranged from 87–93 in 5 mapping populations were comparable. While these populations were slightly small, it's highly improbable that we would observe large LBs in three different indica by japonica crosses while not observing it in two indica by indica crosses. Significant reduction of LB from indica and japonica cross in comparison with that of indica and indica cross suggests that large LB could be due to indica and japonica cross. Further examination of LBs from japonica and japonica cross will help to understand the genetic and molecular bases of LB.

The existence of sexual incompatibility between *indica* and *japonica* rice on chromosome 12 suggest that both *japonica* and *indica* are not as closely related subspecies of cultivated rice *Oryza sativa* as previously thought. It is well known that *japonica* and *indica rice* possess distinct morphological and molecular differences [26]

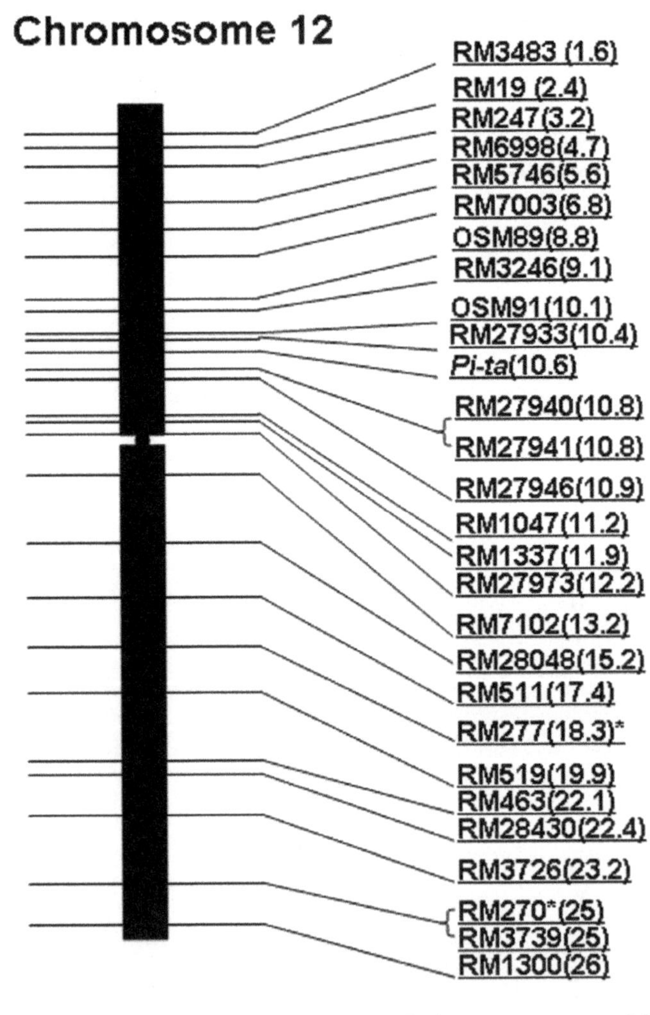

Figure 2. Schematic presentation of chromosome 12 with physical location (MB) of each SSR marker used for this study. Locations of SSR markers were obtained from www.gramene.org. *Physical location unavailable in gramene thus was determined by blasting the primers to the Nipponbare genome.

suggesting unique origins and domestications. One theory is that *O. sativa* was originated from the wild relative *Oryza rufipogon*. However, historical records of the time and geographic separation of *indica* from *japonica* subspecies remains unclear. Furthermore, Xu et al [27] reported sequence information of cultivars and wild relatives of rice suggesting that *indica* is more closely related with wild relative *O. nivara* and *japonica* is closely related with *O. rufipogon* [27]. Contemporary surveys have indicated that *indica* rice is commonly grown in lowland throughout tropical Asia while *japonica* is often cultivated in dry fields in temperate East Asia, upland areas of Southeast Asia and high elevations in South Asia, and the United States of America (US). There is also another theory that *Oryza sativa* rice was domesticated at least twice where *indica* was from eastern India, Myanmar and Thailand; and *japonica* from southern China and Vietnam even though there is archaeological and genetic evidence for a single domestication of rice from the lowlands of China [28,29]. How environmental conditions and domestication contributed to the observed differences of *indica* and *japonica* rice will need to be further investigated.

Three major hypotheses can explain LB at the *Pi-ta* locus: 1) Genes required for blast resistance are clustered together. This hypothesis is supported by the finding that *Pi-ta*-mediated resistance responses require another plant gene *Ptr(t)* and this gene has been mapped nearby the *Pi-ta* gene [25]. The *Ptr(t)* gene was recently fine mapped outside 800 kb of the *Pi-ta* gene (Wang and Jia, unpublished data) further suggesting LB can be reduced without selection for blast resistance. 2) The centromeric region of plant genome often has a significant amount of cross over suppression [30,31], in fact, there is some recombination suppression at the centromeric region on chromosome 8 (Jia, unpublished data). Thus, the sizes of LB observed near the centromere may be larger than that in other regions of chromosome. 3) Incompatibility occurs between *indica* and *japonica* crosses. The fact that recombination suppression occurs in indica by japonica crosses but not indica by indica crosses indicate that large LB at the *Pi-ta* locus is mostly likely due to the incompatibility of *indica* and *japonica*. It would be a priority to investigate the size of LB involved in japonica parents in the present study with other japonica genotypes.

If large LB is due to the incompatibility between *indica* and *japonica*, perhaps genome sequence of parents in the present study should give us some ideas as what genes may be involved. The major homologous recombination gene DMC1 has types A and B [32,33] in the reference genome Nipponbare. One member of the type A is in chromosome 12 [34]. Further sequence analysis of type A and functional validation will be needed to establish if type A plays any role in the observed LB between *indica* and *japonica* crosses. Additionally, genomic comparisons thus far have revealed that there are considerable differences on transposable elements on chromosome 12. Recently, a transposon at the promoter region of *Pi-ta* in resistant rice cultivars was found suggesting its active role in regulation of gene expression [17]. Predicted DNA sequences encode a protein with 844 amino acids with features similar in zinc fingers and transcription factors and with domains commonly found in hAT family dimerization (hATC) of a transposable element [17]. Additionally, a study at chromosome 8 suggested that transposable elements LTR-retrotransposon on chromosome 8 may play role in observed recombination suppression [35]. However, differences in number and type of transposable elements on chromosome 12 are not as great as that on chromosome 8. Further examination of recombination events of the crosses involved in rice cultivars that were not screened for *Pi-ta* mediated blast resistance, and sequence all genes on chromosome 12 in these cultivars may shed more light on the mechanisms of incompatibility between indica and japonica rice.

Understanding of LB on a particular chromosome is important for crop improvement. Rice cultivars with *Pi-ta* have been effective for preventing blast disease over two decades in the southern US [8,9]. This could be due to LB at the *Pi-ta* region where all components including *Ptr(t)* [25] for blast resistance have been inherited together. Thus far, *Pi-ta* and *Pi-ta2* are the only blast *R* genes in that region that are effective at preventing blast disease in the US [10,13,36] although it is unclear if *Pi-ta2* is present in all *Pi-ta* containing cultivars in the southern US. Cloning of *Pi-ta2* should help to address this question. In contrast, LB can be a bottleneck for crop improvement. Genes for desired traits are often embedded with genes for unwanted traits which can be problematic for conventional crop improvement. In fact, historically it has been observed that yield potentials of 11 rice cultivars with *Pi-ta* were not as high as some rice cultivars without *Pi-ta* in Arkansas, suggesting the negative factors for yield may have been inherited from the original donor for *Pi-ta* due to LB at the *Pi-ta* locus. A population with mono- and digenic lines of *Pi-ta, Pi-ks/h* that are segregating for yield components [37] and near isogenic line with *Pi-ta* developed by Jia and Martin [25] should help us to determine epistatic relationship between blast resistance and yield potential in the future.

In summary, using five random mapping populations of rice we demonstrated that LB around a blast resistance gene that was largely due to the incompatibility between *indica* and *japonica* rice. Identified recombinant lines defined recombination suppression will be useful for further sequence, and analysis of molecular differences between *indica* and *japonica* rice in order to understand molecular basis of recombination suppression, a phenomena that has been commonly observed in plant genetics and breeding.

Materials and Methods

Choice of Rice Genotype (Table 2)

"Early" (RU9101001) is an early maturing, cold tolerant tropical *japonica* rice line developed by University of Arkansas (Moldenhauer, unpublished data). This line has been used as an experimental line for marker development and genetic studies. Katy (PI 527707), released in 1990, is a tropical *japonica* cultivar that contains *Pi-ta* [37] and has been a donor for *Pi-ta* for rice breeding programs in the Southern US. Lemont (PI 475833) is a tropical *japonica* cultivar grown from the 1980s to 1990s in the southern US that does not possess *Pi-ta* [38]. Jasmine 85 (PI 595927), a long-grain aromatic *indica* cultivar lacking *Pi-ta* initially developed at the International Rice Research Institute (IRRI), in the Philippines and adapted to the southern US, was officially released in 1989 by the USDA-ARS and the Texas Agricultural Experiment Station [39]. Cocodrie is a tropical *japonica* cultivar with good yield potential and lacks *Pi-ta* [40]. MCR 01-0277 (PI 641932) is a breeding line developed in the US that contains considerable *indica* genome [41]. Zhe733 (PI 629016) is an *indica* cultivar with high yield, early maturity and effective blast resistance genes lacking *Pi-ta* from China. "172" is the *indica* cultivar Amane (*Pi-ta/Pi-ta*) (PI 373335) from Sri Lanka that contains *Pi-ta,* but is susceptible to an avirulent isolate (Jia, unpublished data). It was predicted that Amane contains a defective component [*ptr(t)/ptr(t)*] required for *Pi-ta*-mediated disease resistance [21,24]. The *indica* cultivar 172 contains *Pi-ta,* but is susceptible to the most of common races of blast [15] and was used as a parent crossed to Katy to develop a mapping population for detection of resistance to blast not due to *Pi-ta*, but due to the defect of additional critical component *Ptr(t)* [25]. A

Table 2. Description of rice cultivars and breeding lines used in this study.[1]

Name	Plant Identification	Characteristics	Existence of *Pi-ta*[3]	Reference
Early (RU9101001)	Unregistered	Tropical japonica line, early mature, blast susceptible	–	43
Katy	PI 527707	Tropical japonica cultivar with large introgression of indica from an indica landrace Tetep	+	37
Lemont	PI 475833	Tropical japonica cultivar	–	38
Jasmine 85	PI 595927	Indica	–	39
Cocodrie	PI 606331	Tropical japonica line, early mature		40
MCR01-0277	PI 641932	Indica like breeding line	–	41
Zhe 733	PI 629016	Early mature, blast resistant, high yielding indica	–	46, from China
"172"[2]	PI 37335	Indica	+	From Sri Lanka

[1]Characteristics of rice genotypes were verified using 1536 SNP markers [42].
[2]Cultivar 172 was susceptible to blast although it contains *Pi-ta* because lacking *Ptr(t)* [25].
[3]Indicates the absence, and + indicates the presence of resistant *Pi-ta* allele in each genotype as determined using method described in [23].

previous study genotyped all of these rice germplasm accessions, except MCR 01-0277, with 1536 SNP markers, and genomic contents of *indica* and *japonica* on rice chromosome 12 were verified by Zhao et al [42].

Mapping Populations

In order to reduce negative impact of recombination suppression and polymorphic marker density limitation of some parents, a total of five mapping populations derived from different indica and japonica and indica and indica crosses was used to investigate LB on rice chromosome 12. Five mapping populations, four of which were requested from the Genetic Stocks Oryza (GSOR) collection, USDA-ARS (http://www.ars.usda.gov/spa/dbnrrc/gsor) were used for this study: 1) Early/Katy: 87 individuals of an F_{10} recombinant inbred line (RIL) population derived from a cross of a tropical *japonica* line RU9101001 (*pi-ta/pi-ta*) with a tropical *japonica* cultivar Katy (*Pi-ta/Pi-ta*) [43]. About half (14 megabases) of chromosome 12 in Katy was derived from a landrace *indica* Tetep. 2) LJ: 93 individuals of an F_{10} RIL population derived from a cross of a tropical *japonica* Lemont (*pi-ta/pi-ta*) with an *indica* cultivar Jasmine 85 (*pi-ta/pi-ta*). 3) KZ: 89 individuals of an F_{10} RIL population of a cross of an *indica* Zhe733 (*pi-ta/pi-ta*) and tropical genetic stock *Kaybonnet-lpa1-1* (*Pi-ta/Pi-ta*) with the chromosome 12 identical to that in Katy [44]. 4) K172:94 individuals of an F_2 population derived from a cross of Katy [*Pi-ta/Pi-ta Ptr(t)Ptr(t)*] with '172' [*Pi-ta/Pi-ta ptr(t)ptr(t)*] from Sri Lanka where both contain different *indica* origins for chromosome 12 This population was developed by the Molecular Plant Pathology program of DB NRRC (Jia, unpublished data). 5) MCR: 93 doubled haploid lines of the SB2 mapping population derived from a cross of a tropical *japonica* cultivar Cocodrie (*pi-ta/pi-ta*) and an *indica* derived germplasm MCR 01-0277 (*pi-ta/pi-ta*) [41].

Plant Growth

Rice plants were grown in a greenhouse using sterilized local clay soils under conditions as previously described [45]. For RILs, seeds (5 to 10) from each line were directly sown on top of moistened soil in a small pot to produce 3 plants per line. Small pots with 4 holes at the bottom that allow roots to access water were placed in a tray with tap water. Leaves were bulked from 3 plants at the 3- to 4-leaf stage on ice for DNA extraction. For the F_2 population, seeds were planted into a 64 well flat and tissue was harvested from each plant at the 3- to 4-leaf stage. DNA was prepared as described in Liu et al [46].

SSR Analysis

To obtain dense coverage on chromosome 12, forty SSR markers were used to identify polymorphic markers from both parents and 28 polymorphic markers were identified for this study (Figure 2). The primer sequences and map position of the SSR markers were obtained from the Gramene Version 21 database (http://www.gramene.org). PCR reactions were performed in 25 μL reaction volumes consisting of 20 ng of DNA, 10 mM Tris–HCl pH 8.3, 50 mM KCl, 2.5 mM MgCl2, 300 nM of each primer, 1 U of Taq DNA polymerase (Promega, Madison, WI, USA).

For each marker analysis, the reverse primers were unlabeled to reduce the cost, and the forward primers were labeled with either 6FAM, PET, NED or Hex (Applied Biosystems, Foster City, CA, USA or Integrated DNA Technologies, Coralville, IA, USA). DNA were amplified with MJ Research Tetrad PCR machines (Waltham, MA, USA) under the following conditions: initial denaturation at 94°C for 5 min; then 30 cycles at 94°C for 30 s, 55–67°C (varies dependent on marker) for 30 s, and 72°C for 1 min; 5 min final extension at 72°C. PCR products were pooled based on size of amplified fragments (typically three markers per run along with LIZ-labeled size standard) to reduce cost, and the DNA was denatured by heating samples at 94°C for 5 min. The samples were separated on an ABI Prism 3730 DNA analyzer using the method as described by the manufacturer (Applied Biosystems). SSR fragment size was estimated using the software GeneMapper v 3.7 (Applied Biosystems).

Blast Evaluation and Marker for *Pi-ta* Mediated Blast Resistance

Disease reaction to blast was determined using a modified standard pathogenicity assay as previously described in Jia et al [45]. Specifically, rice seedlings at 3–4 leaf stage in a plastic bag were spray-inoculated with blast spore suspension at $1–5\times10^5$ spores/ML. After inoculation plastic bags were sealed to maintain high humidity for 24 h before removing out from bags. Subsequently, plants were maintained in a greenhouse for additional 6 days to allow the development of clear disease symptoms. The disease reaction was classified as 0–2 as resistance and 3–5 as susceptibility based on visual number and amount of lesions at the next of youngest leaf. In the K172 population $F_{2:3}$ progeny were also used to verify blast reactions [13] and marker segregation. SSR marker RM27941 in Katy was found to co-segregate perfectly with blast resistance [X. Wang and Y. Jia, (unpublished data)]. RM27941 was used as a marker for *Pi-ta* mediated blast resistance in the present study.

Construction of Linkage Maps

JoinMap Version 4.0 was used for linkage analysis [47]. A minimum LOD score of 3.0 was used to identify groups and maps were generated using Maximum likelihood mapping with default settings. The order of the markers on each chromosome was referred to the SSR marker database of Cornell SSR 2001 (Cornell2001) as described in Gramene, and by Conaway-Bormanset et al [48] and Jia et al [23]. SSR marker data were converted for JoinMap and are summarized in the (Table S1, S2, S3, S4, S5, S6, S7, S8, S9, S10, Table S11).

Supporting Information

Table S1 Note to mapping population.

Table S2 Mapping profile for Early Katy mapping population for JoinMap.

Table S3 SSR marker file for Early and Katy mapping population.

Table S4 Mapping profile for LJ mapping population for JoinMap.

Table S5 SSR marker file for LJ mapping population.

Table S6 Mapping profile for Cocodrie and MCR mapping population for JoinMap.

Table S7 SSR marker file for Cocodrie and MCR mapping population.

Table S8 Mapping profile for Katy and172 mapping population for JoinMap.

Table S9 SSR marker file for Katy and 172 mapping population.

Table S10 Mapping profile for KZ mapping population for JoinMap.

Table S11 SSR marker file for Zhe733 and Katy mapping population.

Acknowledgments

The authors thank Michael Lin for developing mapping populations and growing plants in the greenhouse and Eric Green for assisting SSR marker analysis and for other mapping populations that was distributed through GSOR. USDA is an equal opportunity provider and employer.

Author Contributions

Conceived and designed the experiments: YJ MHJ. Performed the experiments: YJ MHJ XW GL. Analyzed the data: MHJ YJ GL. Contributed reagents/materials/analysis tools: YJ GL MHJ XW. Wrote the paper: YJ MHJ GL XW.

References

1. Naveira H, Barbadilla A (1992) The theoretical distribution of lengths of intact chromosome segments around a locus held heterozygous with backcrossing in a diploid species. Genetics 130: 205–209.
2. Remington DL, Thornsberry JM, Matsuoka Y, Wilson LM, Whitt SR, et al. (2001) Structure of linkage disequilibrium and phenotypic associations in the maize genome. Proc Natl Acad Sci USA 98: 11479–11484.
3. Anderson LK, Doyle GG, Brigham B, Carter B, Hooker KD, et al. (2003) High resolution crossover maps for each bivalent of *Zea mays* using recombination nodules. Genetics 165: 849–865.
4. Beadle GW (1932) A possible influence of the spindle fibre on crossingover in *Drosophila*. Proc Natl Acad Sci USA 18: 160–165.
5. Lambie EJ, Roeder GS (1986) Repression of meiotic crossing over by a centromere (CEN3) in *Saccharomyces cerevisiae*. Genetics 114: 769–789.
6. Mather K (1993) Crossing over and heterochromatin in the X chromosome of *Drosophila melanogaster*. Genetics 24: 413–435.
7. Ballini E, Morel JB, Droc G, Price AH, Courtois B, et al. (2008) Genome wide meta analysis of rice blast resistance genes and quantitative trait loci provides new insights into partial and complete resistance. Mol Plant Microbe Interact 21: 859–868.
8. Jia Y, Liu G, Costanzo S, Lee S, Dai Y (2009) Current progress on genetic interactions of rice with rice blast and sheath blight fungi. Front of Agri in China 3: 231–239.
9. RoyChowdhury M, Jia Y, Cartwright RD (2012) Structure, Function and Co-evolution of Rice Blast Resistance Genes. Acta Agronomica Sinica 38: 381–393.
10. Bryan GT, Wu K, Farrall L, Jia Y, Hershey HP, et al. (2000) A single amino acid difference distinguishes resistant and susceptible alleles of the rice blast resistance gene *Pi-ta*. Plant Cell 12: 2033–2045.
11. Orbach MJ, Farrall L, Sweigard JA, Chumley FG, Valent B (2000) A telomeric avirulence gene determines efficacy for the rice blast resistance gene *Pi-ta*. Plant Cell 12: 2019–2032.
12. Jia Y, McAdams SA, Bryan GT, Hershey HP, Valent B (2000) Direct interaction of resistance gene and avirulence gene products confers rice blast resistance. EMBO J 19: 4004–4014.
13. Jia Y, Wang Z, Fjellstrom RG, Moldenhauer KAK, Azam MA, et al. (2004) Rice *Pi-ta* gene confers resistance to the major pathotypes of the rice blast fungus in the United States. Phytopathology 94: 296–301.
14. Zhou E, Jia Y, Correll J, Lee FN (2007) Instability of the *Magnaporthe oryzae* avirulence gene *AVR-Pita* alters virulence. Fungal Genet Biol 10: 1024–1034.
15. Wang X, Fjellstrom R, Jia Y, Yan W-G, Jia MH, et al. (2010) Characterization of *Pi-ta* blast resistance gene in an international rice core collection. Plant Breed 129: 491–501.
16. Huang C, Hwang S, Chiang Y, Lin T (2008) Molecular evolution of the *Pi-ta* gene resistant to rice blast in wild rice (*Oryza rufipogon*). Genetics 179: 1527–1538.
17. Lee S, Costanzo S, Jia Y, Olsen K, Caicedo A (2009) Evolutionary dynamics of the genomic region around the blast resistance gene *Pi-ta* in AA genome *Oryza* species. Genetics 183: 1315–1325.
18. Lee S, Jia Y, Jia MH, Gealy D, Olsen KM, et al. (2011) Molecular evolution of the rice blast resistance gene *Pi-ta* in invasive weedy rice in the USA. PLoS ONE 6: 10.
19. Costanzo S, Jia Y (2009) Alternatively spliced transcripts of *Pi-ta* blast resistance gene in *Oryza sativa*. Plant Sci 177: 468–478.
20. Dai Y, Jia Y, Correll J, Wang X, Wang Y (2010) Diversification and evolution of the avirulence gene *AVR-Pita1* in field isolates of *Magnaporthe oryzae*. Fungal Genet and Biol 47: 974–980.
21. Wang X, Jia Y, Shu Q, Wu D (2008) Haplotype diversity at the *Pi-ta* locus in cultivated rice and its wild relatives. Phytopathology 98: 1305–1311.
22. Jia Y, Wang Z, Singh P (2002) Development of dominant rice blast resistance *Pi-ta* gene markers. Crop Sci 42: 2145–2149.
23. Jia Y, Redus M, Wang Z, Rutger JN (2004) Development of a SNLP marker from the *Pi-ta* blast resistance gene by tri-Primer PCR. Euphytica 138: 97–105.

24. Jia Y (2009) Artificial introgression of a large fragment around the *Pi-ta* rice blast resistance gene in backcross progenies and several elite rice cultivars. Heredity 103: 333–339.
25. Jia Y, Martin R (2008) Identification of a new locus, *Ptr(t)* required for rice blast resistance gene *Pi-ta* mediated resistance. Mol Plant Microbe Interact 21: 396–403.
26. Kovach MJ, Sweeney MT, McCouch SR (2007) New insights into the history of rice domestication. Trends Genet 23: 578–587.
27. Xu X, Liu X, Ge S, Jensen JD, Hu F, et al. (2011) Resequencing 50 accessions of cultivated and wild rice yields markers for identifying agronomically important genes. Nature Biot 30: 105–111.
28. Londo JP, Chiang YC, Hung KH, Chiang TY, Schaal BA (2006) Phylogeography of Asian wild rice, *Oryzae rufipogon*, reveals multiple independent domestications of cultivated rice, *Oryza sativa*". Proc Natl Acad Sci USA 103: 9578–83.
29. Vaughan DA, Lu BR, Tomooka N (2008) The evolving story of rice evolution. Plant Sci 174: 394–408.
30. Henikoff S, Ahmad K, Malik HS (2001) The centromere paradox: stable inheritance with rapidly evolving DNA. Science 293: 1098–1102.
31. Ma J, Wing RA, Bennetzen JL, Jackson SA (2007) Plant centromere organization: a dynamic structure with conserved functions. Trends in Genet 23: 134–139.
32. Metkar S, Sains JK, Mahajan SK (2004) Cloning and characterization of the *DMC1* genes in *Oryza sativa*. Current Sci 87: 353–357.
33. Jain M, Tyagi SB, Thakur JK, Tyagi AK, Khurana JP (2004) Molecular characterization of a light-responsive gene, breast basic conserved protein 1 (*OsiBBC1*), encoding nuclear-localized protein homologous to ribosomal protein L13 from *Oryza sativa indica*. Biochim Biophys Acta 1676: 182–192.
34. Jain M, Tyagi AK, Khurana JP (2008) Constitutive expression of a meiotic recombination protein gene homolog, OsTOP6A1, from rice confers abiotic stress tolerance in *Arabidopsis* plants. Plant Cell Rep 27: 767–778.
35. Tian Z, Yu Y, Lin F, Yu Y, Sanmiguel PJ, et al. (2011) Exceptional lability of a genomic complex in rice and its close relatives revealed by interspecific and intraspecific comparison and population analysis. BMC Genomics 12: 142.
36. Nakamura S, Asakawa S, Ohmido N, Fukui K, Shimizu N, et al. (1997) Construction of an 800-kb contig in the near-centromeric region of the rice blast resistance gene *Pi-ta2* using a highly representative rice BAC library. Mol Gen Genet 254: 611–620.
37. Moldenhauer KAK, Lee FN, Norman RJ, Helms RS, Well RH, et al. (1990) Registration of 'Katy' rice. Crop Sci 30: 747–748.
38. Bollich CN, Webb BD, Marchetti MA, Scott JE (1985) Registration of 'Lemont' rice. Crop Sci 25: 883–885.
39. Marchetti MA, Bollich CN, Webb BD, Jackson BR, McClung AM, et al. (1998) Registration of 'Jasmine 85′ rice. Crop Sci 38: 896.
40. Linscombe SD, Jodari F, Bollich PK, Groth DE, White LM, et al. (2000) Registration of 'Cocodrie' rice. Crop Sci 40: 294.
41. Chu QR, Linscombe SD, Rush MC, Groth DE, Oard J, et al. (2006) Registration of a C/M doubled haploid mapping population of rice. Crop Sci 46: 1417.
42. Zhao K, Wright MK, Kimball JA, Eizenga GC, McClung AM, et al. (2010) Genomic diversity and introgression in *O. sativa* reveal the impact of domestication and breeding on the rice genome. PLoS ONE 5: e10790. Doi10.1371/journal.pone.010780.
43. Jia Y, Moldenhauer KAK (2010) Development of monogenic and digenic rice lines for blast resistance genes *Pi-ta, Pi-kh/Pi-ks*. J of Plant Reg 4: 163–166.
44. Rutger JN, Tai TH (2005) Registration of K/Z mapping population of rice Crop Sci 45: 2671–2672.
45. Jia Y, Valent B, Lee FN (2003) Determination of host responses to *Magnaporthe grisea* on detached rice leaves using a spot inoculation method. Plant Dis 87: 129–133.
46. Liu G, Bernhardt J, Jia MH, Wamishe Y, Jia Y (2008) Molecular characterization of rice recombinant inbred line population derived from a japonica-indica cross. Euphytica 159: 73–82.
47. Van Ooijen JW (2006) Joinmap 4.0, Software for the calculation of genetic linkage map in experimental population. Kyazma BV, Wageningen, Netherlands.
48. Conaway-Bormans CA, Marchetti MA, Johnson CW, McClung AM, Park WD (2003) Molecular markers linked to the blast resistance gene *Pi-z* in rice for use in marker assisted selection. Theor Appl Genet 107: 1014–1020.

8

An International Reference Consensus Genetic Map with 897 Marker Loci Based on 11 Mapping Populations for Tetraploid Groundnut (*Arachis hypogaea* L.)

Bhimana Gautami[1,¶], Daniel Fonceka[2,¶], Manish K. Pandey[1,3,4], Márcio C. Moretzsohn[5], Venkataswamy Sujay[1,6], Hongde Qin[3,7], Yanbin Hong[8], Issa Faye[9], Xiaoping Chen[8], Amindala BhanuPrakash[1], Trushar M. Shah[1], Makanahally V. C. Gowda[6], Shyam N. Nigam[1], Xuanqiang Liang[8], Dave A. Hoisington[1], Baozhu Guo[4], David J. Bertioli[10], Jean-Francois Rami[2], Rajeev K. Varshney[1,8,11]*

1 Center of Excellence in Genomics, International Crops Research Institute for the Semi-Arid Tropics (ICRISAT), Patancheru, India, **2** UMR Développement et Amélioration des plantes, Centre de coopération Internationale en Recherche Agronomique pour le Développement (CIRAD), Montpellier, France, **3** Department of Plant Pathology, University of Georgia (UGA), Tifton, Georgia, United States of America, **4** Crop Protection and Management Research Unit, USDA-Agricultural Research Service, Tifton, Georgia, United States of America, **5** Plant Genetics Lab, EMBRAPA Genetic Resources and Biotechnology, Brasilia, Brazil, **6** Department of Genetics and Plant Breeding, University of Agricultural Sciences (UAS-D), Dharwad, India, **7** Cash Crop Research Institute, Hubei Academy of Agricultural Sciences (HAAS), Wuhan, Hubei, China, **8** Crops Research Institute, Guangdong Academy of Agricultural Sciences (GAAS), Guangzhou, Guangdong, China, **9** Centre National de Recherche Agronomique (CNRA), Institut Sénégalais de Recherches Agricoles (ISRA), Bambey, Sénégal, **10** Genetics Department, University of Brasilia, Brasilia, Brazil, **11** Theme- Comparative and Applied Genomics, CGIAR Generation Challenge Programme (GCP), CIMMYT, Mexico, Mexico

Abstract

Only a few genetic maps based on recombinant inbred line (RIL) and backcross (BC) populations have been developed for tetraploid groundnut. The marker density, however, is not very satisfactory especially in the context of large genome size (2800 Mb/1C) and 20 linkage groups (LGs). Therefore, using marker segregation data for 10 RILs and one BC population from the international groundnut community, with the help of common markers across different populations, a reference consensus genetic map has been developed. This map is comprised of 897 marker loci including 895 simple sequence repeat (SSR) and 2 cleaved amplified polymorphic sequence (CAPS) loci distributed on 20 LGs (a01–a10 and b01–b10) spanning a map distance of 3, 863.6 cM with an average map density of 4.4 cM. The highest numbers of markers (70) were integrated on a01 and the least number of markers (21) on b09. The marker density, however, was lowest (6.4 cM) on a08 and highest (2.5 cM) on a01. The reference consensus map has been divided into 20 cM long 203 BINs. These BINs carry 1 (a10_02, a10_08 and a10_09) to 20 (a10_04) loci with an average of 4 marker loci per BIN. Although the polymorphism information content (PIC) value was available for 526 markers in 190 BINs, 36 and 111 BINs have at least one marker with >0.70 and >0.50 PIC values, respectively. This information will be useful for selecting highly informative and uniformly distributed markers for developing new genetic maps, background selection and diversity analysis. Most importantly, this reference consensus map will serve as a reliable reference for aligning new genetic and physical maps, performing QTL analysis in a multi-populations design, evaluating the genetic background effect on QTL expression, and serving other genetic and molecular breeding activities in groundnut.

Editor: Shu-Biao Wu, University of New England, Australia

Funding: Financial support from Generation Challenge Programme (http://www.generationcp.org) of Consultative Group on International Agricultural Research (CGIAR) is gratefully acknowledged. The funders had no role in study design, data collection and analysis, decision to publish, or preparation of the manuscript.

Competing Interests: The authors have declared that no competing interests exist.

* E-mail: r.k.varshney@cgiar.org

¶ These authors contributed equally to this work.

Introduction

Dense genetic linkage maps are cornerstones for wide spectrum of genetics and breeding applications such as linkage mapping or association analysis based trait mapping, marker-assisted breeding, map-based cloning and physical map alignment. In general, it is possible to map only limited number of molecular markers in a given mapping population due to polymorphism constraints. As a result, several mapping populations are used for developing different genetic maps so that maximum number of marker loci available are mapped in the given crop species. Subsequently, with an objective to increase the number of mapped marker loci, genetic maps developed for different mapping populations are used for developing a consensus map. As compared to individual genetic maps, consensus maps offer several advantages such as: (i) mapping of a large number of marker loci onto a single map, (ii) determining relative position of common markers across the mapping populations, (iii) determining stability of marker locus position across the genome, (iv) provides evidence for chromosomal rearrangements [1,2], gene duplication [2,3] and assists in the assignment of linkage groups to chromosome [1], (v) provides the basic information for comparative genomic studies among related species and subspecies [2–4] and (vi) provides genetic information for greater genomic coverage [5]. Because of above mentioned features, consensus genetic maps have been developed in many crop species like maize [6,7], wheat [8] barley, [9,10], soybean [11,12] and pigeonpea [13].

Groundnut or peanut (*Arachis hypogaea* L.), an economically important oil seed crop, is cultivated mostly in semi-arid regions of the world. It is an allotetraploid ($2n=4x=40$) with a large genome size 2800 Mb/1C. Based on the origin complexity such as polyploidy nature, narrow genetic base with very low DNA polymorphism in cultivated tetraploid groundnuts, initially genetic maps were developed for AA- genome [14–16] and BB- genome [17,18]. Only recently a few mapping populations have been used for developing the genetic maps for cultivated groundnut species [19–22] or based on cross of cultivated and synthetic tetraploid groundnut species [23]. In some cases, consensus genetic maps based on two or three mapping populations have also been developed [24–27]. The most dense consensus genetic map developed so far is based on two mapping populations and is comprised of 324 SSR loci [27]. However because of availability of >4000 SSR markers in *Arachis* species [28], international *Arachis* community has been striving towards developing a consensus genetic map compiling a maximum number of genetic markers especially when efforts have been initiated to sequence the genome of *Arachis* species (http://www.peanutbioscience.com/peanutgenomeproject.html).

Keeping in view of above, this article reports assembling of SSR marker genotyping data for 11 mapping populations including 10 recombinant inbred lines (RILs) and one backcross (BC) population. These genotyping data have been used to develop a consensus genetic map with 895 SSR marker loci and 2 CAPS loci. For enhancing the utility of the consensus genetic map, the map has been divided into 20 cM long 203 BINs and the polymorphism information content (PIC) values for the markers, wherever possible, present in these BINs have also been presented.

Results

High-quality Marker Segregation Data

Marker segregation data were assembled for a total of 1961 markers ranging from 64 markers (RIL-8) to 339 markers (BC-1) per population (Table 1). A chi-square test was performed on marker genotyping data for individual mapping population to test the null hypothesis of segregation ratios of 1:1 at the threshold of $p=0.05$. A variable percentage of distorted markers ranging from 3.45% (RIL-8) to 52.34% (RIL-2) were observed for individual mapping populations.

Individual or Component Genetic Maps

The genotyping data obtained on 11 mapping populations (1961 markers) were used for constructing the component genetic maps for the respective mapping population using MAPMAKER/EXP V 3.0 [29]. All developed component genetic maps can be visualized in CMap database at http://cmap.icrisat.ac.in/cmap/sm/gn/gautami/. The numbers of marker loci ranged from 46 (RIL-8) to 332 (BC-1) per component genetic maps for different mapping populations. Genetic map distance covered from 357.4 cM (RIL-8) to 2208.2 cM (RIL-2) with a range of map density from 2.5 cM (BC-1) to 18.6 cM (RIL-2) (Table 2). As several markers integrated into component maps have segregation distortion, the linkage group (LG)-wise segregation pattern of markers in each mapping population has been shown in Figure S1.

Reference Consensus Genetic Map

A reference consensus genetic map was constructed by integrating all 11 component genetic maps using common markers across different genetic maps using MergeMap program. While integrating component genetic maps, some discrepancies were observed in the names of markers for which genotyping data were available on more than one mapping population. However, to facilitate integration, uniformity in marker nomenclature was maintained for all the markers. For example, 'pPGPseq xx' and pPGSseqxx' were changed to 'seqxx', and 'XIPxx' was changed to 'IPAHMxx' to maintain uniformity in names of marker loci.

Based on the common markers and the comparison between component genetic maps, most of the linkage groups were consistent among the individual maps with few exceptions which can be visually assessed from http://cmap.icrisat.ac.in/cmap/sm/gn/gautami/(also see Table S1). A total of 542 markers were unique markers i.e. mapped only in one mapping population, while the remaining 355 markers were common, i.e. they were mapped in at least two mapping populations (187 markers were common between two maps, 72 between three maps, for four maps 57 are common, 20 markers are common between 5 maps, between 6 maps 13 markers are common, 3 markers are common between 7 maps, 2 markers between 8 maps and 1 marker is common between 9 maps) and served as anchor points for the map integration (Table 3). The grouping of different LGs from component genetic maps to develop the consensus map were given in Table S2. Therefore, in the consensus genetic map, a total of 39.58% (355) markers are anchor markers present on all 20 LGs. The remaining 60.42% (542) markers are the markers which are unique to the individual mapping populations.

Multiple segregating fragments (loci) identified with one SSR primer pair were designated with one lower case letter "a", "b" or "c" as suffix with the name of marker. For example two loci mapped on LG_AhVII and LG_AhXVII by using the same marker (RIL-1), were designated as IPAHM108a and IPAHM108b loci (Table S1).

Seventy homeologous loci were identified on "a" and "b" linkage groups (Figure 1), which facilitate the detection of ten homeologous pair named from a01 to a10 and b01 to b10 based on the same loci detected on the framework map (BC-1 in the present study) developed by Foncéka et al. [23]. Out of these 70 homeologous loci, 11 loci were located between the group a01–b01 and a03–b03, followed by eight loci between a02–b02 and a04–b04 and four loci between a09–b09. Except for the groups between a01–b01, a03–b03 and a04–b04 markers order and inter-loci map distance were well conserved between homeologous groups (Figure 1).

In some cases, the same marker mapped single locus on different linkage groups in different mapping populations, these loci were not considered as the same loci and were included as unique loci (with the same name) in the reference consensus genetic map. Twenty nine (26.13%) primer pairs detected duplicated non-homeologous loci between linkage groups (e.g., seq12F07 detected two loci, one on a01 and one on a10; IPAHM524 detected two loci, one on b02 and one on b06 and IPAHM171 detected three loci on a06, b01 and b08) (Figure 1, Table S1).

Although it was planned to include only SSR marker loci in the reference consensus genetic map, two CAPS (cleaved amplified

Table 1. Source of marker data used for constructing the reference consensus genetic map.

Mapping populations	Population type	Population size	Genotyping data assembled (no. of loci)	Trait segregation
International Crops Research Institute for the Semi-Arid Tropics (ICRISAT), Patancheru, India				
TAG 24 × ICGV 86031 (RIL-1)	RIL (F_8)	318	211	Drought tolerance related traits
ICGS 76 × CSMG 84–1 (RIL-2)	RIL (F_9)	176	128	Drought tolerance related traits
ICGS 44 × ICGS 76 (RIL-3)	RIL (F_8)	188	87	Drought tolerance related traits
University of Agricultural Sciences-Dharwad (UAS-D), Dharwad, India				
TAG 24 × GPBD4 (RIL-4)	RIL (F_7)	266	209	Late leaf spot and rust resistance
TG 26 × GPBD 4 (RIL-5)	RIL (F_7)	146	209	Late leaf spot and rust resistance
Guangdong Academy of Agricultural Sciences (GAAS), China				
Yueyou 13 × Zhen Zhuhei (RIL-6)	RIL ($F_{4:6}$)	142	146	Protein content
Yueyou 13 × FU 95–5 (RIL-7)	RIL ($F_{4:6}$)	84	124	Oil content
Yueyou 13 × J 11 (RIL-8)	RIL ($F_{4:6}$)	136	64	Resistance to *Aspergillus flavus* and aflatoxin contamination
University of Georgia (UGA), Tifton, USA				
Tifrunner × GT-C20 (RIL-9)	RIL ($F_{2:6}$)	248	261	Tomato spotted wilt virus (TSWV) resistance and several agronomic traits
SunOleic 97R × NC94022 (RIL-10)	RIL ($F_{2:6}$)	352	197	Tomato spotted wilt virus (TSWV) resistance and several agronomic traits
Centre de coopération Internationale en Recherche Agronomique pour le Développement (CIRAD), Montpellier, France				
Fleur11 × AiAd (synthetic amphidiploid) (BC-1)	BC_1F_1	88	339	Several agronomic traits

polymorphic sequence) markers i.e., ahFAD2A and ahFAD2B, due to their association with high oleic acid to linoleic acid ratios (high O/L) [30], very important trait in groundnut, were also integrated in the reference consensus genetic map.

In summary, the reference consensus map is comprised of 895 SSR and 2 CAPS loci distributed over 20 LGs. Nomenclature of LGs in the reference consensus map was given in the same way as in the framework map (BC-1 in the present study) developed by Foncéka et al [23]. The map density in the reference consensus map ranged from 2.5 cM (a01) to 6.4 cM (a08) with an average of 4.3 cM per marker. The inter-locus gap distance ranged from 1.5 cM (a01) to 5.4 cM (a08), with a mean value of 4.5 cM per marker (Table 4). Among the 20 LGs, a01 possess maximum marker loci with 70 loci followed by a03, a05 and b03 with 65, 61 and 60 loci respectively, while a02 and b09 have only 23 and 21 loci, respectively (Figure 1, Table 4). The low number of SSR loci mapped on a02 and b10 may be related to the lack of polymorphism on these two LGs. For example, the consensus LG a02 is built with seven LGs of the different component genetic maps, among which four LGs have only two mapped loci. For these small LGs additional genetic markers are needed for increasing the map density.

Table 2. Features of the component and reference consensus genetic maps.

Maps	Linkage groups	Mapped loci	Map length (cM)	Map density (cM)	Inter-locus gap distance (cM)	References
RIL-1	22	191	1785.4	9.4	9.4	Varshney et al. 2009b, Ravi et al. 2011
RIL-2	20	119	2208.2	18.6	18.7	Gautami et al. 2012
RIL-3	15	82	831.4	10.1	10.3	Gautami et al. 2012
RIL-4	20	188	1922.4	10.2	10.3	Khedikar et al. 2010, Sujay et al. 2012
RIL-5	21	181	1963.0	10.8	10.9	Sarvamangala et al. 2011, Sujay et al. 2012
RIL-6	19	133	793.1	6.0	6.1	Hong et al. 2010
RIL-7	21	109	503.1	4.6	4.7	Hong et al. 2010
RIL-8	13	46	357.4	7.7	7.9	Hong et al. 2010
RIL-9	26	233	1304.9	5.6	5.6	Qin et al. 2012
RIL-10	22	193	917.5	5.3	5.4	Qin et al. 2012
BC-1	21	332	847.4	2.5	2.6	Foncéka et al. 2009
Reference consensus genetic map	20	897	3863.6	4.4	4.5	–

Table 3. Summary of number of loci common between genetic maps for different mapping populations.

S.No	Mapping population	No. of mapped loci	No. of mapped loci used in consensus map	Number of markers in common with *n* other mapping populations								
				n=0	*n*=1	*n*=2	*n*=3	*n*=4	*n*=5	*n*=6	*n*=7	*n*=8
1	RIL-1	191	178	55	36	35	27	11	8	3	2	1
2	RIL-2	119	81	39	12	7	11	5	4	2	1	0
3	RIL-3	82	72	18	14	16	8	9	4	1	1	1
4	RIL-4	188	176	19	67	28	32	16	9	2	2	1
5	RIL-5	181	168	17	72	23	31	12	8	3	1	1
6	RIL-6	133	114	27	28	18	17	10	9	3	1	1
7	RIL-7	109	96	12	30	14	20	7	9	2	1	1
8	RIL-8	46	36	10	4	7	6	4	3	0	1	1
9	RIL-9	233	194	85	43	19	23	12	7	2	2	1
10	RIL-10	193	145	51	40	19	18	7	7	1	2	0
11	BC-1	332	324	209	28	30	35	7	10	2	2	1
	Total			542	187	72	57	20	13	3	2	1

In the consensus map, some gaps are observed on the distal ends of the a02, b02, a03, a05, b05, a08, a09, b09 and a10. Of the 897 mapped loci, 32.33% (290 loci) of the marker intervals were smaller than 1 cM while 41.14% (369 loci) marker intervals were between 1–5 cM, 15.94% (143 loci) between 5–10 cM, 7.36% (66 loci) between 10–20 cM, and 3.23% (29 loci) marker intervals were greater than 20 cM.

Added Value Features of the Reference Consensus Genetic Map

As SSR markers are the marker of choice in breeding applications, an attempt was made to understand the distribution of different SSR motifs as well as the polymorphism information content (PIC) values for these markers.

Out of 895 SSR loci integrated into the reference consensus map, information on repeat motifs was available for 788 SSR loci. Of the 788 SSRs, 612 SSR loci represent simple repeat motifs and 176 SSR loci contain compound repeat motifs. Among simple repeat motifs contained SSR loci, 375 SSR loci (47.58%) are comprised of di- (NN) repeats followed by 226 (28.70%) tri-nucleotides (NNN) repeats. The longer repeat classes, i.e. tetra- (NNNN, 8 loci) and hexa-nucleotide (NNNNNN, 3 loci) represented 1.39% of the SSR loci (Table S3). In the case of the compound repeats containing SSR loci, 93 loci were comprised of NN repeats and the remaining 83 loci comprised with mixed repeats.

Of the 897 mapped marker loci, the information on PIC values was available for 526 SSR marker loci from the studies in which the corresponding SSR loci were mapped (Table S3). Based on genotypes surveyed in those earlier studies, 144 marker loci have >0.50 PIC value. Majority of the loci (181) have 0.31–0.40 PIC value (Figure S2). Average PIC value of individual LGs varied from 0.55 (a02) to 0.81 (a01).

For making the consensus map more informative, an attempt has been made to divide the genetic map in 20 cM long BINs. As a result, the reference groundnut genetic map has a total of 203 BINs ranging from 5 (a02 and b06) to 16 (b01) with an average of 4 per linkage group. These BINs carry 1 (a10_02, a10_08 and a10_09) to 20 (a10_04) with an average of 4.41 marker per BIN. In terms of selecting highly informative SSR markers based on available PIC values, 36 BINs have at least one marker that has >0.70 PIC value and 111 BINs carry at least one marker with >0.50 PIC value. 166 BINs have the marker loci with <0.50 PIC value and 23 BINs do not have the information available on PIC values. 13 BINs do not have any marker.

Finally, for deciphering the relationships between LGs of the different component maps, we have identified a total of 58 genome specific SSR markers. These markers will be of great interest for subgenome assignment of SSR loci in cultivated x cultivated mapping studies. These markers could also be used in diversity analysis as they give access to the diversity at the diploid genome level allowing differentiating the structural heterozygosity linked to polyploidy from true heterozygosity.

Relationships of the Reference Genetic Map and Component Maps

As the reference map was developed based on the common marker loci mapped in the different component genetic maps using the same nomenclature of LGs, there was a good congruence except a few exceptions between marker orders and positions among component maps and the reference consensus map (http://cmap.icrisat.ac.in/cmap/sm/gn/gautami/and also see in Table S1). Comparison of a03 and b08 for all the component genetic maps and the reference consensus map, for example, has been shown in Figure 2.

Comparison with Diploid Genetic Maps

The results of the reference consensus genetic map were compared with the diploid AA and BB maps published earlier [15,18]. The LGs of the reference consensus map in this study are named according to the LGs named in Foncéka et al. [23] (a01 to a10 and b01 to b10). In these maps, LGs of AA and BB genome map were named as Group 1 to Group 11 and B1 to B10 respectively. The synteny study between the reference consensus map and AA map assessed 68 common SSR marker loci and between BB map assessed 43 common SSR marker loci (Table S4). However, for all the ten LGs of the present constructed reference consensus genetic maps, overall good

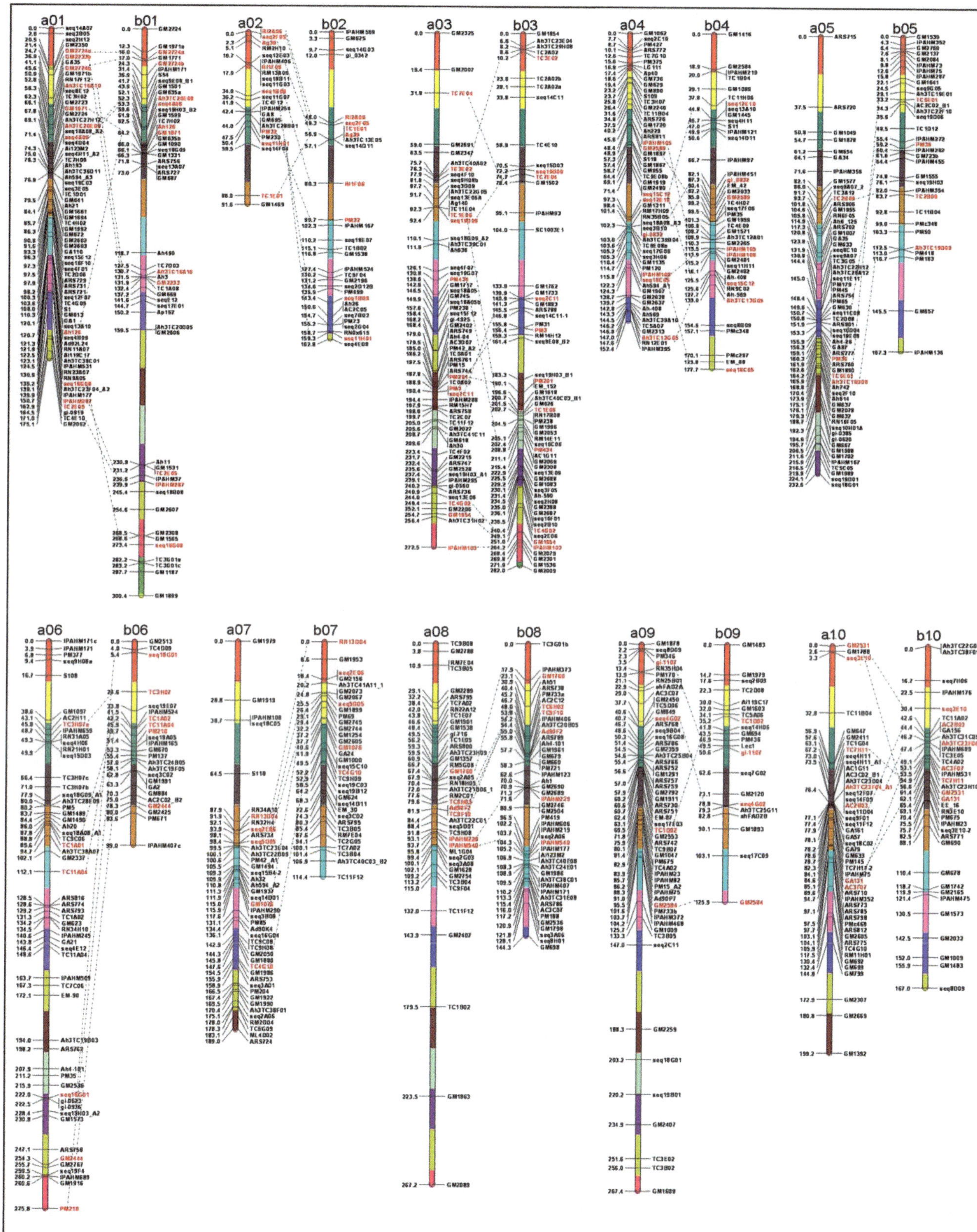

Figure 1. A microsatellite consensus genetic map comprising 897 marker loci based on 11 mapping populations. Markers are shown on *right* side of the LG while map distances are shown on the *left* side. Each LG has been divided into 203 BINs of 20 cM each. The homoeologous loci between the corresponding LGs in the reference consensus map are indicated in red colour.

Table 4. Features of the reference consensus genetic map.

LGs	No. of mapped loci	Map distance (cM)	Map density (cM)	Inter-locus gap distance (cM)
a01	70	175.1	2.5	2.5
b01	51	300.4	6.0	6.0
a02	23	91.6	4.0	4.2
b02	30	162.8	5.4	5.6
a03	65	272.5	4.1	4.3
b03	60	282.0	4.7	4.8
a04	56	152.4	2.7	2.8
b04	42	177.7	4.2	4.3
a05	61	232.6	3.8	4.0
b05	33	167.3	5.1	5.2
a06	57	275.8	4.8	5.0
b06	24	99.0	4.1	4.3
a07	43	189.0	4.4	5.0
b07	34	114.4	3.4	3.5
a08	42	267.2	6.4	6.5
b08	47	144.3	3.1	3.1
a09	56	267.4	4.8	5.0
b09	21	125.9	6.0	6.3
a10	47	199.2	4.2	4.3
b10	35	167.0	4.8	5.0
Total	897	3863.6	–	–
Mean	45	193.2	4.4	4.5

collinerity was observed for the corresponding LGs of the two diploid maps, with a few exceptions in some marker positions. The salient features of the comparison of the reference consensus genetic map with AA and BB maps for six LGs each are shown in the Figure 3. The number of common SSR

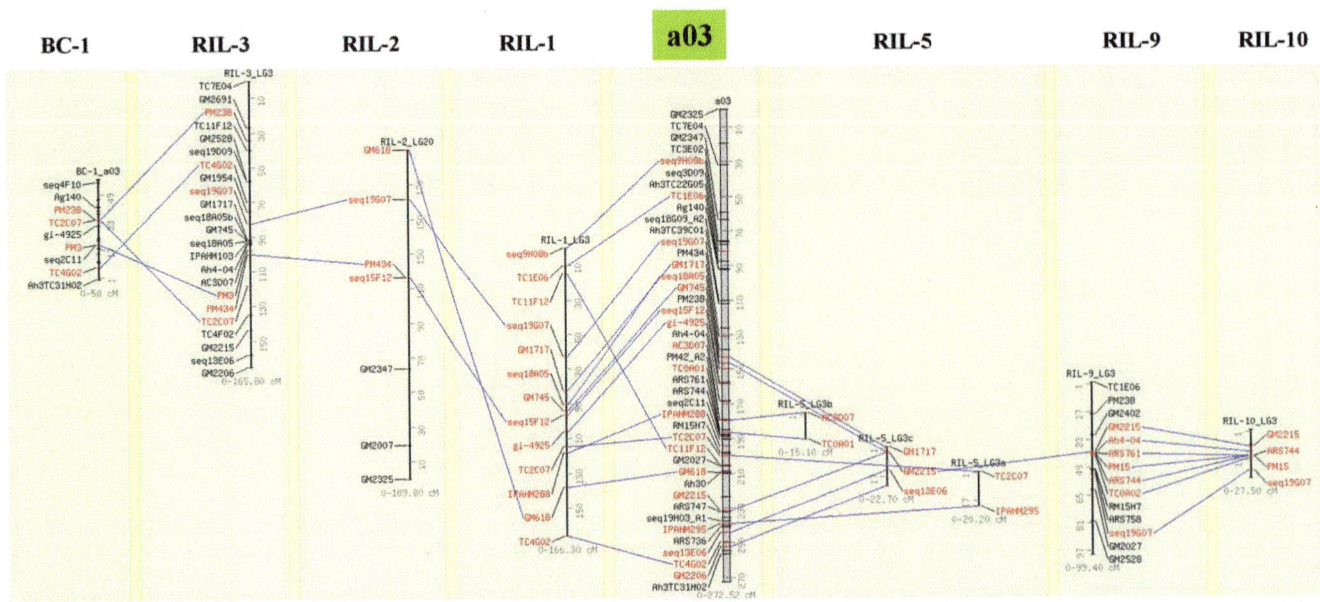

Figure 2. A marker based correspondence for a03 among reference consensus and individual genetic maps. Only common markers i.e. landmarks are included to visually asses the co-linearity of marker orders and marker positions. LGs are aligned together using comparative mapping program CMap version 1.01. Figure can also be accessed from http://cmap.icrisat.ac.in/cmap/sm/gn/gautami/.

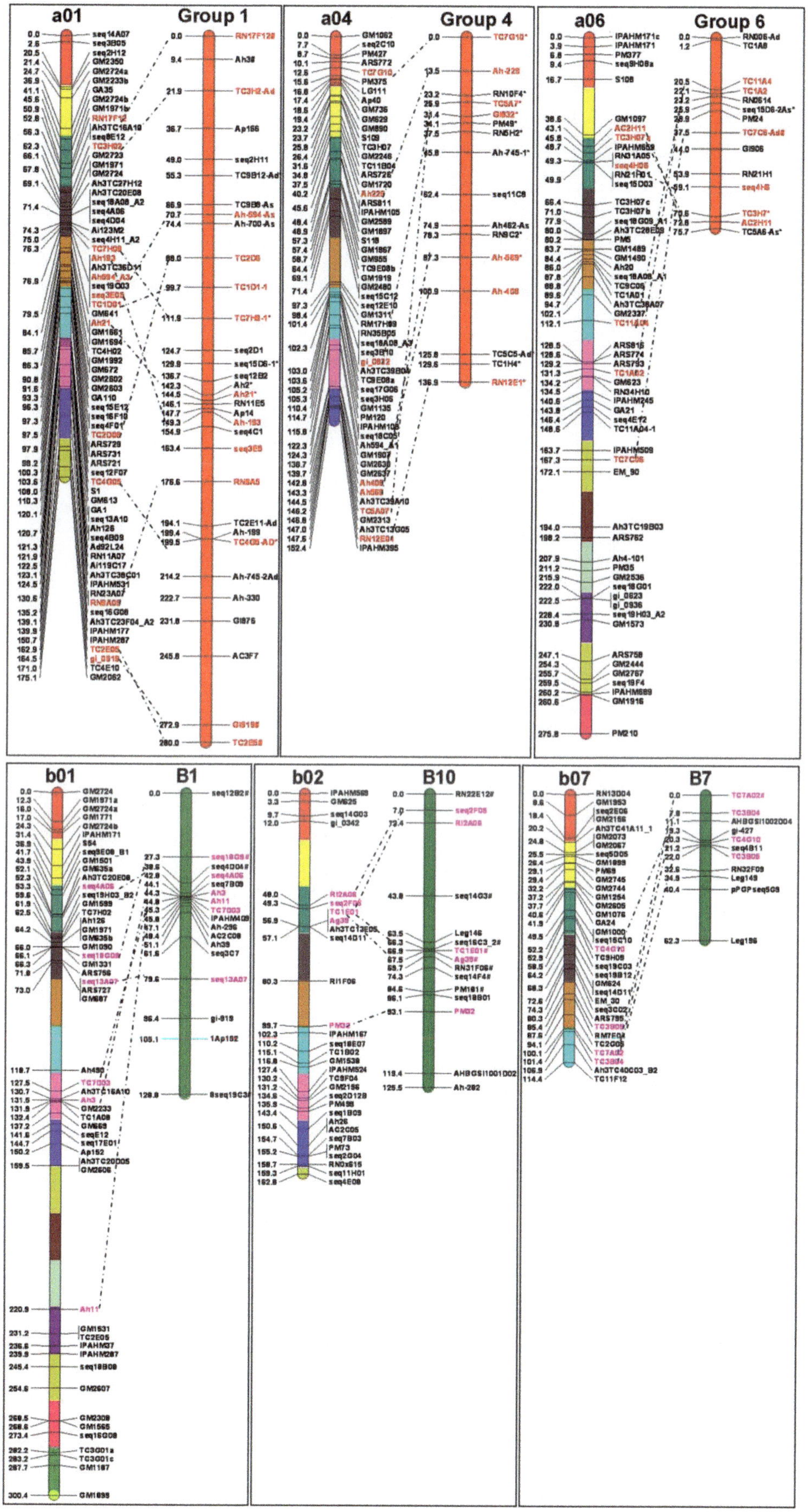
a01
Group 1
a04
Group 4
a06
Group 6
b01
B1
b02
B10
b07
B7

Figure 3. Comparison between the LGs of the reference consensus map and the diploid AA and BB maps. The LGs of the reference consensus map are represented as a01 to a10 and b01 to b10. The LGs of AA map are named as Group 1 to Group 11 and for BB map as B1 to B10 respectively (published by Moretzsohn et al 2005, 2009). The AA map was represented by a red bar and the BB map with green colour. The common markers between corresponding LGs in the reference consensus map and AA map are indicated in red colour and pink colour with BB map.

marker loci per homologous LGs varied between 2 and 10 with AA map and with BB map between 1 and 9.

Discussion

Significant progress has been made during last decade in high throughput genotyping and various linkage mapping tools to place a large number of marker loci on genetic maps in several crop species [8,13,31–33]. In the case of tetraploid groundnut, genetic mapping efforts have been initiated only recently and few genetic maps with 46–332 marker loci have been developed [34]. To enhance the marker density, a few consensus maps have also been developed using 2–3 mapping populations and the mapped marker loci on these maps has not gone beyond 324 loci. The major objective of this study, therefore, was the development of a dense consensus genetic map that can be used as a reference map by the international groundnut community.

Dense genetic maps can be developed mainly by using two approaches: (a) map maximum number of marker loci using highly diverse population, (b) merge available genetic maps using common markers mapped across the populations. While the first approach is quite challenging and laborious but precise, the second approach was used in the present study. In this context, segregation data for a total of 1961 marker loci generated for 11 (10 RIL and 1 BC) populations were assembled from different organizations. As a first step, component genetic maps were developed for all 11 populations. While comparing the component genetics maps developed in this study with the ones published by the source laboratory, all mapped marker loci could not be integrated into component genetic maps in this study. One of the main reasons for this may be use of a stringent and common approach to develop all the component genetic maps.

Building a consensus map is not possible without common or bridge loci present on each LG [9]. A bridge marker was considered as such when it had an identical name and should have a similar position in different mapping populations that are underpinned. Markers with the same name that mapped to different positions in different populations were not considered to be common or bridge markers. However a minimum of three common markers per linkage group should be considered while, in the present study, at least one common marker per LG is also taken into consideration in some LGs because of availability of lower number of markers in some LGs.

During the process of construction of consensus map, the major emphasis was given towards obtaining a general order and distance because as a known fact, groundnut is polyploid with a large genome size (2800 Mb/C), and has a narrow genetic base with very low DNA polymorphism. Slight discrepancies in marker orders as well as positions observed in some LGs (http://cmap.icrisat.ac.in/cmap/sm/gn/gautami/and Table S1) among different component genetic maps may be due to (i) different mapping population sizes used (ii) different type of mapping populations used and (iii) genotyping errors [35] or sometimes these small differences might be due to mapping- imprecision rather than real rearrangements.

Developed consensus map integrates a total of 897 marker loci including 895 SSR and 2 CAPS loci with an average map density of 4.3 cM. This map is the most dense and community map and, therefore, is proposed as a reference consensus map. Despite of dense placing of markers on various LGs, some gaps were observed on the distal ends of some LGs e.g. a02, b02, a03, a05, b05, a08, a09, b09 and a10. These regions may be high recombination prone regions and some of them were also observed in other mapping studies also [19,21,24–27]. Another reason for these gaps may be due to under-representation or deficiency of marker loci from these genomic regions in the dataset used for developing the reference consensus map [9,13,19,21].

In present mapping protocol, the homologous LGs taking into consideration of homeologous relationship were used to generate consensus map LGs one at a time using MergeMap to establish marker orders (see materials and methods). Therefore, the marker orders in the consensus map are consistent throughout most of the linkage groups with few exceptions where the marker orders are in opposite orientation. Moreover, maximum markers were mapped onto the consensus map in their original orders similar to the individual maps, but small number of markers were joined with order changes, which could be caused by computational variation resulting from (i) recombination heterogeneity between different populations, (ii) weak linkages existing in the various LGs of maps, (iii) missing or poor quality data, (iv) different mapping programmes being used for constructing the individuals and the consensus maps and, (v) different thresholds statistics being applied for creating the consensus map and the original maps [36].

While utmost precautions were taken in preparing this consensus map, there could be some disagreement in order of closely linked markers between the individual maps within some LGs intervals. Such a disagreement may be due to the quality as well as the quantity and distribution along the LGs of the bridge (common) markers used for preparing the consensus map, or to mapping populations, algorithm and stringency criteria of computer programme [9,24,36]. For example, the mapping populations from which the consensus map has been prepared have different numbers and different types of progeny lines. In smaller populations, the chance that informative recombinant progeny lines are present in the population to accurately position markers is lower than in larger populations [9,36]. Further, even for a given mapping population, different markers were mapped using different subsets of progeny lines in different laboratories. Therefore, the users of the consensus SSR map must consider that the marker order is conditioned by several factors like the progeny lines used and the position of cross over along chromosome within the progeny lines. The precise fine markers order may slightly differ in other population and users may need to verify the order of closely linked markers in their mapping and breeding populations.

This reference consensus map integrated almost all types of SSR motifs, however di- and tri-nucleotide microsatellites at 47.58% and 28.70%, respectively, are present in higher proportions than the compound (22.33%) and other types of SSRs (1.39%). The underlying reason may be that the majority of SSR loci integrated in the consensus map were derived from the genomic DNA libraries that had been enriched for dinucleotide and trinucleotide SSR probes [28,34]. Therefore, the availability of different types of SSR loci in a given region will facilitate selection of the SSR repeat motifs of choice in a particular region of interest. Availability of the primer sequences for a total of 885 SSR loci, approximately 90% of all loci integrated in the consensus map, at one place should accelerate the use of SSR markers in groundnut breeding activities. Similarly, the genotyping data has been made available for all the mapped SSR loci in the present study and this will allow the community to extend the dataset with their own data set in future.

Another feature of the developed reference consensus map is the defining of the groundnut genetic map in 203 BINs. Furthermore marker loci present in these BINs are associated with the PIC values. One marker from each of such BIN with higher PIC value has also been identified. Using this criteria, 36 BINs have been identified that have at least one marker with >0.70 PIC value and 111 BINs carry at least one marker >0.50 PIC value. This information will be very useful to select the genome-wide markers that will have higher probability of showing polymorphism in the parental genotypes of the mapping populations or germplasm collections to be analyzed. It is also important to mention that primer sequence information also has been provided here for 885 markers.

Materials and Methods

Assembling Marker Segregation Data

SSR marker segregation data available on ten recombinant inbred lines (RILs) and one backcross (BC) mapping populations were assembled from collaborators as mentioned in Table 1. The populations, for which marker segregation data were assembled, for the convenience of referring in this article, have been referred as RIL-1 to RIL-10 and BC-1.

Three mapping populations (RIL-1, RIL-2, and RIL-3), developed at ICRISAT, segregate for drought tolerance related traits [25], two mapping populations (RIL-4 and RIL-5), developed at UAS-Dharwad, segregate for foliar disease resistance [26] and two populations (RIL-9 and RIL-10), developed at UGA and HAAS, segregate for tomato spotted wilt virus (TSWV). In the case of RIL-6, RIL-7 and RIL-8, developed at GAAS, Yueyou 13 (Y13), a Spanish type with high yield was the common female parent. While the RIL-6 segregates for oil content, the RIL-7 segregates for protein content and the RIL-8 segregates for resistance to Aspergillus flavus and aflatoxin contamination [24]. The remaining BC-1 population (BC1F1) was developed using a wild tetraploid AABB amphidiploid (*A. ipaënsis* KG30076 × *A. duranensis* V14167), called AiAd [37] and a cultivated tetraploid AABB variety (Fleur 11). This population segregates for several agro-morphological and drought related traits [38].

In brief, the segregation data for 211 marker loci on RIL-1 [19,21], 128 and 87 markers loci on RIL-2 and RIL-3, respectively [25] and 209 marker loci each on RIL-4 and RIL-5 populations [20,22,26] were obtained. RIL-6, RIL-7 and RIL-8 contributed marker data for 146, 124 and 64 marker loci respectively [24]. Segregation data were obtained for 261 and 193 marker loci on RIL-9 and RIL-10, respectively [27]. The lone BC-1 population contributed segregation data of maximum number (339) of marker loci [23]. Genotyping data as mentioned above have been provided in Table S5.

Construction of Component Genetic Maps

Segregation data for 1961 markers obtained on all the 11 mapping populations were subjected to chi-square ($\chi 2$) test to examine distortion from the expected 1:1 segregation using "Locus genotype frequency" function of JoinMap V 3.0 [39]. Individual or component genetic maps were constructed using MAPMAKER/EXP [29] and Kosambi mapping function [40] to assemble linkage groups by maximum-likelihood for respective mapping populations. Marker clusters were identified using a minimum LOD score of 5.0 and a maximum recombination fraction (θ) of 0.35. The most likely marker order within each LG was estimated by comparing the log-likelihood of the possible orders of markers using multipoint analysis "Compare" command. The "Try" command was also used to determine the most likely placement of the unlinked markers, and subsequent orders were tested using the "Ripple" command with "Error Detection" and "Use Three Points" options enabled. The distance between neighboring markers were calculated using the multipoint analysis implemented in the "Map" command.

Construction of Reference Consensus Genetic Map

A reference consensus genetic map was constructed using the markers mapped in ten RILs and one BC mapping populations. As peanut is an allotetraploid, deciphering the homologous versus homeologous relationships between LGs of the different component maps was necessary before constructing the consensus map. We first identified the sub-genome origin of each LG of the different component maps using a set of 58 single dose SSR markers (Table S6) that consistently amplified only one locus on the A or B sub-genomes. We then merged all LGs belonging to the same homology group with the software MergeMap [41]. In brief, LGs belonging to the same group of homology were first converted to direct acyclic graphs (DAG), which were then merged into a consensus graph on the basis of their shared vertices. Subsequently, efforts were made to resolve conflicts among the individual LGs by deleting a minimum set of marker occurrences. The result of the conflict-resolution step was a consensus DAG, which was then simplified and linearised to produce the consensus map. The final map was drawn with the help of Mapchart V 2.2 [42].

For efficient visualization of individual and consensus maps as well as their comparison, mapping data were put in the comparative mapping programme CMap version 1.01 http://www.gmod.org/cmap. This helped in assessing the congruency of marker positions and order by making a pairwise comparison among different genetic maps. Considering only the common loci existing among various genetic maps, highly conserved marker order was manifested. Subsequently, all the developed 11 individual genetic maps and the reference consensus map were aligned together in CMap.

Conclusion

This article reports the first dense reference consensus map of the international groundnut community for wider applications in groundnut research. The consensus map provides the marker order for a maximum number of markers available in groundnut, which will be very helpful for aligning any new genetic map as well as anchoring genetic map to the future physical map. Furthermore, the reference consensus map now offers the possibility to select desirable set of markers with appropriate repeat motifs as well as PIC value that are uniformly distributed throughout the genome. In addition, marker segregation and mapping data as well as primer sequence information for as many as markers as possible have also been provided as supplementary tables that will be very useful for the groundnut community for future genetics research and breeding applications.

Supporting Information

Figure S1 LG wise segregation patterns of markers in each population. In the scatter plot, markers from component mapping populations viz. RIL-1, RIL-2, RIL-3, RIL-4, RIL-5, RIL-6, RIL-7, RIL-7, RIL-8, RIL-9, RIL-10 and BC-1 are shown by blue hexagon, red square, pink triangle, orange circle sea green hexagon, bright square, plum triangle, blue circle, yellow hexagon, lavender square and violet triangle respectively.

Figure S2 Classification of polymorphic markers into different ranges of PIC values. This figure provides frequency distribution of mapped markers with variable range of PIC values.

Table S1 Details of the component and the reference consensus genetic maps with marker distances. This table provides the comparative details on mapped loci and their map distance in each LGs of component genetic maps (RIL-1, RIL-2, RIL-3, RIL-4, RIL-5, RIL-6, RIL-7, RIL-8, RIL-9, RIL-10 and BC-1) and the reference consensus map.

Table S2 Summary of consensus map based on ten RILs and one BC mapping populations. This table provides information about different homologous linkage groups used from the 11 component genetic maps for constructing the reference consensus genetic map.

Table S3 Details of reference consensus genetic map with 897 marker loci based on 11 mapping populations. This table provides BIN wise information for integrated markers along with repeat motifs, PIC values and primer sequence information.

Table S4 Summary of comparative information between tetraploid cultivated reference map and with diploid groundnut maps. This table provides comparative information on common markers between the tetraploid cultivated reference groundnut map with A-genome (Moretzsohn et al. 2005) and B-genome (Moretzsohn et al. 2009) diploid maps.

Table S5 Genotyping scores for ten RILs and one BC mapping populations used to construct the reference consensus genetic map. This table provides detailed genotyping data for all the 11 mapping populations used for the construction of reference genetic map.

Table S6 List of the "A" and "B" genome specific markers mapped in the reference consensus genetic map. This table provides list of SSR markers identified specific to A and B genomes of the tetraploid groundnut.

Acknowledgments

Authors are thankful to Ms. M. Sriswathi for extending her help while preparing this MS.

Author Contributions

Conceived and designed the experiments: RKV. Performed the experiments: B. Gautami DF MKP. Analyzed the data: B. Gautami MCM DJB JFR RKV. Contributed reagents/materials/analysis tools: VS HQ YH IF XC AB TMS MVCG SNN XL DAH B. Guo DJB JFR RKV. Wrote the paper: RKV B. Gautami MKP DF MCM DJB JFR.

References

1. Beavis WD, Grant D (1991) A linkage map based on information from four F_2 populations of maize (*Zea mays* L.). Theor Appl Genet 82:636–644.
2. Kianian SF, Quiros CF (1992) Generation of a *Brassica oleracea* composite RFLP map: linkage arrangements among various populations and evolutionary implications. Theor Appl Genet 84:544–554.
3. Gentzbittel L, Vear F, Zhang YX, Breville A, Nicolas P (1995) Development of a consensus linkage RFLP map of cultivated sunflower (*Helianthus annus* L.). Theor Appl Genet 90:1079–1086.
4. Hauge BM, Hanley SM, Cartinhour S, Cherry JM, Goodman HM (1993) An integrated genetic/RFLP map of the *Arabidopsis thaliana* genome. Plant J 3:745–754.
5. Sewell MM, Sherman BK, Neale DB (1999) A consensus map for loblolly pine (*Pinus taeda* L.). I. construction and integration of individual linkage maps from two outbreed three-generation pedigrees. Genetics 151:321–330.
6. Sharopova N, McMullen MD, Schultz L, Schroeder S, Sanchez-Villeda H, et al. (2002) Development and mapping of SSR markers for maize. Plant Mol Biol 48:463–481.
7. Falque M, Décousset L, Dervins D, Jacob AM, Joets J, et al. (2005) Linkage mapping of 1454 maize candidate gene loci. Genetics 70:1957–1966.
8. Somers DJ, Isaac P, Edwards K (2004) A high-density microsatellite consensus map for bread wheat (*Triticum aestivum* L.). Theor Appl Genet 109:1105–1114.
9. Varshney RK, Marcel TC, Ramsay L, Russell J, Röder MS, et al. (2007) A high density barley microsatellite consensus map with 775 SSR loci. Theor Appl Genet 114:1091–1103.
10. Marcel TC, Varshney RK, Barbieri M, Jafary H, de Kock MJ, et al. (2007) A high-density consensus map of barley to compare the distribution of QTLs for partial resistance to *Puccinia hordei* and of defence gene homologues. Theor Appl Genet 114:487–500.
11. Song QJ, Marek LF, Shoemaker RC, Lark KG, Concibido VC, et al. (2004) A new integrated genetic linkage map of the soybean. Theor Appl Genet 109:122–128.
12. Choi IY, Hyten DL, Matukumalli LK, Song Q, Chaky JM, et al. (2007) A soybean transcript map: gene distribution, haplotype and single- nucleotide polymorphism analysis. Genetics 176:685–696.
13. Bohra A, Saxena RK, Gnanesh BN, Saxena K, Byregowda M, et al. (2012) An intra-specific consensus genetic map of pigeonpea [*Cajanus cajan* (L.) Millspaugh] derived from six mapping populations. Theor Appl Genet (accepted).
14. Halward T, Stalker HT, Kochert G (1993) Development of an RFLP Linkage Map in Diploid Peanut Species. Theor Appl Genet 87:379–384.
15. Moretzsohn MC, Leoi L, Proite K, Guimarães PM, Leal-Bertioli SCM, et al. (2005) A microsatellite-based, gene-rich linkage map for the AA genome of *Arachis* (*Fabaceae*). Theor Appl Genet 111:1060–1071.
16. Leal-Bertioli SC, José AC, Alves-Freitas DM, Moretzsohn MC, Guimarães PM, et al. (2009) Identification of candidate genome regions controlling disease resistance in *Arachis*. BMC Plant Biol 9:112.
17. Gobbi A, Teixeira C, Moretzsohn M, Guimarães P, Leal-Bertioli S, et al. (2006) Development of a linkage map to species of B genome related to the peanut (*Arachis hypogaea* – AABB). *In: Plant and Animal Genomes (PAG) XIV Conference*, San Diego, CA, USA. p 679.
18. Moretzsohn MC, Barbosa AVG, Alves-Freitas DMT, Teixeira C, Leal-Bertioli SCM, et al. (2009) A linkage map for the B-genome of *Arachis* (Fabaceae) and its synteny to the A-genome. BMC Plant Biol 9:40.
19. Varshney RK, Bertioli DJ, Moretzsohn MC, Vadez V, Krishnamurthy L, et al. (2009) The first SSR- based genetic linkage map for cultivated groundnut (*Arachis hypogaea* L.). Theor Appl Genet 118:729–739.
20. Khedikar YP, Gowda MVC, Sarvamangala C, Patgar KV, Upadhyaya HD, et al. (2010) A QTL study on late leaf spot and rust revealed one major QTL for molecular breeding for rust resistance in groundnut (*Arachis hypogaea* L.). Theor Appl Genet 121:971–984.
21. Ravi K, Vadez V, Isobe S, Mir RR, Guo Y, et al. (2011) Identification of several small main-effect QTLs and a large number of epistatic QTLs for drought tolerance related traits in groundnut (*Arachis hypogaea* L.). Theor Appl Genet 122:1119–1132.
22. Sarvamangala C, Gowda MVC, Varshney RK (2011) Identification of quantitative trait loci for protein content, oil content and oil quality for groundnut (*Arachis hypogaea* L.). Field Crops Res 122:49–59.
23. Fonceka D, Hodo-Abalo T, Rivallan R, Faye I, Ndoye Sall M, et al. (2009) Genetic mapping of wild introgressions into cultivated peanut: a way toward enlarging the genetic basis of a recent allotetraploid. BMC Plant Biol 9:103.
24. Hong Y, Chen X, Liang X, Liu H, Zhou G, et al. (2010) A SSR-based composite genetic linkage map for the cultivated peanut (*Arachis hypogaea* L.) genome. BMC Plant Biol 10:17.
25. Gautami B, Pandey MK, Vadez V, Nigam SN, Ratnakumar P, et al. (2012) Quantitative trait locus analysis and construction of consensus genetic map for drought tolerance traits based on three recombinant inbred line populations in cultivated groundnut (*Arachis hypogaea* L.). Mol Breed DOI 10.1007/s11032-011-9660-0.
26. Sujay V, Gowda MVC, Pandey MK, Bhat RS, Khedikar YP, et al. (2012) Quantitative trait locus analysis and construction of consensus genetic map for foliar disease resistance based on two recombinant inbred line populations in

cultivated groundnut (*Arachis hypogaea* L.). Mol Breed DOI 10.1007/s11032-011-9661-z.

27. Qin H, Feng S, Chen C, Guo Y, Knapp S, et al. (2012) An integrated genetic linkage map of cultivated peanut (*Arachis hypogaea* L.) constructed from two RIL populations. Theor Appl Genet 124:653–664.
28. Pandey MK, Gautami B, Jayakumar T, Sriswathi M, Upadhyaya HD, et al. (2012) Highly informative genic and genomic SSR markers to facilitate molecular breeding in cultivated groundnut (*Arachis hypogaea* L.). Plant Breed 131:139–147.
29. Lander ES, Green P, Abrahamson J, Barlow A, Daly MJ (1987) MAPMAKER: an interactive computer package for constructing primary genetic linkage maps of experimental and natural populations. Genomics 1:174–181.
30. Chu Ye, Holbrook CC, Ozias-Akins P (2009) Two alleles of *ahFAD2B* control the high oleic acid trait in cultivated peanut. Crop Sci 49:2029–2036.
31. Langridge P, Karakousis A, Collins N, Kretschmer J, Manning S (1995) A consensus linkage map of barley. Mol Breed 1:389–395.
32. Mace ES, Rami JF, Bouchet S, Klein PE, Klein RR, et al. (2009) A consensus genetic map of sorghum that integrates multiple component maps and high-throughput Diversity Array Technology (DArT) markers. BMC Plant Biol 9:13.
33. Hyten DL, Choi IY, Song Q, Specht JE, Carter TE, et al. (2010) A high density integrated genetic linkage map of soybean and the development of a 1536 universal soy linkage panel for quantitative trait locus mapping. Crop Sci 50:960–968.
34. Pandey MK, Monyo E, Ozias-Akins P, Liang X, Guimaraes P, et al. (2012): Advances in *Arachis* genomics for peanut improvement. Biotech Adv 30:639–651.
35. Feltus FA, Hart GE, Schertz KF, Casa AM, Kresovich S, et al. (2006) Alignment of genetic maps and QTLs between inter- and intraspecific sorghum populations. Theor Appl Genet 112:1295–1305.
36. Gustafson JP, Ma XF, Korzun V, Snape JW (2009) A consensus map of rye integrating mapping data from five mapping populations. Theor Appl Genet 118:793–800.
37. Fávero AP, Simpson CE, Valls JF, Vello NA (2006) Study of the evolution of cultivated peanut through crossability studies among *Ara chis ipaensis*, *A. duranensis*, and *A. hypogaea*. Crop Sci 46:1546–1552.
38. Foncéka D, Hodo-Abalo T, Rivallan R, Vignes H, Faye I, et al. (2012) Fostered and left behind alleles in peanut: interspecific QTL mapping reveals footprints of domestication and useful natural variation for breeding. BMC Plant Biol 12:26.
39. Stam P (1993) Construction of integrated genetic linkage maps by means of a new computer package JoinMap. Plant J 3:739–744.
40. Kosambi DD (1994) The estimation of map distance from recombination values. An Eugen 12:172–175.
41. Wu Y, Close TJ, Lonardi S (2008) On the accurate construction of consensus genetic maps. In: Proceedings of LSS Computational Systems. *Bioinformatics Conference*, 26–29 August 2008, Stanford Edited by: Peter Markstein p285–296.
42. Voorrips RE: Mapchart (2002) software for the graphical presentation of linkage maps and QTLs. J Hered 93:77–78.

9

Increasing Strawberry Fruit Sensorial and Nutritional Quality Using Wild and Cultivated Germplasm

Jacopo Diamanti[1], Franco Capocasa[1], Francesca Balducci[1], Maurizio Battino[2], Jim Hancock[3], Bruno Mezzetti[1]*

1 Dipartimento di Scienze Agrarie, Alimentari e Ambientali, Università Politecnica delle Marche, Ancona, Italy, 2 Dipartimento di Scienze Cliniche Specialistiche ed Odontostomatologiche, Università Politecnica delle Marche, Ancona, Italy, 3 Department of Horticulture, Michigan State University, East Lansing, Michigan, United States of America

Abstract

Background: Increasing antioxidant levels in fruit through breeding is an important option to support higher antioxidant intake particularly when fruit consumption is low. Indeed, if nutritional components are also combined with a high standard of sensorial fruit quality, the perspective for consumer health can be further improved by encouraging more fruit consumption. Wild species are valued by strawberry breeders as sources of novel traits, especially for pest resistance and abiotic stress tolerance. Furthermore, previous investigations have shown improvements in fruit nutritional quality in breeding material that originated from *Fragaria virginiana* ssp. *glauca* (FVG) inter-species crosses. Recently, commercial varieties of strawberries have also shown interesting variability in fruit nutritional quality.

Results: Strawberry fruit sensorial and nutritional qualities generated by *Fragaria* inter-species and intra-species crosses were evaluated on 78 offspring derived from 8 families: two that originated from *F.* × *ananassa* intra-species crossing; three from back-crossing of F1- FVG × *F.* × *ananassa*; and three from back-crossing of BC1- FVG × *F.* × *ananassa*. The genetic variability from the three types of cross combinations was analyzed by calculation of the correlations among the fruit sensorial and nutritional parameters. The results obtained show that two subsequent back-crossing generations from an inter-species crossing combination with *F. virginiana* ssp. *glauca* provides useful improvement of the fruit nutritional and sensorial qualities that is combined with agronomic standards that are close to those requested at the commercial level. Improvements of these traits can also be achieved by programming *F.* × *ananassa* intra-species crosses and producing progeny with productivity traits more similar to those of the commercial cultivars.

Conclusions: The two types of combination programs (inter-species back-crosses, and intra-species crosses) can be used to improve strawberry nutritional quality.

Editor: Cameron Peace, Washington State University, United States of America

Funding: The research work was carried out thanks to the project on 'Valorization of Red Fruit Genetic Resources' supported by the Italian Ministry of Environment and for the last analyses to the EU FP7 EUBerry project No. 265942. The funders had no role in study design, data collection and analysis, decision to publish, or preparation of the manuscript.

Competing Interests: The authors have declared that no competing interests exist.

* E-mail: b.mezzetti@univpm.it

Introduction

Greater consumption of fruit and vegetables is considered as one way of increasing the intake of antioxidants, and like other berries, strawberry represents one of the most important sources of bioactive compounds with high antioxidant capacity [1–6]. Accordingly, an increase in the consumption of berries richer in 'healthy compounds' is seen as an appropriate strategy for improving human health.

Increasing the antioxidant levels in fruit through breeding and/or biotechnology is an important option to maintain a higher antioxidant intake, particularly when fruit consumption is low. Indeed, if nutritional bioactive components are combined with a high standard of sensorial fruit quality, consumer fruit consumption can be encouraged, with much greater positive health effects [3], as would be the case for strawberry fruit [7].

The breeding of more nutritious, better-tasting cultivars can be successful if the variability and heritability of the bioactive compounds, which is defined as the total antioxidant capacity (TAC), indicate the possibility of achieving breeding progress. It is well know that the availability of genetic diversity within compatible species of any given crop will enhance the extent of any improvement [8]. The biotechnological approach is also an option to supplement this improvement, through modifications of specific biosynthetic pathways [9]. However, the success of both breeding and biotech approaches is dependent on deep knowledge of the sources of the most useful wild and cultivated genetic diversity to be used in genetic and genomic studies.

The effect of genotype on the nutritional quality of strawberry is well known [10–13]. The levels of antioxidants and the TAC in strawberry extracts from whole fruits vary considerably among genotypes [4,12,14], although few genotypes have been well characterized for these important features. Furthermore, limited knowledge is available on the possibility of improving strawberry nutritional traits by breeding [15–16]. Some results have been obtained for blueberries and raspberries, with a moderate heritability for TAC, total phenols content (TPH) and anthocyanins content (ACY) demonstrated [17–21].

Accessions of the progenitor wild species are valued by strawberry breeders as sources of novel traits, and especially for pest resistance and abiotic stress tolerance [22]. Furthermore, previous investigations [4,5,15,16] have shown improvements in fruit nutritional quality in breeding material that originated from *Fragaria virginiana* ssp. *glauca* (FVG) in inter-species crosses. This result is supported by a study carried out by Wang *et al.* [14] that showed that the fruit of *F. virginiana* accessions have significantly higher TAC, TPH, and total ACY than the fruit from different lines of *F. chiloensis* and *F.* × *ananassa.*

Since 2006, a breeding program based on the evaluation and comparison of different populations of *F.* × *ananassa* crosses and FVG inter-species back-crosses has been conducted by the *Università Politecnica delle Marche*, with the objective being to generate new selections with increased fruit nutritional quality.

With the aim of studying strawberry fruit sensory and nutritional quality generated by *Fragaria* inter-species and intra-species crosses, 78 offspring derived from 8 families were evaluated, two originating from *F.* × *ananassa* intra-species crosses; three from back-crossing of F1– FVG × *F.* × *ananassa*; and three from back-crossing of BC1– FVG × *F.* × *ananassa* (Figure 1). The selection AN94,414,52 (F1 FVG) used in the breeding program was originated by the inter-species cross of *F. virginiana* ssp. *glauca* "FVG22" × *F.* × *ananassa* selection 91,143,5. FVG22 is a clone of *F. virginiana* ssp. *glauca* that is maintained in the National Clonal Germplasm Repository in Corvallis (OR, USA), while the *F.* × *ananassa* selection 91,143,5 originated from the breeding program of the Italian Experimental Agriculture Research Council, Forlì Fruit Research unit (CRA-FRF, Italy).

Fruit sensorial and nutritional parameters of all of the offspring were evaluated in 2009 from single seedling plants, and in 2010 on vegetatively propagated selections. For both years, all of the offspring and the corresponding parents were evaluated for their fruit nutritional qualities (fruit TAC, TPH and ACY), while in the second year, they were evaluated for their production (fruit weight and total plant production, commercial, misshapen, undersized and rotten fruit production, per plant) and sensorial (fruit firmness, color, solid soluble [SS] content and total acidity [TA]) parameters.

The genetic variability that originated from the three types of cross-combinations (BC1 back-cross, BC2 back-cross, and intra-species cross) was analyzed by correlating fruit sensorial and nutritional parameters. With such an analysis, we were able to evaluate the relative effects of intra-species and inter-species back-cross populations.

Results

Fruit nutritional and sensorial parameters of the different breeding populations were evaluated by two-way nested ANOVA, which revealed significant differences between the inter-species back-crosses and the intra-species crosses (Table 1).

SS content, TA, TAC, TPH and ACY were significantly higher in the BC1 and BC2 back-crossed populations that contained genes from *Fragaria virginiana* ssp *glauca* (Table 2) than in the *F.* × *ananassa* intra-species crosses, while the intra-species crosses were firmer and more darkly colored (Table 2). For all of the parameters examined, there were no significant differences between the BC1 and BC2 back-crosses, although the individual populations were significantly different.

Selections of the BC1 Populations

The two-way nested ANOVA evaluated for the BC1 back-cross populations showed significant differences for fruit TA, TAC, TPH and ACY, but not for SS content, firmness, Chroma and L* (Table 1).

The highest mean values of the fruit TA were seen for AN07,004 and AN07,005, while the AN07,003 cross-combination showed a significantly lower TA than for AN07,004, although not a significantly different TA than for AN07,005 (Table 2). The highest fruit TAC was seen for the fruit of AN07,004, while the lowest was for the fruit of AN07,005; these TACs were significantly different. Population AN07,004 also showed the highest fruit TPH, which was significantly different from the AN07,003 back-cross population with the lowest TPH (Table 2).

In contrast to these results, population AN07,003 showed the highest fruit ACY, which was significantly higher than for AN07,004 (Table 2).

The BC1 populations also differed significantly in their plant production parameters. Indeed, AN07,004 had the highest mean for fruit weight and commercial production, and the lowest for misshapen fruit per plant (Table 3). The lowest mean fruit weight was seen for AN07,003, while the lowest commercial production was for AN07,005. The lowest amount of rotted fruit was seen for BC1 AN07,004 (Table 3).

Fruit sensorial and nutritional parameters of the BC1 populations had different levels of correlation (Table 4). Fruit TA was negatively correlated with fruit SS content (−0.24), firmness (−0.50) and ACY (−0.30). A high negative correlation was also seen between fruit firmness and TAC (−0.34). The Chroma index had a high negative correlation with TAC (−0.29) and ACY (−0.46). Finally, L* grade showed a high negative correlation with fruit ACY (−0.47) (Table 4).

Positive correlations were seen between SS content and fruit ACY (0.26), Chroma index and L* grade (0.83), and fruit TAC and TPH (0.46) and ACY (0.33) (Table 4).

These data show the positive effects of using *F. virginiana glauca* in strawberry breeding, in terms of improving the fruit nutritional parameters, although there was a negative effect on fruit sensorial parameters, as shown by the correlation matrix for the BC1 FVG back-crosses. Indeed, the increase in fruit nutritional parameters was associated with a large increase in fruit TA, which was linked with a decrease in fruit SS content, firmness and ACY. In particular, fruit TAC was negatively correlated with fruit firmness and Chroma index. This indicates that in BC1 progenies, the higher content of bioactive compounds that determine the fruit TAC is associated with negative effects on fruit firmness, color (darker fruit) and ACY. The correlation matrix of the same BC1 progenies reveals negative correlation between ACY and both colorimetric parameters, while in the BC1 population, ACY was positively correlated with an increase in fruit SS content. Among the nutritional parameters, the positive correlations between fruit TAC and TPH, and fruit TAC and ACY were consistent with what was reported by Scalzo et al. (2005).

Principal component analysis (PCA) bi-plots of the BC1 families showed strong associations between the TAC and TPH nutritional parameters for AN07,004, and in particular for AN07,004,51 and AN07,004,54 (Figure 2b), confirming the data for the nutritional parameters in Table 4. AN07,003 showed the highest fruit ACY (Table 2) and PCA analysis confirmed this trend (Figure 2b). The PCA bi-plots combining the data of the fruit TAC, TPH, ACY, and weight, and the commercial production of the selections and

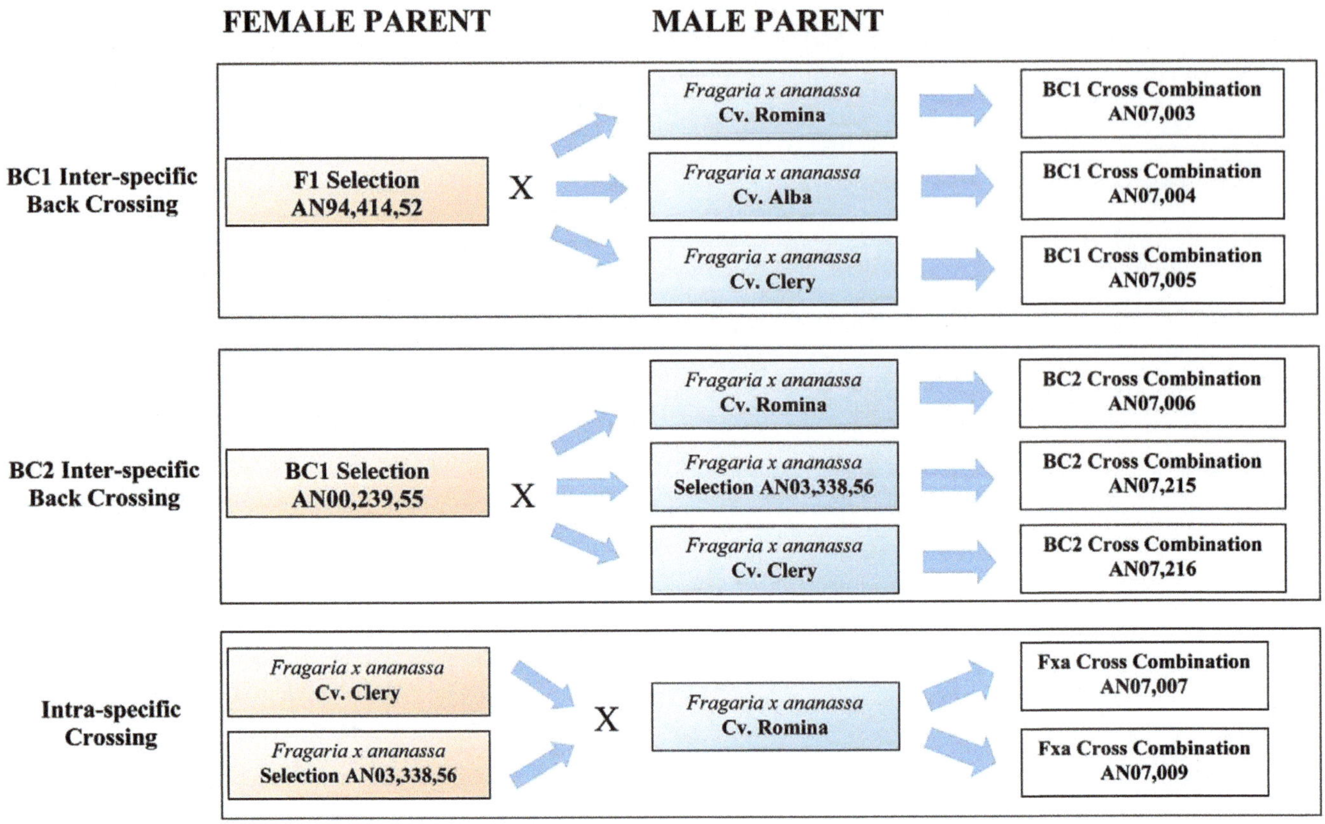

Figure 1. Crossing combination scheme. Scheme of the crossing combinations, including *F.* × *ananassa* and *F. virginiana glauca* back-crossing, and *F.* × *ananassa* intra-species crossing.

corresponding parents of the BC1 populations (Figure 2a, vector distribution plot; Figure 2b, case distribution plot), highlight the selections AN07,004,51, AN07,004,52, AN04,007,54 and AN07,003,60 for their positions in the upper right-hand section of the plot, corresponding to high values of the fruit ACY, TPH and TAC variables. Thus, these selections have the highest fruit nutritional value, although they still have some deficiencies in fruit fresh weight and the levels of commercial production. Conversely, selections AN07,004,57 and AN07,005,58 are located in the upper left-hand quadrant of the plot, located towards the vectors of fruit weight and commercial production. All of the other offspring are located between the positions of the parents, which are in the central area of the plot. The male parents (cv. Alba and Clery, selection Romina) are situated on the left-hand side of the plot, again, corresponding to the vectors of fruit weight and commercial production, while the common female parent, the F1 selection AN94,414,52, is located on the right-hand side of the plot, corresponding to the vectors of fruit TAC, TPH and ACY. This confirms the offspring improvement in fruit size and plant production with regard to the FVG mid-parent [22].

The PCA bi-plots of the BC1 families for fruit sensorial parameters (SS content, TA) combined with fruit weight and commercial production variables (Figure 3a) locates most of the selections towards the SS content vector instead of the fruit weight and commercial production vectors, revealing an improvement in the sensorial quality of the progenies, with respect to their parents (Figure 3b). With all three of the families more towards the SS content vector, AN07,003,60 is the selection that is most oriented towards the fruit SS content vector.

In the PCA bi-plots of the BC1 families for Chroma index, fruit ACY and TPH (Figure 4a, vector distributions), most of the offspring and parents are distributed in the left-hand central area of the plot, where the Chroma index vector is located (Figure 4b, case distributions). The only exceptions are selections AN07,003,60 and AN07,004,54: AN07,003,60 is located on the right-hand side of the plot, towards the vector of the fruit ACY variable, and AN07,004,54 is located in the upper right-hand quadrant, in the direction of the fruit TPH.

These data on the fruit sensorial and nutritional parameters demonstrate the importance of using the F1 FVG parent in the breeding, as it positively influences the fruit nutritional characteristics of the progeny seedlings.

Selections of the BC2 Populations

The ANOVA analysis of the BC2 population in 2010 showed significant differences for fruit TA, Chroma index, L* grade, TAC and TPH, while there were no significant differences for fruit SS content, firmness and ACY (Table 1).

The significant highest mean values for fruit TA, Chroma index, TAC and TPH were seen for AN07,215, while the significant highest values for SS content and L* were in AN07,006 (Table 2). Fruit firmness and ACY did not show significant differences among the BC2 populations; however, the lowest fruit TA, Chroma, L* and TAC were seen in AN07,216 (Table 2).

The highest means for fruit weight and commercial production were seen in the AN07,216 and AN07,006 populations, although these also had the highest levels of misshapen and undersized fruit per plant (Table 3). The lowest mean commercial production was

Table 1. Two-way nested ANOVA for the type of cross combination and the individual BC1, BC2 and intra-species populations for the fruit sensorial and nutritional parameters.

Factor	DF	SS content (°Brix)		TA (meq NaOH/100 g FW)		Firmness (g)		L*		Chroma		TAC (mmol TE/kg FW)		TPH (mg GA/kg FW)		ACY (mg PEL-3-GLU/kg FW)	
		MS	F	MS	F	MS	F	MS	F	MS	F	MS	F	MS	F	MS	F
Type of cross	2	7,.1	10.9	52.8	16.1	21363	9,7	66.2	17.4	87.3	8.8	338.1	65.1	299300	3.6	103216	9.3
Error	7	0.7		3.3		2195		3.8		9.9		5.2		83518		11076	
BC1 POP	2	1.1	1.5	7.2	1.8	1768.0	0.8	1.7	0.4	12.4	1.0	138.4	16.5	198108.5	1.4	12373.1	1.1
Error	28	0.7		4.0		2139.4		4.2		11.8		8.4		144708.2		11218.8	
BC2 POP	2	1.5	2.5	17.9	5.6	1491.7	1.0	17.5	5.8	36.3	4.2	313.6	81.2	347355.1	12.4	10471.7	0.9
Error	22	0.6		3.2		1442.0		3.0		8.6		3.9		27991.4		11191.3	
IS POP	1	0.5	0.8	13.7	8.3	5873.3	2.4	79.6	16.7	158.2	16.6	4.7	2.5	546376.7	9.7	57935.4	5.4
Error	18	0.7		1.7		2449.8		4.8		9.5		1.9		56198.9		10712.1	

ANOVA analysis was calculated between the types of crossing and within each cross combination. For each factor of analysis, the degree of freedom (DF), mean square (MS) and F value (F) for the sensorial and nutritional parameters were evaluated and are reported. IS, intra-species; POP, population.

Table 2. Fruit sensorial and nutritional parameters of the clonal populations of selected seedlings for each of the individual cross combinations within the BC1 and BC2 back-crosses and the intra-species cross populations.

Type of cross	Cross code	SS content (°Brix)	TA (meq NaOH/100 g FW)	Firmness (g)	Chroma	L*	TAC (mmol TE/kg FW)	TPH (mg GA/kg FW)	ACY (mg PEL-3-GLU/kg FW)
BC1	AN07,003	9.32 ns (0.24)	11.75 b (0.42)	338.85 ns (9.86)	45.55 ns (0.47)	34.49 ns (0.39)	20.91 b (0.53)	1604.53 b (70.54)	494.6 a (19.89)
BC1	AN07,004	9.07 ns (0.23)	13.37 a (0.4)	321.51 ns (9.57)	44.29 ns (0.46)	34.41 ns (0.37)	23.54 a (0.52)	1875.13 a (68.49)	429.08 b (17.37)
BC1	AN07,005	9.79 ns (0.24)	12.01 ab (0.42)	349.33 ns (9.86)	43.51 ns (0.47)	33.85 ns (0.39)	16.17 c (0.53)	1650.77 ab (70.54)	457.77 ab (17.89)
BC2	AN07,006	8.93 ns (0.22)	12.48 b (0.39)	341.07 ns (0.37)	42.88 ab (0.45)	35.60 a (8.98)	22.88 b (0.4)	1558.72 b (37.34)	498.49 ns (21.39)
BC2	AN07,215	9.44 ns (0.26)	15.38 a (0.45)	312.41 ns (0.42)	44.81 a (0.52)	35.54 a (10.31)	27.21 a (0.46)	1964.33 a (42.87)	446.10 ns (24.55)
BC2	AN07,216	8.53 ns (0.22)	12.71 b (0.39)	336.47 ns (0.36)	41.09 b (0.44)	33.62 b (8.83)	15.12 c (0.39)	1657.34 b (36.71)	505.69 ns (21.03)
IS	AN07,007	8.42 ns (0.23)	11.4 a (0.31)	363.58 ns (10.88)	49.65 a (0.59)	39.69 a (0.43)	14.78 ns (0.43)	1690.76 a (39.06)	288.94 b (16.33)
IS	AN07.009	8.78 ns (0.25)	9.74 b (0.29)	398.71 ns (10.20)	44.03 b (0.54)	35.73 b (0.46)	13.80 ns (0.40)	1363.16 b (34.63)	393.35 a (17.64)

Data are means and standard errors (SE) of the offspring selections for each family. Means followed by different letters are significantly different. SNK Test $P \leq 0.05$. IS, intra-species.

Table 3. Plant production parameters of the selections from the BC1 and BC2 back-cross and the intra-species cross populations.

Type of cross	Cross code	Fruit weight (g/20 fruits)	Misshapen fruit (g/plant)	Undersized fruit (g/plant)	Rotten fruit (g/plant)	Commercial production (g/plant)	Total production (g/plant)
BC1	AN07,003	11.7 (0.6)	18.5 (5.4)	110.5 (19.1)	78.8 (16.6)	349.0 (26.7)	556.8 (39.2)
BC1	AN07,004	13.9 (1.0)	10.4 (2.9)	116.7 (20.2)	48.4 (7.8)	356.7 (52.7)	532.1 (54.0)
BC1	AN07,005	12.9 (0.1)	15.0 (3.5)	190.8 (38.7)	62.4 (12)	300.5 (41.2)	568.6 (67.9)
BC2	AN07,006	14.4 (0.8)	15.1 (3.9)	162.1 (21.9)	77.2 (7.7)	401.6 (29.3)	656.1 (41.9)
BC2	AN07,215	13.6 (0.5)	3.9 (1.5)	97.8 (31.3)	135.6 (34.8)	265.6 (35.3)	502.2 (40.1)
BC2	AN07,216	14.7 (0.9)	10.7 (4.1)	110.5 (28.2)	51.1 (11.5)	376.3 (48.9)	548.6 (76.3)
IS	AN07,007	16.3 (1.0)	27.5 (5.1)	103.0 (11.5)	182.0 (32)	464.3 (38.9)	776.8 (57.7)
IS	AN07,009	18.0 (0.6)	22.1 (5.7)	76.3 (14.8)	190.9 (36.4)	431.1 (38.2)	720.4 (80.3)

Data are means and standard errors (SE) of the offspring selections for each family.

seen for AN07,215, together with the lowest level of misshapen and undersized fruit production per plant, although also with the highest value for rotten fruit production (Table 3). Few misshapen fruit were seen in the BC2 populations, in comparison with the intra-species cross population (Table 3).

The Pearson matrix for the fruit of the BC2 populations showed a number of negative correlations among the various sensorial and nutritional parameters (Table 4). Indeed, significant negative correlations were seen between fruit firmness and fruit SS content (−0.22), TA (−0.24), L* grade (−0.26) and fruit antioxidant capacity (−0.21). Most significantly, negative correlations were seen between fruit ACY and Chroma index (−0.62) and L* grade (−0.65) (Table 4).

Significant positive correlations were seen between TA and fruit TAC (0.45), TA and TPH (0.44), fruit TAC and TPH (0.47), and Chroma index and L* grade (0.82); also, stronger significant positive correlations were seen between SS content and TAC (0.29), and between SS content and TPH (0.29). Finally, firmness and ACY were positively correlated (0.24) (Table 4).

The results from the correlation matrix of the BC2 families show that in moving ahead by one back-cross generation there is a substantial improvement, seen as favorable correlations between sensorial and nutritional parameters. Indeed, in the BC2 populations, high fruit TAC and TPH content correspond to an increase in the fruit SS content and TA, and an increase in TAC corresponds to an increase in fruit brightness (L*) (Tables 2, 4). There are still some negative correlations, mainly for fruit firmness, which is negatively correlated with SS content, TA, L* grade and fruit TAC. So, highlighting these effects, a second back-cross generation can have positive effects for the improvement of fruit sensorial (SS content, TA, and L*) and nutritional (TAC) quality, but not of fruit firmness (Table 4). Negative correlations also remain between the fruit color parameters and ACY; indeed, in the BC2 families, high fruit ACY corresponds to a darker and more opaque fruit (Tables 2, 4).

The PCA bi-plots of fruit TAC, TPH, ACY, fruit weight and plant commercial production for the offspring and parents of these BC2 back-cross combinations (male parents: Clery, Romina and AN03,338,56; female parent: AN00,239,55) (Figure 5a) show a unique distribution for each cross combination; indeed, the seedlings of AN07,215 were almost all located in the lower right-hand quadrant of the plot, where the vectors for the fruit TAC and TPH variables are seen, in particular for the selections AN07,215,55 and AN07,215,57 (Figure 5b). The other two cross combinations are located mainly on the mid left-hand side of the plot (Figure 5b), where the vector for plant commercial production is located (Figure 5a), showing their lower affinity for the nutritional parameters. The selection AN07,216,60 here deserves special attention, as it is located in the upper right-hand quadrant where the vector for fruit ACY is positioned (Figure 4a), together with the female parent of the BC1 selection AN00,239,55.

The PCA plots for fruit sensorial parameters (SS content, TA) and productive parameters (fruit weight, commercial production) (Figure 6a) show wider distributions of the offspring for the cases plot (Figure 6b), in the offspring that are characterized by higher sensorial parameters and the offspring that are characterized by higher productive parameters.

The selections AN07,215,55 and AN07,215,53 are in the lower left-hand quadrant of the plot (Figure 6b), where the SS content and TA vectors are located (Figure 6a). The selections AN07,006,62 and AN07,216,55 and the male parents (Clery and AN03,338,56) (Figure 6b) are located in the lower right-hand quadrant of the plot, showing more affinity with the fruit weight and commercial production vectors (Figure 6a).

Table 4. Pearson's correlation matrix of the fruit sensorial and nutritional parameters for type of crossing evaluated on clonal populations of selected seedling.

BC1 populations		SS content	TA	Firmness	Chroma	L*	TAC	TPH
	SS content	–						
	TA	−0.24**	–					
	Firmness	0.03	$-0.50^{\$}$	–				
	Chroma	0.03	−0.08	0.05	–			
	L*	0.16	−0.11	−0.04	$0.83^{\$}$	–		
	TAC	0.02	0.06	$-0.34^{\$}$	$-0.29^{\$}$	−0.14	–	
	TPH	0.06	−0.12	−0.15	−0.10	−0.07	$0.46^{\$}$	–
	ACY	0.26**	−0.30**	−0.07	$-0.46^{\$}$	$-0.47^{\$}$	$0.33^{\$}$	0.18
BC2 populations	SS content	–						
	TA	0.18	–					
	Firmness	−0.22*	−0.24*	–				
	Chroma	−0.02	0.01	−0.09	–			
	L*	0.04	0.06	−0.26*	$0.82^{\$}$	–		
	TAC	0.29**	$0.45^{\$}$	−0.21*	0.15	0.20*	–	
	TPH	0.29**	$0.44^{\$}$	−0.17	0.08	0.15	$0.47^{\$}$	–
	ACY	0.02	0.04	0.24*	$-0.62^{\$}$	$-0.65^{\$}$	0.04	0.07
Intra-species cross populations	SS	–						
	TA	−0.16	–					
	Firmness	−0.06	−0.19	–				
	Chroma	0.21	0.28*	$-0.48^{\$}$	–			
	L*	0.26*	0.12	$-0.44^{\$}$	$0.75^{\$}$	–		
	TAC	0.03	0.27*	−0.05	0.07	0.15	–	
	TPH	0.02	0.34**	−0.10	0.23*	0.30**	$0.83^{\$}$	–
	ACY	−0.08	−0.22*	0.28*	$-0.64^{\$}$	$-0.53^{\$}$	$0.37^{\$}$	0.18

*, **, $ correlation levels significant at $p \leq 0.05$; $p \leq 0.01$; $p \leq 0.001$ respectively.

The other selections of BC2 are positioned in the central area of the plot, or in the opposite quadrants from where the vectors are located. From this distribution, it can be noted that most of the BC2 offspring are more oriented towards the commercial production and fruit weight vectors, in comparison with the BC1 parent AN00,239,55 (Figure 6b). This confirms the greater improvement of this trait in the BC2 genotypes than in their FVG parent [22].

The PCA plot for the fruit Chroma index, TPH and ACY variables (Figure 7a) shows the distribution of the offspring (Figure 7b) around the central area of the plot, with only the selection AN07,215,55 differing from the others; indeed, it is located in the lower left-hand quadrant of the plot (Figure 7b), where the TPH vector is located. The location of the female parent AN00,239,55 is in the lower right-hand quadrant and the location of the male parent AN03,338,56 is in mid-left-hand area of the plot (Figure 7b), following the orientations of the ACY vector and the Chroma index vector, respectively (Figure 7b).

The offspring AN07,215,55 showed high affinity for fruit nutritional quality (Figure 5b), and the PCA bi-plot also revealed its high fruit sensorial quality (Figure 6b), which indicates a good equilibrium between fruit SS content and TA. Also, the other offspring of the AN07,215 back-cross population showed a good combination of fruit nutritional (Figure 5b) and sensorial (Figure 6b) qualities, with respect to the other BC2 populations. These selections can be considered as an important improvement from the original FVG inter-species cross combination, and to be very close to the characteristics now requested for a commercial variety.

Selections from the Intra-species Crosses

The selections from the intra-species cross populations evaluated in 2010 differed significantly for fruit TA, Chroma index, brightness (L*), TPH and ACY, while the SS content, fruit firmness and TAC remained more stable within the populations (Table 1) and showed no significant differences.

The intra-species cross population AN07,007 (Romina × AN03,338,56) showed the most significant values for fruit TA, Chroma index, brightness (L*) and fruit TPH, while the highest fruit firmness and ACY were seen for the AN07,009 intra-species cross combination (Romina × Clery) (Table 2).

For the production parameters, the highest fruit weight was seen for AN07,009, which also had the lowest values for undersized and misshapen fruit per plant (Table 3), while AN07,007 showed the highest values of commercial production and total production (Table 3). The rotten fruit per plant were remarkable for both of these cross combinations, as this was almost half of that seen for the commercial production (Table 3).

For the Pearson's correlations, the Chroma index showed a high level of negative correlation with firmness (−0.48) and fruit ACY

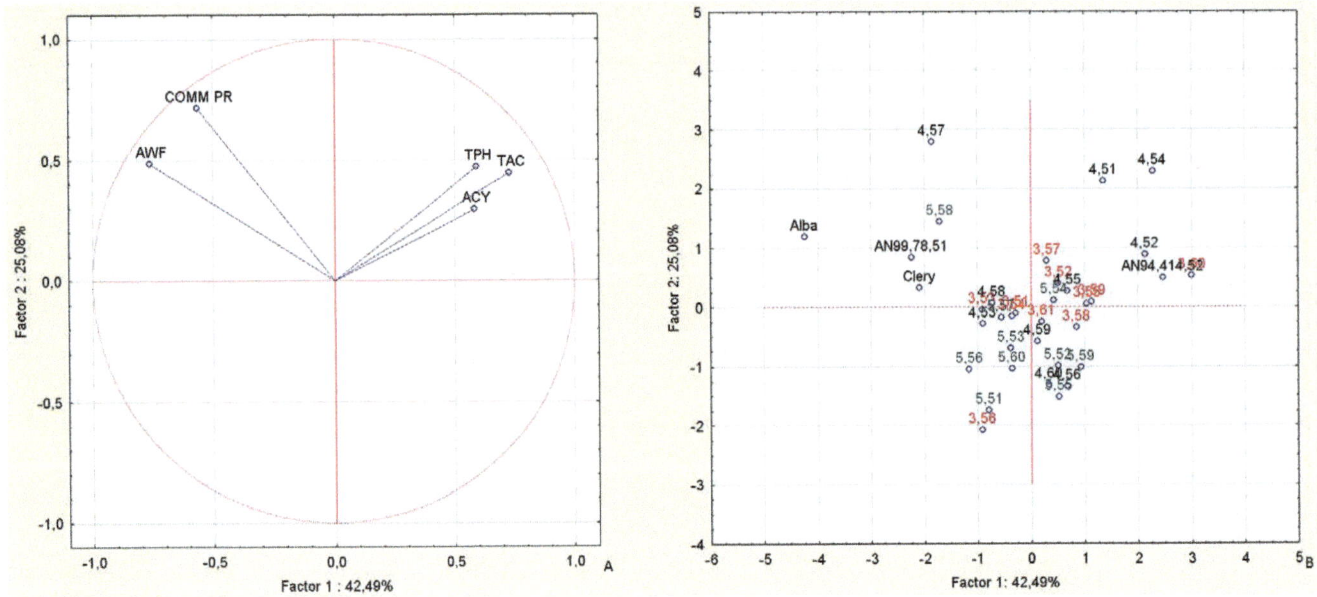

Figure 2. Principal component analysis. Bi-plot of the fruit weight, commercial production, TAC, TPH and ACY parameters of the selections and their corresponding parents, of the BC1 back-cross populations. Factor 1 and Factor 2 explain 67.57% of the data variation. a: Variable vector distributions; b: Case distributions. The seedlings reported in the PCA are identified only with the codes related to the cross combinations and the progressive seedling numbering.

(−0.64), and even the L* grade showed high negative correlation with fruit firmness (−0.44) and ACY (−0.53). Finally, negative correlation was also seen between fruit TA and ACY (−0.22) (Table 4).

Positive correlations were seen for fruit TA and Chroma index (0.28), TAC (0.27) and TPH (0.34). Fruit TPH also showed positive correlation with Chroma index (0.23) and L* grade (0.30), and high positive correlation with TAC (0.83). Fruit ACY also shown positive correlation with firmness (0.28) and high correlation with fruit TAC (0.37). Finally, fruit SS content showed positive correlation with L* grade (0.26) (Table 4).

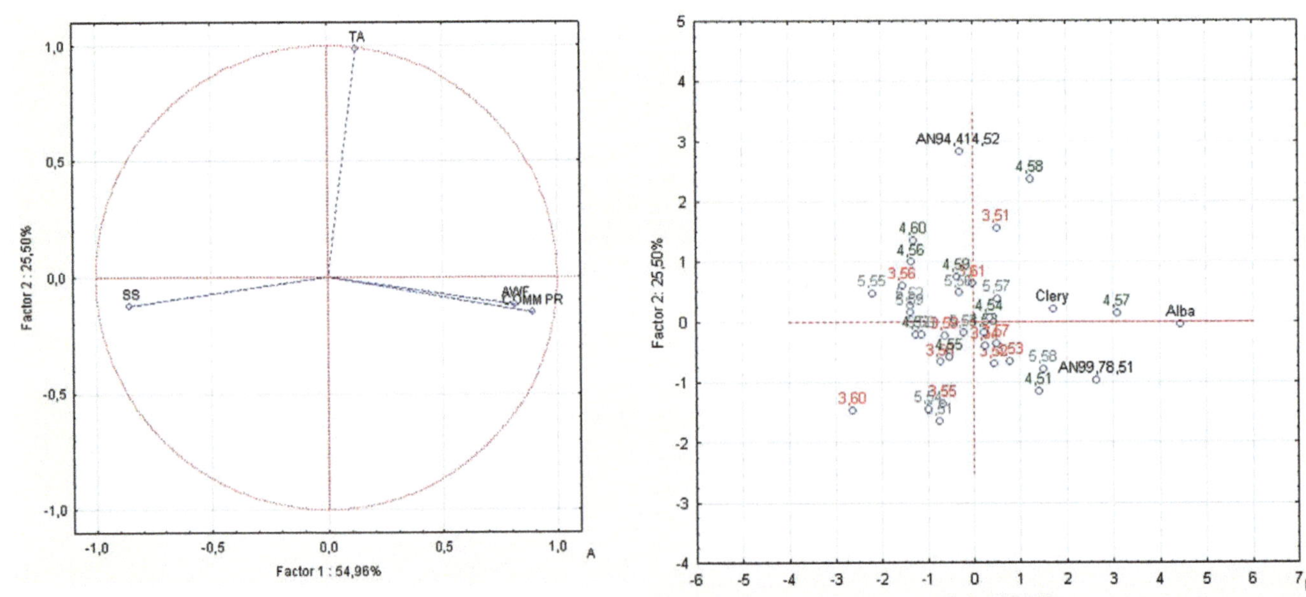

Figure 3. Principal component analysis. Bi-plot of fruit weight and commercial production *versus* SS content and TA for the offspring and parents of the BC1 back-cross populations. Factor 1 and Factor 2 explain 80.46% of the data variation. a: Variable vector distributions; b: Case distributions. The seedlings reported in the PCA are identified only with the codes related to the cross combinations and the progressive seedling numbering.

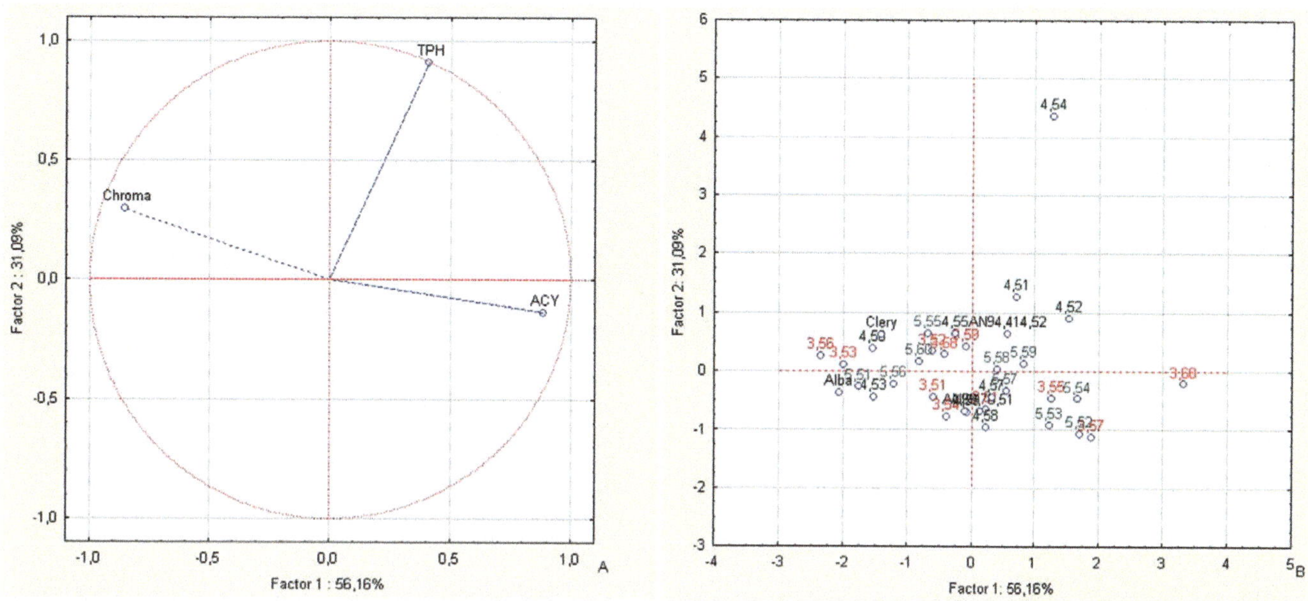

Figure 4. Principal component analysis. Bi-plot of the ACY, TPH and Chroma index components for the offspring and the parents of the BC1 back-cross populations. Factor 1 and Factor 2 explain 87.25% of the data variation. a: Variable vector distributions; b: Case distributions. The seedlings reported in the PCA are identified only with the codes related to the cross combinations and the progressive seedling numbering.

For the intra-species cross populations, positive correlations were seen between the fruit sensorial TA and Chroma index parameters and the TAC and TPH nutritional parameters (Table 4). AN07,007 had the highest TA, Chroma index, TPH and TAC, even if the TAC was not significantly different from that detected for fruit of the AN07,009 family (Table 2). In addition, positive correlation was seen for TPH and both of the colorimetric parameters, for both families, even if the fruits of the AN07,007 cross combination showed significantly higher TPH content, Chroma index and L* grade, in comparison with the fruit of AN07,009 (Table 2). In both of these cross combinations, Romina was used as the common male parent (Table 5). Romina is a new variety that was released by the *Università Politecnica delle Marche*, and it is characterized by improved fruit sensorial and nutritional quality. It was crossed with the selection AN03,338,56 from the University breeding program, which was chosen because of its

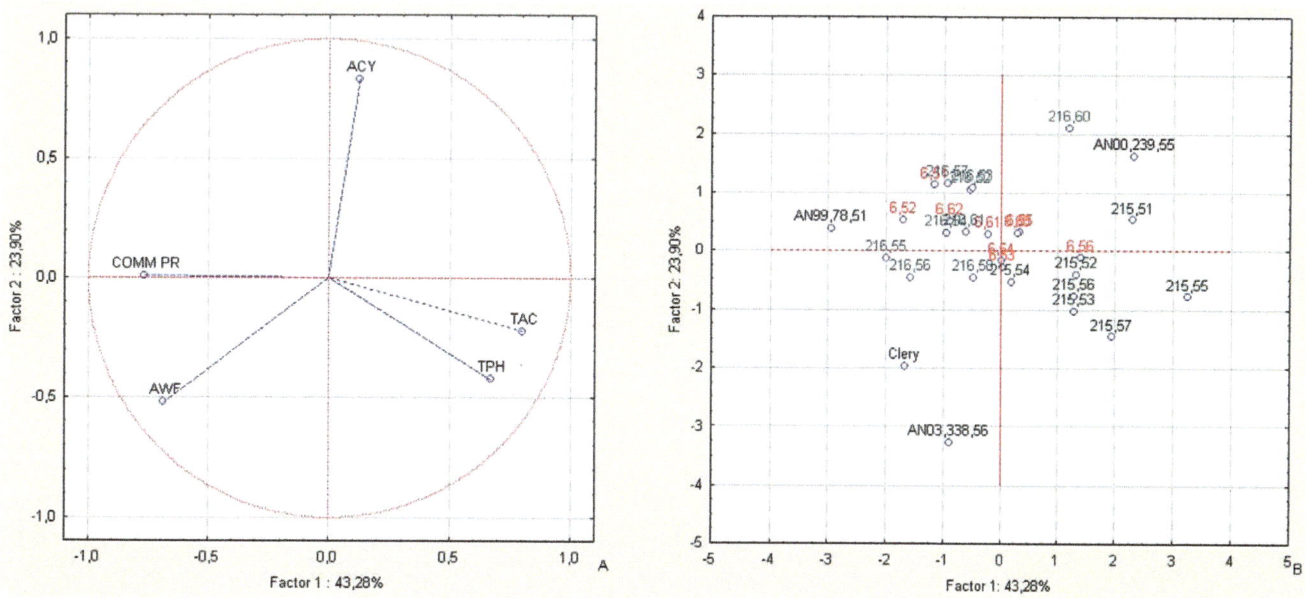

Figure 5. Principal component analysis. Bi-plot of the fruit weight, commercial production, TAC, TPH and ACY parameters of the selections and the corresponding parents of the BC2 back-cross populations. Factor 1 and Factor 2 explain 67.18% of the data variation. a: Variable vector distributions; b: Case distributions. The seedlings reported in the PCA are identified only with the codes related to the cross combinations and the progressive seedling numbering.

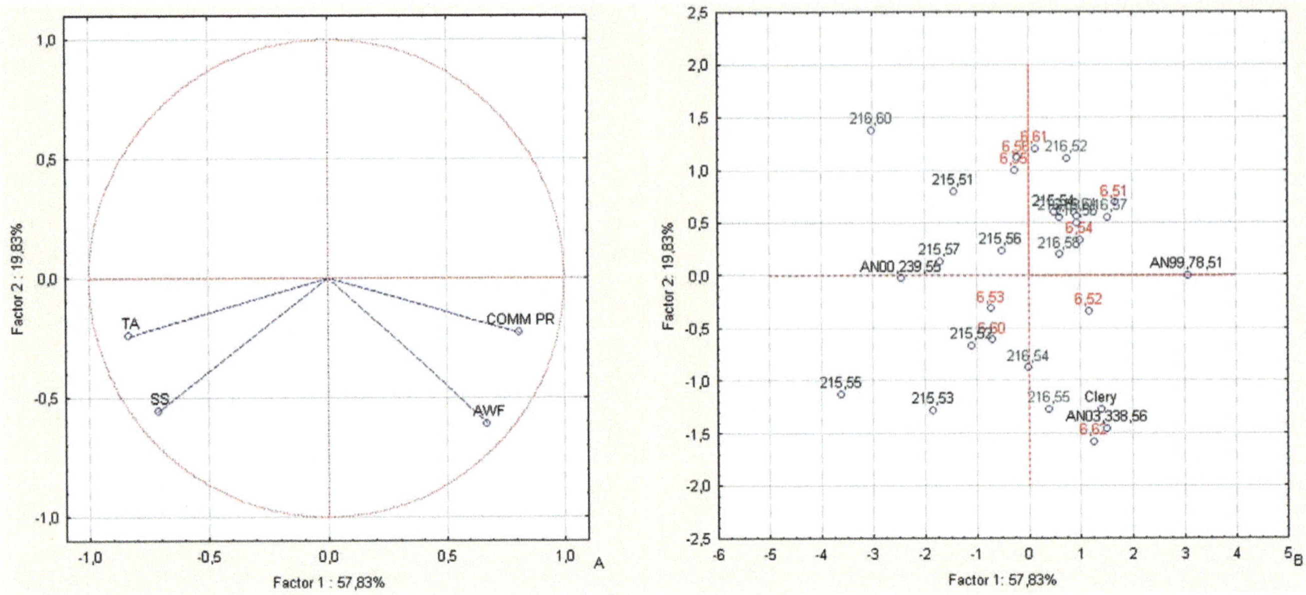

Figure 6. Principal component analysis. Bi-plot of the fruit weight, commercial production, TA and SS content parameters of the selections and the corresponding parents of the BC2 back-cross populations. Factor 1 and Factor 2 explain 78.66% of the data variation. a: Variable vector distributions; b: Case distributions. The seedlings reported in the PCA are identified only with the codes related to the cross combinations and the progressive seedling numbering.

interesting sensorial and nutritional parameters, and with Clery, a well-know commercial variety on the European Union market. The differences detected for these two intra-species families reveal the influence of the Romina genetic background in the production of a high combination of fruit sensorial and nutritional parameters, and indicate that these values can be improved even more if further genetic material identified for additional improved fruit sensorial and nutritional values can be used.

These two intra-species cross populations also showed negative correlations between both of the color parameters, Chroma index and L* grade, and the fruit firmness and ACY, which means that the high color parameters correspond to a softer fruit with lower ACY (Table 4). Indeed, particularly for the fruits of the AN07,007 family, the significantly higher values of both of the color parameters correspond to significantly lower fruit firmness and ACY.

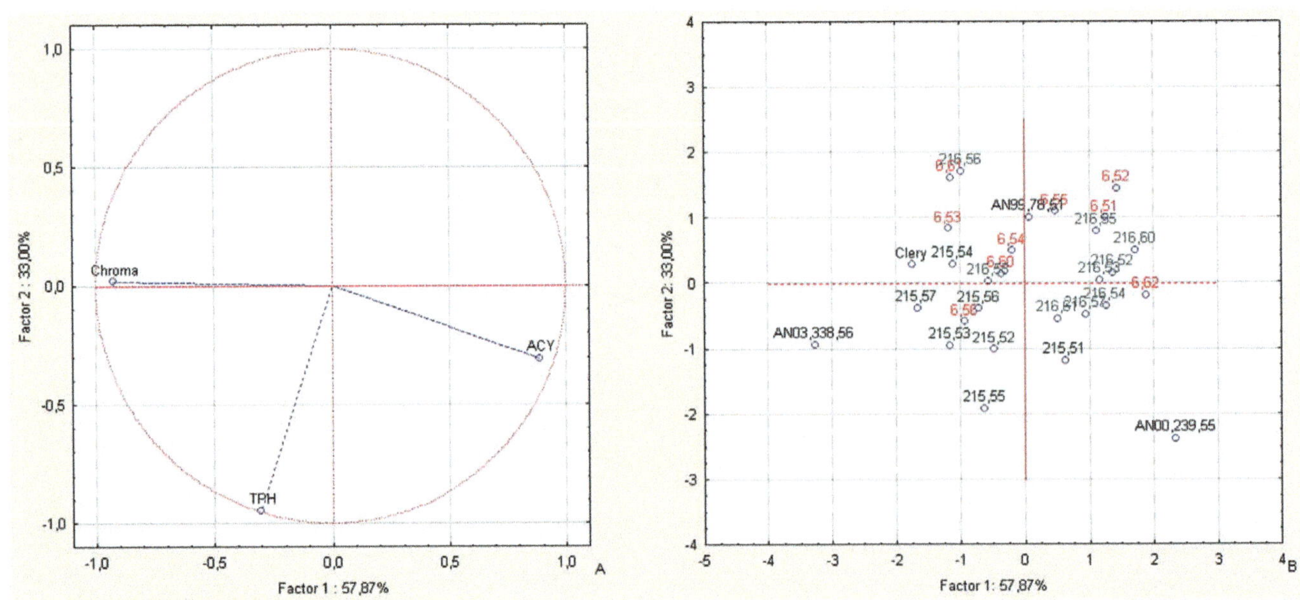

Figure 7. Principal component analysis. Bi-plot of the Chroma, TPH and ACY parameters of the selections and the corresponding parents of the BC2 back-cross populations. Factor 1 and Factor 2 explain 90.87% of the data variation. a: Variable vector distributions; b: Case distributions. The seedlings reported in the PCA are identified only with the codes related to the cross combinations and the progressive seedling numbering.

Table 5. The families with the corresponding cross and back-cross (BC) combination, as used in the present study.

Families		Progenitors Mother		Father	N° of offspring	Type of cross
A	AN07,003	AN94,414,52	×	Romina	11	BC1
	AN07,004	AN94,414,52	×	ALBA	10	BC1
	AN07,005	AN94,414,52	×	CLERY	10	BC1
B	AN07,006	AN00,239,55	×	Romina	8	BC2
	AN07,215	AN00,239,55	×	AN03,338,56	7	BC2
	AN07,216	AN00,239,55	×	CLERY	9	BC2
C	AN07,007	AN03,338,56	×	Romina	10	Intra-species
	AN07,009	CLERY	×	Romina	10	Intra-species

The corresponding parents and number of offspring evaluated for their fruit nutritional quality of the half-sib BC1 FVG, half-sib BC2 FVG and half-sib intra-species crossing. AN94,414,52 is a cross of the *Fragaria* × *ananassa* cultivar Don × *F. virginiana* ssp. *glauca* selection FVG 22 from USDA Corvallis germplasm repository (USA). AN00,239,55 (BC1) is an elite selection from the cross of selection AN94,414,52 (F1) × *F.* × *ananassa* selection 91,143,5. Romina and AN03,338,56 are elite selections from the Università Politecnica delle Marche breeding program, containing no wild germplasm.

The PCA bi-plot of fruit nutritional parameters showed a clear distribution of the populations on the plot that highlights the selection AN07,007,55 as having the high values of the fruit nutritional parameters (TAC and TPH).

The PCA bi-plot of fruit weight, plant commercial production, and fruit nutritional parameters (TAC, TPH and ACY) for the offspring and parents of the intra-species cross combinations (Figure 8a) showed distinct distributions for the two populations (Figure 8b). Indeed, the offspring of AN07,007 were mainly positioned in the central-right-hand area of the plot, where the TPH and TAC vectors are located, while only AN07,007,57 is located in the lower right-hand quadrant. The offspring of AN07,009 are mainly located in the lower quadrant (Figure 8b), where the vectors of the fruit weight and commercial production variables are located (Figure 8a). The female parents, Clery and AN03,338,56, are located on the right-hand side of the plot (Figure 8b), where the TPH and TAC vectors are located (Figure 8a), while the common male parent, Romina, is located on the lower part of the plot, where the vectors for fruit weight, commercial production and ACY are located.

The PCA bi-plot for fruit weight and commercial production combined with the sensorial parameters of SS content and TA (Figure 9a) distributes the offspring and parents widely in all four of the quadrants of the plot (Figure 9b). This distribution makes the different affinities of the two cross combinations towards the TA sensorial parameter clear. Indeed, the fruit of AN07,007 are located mainly towards the sensorial parameters vector TA (Figure 9b), while the fruit of the AN07,009 cross combination show a distribution that is mainly in the lower quadrants, on the opposite side of the TA vectors. From the distribution of the offspring, it can be noted that AN07,009,54, AN07,009,56 and AN07,009,60 are located in the lower left-hand quadrant, just in the opposite direction of the vectors that identify the TA variable, and also AN07,009,53, AN07,009,52 and AN07,009,51 are located on the lower part of the plot (Figure 9b), more oriented towards the SS content vector (Figure 9a). Half of the offspring of AN07,007, together with the female parents, AN03,338,56 and Clery, are located in the upper part of the plot (Figure 9b), between the vectors for the production variables and the fruit TA (Figure 9a). The offspring AN07,007,57 and AN07,009,57, together with the male parent (Romina), are located on the left-hand side of the plot (Figure 9b), where the production variables are located (Figure 9a). The selections AN07,007,53 and AN07,009,51, are located towards the SS content vector.

The PCA bi-plot for the fruit Chroma index, TPH and ACY parameters (Figure 10a) shows clear-cut distributions of the parents and offspring (Figure 10b). Indeed, the female parents together with most of the AN07,007 offspring are located on the left-hand side of the plot (Figure 10b), where the Chroma vector is located (Figure 10a), while the male parents and most of the AN07,009 offspring are on the right-hand side of the plot, which shows the separation between these two populations. The AN07,007 family is mainly located in the mid-upper part of the plot (Figure 10b), just on the opposite side of the TPH vector (Figure 10a).

The selection AN07,007,55 is located between the ACY and TPH vectors, while the selection AN07,009,56 is located towards the ACY vector, and the other offspring are not influenced by these variables (Figure 10b).

Discussion

To date, strawberry breeding programs have been mainly focused on the improvement of agronomic and commercial traits [23], although more recently the major breeding targets have shifted more towards the sensorial and nutritional qualities of the strawberry fruit [24–25]. Very recent studies that have focused on the sensorial [26–27] and nutritional [4–7,13–15,24–25,27] parameters in strawberry fruit have given more information and have considered further the concepts relating to their origins in plants and their importance for human health.

The major aim of the present study was to compare different breeding approaches to increase the narrow genetic base of the cultivated strawberry, and in particular, to increase the strawberry fruit sensorial and nutritional values. To achieve this, the contributions of *Fragaria virginiana* ssp. *glauca,* a wild octoploid strawberry, were tested in inter-species crosses and through subsequent back-cross cycles. The use of wild strawberry in this breeding program, which is aimed at producing new commercial varieties, is motivated by the need to increase the genetic variability of *Fragaria* × *ananassa* [28–29], because of the breeding selection process that has developed since the domestication of the first strawberry hybrids. The usefulness of FVG as an original source of fruit nutritional quality has already been reported [14]; however, long-term back-cross generations are necessary to provide new genotypes with high commercial value [5].

From the study of the inter-species back-cross populations, the offspring of BC1 and BC2 from FVG showed overall high fruit sensorial and nutritional qualities, as seen in the original F1 inter-species crosses [5]. These were combined with improvements in the other important commercial traits (mostly for fruit weight and commercial production), which were lost in the F1 lines. Indeed, the lack of association between the productivity and quality traits is well represented by the different PCA plots (Figures 2, 3, 5 and 6). These results highlight the independent genetic control of these important plant and fruit traits.

The use of FVG was also particularly efficient for the creation of greater variability, as seen with the production of a relatively high number of transgressive segregants. This result is considered to be particularly relevant in comparison with what was achieved from the *F.* × *ananassa* intra-species crosses included in the present study and in other studies [24]. Indeed, the FVG-derived populations

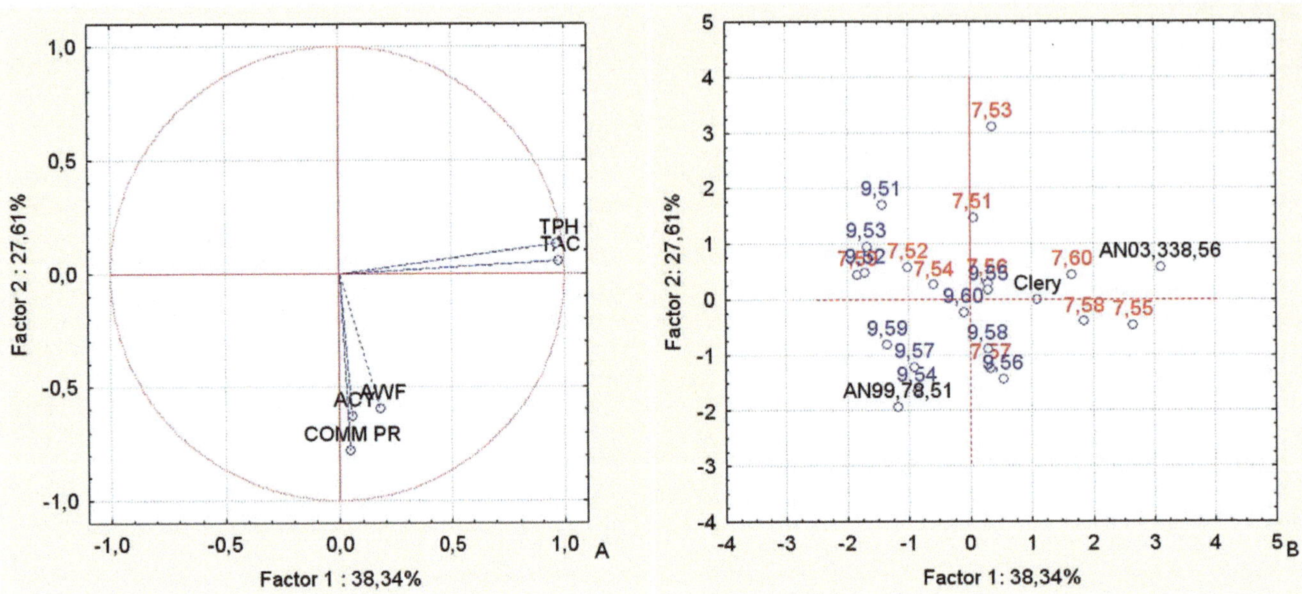

Figure 8. Principal component analysis. Bi-plot of the fruit weight, commercial production, TAC, TPH and ACY paraemters of the selections and the corresponding parents of the intra-species cross populations. Factor 1 and Factor 2 explain 65.95% of the data variation. a: Variable vector distributions; b: Case distributions. The seedlings reported in the PCA are identified only with the codes related to the cross combinations and the progressive seedling numbering.

had a higher number of new transgressive segregants, and also higher values for each of the sensorial and nutritional traits. This achievement is probably the result of an increased number of alleles that can interact epistatically [30], and also dominance/overdominance effects might have contributed to the presence of transgressive segregants for fruit quality [31].

The increase in fruit ACY is reflected phenotypically by the darker red coloration of the berry [24]. Fruit color is controlled at multiple levels that involve transcription factors and structural genes [32–33]. Quantitative trait loci for anthocyanin and fruit color have been evaluated for the *flavanone 3-hydroxylase, chalcone isomerase, flavonol synthetase* and *dihydroflavonol reductase* genes, which are involved in anthocyanin biosynthesis, but no gene co-location

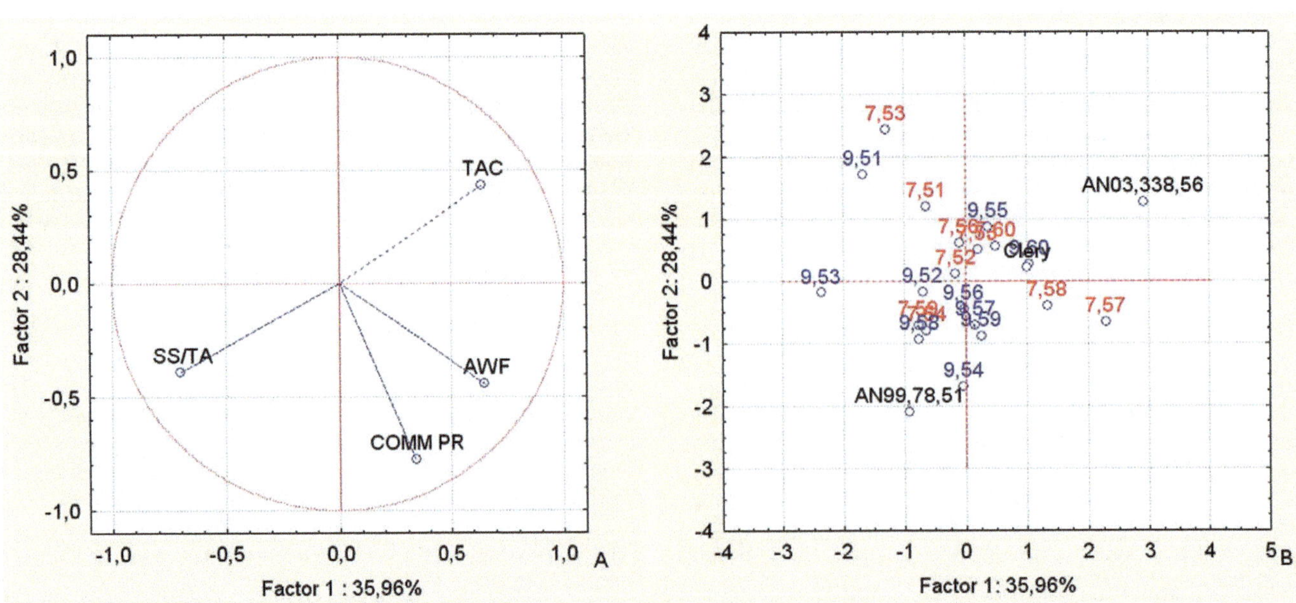

Figure 9. Principal component analysis. Bi-plot of the AWF, commercial production, SS conent and TA parameters of the selections and the corresponding parents of the intra-species cross populations. Factor 1 and Factor 2 explain 75.00% of the data variation. a: Variable vector distributions; b: Case distributions. The seedlings reported in the PCA are identified only with the codes related to the cross combinations and the progressive seedling numbering.

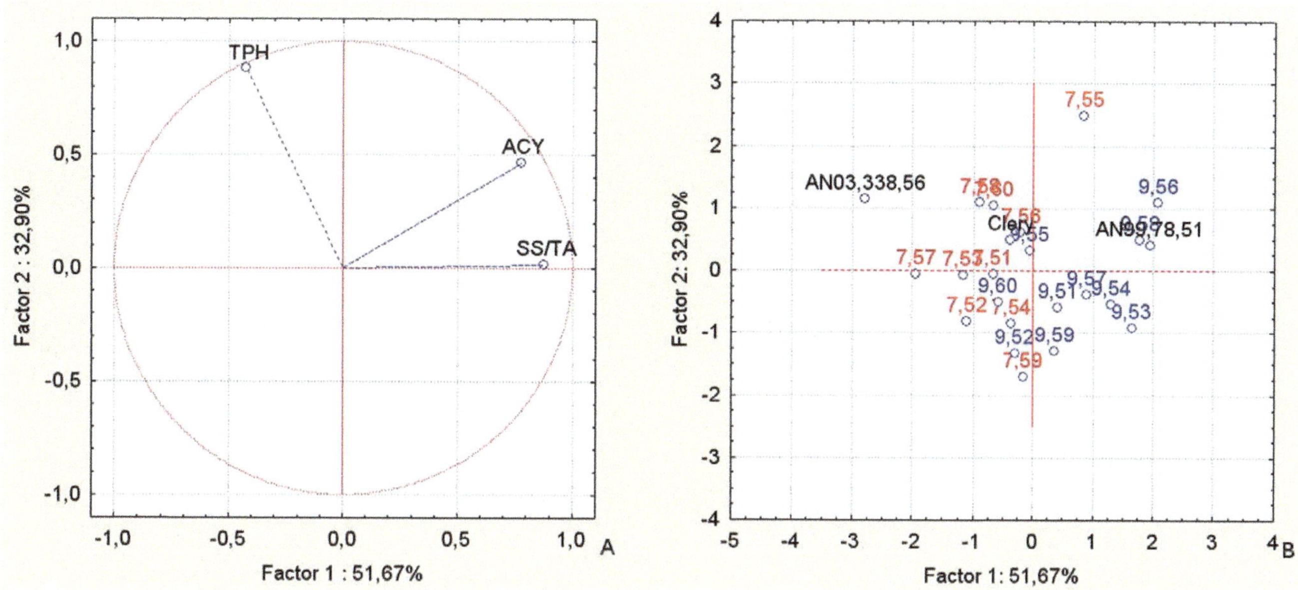

Figure 10. Principal component analysis. Bi-plot of the Chroma, TPH and ACY parameters of the selections and the corresponding parents of the intra-species cross populations. Factor 1 and Factor 2 explain 90.87% of the data variation. a: Variable vector distributions; b: Case distributions. The seedlings reported in the PCA are identified only with the codes related to the cross combinations and the progressive seedling numbering.

with fruit color or ACY was seen [25]. Improvements in the nutritional values in the BC1 and BC2 FVG populations, which were due in particular to increased ACY (Table 6), were associated with a decrease in the Chroma index. This was supported by the correlation matrix (Table 4), where these parameters were highly negatively correlated, both in the BC and intra-species offspring populations (Table 4) [24]. While for the intra-species populations, these were the only ones where a high negative correlation between colorimetric parameters and phenol content was confirmed [5,15]. The PCA plots for the BC1 populations (Figure 4) show a high number of transgressive segregants for fruit color and ACY, while the BC2 populations do not show segregants. Hence, these data confirm the concept that fruit color is controlled by epistatic and dominance/overdominance effects [30–31], as seen for the nutritional compounds.

The importance of *F. virginiana*, and in particular ssp. *glauca*, for increases in pathogen resistance of the hybrid progenies was evaluated in a breeding program that was designed to widen the genetic base of the strawberry germplasm [22]. The BC1 and BC2 populations showed marked reductions in rotten fruit production in comparison to the *F.* × *ananassa* populations. This lower rotten fruit production was accompanied by a higher content of bioactive compounds, and specifically of anthocyanins and phenols, which have active roles in the protection of plants and fruit against pathogens [34–36]. Hence, the increases in the bioactive compounds in the BC1 and BC2 populations, and the data showing lower levels of rotten fruit production, can be considered as confirmation that the increased content of these bioactive compounds can have effects on increased tolerance to the major strawberry diseases [22]. Therefore, these achievements underline the importance of the wild genetic resources as sources of different important genes and the need for continued evaluation of back-cross generations for the creation of new genotypes with both improved nutritional compound content and disease resistance. Moreover, studies performed by Tulipani et al. [37] also showed that wild species can contribute to increases in not only interesting nutritional features in cultivated strawberry, but also in non-nutritional compounds, which might have implications for human consumption. From these findings, it is clear that intra-species crosses induce low increases in allergens contents, while the fruit from the F1 and BC1 selections had much higher contents of allergens.

In general, the performances of the BC1 and BC2 FVG populations can be further improved by additional back-cross generations, so as to be closer to the overall commercial quality that is required for any new commercial variety. By the third back-cross generation, the plant and fruit commercial standards can be very near to those requested for commercial cultivars [8].

The results for the *F.* × *ananassa* intra-species crosses show that there is a genetic background in the cultivated species that can induce an increase in the fruit sensorial and nutritional qualities. Even if these are of lower value in comparison with the FVG back-cross program, they are already combined with a high standard of commercial performance. However, this positive increase is strictly related to the genetic background of the selected parents. Indeed, the parents used in the cross combinations were chosen because they had already been identified as having high fruit nutritional quality in comparison with other cultivars [5,12,15].

Conclusions

The present study has demonstrated that these two types of combination programs (inter-species back-crosses and intra-species crosses) can be used for the improvement of strawberry nutritional quality, although the success of the program is strictly related to the combining attitude of the different parents. For this reason, this type of experiment can be considered important for the production of new genotypes with improved sensorial and nutritional qualities, and for the identification of cultivars and selections that can perform in the same way as the most useful parents. These can be used in strawberry breeding programs that are aimed at the production of new improved varieties, for the better combination of the agronomic and overall fruit quality traits.

Table 6. Fruit sensorial and nutritional parameters of the BC1 and BC2 back-cross combinations and the intra-species cross combinations (evaluated in 2010 on clonal population of selected seedlings).

Type of Cross	SS content (°Brix)	TA (meq NaOH/100 g FW)	Firmness (g)	Chroma	L*	TAC (mmol TE/kg FW)	TPH (mg GA/kg FW)	ACY (mg PEL-3-GLU/ kg FW)
BC1	9.40 a (0.15)	12.28 b (0.37)	336.00 b (8.26)	44.48 b (0.62)	34.24 b (0.36)	20.25 a (0.74)	1707.46 a (69.16)	464.16 a (19.09)
BC2	8.99 a (0.17)	13.55 a (0.42)	330.79 b (7.61)	42.67 b (0.66)	34.75 b (0.41)	21.42 a (1.09)	1708.72 ab (46.74)	484.34 a (21.1)
IS	8.45 b (0.18)	10.49 c (0.34)	381.77 a (11.47)	46.84 a (0.93)	37.54 a (0.66)	14.37 b (0.32)	1517.32 b (64.03)	353.97 b (25.69)

Data are means and standard errors (SE) of offspring selections for each cross combination type. Means followed by different letters are significantly different. SNK Test P≤0.05. IS, intra-species.

These evaluations carried out on the genetic pools that originated from the *F. virginiana* ssp. *glauca* inter-species back-cross and the *F.* × *ananassa* intra-species cross confirm the importance of using FVG, or genotypes with an FVG genetic background, in breeding programs aimed at increasing the fruit nutritional (TAC, TPH and ACY) and sensorial (SS content, TA, Chroma, L*, firmness) qualities. Indeed, both the BC1 and BC2 populations showed these characteristics, thus providing useful improvements in the fruit nutritional and sensorial qualities combined with agronomic standards that are closer to the those requested at the commercial level with respect to their respective FVG hybrid parents.

The results obtained in the present study show that two sequential back-crossing generations (BC1 and BC2) from an inter-species hybrid with a wild genotype of *F. virginiana* ssp. *glauca* can improve sensorial and nutritional traits. These data confirm that at least two back-cross generations are needed to be close to the agronomic values requested at the commercial level [8].

This study has created new genotypes with the FVG genetic background introduced by inter-species crosses, thus increasing the genetic variability in cultivated strawberry, particularly for their specific nutritional traits [5]. This helps to counteract the narrow germplasm base that is used for breeding, which has caused deleterious effects of inbreeding and genetic vulnerability [27]. From our data, it is quite clear which parameters have to be increased in a back-crossing program. Fruit firmness, size, acidity and color are the main traits that can be recovered in the back-crossing generations. Clearly, the high positive values detected for some of the parameters in the F1 genotypes (TAC, polyphenols and folate) are reduced in the back-crossing generations, although they always remain higher than those in the commercial *F.* × *ananassa* genotypes. These findings suggest that controlled crosses of wild species with cultivated strawberry will contribute to introgression of interesting nutritional features; however, this can also lead to an increase in unfavorable compounds that might have health implications for humans, and that thus need specific evaluation.

Successive breeding studies using wild *Fragaria* octoploid species can introduce new genetic variability for nutritional, flavor and disease-resistance traits, and can contribute to our understanding of their heritability ratios and the hypothetical correlations with nutritional parameters [27]. Indeed, further studies on wild genotypes have evaluated the importance of FVG [22], and also of other wild genotypes with different ploidy levels [38–40], in terms of their flavor and resistance parameters, using gene expressions studies [41].

Improvements in sensorial and nutritional traits can also be achieved by the programming of *F.* × *ananassa* intra-species crosses and the production of progeny with productivity traits that are closer to those of the commercial cultivars, even if these improvements are fewer in number and quality with respect to the FVG back-cross populations.

Materials and Methods

Cross Combinations and Seedling Evaluations

A breeding program was carried out to obtain eight half-sib cross combinations originated using the following six different strawberry inter-species back-cross combinations and two intra-species cross combinations (Table 5, Figure 1):

a) Three half-sib cross combinations originated as back-cross populations (BC1) with the F1 selection AN94,414,52, and an inter-species cross of *F.* × *ananassa* (Don) × *F. virginiana glauca* (FVG22) as the common parent;

b) Three half-sib cross combinations originated as back-cross populations (BC2) with AN00,239,55, a BC1 selection derived from the back-crossing of selection F1 AN94,414,52 with 91,143,5 (*F.* × *ananassa* advanced selection) as the common parent;

c) Two half-sib cross combinations originated by an intra-species cross with the selection Romina as the common parent.

Seedlings from all of these cross combinations were planted in 2008 in non-fumigated soil at the 'P. Rosati' University experimental farm, Agugliano (Ancona, Italy), and grown under open-field conditions using the plastic hill culture production system. For each cross combination of about 50 seedlings, seven to 11 elite seedlings were selected, based on a subjective phenotypic evaluation of the major plant vegetative and fruit sensorial traits, which was performed at the stage of fruit ripening. All of the seedlings of the different families were identified by code number, including AN as the Ancona site of the crossing, followed by 07 as the year of the crossing, with the code for the type of crossing, followed by progressive numbering (starting from 51) for each selected seedling. From this subjective selection, a total of 80 elite seedlings were identified and propagated by runners to create clonal population of the selected seedling. After the 2009 evaluation, 78 elite seedlings (two were lost) were planted as fresh plants in July 2009, in single plots of six plants each. They were then harvested and evaluated in May 2010, as selections from clonal population of selected seedlings.

This second evaluation was performed under the same open-field conditions in non-fumigated soil at the 'P. Rosati' University experimental farm, Agugliano (Ancona, Italy), again using the plastic hill culture production system. In the same field, under the same conditions, the corresponding parents of each of the cross combinations were also grown. Fruit samples of the fully red berries were harvested at the second, third and fourth main pickings. The evaluations took into account plant productivity (fruit weight and total plant production), and fruit nutritional (fruit TAC, phenols and ACY) and sensorial (fruit firmness, color, SS content and TA) parameters.

Chemicals

The 99% methanol (ACS-ISO) for the analysis, and the bromothymol blue were purchased from Carlo Erba Reagenti (Milan, Italy). Folin-Ciocalteu's reagent anhydrous, sodium carbonate, potassium chloride, sodium acetate, chloridric acid, glacial acetic acid, dihydrogen potassium phosphate, dipotassium hydrogen phosphate, 2,2′-azino-bis (3-ethylbenzothiazoline-6-sulphonic acid) diammonium salt (ABTS•+), 6-hydroxy-2,5,7,8-tetramethylchromane-2-carboxylic acid (Trolox), potassium persulfate, 3,4,5-trihydroxybenzoic acid (gallic acid), and sodium hydroxide, were purchased from Sigma-Aldrich (Milan, Italy).

Strawberry Production Parameters

In 2010, the strawberry plant production was evaluated in the single plot of all of the selections at each harvest, by evaluating the following parameters:

a) Total fruit production (g/plant).

b) Commercial production: total weight (g/plant) of ripe fruit with a commercial diameter (Ø≤0.22 mm) and without injuries.

c) Fruit weight: average weight (g) of 20 commercial fruits from each harvest.

d) Undersized fruit: total weight (g/plant) of ripe fruits with diameter Ø ≥0.22 mm, without injuries.

e) Misshapen fruit: total weight (g/plant) of misshapen fruits.

f) Rotten fruit: total weight (g/plant) of rotten fruits.

Strawberry Instrumental Sensorial Parameters

Fruit sensorial quality of the selections and the parents was analyzed on the harvesting day, at each harvest (2nd, 3rd, 4th), on 10 commercial fruits, taking in account the following parameters:

a) SS content: determined using a hand-held refractometer (ATAGO), with results expressed as °Brix.

b) TA: determined from 10 mL juice diluted with distilled water (1/2, v/v) and titrated with 0.1 N NaOH solution to pH 8.2, and expressed as mEQ of NaOH per 100 g fruit (fresh weight; FW).

c) Fruit color: determined as the Chroma index using a Minolta Chromameter CR 400, for two sides of 10 sound, ripe, undamaged and uniform fruits. The instruments included three parameters L* (luminescence), a* (red tone), b* (yellow tone). The Chroma index was evaluated from a and b $[(a^{*2}+b^{*2})^{½}]$, with a higher Chroma index representing pale fruit and a low Chroma index representing dark fruit.

d) Fruit firmness: measured using a penetrometer 327 (Effegì, Ravenna, Italy), with the results expressed in g.

Strawberry Nutritional Parameters

Fruit extraction method. From a pool of fruit for each genotype harvested in 2010 and stored at −20°C, 10 ripe commercial fruits were sampled, and from each fruit two slices from opposite sides were cut and milled into small pieces. Ten gram of this blend was weighed and placed into a tube for extraction with methanol (1:4, fruit: methanol, w/w), including two subsequent steps. The first step consisted of homogenization with an Ultraturrax T25 homogenizer, for a fruit blend in 20 mL methanol (Janke and Kunkel, IKA Labortechnik, Staufen, Germany). The homogenized suspension was continuously agitated for 30 min in the dark. The suspension was centrifuged at 4,500×*g* for 10 min (Centrifuge Rotofix32, Hettich Zentrifugen, Tuttlingen, Germany), and the supernatant was collected and stored in three amber vials, of 1 mL each. For a more complete and extensive extraction, the pellet of the fruit was extracted a second time by adding another 20 mL methanol and repeating the procedure described above. The second supernatant was collected and added to the first one, always using 1 mL for each of the 3 amber vials per sample, and then stored at −20°C.

TAC. The fruit extract from each genotype was used to analyze the fruit TAC using the ABTS assay, according to a previously validated procedure [42–43]. ABTS is a chromogen and a colorless substance; it changes into its colored monocationic radical form (ABTS•+) with exposure to an oxidative agent. Addition of antioxidants reduces ABTS•+ into its colorless form. The extent of decolorization as a percentage of inhibition of ABTS•+ was determined as a function of the concentration and was calculated relative to the reactivity of Trolox, a water-soluble vitamin E analog. The antioxidant activity is expressed as mg Trolox equivalent per kg fresh pulp weight. The calibration was calculated as the linear regression from the dose-response of the Trolox standard.

TPH. The fruit TPH was evaluated on fruit extracts using the Folin-Ciocaltou's reagent method [44], with gallic acid as the standard for the calibration curve. Briefly, a glass test-tube was filled with 7.0 mL water. Afterwards, 1 mL of the diluted sample (1:20) was added, followed by the addition of 500 μL Folin-Ciocalteu-Reagent and vortexing. After 3 min, 1.5 mL sodium carbonate (0.53 mol/L) was added, and the tube was mixed once more and then stored in the dark for 60 min. The absorbance of the sample was measured at 760 nm after exactly 60 min. The data were calculated and expressed as mg gallic acid per kg fresh fruit, using a standard curve with the range of standard values of 10 to 50 mg gallic acid/L.

ACY. Total fruit ACY was measured on the fruit extracts using the pH differential shift method [45]. This assay is based on the anthocyanins characteristic of a change in intensity of hue depending on the pH shift. Briefly, the samples were diluted to a ratio of 1:10 with potassium chloride (pH 1.00) and with sodium acetate (pH 4.50), and then the corresponding maximum absorbance for both of the solutions was measured at $\lambda = 500$ nm and at $\lambda = 700$ nm, respectively. The data were expressed as mg pelargonidin-3-glucoside (the anthocyanin more representative in strawberry) per kg fresh weight (mg Pel-3-Glu/kg FW).

Experimental Trial and Statistical Analysis

In the 2010 harvest season, the fruit sensorial (SS content, TA, firmness, Chroma, L*) and nutritional (TAC, TPH, ACY) parameters were analyzed in triplicate for each clonal population of selected seedlings identified in 2009, and for their corresponding parents.

All of these data were subjected to the two-way nested analysis of variance, MANOVA test. The analysis of variance was performed for single cross types and single cross combinations as random effects. The data on fruit sensorial and nutritional parameters are reported as means±standard error (SE). Multiple comparisons were calculated according to the SNK test and were considered significant at $p \leq 0.05$.

Correlations among the sensorial (SS content, TA, firmness, Chroma, L*) and nutritional (TAC, TPH and ACY) parameters were analyzed using Pearson's correlations. PCA was also used to evaluate the levels of association among the various production, sensorial and nutritional parameters of the selections that originated from the different cross combinations and their corresponding parents, calculated on a genotype-mean basis. Based on the theoretical arguments of PCA described by [46], the significant factor loading values ≥0.7 were used to identify the most important variables and observations in each dimension (PC). The factor loading values are the correlations of each variable with the PC. They are represented as vectors (positions) in the space represented by the axes of the PCA bi-plots (Figures 2a to 10a). In the graphs, the variables (Figures 2a to 10a) and observations (Figures 2b to 10b) that are closest to each other in the same geometric plane of the bi-plot are considered as interrelated, and as a result they are distant from the variables and observations to which they are not related, or are negatively related. The greater the distance of a vector from the origin of the axis, the higher the correlation of the variable with the PC represented in that dimension (axis). All of the analyses were performed using STATISTICA (Statsoft Inc., Tulsa, OK, USA).

Acknowledgments

Our thanks are expressed to Dr. Enrico Muzzi for assistance in planning and discussing statistical analyses.

Author Contributions

Conceived and designed the experiments: BM FC JD. Performed the experiments: JD FC FB. Analyzed the data: JD FC JH BM. Contributed reagents/materials/analysis tools: JD BM. Wrote the paper: JD MB JH BM.

References

1. Meyers KJ, Watkins CB, Pritts MP, Liu RH (2003) Antioxidant and antiproliferative activities of strawberries. J Agric Food Chem 51: 6887–6892.
2. Prior RL, Cao G, Martin A, Sofic N, McEwen J, et al. (1998) Antioxidant capacity as influenced by total phenolic and ACY, maturity and variety of *Vaccinium* species. J Agric Food Chem 46: 2686–2693.
3. Proteggente AR, Pannala AS, Pagana G, Van Buren L, Wagner E, et al. (2002) The antioxidant activity of regular consumed fruit and vegetables reflects their phenolic and vitamin C composition. Free Radic Res 36: 217–233.
4. Shiow Y, Wang L, Kim S (2007) Antioxidant capacity and flavonoid content in wild strawberries. J Amer Soc of Hort Sci 132: 629–637.
5. Tulipani S, Mezzetti B, Capocasa F, Bompadre S, Beekwilder J, et al. (2008) Antioxidants, phenolic compounds, and nutritional quality of different strawberry genotypes. J Agric Food Chem 2008, 56: 696–704.
6. Battino M, Beekwilder J, Denoyes-Rothan B, Laimer M, McDougall JG, et al. (2009) Bioactivities of berries relevant to human health. Nutr Rev 67: 145–150.
7. Tulipani S, Romandini S, Busco F, Bompadre S, Mezzetti B, et al. (2009) Ascorbate, not urate, modulates the plasma antioxidant capacity after strawberry intake. Food Chem 117: 181–188.
8. Bringhurst RS & Voth V (1978) Origin and evolutionary potentiality of the day-neutral trait in octoploid. Fragaria Genetics 90: 510.
9. Della Penna D (2001) Plant metabolic engineering. Plant Physiology 125: 160–163.
10. Azodanlou R, Darbellay C, Luisier J, Villettaz J, Amadò R (2003) Quality assessment of strawberry (*Fragaria* species). J Agric Food Chem 51: 715–721.
11. Olsson ME, Ekvall J, Gustavsson K, Nilsson J, Pillai D (2004) Antioxidants, low molecular weight carbohydrates, and TAC in strawberry (*Fragaria* x *ananassa*). J Agric Food Chem 52: 2490–2498.
12. Scalzo J, Politi A, Mezzetti B, Battino M (2005) Plant genotype affects TAC and phenolic contents in fruit. Nutrition 21: 207–213.
13. Wang SY, Zheng W, Galletta GJ (2002) Cultural system affects fruit quality and antioxidant capacity in strawberries. J Agric Food Chem 50: 6534–6542.
14. Wang SY, Lewers KS (2007) Antioxidant capacity and flavonoid content in wild strawberries. J Amer Soc Hort Sci 132: 629–637.
15. Capocasa F, Diamanti J, Tulipani S, Battino M, Mezzetti B (2008) Breeding strawberry (*Fragaria* X *ananassa* Duch) to increase fruit nutritional quality. Bio Factors 34: 67–72.
16. Capocasa F, Bordi M, Mezzetti B (2009) Comparing frigo and fresh plant in not fumigated and heavy soil: the response of ten strawberry genotypes. Proc VIth International Strawberry Symposium Acta Hortic (ISHS) 842: 129–133.
17. Connor AM, Luby JJ, Tong CBS, Finn CE, Hancock JF (2002) Variation and heritability estimates for antioxidant activity, total phenolic content and ACY in blueberry progenies. J Amer Soc Hort Sci 1: 82–88.
18. Connor AM, Stephens MJ, Hall HK, Alspach PA (2005) Variation and heritabilities of antioxidant activity and total phenolic content estimated from a red raspberry factorial experiment. J Amer Soc Hort Sci 3: 130.
19. Deighton N, Brennan R, Finn C, Davies HV (2000) Antioxidant properties of domesticated and wild *Rubus* species. J. Sci. Food Agric. 80: 1307–1313.
20. Wang SY, Lin HS (2000) Antioxidant activity in fruits and leaves of blackberry, raspberry, and strawberry varies with cultivar and developmental stage. J Agric Food Chem 48: 140–146.
21. Zhen W, Wang SY (2003) Oxygen radical absorbing capacity of phenolics in blueberries, cranberries, chokeberries, and lingonberries. J Agric Food Chem 51: 502–509.
22. Hancock JF, Luby JJ, Dale A, Callow PW, Serce S, et al. (2002) Utilizing wild *Fragaria virginiana* in strawberry cultivar development: inheritance of photoperiod sensitivity, fruit size, gender, female fertility and disease resistance. Euphytica 126: 177–184.
23. Darrow GM (1966) The strawberry: history, breeding and physiology. Holt, Rinehart, and Winston, New York.
24. Lerceteau-Köhler E, Moing A, Guérin G, Renaud C, Petit A, et al. (in press) Genetic dissection of fruit quality traits in the octoploid cultivated strawberry highlights the role of homoeo-QTL in their control. Theor Appl Gen DOI: 10.1007/s00122-011-1769-3.
25. Zorrilla-Fontanesi Y, Cabeza A, Domínguez P, Medina JJ, Valpuesta V, et al. (2011) Quantitative trait loci and underlying candidate genes controlling agronomical and fruit quality traits in octoploid strawberry (*Fragaria* × *ananassa*). Theor Appl Gen 123: 755–778.

26. Kader AA (1991) Quality and its maintenance in relation to the post-harvest physiology of strawberry. In: Luby JJ, Dale A (eds) Quality and its maintenance in relation to the post-harvest physiology of strawberry. Timber press, Portland, 145–152.
27. Aharoni A, Giri AP, Verstappen FWA, Bertea CM, Sevenier R, et al. (2004) Gain and loss of fruit flavor compounds produced by wild and cultivated strawberry species. Plant Cell 16: 3110–3131.
28. Sjulin TM, Dale A (1987) Genetic diversity of North American strawberry cultivars. J Am Soc Hortic Sci 112: 375–385.
29. Hancock J, Luby J (1995) Adaptative zones and ancestry of the most important North American strawberry cultivars. Fruit Varieties J 49: 85–90.
30. Coelho CM, Wu S, Li YC, Hunter B, Dante RA, et al. (2007) Identification of quantitative trait loci that affect endoreduplication in maize endosperm. Theor Appl Genet 115: 1147–1162.
31. Flagel LE, Wendel JF (2009) Gene duplication and evolutionary novelty in plants. New Phytol 183: 557–564.
32. Carbone F, Preuss A, De Vos RC, D'Amico E, Perrotta G, et al. (2009) Developmental, genetic and environmental factors affect the expression of flavonoid genes, enzymes and metabolites in strawberry fruits. Plant Cell Environ 32: 1117–1131.
33. Allan AC, Hellens RP, Laing WA (2008) MYB transcription factors that colour our fruit. Trends Plant Sci 13: 99–102.
34. Karjalainen R, Lehtinen A, Hietaniemi V, Pihlava JM, Jokinen K, et al. (2002) Benzothiadiazole and glycine betaine treatments enhance phenolic compound production in strawberry. Acta Hort. 567: 353–356.
35. Hukkanen AT, Kokko HI, Buchala AJ, McDougall GJ, Stewart D, et al. (2007) Benzothiadiazole induces the accumulation of phenolics and improves resistance to powdery mildew in strawberries. J. Agric. Food Chem. 55: 1862–1870.
36. Hebert C, Charles MT, Gauthier L, Willemot C, Khanizadeh S, et al. (2002) Strawberry proanthocyanidins: biochemical markers for *Botrytis cinerea* resistance and shelf-life predictability. Acta Hort. 567: 659–661.
37. Tulipani S., Marzban G., Herndl A., Laimer M., Mezzetti, et al. (2011). Influence of environmental and genetic factors on health-related compounds in strawberry. Food Chem 124: 906–913.
38. Amil-Ruiz F, Blanco-Portales R, Muñoz-Blanco J, Caballero JL (2011) The strawberry plant defense mechanism: a molecular review. Plant Cell Physiol 52: 1873–1903.
39. Gooding HJ, Mcnicol RJ, Macintyre D (1981) Methods of screening strawberries for resistance to *Sphaerotheca macularis* (wall ex Frier) and *Phytophthora cactorum* (Leb. and Cohn). J. Hort. Sci. Biotechnol. 56: 239–245.
40. Harland SC, King E (1957) Inheritance of mildew resistance in *Fragaria* with special reference to cytoplasmatic effects. Heredity 11: 257.
41. Maas J (1998) Compendium of Strawberry Diseases. American Phytopathological Society, St Paul, MN.
42. Miller NJ, Rice-Evans C, Davis MJ (1993) A novel method for measuring antioxidant capacity and its application to monitoring the antioxidant status in premature neonates. Clinical Science 83: 407–412.
43. Re R, Pellegrini N, Proteggente A, Pannala A, Yang M (1999) Antioxidant activity applying an improved ABTS radical cation decolorization assay. Free Radic Biol & Med 26: 1231–1237.
44. Slinkard K, Singleton VL (1977) Total phenol analysis: automation and comparision with manual methods. Amer J Enol and Vitic 28: 49–55.
45. Giusti M, Wrolstad RE (2001) Characterization and Measurement of Anthocyanins by UV-visible Spectroscopy. In Current Protocols in Food Analytical Chemistry. Hoboken: John Wiley and Sons F1.2.1–F1.2.13.
46. Hair F, Anderson J, Tatham L, Black C (1998) Multivariate data analysis, 5th edition. Prentice Hall.

10

Vip3A Resistance Alleles Exist at High Levels in Australian Targets before Release of Cotton Expressing This Toxin

Rod J. Mahon[1]*, Sharon J Downes[2], Bill James[1]

1 Ecosystem Sciences Division, The Commonwealth Scientific and Industrial Research Organisation, Black Mountain Laboratories, Canberra, Australian Capital Territory, Australia, **2** Ecosystem Sciences Division, The Commonwealth Scientific and Industrial Research Organisation, Australian Cotton Research Institute, Narrabri, New South Wales, Australia

Abstract

Crops engineered to produce insecticidal crystal (Cry) proteins from the soil bacterium *Bacillus thuringiensis* (Bt) have revolutionised pest control in agriculture. However field-level resistance to Bt has developed in some targets. Utilising novel vegetative insecticidal proteins (Vips), also derived from Bt but genetically distinct from Cry toxins, is a possible solution that biotechnical companies intend to employ. Using data collected over two seasons we determined that, before deployment of Vip-expressing plants in Australia, resistance alleles exist in key targets as polymorphisms at frequencies of 0.027 (n = 273 lines, 95% CI = 0.019–0.038) in *H. armigera* and 0.008 (n = 248 lines, 0.004–0.015) in *H. punctigera*. These frequencies are above mutation rates normally encountered. Homozygous resistant neonates survived doses of Vip3A higher than those estimated in field-grown plants. Fortunately the resistance is largely, if not completely, recessive and does not confer resistance to the Bt toxins Cry1Ac or Cry2Ab already deployed in cotton crops. These later characteristics are favourable for resistance management; however the robustness of Vip3A inclusive varieties will depend on resistance frequencies to the Cry toxins when it is released (anticipated 2016) and the efficacy of Vip3A throughout the season. It is appropriate to pre-emptively screen key targets of Bt crops elsewhere, especially those such as *H. zea* in the USA, which is not only closely related to *H. armigera* but also will be exposed to Vip in several varieties of cotton and corn.

Editor: Alejandra Bravo, Universidad Nacional Autonoma de Mexico, Instituto de Biotecnologia, Mexico

Funding: The authors received financial support from the Commonwealth Scientific and Industrial Research Organisation and the Cotton Research and Development Corporation to conduct this research. The funders had no role in study design, data collection and analysis, decision to publish, or preparation of the manuscript.

Competing Interests: The authors received funding from a commercial source (Cotton Research and Development Corporation).

* E-mail: rod.mahon@csiro.au

Introduction

New beneficial traits in populations can arise from selection of existing rare gene polymorphisms (pre-adaptation) or new mutations generated in response to a selection pressure [1]. The extent to which either of these processes contributes to evolution can have different consequences for the rate of adaptation [2]. For instance, mutations of esterase 3 confer malathion resistance on contemporary Australasian *Lucilia cuprina* and PCR analysis of pinned specimens collected before the release of insecticides reveals several cases of changes that conferred resistance. Thus, the early outbreak of resistance in this species can be explained by the pre-existence of mutant alleles encoding malathion resistance [2].

The soil bacterium *Bacillus thuringiensis* produces crystalline inclusions during sporulation which contain insecticidal δ-endotoxins or Cry toxins. Specific *cry* genes engineered into plants to express the proteinaceous toxins, have been the mainstay of transgenic insecticidal crops since the mid-1990's. In several target pests major gene polymorphisms for resistance to Cry toxins were detected before populations were exposed to Bt crops [3–6]. Incipient resistance to Bt was reported in at least one of these targets within several years of exposure to crops [7]. Field level resistance to Bt-crops has been reported in at least 4 target pests [8,9] but the pre-existing levels of gene polymorphisms for resistance are unknown for these systems.

The usual target pests for Bt crops are Lepidoptera. In general this group are only susceptible to toxins in the Cry1 (e.g., Cry1Ac, Cry1Ab, Cry1F) and Cry2 (e.g., Cry2Ab, Cry2Aa, Cry2Ae) classes, several of which are exploited in existing transgenic crops [8]. Within the Cry1 class, insects which are resistant to one toxin are often, but not inevitably, cross resistant to others [10]. Less is known about cross resistance within the Cry2 class, however Cry2Ab resistant *H. armigera* are resistant to Cry2Aa [11] and Cry2Ab resistant *H. armigera* and *H. punctigera* are resistant to Cry2Ae [12]. Therefore, it is likely for most systems that if resistance emerges to a toxin in the Cry1 or Cry 2 class, there are limited alternative Cry toxins for plant breeders to exploit.

Cry toxins are produced during sporulation [13]. In contrast, vegetative insecticidal proteins (Vips), are secreted during the growth stages before sporulation [14]. Vip proteins are toxic to a taxonomically diverse group of Lepidoptera [14] and show no sequence or structural homology with the δ-endotoxins. Thus, Vip toxins could be efficacious against insects that are resistant to Cry toxins, and provide a third Bt class that could be exploited in transgenic crops. In particular, Vip proteins could be added to crops that also express Cry toxins to produce exceptionally robust pyramids. The value of pyramids stem from the rarity of insects that are phenotypically resistant to more than one toxin. To obtain the full value of any pyramid, the toxins to be 'stacked' should require targets to possess separate gene-based mechanisms to overcome each toxin. However the resilience of a pyramid depends on the efficacy of, and levels of pre-existing resistance to, each toxin [15,16].

In 2009/10 and 2010/11 we used F_2 screens, which ensure that a proportion of individuals in the second generation of isofemale lines are homozygous for alleles carried by field collected parents [17]. This enabled the establishment of baseline frequencies of alleles conferring resistance to Vip3A in Australian populations of *H. armigera* and *H. punctigera*. Not only did we isolate from field populations the first examples in any insect of alleles conferring resistance to a vegetative insecticidal protein, we also determined that in both target species, baseline frequencies of Vip3A resistance alleles are higher than mutation rates normally encountered. Herein we report for both species the Vip3A resistance allele frequencies, as well as partial characterisation of the isolated resistance, including information on allelism, dominance, genetic basis, and cross-resistance to commercially relevant Cry toxins. We consider the more general consequences of our findings to the potential usefulness of Vip-expressing crops for countering resistance to Bt crops in Lepidopteran pests worldwide.

Results

For *H. armigera* there was no statistically significant difference between the *r* frequencies obtained for Vip3A in the two study years (2009/10 = 0.029 [11 positive lines, 108 tested lines], 2010/11 = 0.028 [17 positive lines, 165 tested lines]; Fisher's Exact Test, P = 0.9). Pooling the data across the two study years yields an *r* frequency for *H. armigera* of 0.027 with a 95% CI between 0.019 and 0.038. This frequency is based on 273 tested lines, 28 of which scored positive for resistance to Vip3A.

For *H. punctigera* there was no statistically significant difference between the *r* frequencies obtained for Vip3A in the two study years (2009/10 = 0.008 [5 positive lines, 197 tested lines], 2010/11 = 0.015 [2 positive lines, 51 tested lines]; Fisher's Exact Test, P = 0.6). Pooling the data across the two study years yields an *r* frequency for *H. punctigera* of 0.008 with a 95% CI between 0.004 and 0.015. This frequency is based on 248 tested lines, 7 of which scored positive for resistance to Vip3A.

Complementation tests involving crosses of the *H. armigera* isolates SP85 and SP477 and the *H. punctigera* isolates Hp8–48 and Hp9–5442 demonstrated that the F_1 progeny were also resistant to Vip3A (no. larvae at the same stage as parental controls after 7 days/no. larvae tested: *H. armigera*, 44/45; *H. punctigera*, 68/71). This result implies that within each species the resistance in both isolates is due to alleles at a common locus.

Individuals from the SP85, SP477 (*H. armigera*) and Hp8–48 (*H. punctigera*) colonies survived the maximum concentration of Vip3A toxin that could be practically delivered in a surface treatment assay (220 $\mu g/cm^2$) (in all cases, control corrected mortality at this dose was less than 2.4%) and larvae developed at the same rate as siblings reared on non-treated diet. For both species (based on assays with SP85 and Hp8–48), resistance was found to be essentially recessive, with heterozygotes exhibiting concentration-response characteristics that are similar to those of susceptible insects (Table 1). However on 4 out of the 5 occasions tested, the heterozygotes generated from the cross between the resistant (SP85) female and the susceptible (GR) male were more tolerant of Vip3A than the susceptible colony GR or heterozygotes from the reciprocal cross. While this difference was subtle, and generated a resistance ratio of only 3, the difference between the LC_{50} of GR and that of the SP85♀ X GR♂ F_1's was statistically significant ($X^2 = 58.51$, DF 2, P<0.001), as was the comparison with the reciprocal cross ($X^2 = 51.40$, DF 2, P<0.001). In contrast, there was no significant difference in response to Vip3A between GR and the heterozygotes produced by the GR♀ X SP85♂ cross ($X^2 = 0.12$, DF 2, P = 0.942). The situation in *H. punctigera* is less clear cut. While heterozygotes were generated on 4 occasions, only the first attempt yielded useful data for both reciprocal crosses (Table 1).

In both species, reciprocal backcrosses of heterozygotes to resistant colonies produced results for concentration-response assays which confirmed that resistance is essentially recessive – that is, 50% of offspring are homozygous resistant while the remainder are heterozygous and thus phenotypically susceptible (see Figures 1 and 2). These data are also consistent with the hypothesis that resistance in both species is conferred by a single gene.

Resistant and susceptible insects were exposed to ranges of concentrations of toxin incorporated into the diet. In two assays, *H. armigera* (SP85) and GR received a maximum of 1200 μg of Vip3A per g of diet. While SP85 were unaffected at this dose (no mortality and 21 of 23 larvae reached 3rd instar by day 7), no GR survived. GR survivors at lower doses grew more slowly with only one of the 20 survivors among the 45 GR exposed to 100 μg toxin per g of diet reached third instar whereas 61 of 64 control larvae reached that stage. Similarly on four occasions the *H. punctigera* resistant colony Hp8–48 and the susceptible HPM colony received a maximum of 800 μg of Vip3A per g of diet. There was no mortality among a total of 86 resistant insects at that dose and 91, 100, 70 and 80% reached third instar by day 7. In contrast, no HPM survived that dose and only 4 of the 23 susceptible *H. punctigera* insects (HPM) exposed to 100 μg per g diet were alive after 7 days and all were first instar. The estimated LC_{50}±95% CI for susceptible *H. armigera* and *H. punctigera* was 68.31±35.22–104.25 and 55.173±43.89–70.28 μg per g of diet respectively.

The Vip3A resistant colonies SP85, SP477 (*H. armigera*), Hp8–48 and Hp9–5442 (*H. punctigera*) were established from F_2 larvae that survived a discriminating-concentration of Vip3A toxin. Additional screens against Cry1Ac or Cry2Ab were conducted with sub-samples of the same cohorts of F_2 larvae that were exposed to Vip3A; these proved negative suggesting that Vip3A-resistant insects were not cross resistant to additional toxins. This result was confirmed using concentration-response assays where Vip3A resistant insects proved to be as susceptible, or more so, to Cry1Ac and Cry2Ab toxins (Table 1).

Discussion

Whether in response to recent evidence of emerging resistance to existing toxins [7,18] or as planned improvements, leading companies that produce Bt crops recently received access to Vip3A toxin [19]. Monsanto's third generation Bt-cotton, to be traded as 'Genuity Bollgard III', will produce Vip3A as well as the Cry1Ac and Cry2Ab toxins present in the second generation dual-gene Bt cotton known as Bollgard II. Genuity Bollgard III is likely to be available in the USA, and begin a phased entry in Australia, around 2016. It will be especially valuable in Australia where Bollgard II is deployed on at least 90% of the cropping area and alleles that confer Cry2Ab resistance in the target pests *H. armigera* and *H. punctigera* occur at frequencies that are of concern [3,20,21]

Table 1. Susceptibility of various strains of *H. armigera* and *H. punctigera* to Vip3A, Cry2Ab and Cry1Ac.

Species	Vip3A genotype	Colony	Toxin	Slope	LC_{50} ug/cm^2±95% CI
H. armigera	Susceptible (SS)	GR	Vip3A	1.82	0.551±0.37–0.765
			Cry2Ab	1.35	0.030±0.019–0.049
			Cry1Ac	1.42	0.010±0.0064–0.015
	Resistant (RR)	SP477	Vip3A	*	*
			Cry2Ab	1.52	0.0053±0.0033–0.0073
			Cry1Ac	1.64	0.0166±0.0104–0.0254
		SP85	Vip3A	*	*
			Cry2Ab	0.95	0.0077±0.0029–0.0145
			Cry1Ac	2.03	0.0087±0.0059–0.0196
	Heterozygote (RS)	GR♀ X SP85♂	Vip3A	1.92	0.566±0.414–0.736
		SP85♀ X GR♂	Vip3A	2.10	1.473±1.192–1.816
H. punctigera	Susceptible (SS)	LHP	Vip3A	1.77	0.594±0.195–0.984
			Cry2Ab	1.38	0.0271±0.0059–0.0588
			Cry1Ac	0.95	0.0362±0.0103–0.0825
	Resistant (RR)	Hp8–48	Vip3A	*	*
			Cry2Ab	2.17	0.0456±0.0022–0.0724
			Cry1Ac	2.29	0.1172±0.0814–0.1601
	Heterozygote (RS)	LHP♀ X Hp8–48♂	Vip3A	1.56	1.340±0.26–2.396
		Hp8–48♀ X LHP♂	Vip3A	1.76	1.382±0.694–2.085

*LC_{50} and slope could not be calculated as there was no increase in mortality at the maximum titre tested (128 ug/cm^2).

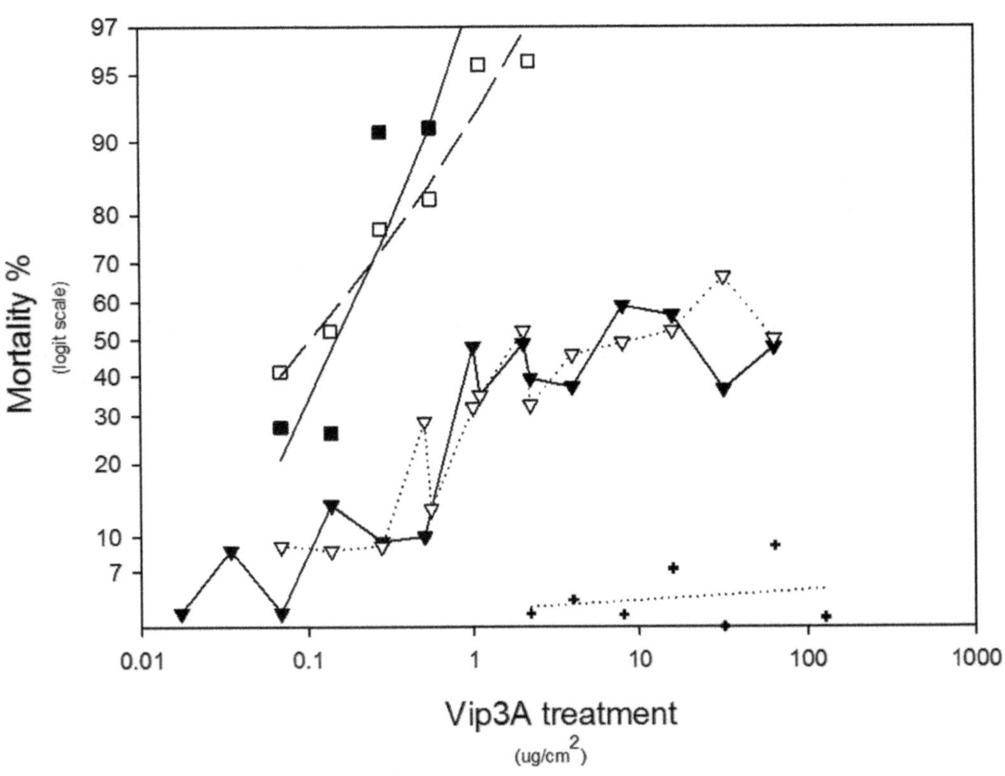

Figure 1. Dose responses of *H. punctigera* genotypes and backcrosses to the resistant colony. Solid squares show data for the homozygous susceptible colony (LHP), open squares show data for the heterozygotes, open triangles show data for the offspring from mating between heterozygous females and resistant males, closed triangles show data for the offspring from mating between heterozygous males and resistant females, and the crosses show data for the homozygous resistant colony (Hp8–48).

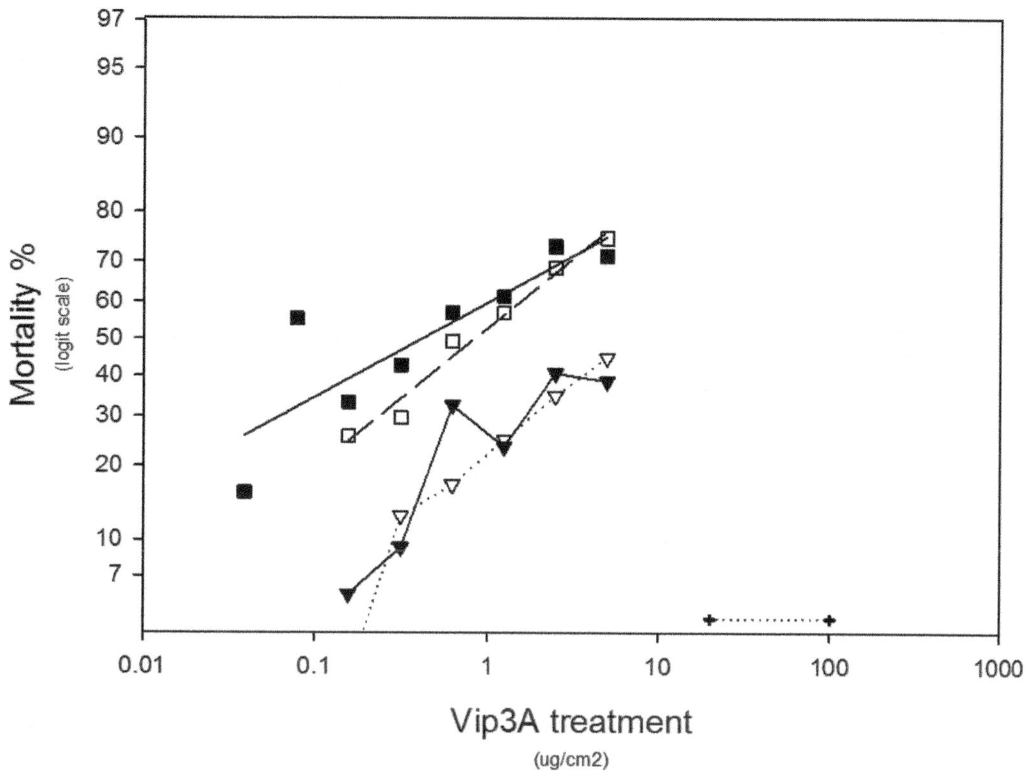

Figure 2. Dose responses of *H. armigera* genotypes and backcrosses to the resistant colony. Solid squares show data for the homozygous susceptible colony (GR), open squares show data for the heterozygotes, open triangles show data for the offspring from mating between heterozygous females and resistant males, closed triangles show data for the offspring from mating between heterozygous males and resistant females, and the crosses show data for the homozygous resistant colony (SP85).

and in *H. punctigera* have increased over time [7,20]. For the USA market, Dow is enhancing Widestrike® to produce Vip3A in addition to the existing Cry1Ac and Cry1F toxins [22]. Bayer plans to licence from Syngenta a Bt-cotton containing Cry1b and Vip3A that trades as Vipcot® [23]. In addition, corn expressing Vip3A and Cry1b (marketed as Agrisure Viptera™) is expected to be available soon [23].

From 2001 to 2003, small trial plots of cotton expressing Vip3A were grown in Australia which proved largely efficacious against Australian *Helicoverpa* spp. [24]. The limited potential exposure of insects to these plots makes it virtually impossible that selection by Bt plants is responsible for the high frequencies of Vip3A resistance alleles that we detected almost 10 years after these trials. We conclude that alleles conferring resistance to Vip3A exist in Australian populations of *H. armigera* and *H. punctigera* as natural polymorphisms. To our knowledge this is the first report of isolation from field populations of alleles conferring resistance to a vegetative insecticidal protein in any insect. Moreover, our data suggest that these pre-existing mutations occur at a relatively high frequency of 0.027 (95% CI, 0.019–0.038) and 0.008 (95% CI, 0.004–0.015) for *H. armigera* and *H. punctigera*, respectively.

The only frequency of resistance to a Bt toxin that is of similar magnitude to the 0.027 for Vip3A alleles in Australian *H. armigera* is the initial estimate of 0.158 for Cry1Ac resistance alleles in *Pectinophora gossypiella* [25]; however, recent estimates are significantly lower at 0.004 [18]. Work on Cry2Ab resistance in *H. armigera* and *H. punctigera* demonstrates that F_2 screens may underestimate frequencies in natural populations by up to 6-fold [26]. Thus, importantly, the frequency of Vip3A resistance alleles in both Helicoverpa species may be well above the mutation rate of 0.0001 assumed in most models of the evolution of resistance [15]. This suggests that another agent is selecting for an advantage against this toxin (see also [3]) and/or there is a high rate of mutation which introduces resistance alleles that are not selected against [27].

On seeking registration for Vip-producing cotton in the USA, Syngenta cited a maximal concentration of toxin in cotton plant tissues of 23.75 μg Vip3A per g fresh weight [28]. Homozygous resistant neonates of *H. armigera* and *H. punctigera* exposed to concentrations of Vip3A incorporated into diet that were 34–51 times that level (800–1200 μg of Vip3A/g diet respectively) exhibited no increased mortality or reduced growth rate relative to control insects. Cotton tissues contain secondary compounds that are not present in artificial diet, and may interact with Bt toxin to increase its efficacy. For instance, Llewellyn et al. [29] showed that Vip-expressing cotton largely controlled Helicoverpa spp., yet the reported titres of toxin in plants is lower than the LC_{50} we found for susceptible insects in our laboratory assays. In addition, the Vip3A toxin employed here differs subtly from that intended to be used in cotton plants – the sequence of the protein differs by one amino acid and the protein includes a HIS tag. Nevertheless, in our Cry2Ab-resistant *H. armigera* and *H. punctigera* colonies there is a good correlation between the results from diet assays and those using field-grown Cry2Ab cotton plants [S. Downes, unpublished data).

While the frequency of Vip3A resistance alleles in *H. armigera* and *H. punctigera* is uncomfortably high, it is essentially recessive in both species. *H. armigera* heterozygotes created by a cross between male susceptible insects and a homozygous Vip3A resistant female were subtly more tolerant of toxin than susceptible insects.

However the difference yields a small resistance ratio of approximately 3. Possible explanations include the presence of a maternal effect, or sex linkage of a gene or genes that subtly modifies the vigour and survival of heterozygotes. Our data demonstrate that Vip3A resistant insects are susceptible to both Cry1Ac and Cry2Ab.

The lack of significant levels of dominance and the absence of cross resistance means that resistance to Vip3A in Helicoverpa should be manageable if Genuity Bollgard III expresses all toxins optimally. Unfortunately, protein levels of Cry1Ac are variable in existing Bt crops, especially later in the season [30,31]. Llewellyn et al. [29] reported that efficacy of Vip3A declined as the season progressed however it did not do so as markedly as Cry1Ac. Additionally, particularly post flowering, occasional fields of Bollgard II support larvae that are susceptible to Bt and can survive to pupation [32]. Presumably during these episodes the expression of toxin(s) declines, and since Cry2Ab expression is more stable than Cry1Ac [33] there are likely to be periods where only Cry2Ab is effective, thereby selecting for resistance (see [32,34]). Thus there is considerable merit in protecting the susceptibility of Helicoverpa spp. to Cry2Ab until Bollgard III becomes available, otherwise Vip3A may be exposed to selection in a similar fashion to what we assume currently occurs for Cry2Ab in Bollgard II.

As we have no knowledge of the evolutionary forces maintaining Cry2Ab or Vip3A resistance polymorphisms in populations prior to the release of Bt-crops expressing these proteins, it is difficult to predict their frequency in pests elsewhere in the world. It is possible that the environmental-ecological conditions needed to generate or maintain these polymorphisms occur only in Australia. *H. armigera* and *H. punctigera* are only distantly related, yet both exhibit evidence of such polymorphisms prior to use of the toxins in Bt plants. Either the environmental-ecological conditions needed to generate or maintain these polymorphisms occur only in Australia or, such polymorphisisms are likely to be encountered in other Helicoverpa pests. Furthermore, the New World congener *H. zea* is very closely related to *H. armigera*, thus resistance to Vip3A in this species is of special interest.

Vip-expressing crops may soon be grown widely. It may be wise for organisations responsible for developing resistance management programs for Bt toxins, particularly those targeting species of Helicoverpa, to survey for Vip3A resistance alleles prior to the widespread use of this toxin. While targeted pest species will be naive to the Vip3A present in plants, they, like Australian Helicoverpa, may already possess the genetic variability to quickly respond to this new agent.

Materials and Methods

A Vip3A clone in *E. coli* was used as a source of toxin. The methods employed to produce the Vip3Aa toxin generally followed those described in [35]. The *vip3A* gene was modified to contain a HIS tag sequence at the C terminus of the protein to facilitate purification. *E. coli* cells were grown at 37°C in a shaking incubator overnight in Luria-Bertani medium. Expression of Vip3A protein was induced by the addition of isopropyl-D-thiogalactopyranoside (IPTG) to a final concentration of 1 mM at $O.D._{600}$ of 1.2 and cells were further cultivated at 28°C overnight. Cells were harvested by centrifugation for 15 minutes at 5000 g and then resuspended and sonicated in PBS (137 mM NaCl, 2.7 mM KCl, 4.3 mM Na_2HPO_4, 1.5 mM KH_2PO_4, pH 7.4) before being partitioned into aliquots and frozen. Purified Vip3A was prepared from a portion of the sonicated cell lysate using a HIS SELECT 1 ml cartridge (Sigma) and was examined for purity and stability by sodium dodecyl sulfate-polyacrylamide gel electrophoresis (SDS-PAGE) analysis. Protein concentration of the purified sample was quantified by using the Bradford method with BSA as a standard. Parallel surface contamination bioassays using either purified toxin or sonicated cell lysate were performed to calibrate the toxicity of the crude solution which was used as a source of toxin in all further assays.

Response to increasing doses of Vip3A toxin were examined in a Bt-susceptible colony of *H. armigera* (designated 'GR') and *H. punctigera* (designated 'HPM') to establish appropriate concentrations to identify likely instances of resistance in F_2 tests. We have found that the susceptibility of the two colonies is typical of the vast majority of field collected insects (S. Downes, unpublished data). Based on these results, discriminating assays were established for the two species. For *H. armigera*, 10 $\mu g/cm^2$ of toxin was applied to the surface of cooled diet while for *H. punctigera* a lower application rate of 2 $\mu g/cm^2$ was employed.

To detect Vip3A resistance, we employed the F_2 techniques previously described for *H. armigera* and *H. punctigera* to screen against Cry1Ac and Cry2Ab [4,11]. Indeed, the only change to our regular routine was to expose an additional sample of approximately 90, F_2 larvae to Vip3A toxin whereas previously we tested only for the two Cry toxins. We followed Baysian inference statistical approaches [17] to determine expected frequency of resistance alleles in the sampled populations ($E[p_R]$) and the 95% credibility intervals for our estimated frequencies.

To examine the characteristics of Vip3A resistance, we studied the *H. punctigera* Vip-resistant colony Hp8–48 which was isolated during the summer of 2008–09 from individuals collected as eggs on cotton in the Gwydir valley. To perform complementation assays we also used the *H. punctigera* Vip-resistant colony Hp9–5442 which was isolated during the summer of 2009–10 from individuals collected as eggs on pigeon pea in the Macintyre valley. To determine the characteristics of Vip3A resistance in *H. armigera*, we examined the Vip-resistant colonies, SP85 and SP477 which were isolated during the summer of 2009–10 from individuals collected as eggs on non-Bt cotton from St. George, Qld (SP85) and the Gwydir Valley, NSW (SP477).

The majority of tests involved surface treatment assays as described in [11] and were repeated at least 3 times using different cohorts of the same colony. Briefly, the surface of aliquots of cooled diet in wells of a 45-well plastic tray was contaminated with a solution of toxin and allowed to air dry. When dry, an individual neonate was added to each well and sealed in place with a perforated lid. After incubation at 26°C for 7 days, assays were scored for 'dead' (incapable of coordinated movement) and live insects, and the instar was recorded for the latter. Such assays were used to conduct complementation tests by exposing to toxin the offspring of reciprocal crosses between (a) the SP85 and SP477 *H. armigera* strains, and (b) the Hp8–48 and 9–5442 *H. punctigera* strains. We also used these assays to examine possible dominance present among neonate heterozygotes produced by crossing the resistant and susceptible strains. Finally, surface treatment assays were used to examine the response of the Vip3A resistant strains to Cry toxins Cry2Ab and Cry1Ac.

We also used diet incorporation assays to further investigate the extent of resistance to Vip3A by the resistant strains. Serial doses of Vip3A toxin were mixed with diet cooled to 45°C, and then poured into 100 mm petri dishes. Once cooled, portions of diet approximately 6 mm^3 were placed in wells of our standard 45-well assay plates. One neonate was placed in each well which was then heat sealed.

No specific permits were required to collect the insect material used in the described research program from commercial farms. In all cases, permission to access and collect insects from commercial fields was obtained directly from the landowners. None of the accessed commercial farms were protected in any way, and our field studies did not involve endangered or protected species.

Acknowledgments

The authors greatly appreciate the support of cotton growers who permit access to crops to collect eggs used in our research. We also thank Karen Stanford, Donna Jones, Wendy Finn, and Norm Winters, and the Crop Consultants of Australia for collecting eggs for the work. Expert technical assistance was provided by our colleagues, Joel Armstrong, Janine Gascoyne, Su Young and Tracey Parker. Two anonymous reviewers provided comments on an earlier draft of this manuscript that improved its clarity.

Author Contributions

Conceived and designed the experiments: RM SD BJ. Performed the experiments: BJ SD RM. Analyzed the data: RM SD. Contributed reagents/materials/analysis tools: RM SD BJ. Wrote the paper: RM SD BJ.

References

1. McKenzie JA (1996) Ecological and evolutionary aspects of insecticide resistance. Melbourne: Academic Press.
2. Hartley C, Newcomb R, Russell R, Yong C, Stevens J, et al. (2006) Amplification of DNA from preserved specimens shows blowflies were preadapted for the rapid evolution of insecticide resistance. Proc Natl Acad Sci U S A 103: 8757–8762.
3. Mahon R, Olsen K, Downes S, Addison S (2007) Frequency of alleles conferring resistance to the Bt toxins Cry1Ac and Cry2Ab in Australian populations of *Helicoverpa armigera* (Lepidoptera: Noctuidae). J Econ Entomol 100: 1844–1853.
4. Downes S, Parker T, Mahon R (2009) Frequency of alleles conferring resistance to the *Bacillus thuringiensis* toxins Cry1Ac and Cry2Ab in Australian populations of *Helicoverpa punctigera* (Lepidoptera: Noctuidae) from 2002 to 2006. J Econ Entomol 102: 733–742.
5. Tabashnik BE, Dennehy TJ, Carriere Y (2005) Delayed resistance to transgenic cotton in pink bollworm. Proc Natl Acad Sci U S A 102: 15389–15393.
6. Genissel A, Augustin S, Courtin C, Pilate G, Lorme P, et al. (2003) Initial frequency of alleles conferring resistance to *Bacillus thuringiensis* poplar in a field population of *Chrysomela tremulae*. Proc Roy Soc Lon B 270: 791–797.
7. Downes S, Parker T, Mahon R (2010) Incipient resistance of *Helicoverpa punctigera* to the Cry2Ab Bt toxin in Bollgard II cotton. PLoS One 5: e12567.
8. Tabashnik BE, Van Rensburg JBJ, Carriere Y (2009) Field-evolved insect resistance to Bt Crops: definition, theory, and data. J Econ Entomol 102: 2011–2025.
9. Gassmann A, Petzold-Maxwell J, Keweshan R, Dunbar M (2011) Field-evolved resistance to Bt maize by Western Corn Rootworm. PLoS One 6: e22629.
10. Ferre J, Van Rie J (2002) Biochemistry and genetics of insect resisitance to *Bacillus thuringiensis*. Annu Rev Entomol 47 501–533.
11. Mahon R, Olsen K, Garsia K, Young S (2007) Resistance to *Bacillus thuringiensis* toxin Cry2Ab in a strain of *Helicoverpa armigera* (Lepidoptera: Noctuidae) in Australia. J Econ Entomol 100: 894–902.
12. Caccia S, Hernández-Rodríguez C, Mahon R, Downes S, James W, et al. (2010) Target site alteration is responsible for field-derived resistance to *Bacillus thuringiensis* Cry2A insecticidal proteins in *Helicoverpa* spp. PLoS One. 5: e9975.
13. Schnepf E, Crickmore N, Van Rie J, Lereclus D, Baum J (1998) *Bacillus thuringiensis* and its pesticidal crystal proteins. Microbiol. Mol Biol Rev 62: 775–806.
14. Estruch J, Warren G, Mullins M, Nye G, Craig JK (1996) Vip3A, a novel *Bacillus thuringiensis* vegetative insecticidal protein with a wide spectrum of activities against lepidopteran insects. Proc Natl Acad Sci U S A 93: 5389–5394.
15. Roush RT (1998) Two-toxin strategies for management of insecticidal transgenic crops: can pyramiding succeed where pesticide mixtures have not? Phil Trans R Soc Lond B 353: 1777–1786.
16. Zhao JZ, Cao J, Li YX, Collins HL, Roush RT, et al. (2003) Transgenic plants expressing two *Bacillus thuringiensis* toxins delay insect resistance evolution. Nature Biotechnol 21: 1493–1497.
17. Andow DA, Alstad DN (1998) F_2 screen for rare resistance alleles. J Econ Entomol 91: 572–578.
18. Tabashnik BE, Gassmann AJ, Crowder DW, Carriere Y (2008) Insect resistance to Bt crops: evidence versus theory. Nature Biotechnol 26: 199–202.
19. Laws F (2010) VipCot nearing end of 'interesting journey'? http://farmindustrynews/corn-hybrids/vipcot-nearing-end-interesting-journey.
20. Downes S, Mahon R (2012) Evolution, ecology and management of resistance in *Helicoverpa* spp. to Bt cotton in Australia. J Invertebr Pathol, In press.
21. Downes S, Mahon R (2012) Successes and challenges of managing resistance in *Helicoverpa armigera* to Bt cotton in Australia. GM Crops, In press.
22. Anon (2010) Syngenta, Dow Agrosciences sign cotton technology licensing agreement. http://www2.syngenta.com/en/media/mediareleases/en_100105.html.
23. Anon (2010) Syngenta and Bayer CropScience sign global cotton technology licencing agreement. http://www2.syngenta.com/en/media/mediareleases/en_100714.html.
24. Llewellyn DJ, Mares CL, Fitt GP (2007) Field performance and seasonal changes in the efficacy against *Helicoverpa armigera* (Hubner) of transgenic cotton expressing the insecticidal protein Vip3A. Ag Forest Entomol 9: 93–101.
25. Tabashnik BE, Patin AL, Dennehy TJ, Liu YB, Carriere Y, et al. (2000) Frequency of resistance to *Bacillus thuringiensis* in field populations of pink bollworm. Proc Natl Acad Sci U S A 97: 12980–12984.
26. Mahon RJ, Downes S, James W, Parker T (2010) Why do F_1 screens estimate higher frequencies of Cry2Ab resistance in *Helicoverpa armigera* (Lepidoptera: Noctuidae) than do F_2 screens? J Econ Entomol 103: 472–481.
27. Clark A (1998) Mutation-selection balance with multiple alleles. Genetica 102–103: 41–47.
28. Artim L (2003) Petition 03–155–01p for the determination of non-regulated status: Lepidopteran insect protected Vip3A cotton transformation event Cot102. http://cera-gmc.org/docs/decdocs/05-209-018.pdf.
29. Llewellyn D, Mares C, Fitt G (2007) Field performance and seasonal changes in the efficacy against *Helicoverpa armigera* (Hübner) of transgenic cotton expressing the insecticidal protein Vip3A. Ag Forest Entomol 9 93–101.
30. Fitt G, Daly JC, Mares CL, Olsen K (1998) Changing efficacy of transgenic Bt cotton - patterns and consequences. Pest Management - Future Challenges, Vols 1 and 2, Proceedings of the Australian Cotton Growers Research Association Conference: 189–196.
31. Rochester IJ (2006) Effect of genotype, edaphic, environmental conditions, and agrononmic practices on Cry1Ac protein expression in transgenic cotton. J Cotton Sci 10: 252–262.
32. Lu B, Downes S, Wilson L, Gregg P, Knight K, et al. (2011) Preferences of field bollworm larvae for cotton plant structures: impact of Bt and history of survival on Bt crops. Entomol Exper Et Appl 140: 17–27.
33. Adamczyk J, Adams LC, Hardee DD (2001) Field efficacy and seasonal expression profiles for terminal leaves of single and double *Bacillus thuringiensis* toxin cotton genotypes. Econ Entomol 94: 1589–1593.
34. Mahon R, Olsen K (2009) Limited survival of a Cry2Ab-resistant strain of *Helicoverpa armigera* (Lepidoptera: Noctuidae) on Bollgard II. J Econ Entomol 102: 708–716.
35. Sena JAD, Hernandez-Rodriguez CS, Ferre J (2009) Interaction of *Bacillus thuringiensis* Cry1 and Vip3A proteins with *Spodoptera frugiperda* midgut binding sites. Appl Environ Microbiol 75: 2236–2237.

11

Geminivirus-Mediated Delivery of Florigen Promotes Determinate Growth in Aerial Organs and Uncouples Flowering from Photoperiod in Cotton

Roisin C. McGarry, Brian G. Ayre*

Department of Biological Sciences, University of North Texas, Denton, Texas, United States of America

Abstract

Background: Plant architecture and the timing and distribution of reproductive structures are fundamental agronomic traits shaped by patterns of determinate and indeterminate growth. Florigen, encoded by *FLOWERING LOCUS T* (*FT*) in Arabidopsis and *SINGLE FLOWER TRUSS* (*SFT*) in tomato, acts as a general growth hormone, advancing determinate growth. Domestication of upland cotton (*Gossypium hirsutum*) converted it from a lanky photoperiodic perennial to a highly inbred, compact day-neutral plant that is managed as an annual row-crop. This dramatic change in plant architecture provides a unique opportunity to analyze the transition from perennial to annual growth.

Methodology/Principal Findings: To explore these architectural changes, we addressed the role of day-length upon flowering in an ancestral, perennial accession and in a domesticated variety of cotton. Using a disarmed *Cotton leaf crumple virus* (CLCrV) as a transient expression system, we delivered *FT* to both cotton accessions. Ectopic expression of *FT* in ancestral cotton mimicked the effects of day-length, promoting photoperiod-independent flowering, precocious determinate architecture, and lanceolate leaf shape. Domesticated cotton infected with *FT* demonstrated more synchronized fruiting and enhanced "annualization". Transient expression of *FT* also facilitated simple crosses between wild photoperiodic and domesticated day-neutral accessions, effectively demonstrating a mechanism to increase genetic diversity among cultivated lines of cotton. Virus was not detected in the F_1 progeny, indicating that crosses made by this approach do not harbor recombinant DNA molecules.

Conclusions: These findings extend our understanding of FT as a general growth hormone that regulates shoot architecture by advancing organ-specific and age-related determinate growth. Judicious manipulation of *FT* could benefit cotton architecture to improve crop management.

Editor: David Edward Somers, Ohio State University, United States of America

Funding: This work was supported by Cotton Incorporated (Cooperative Agreement 10-768 to BGA), a not-for-profit producers organization. The funders had no role in study design, data collection and analysis, decision to publish, or preparation of the manuscript.

Competing Interests: This work was supported by Cotton Incorporated (Cooperative Agreement 10-768 to BGA), a not-for-profit producers organization. The authors declare there are no other relevant declarations relating to employment, consultancy, patents, products in development or marketed products etc.

* E-mail: bgayre@unt.edu

Introduction

Plant architecture is fundamental to agricultural productivity and artificial selection of desired growth habits is prominent in the earliest domestication of exotics into crops, the yield enhancements of the "green revolution", and in modern crop improvement. Shoot architecture is determined by the fate of the apical meristems being indeterminate or determinate, the strength of apical dominance, branching pattern of lateral growth, and the timing and placement of reproductive growth. Indeterminate shoot apical meristems retain a population of vegetative stem cells indefinitely with tissue and organ differentiation occurring below and on the flanks. Because of this single point of continued growth, shoots derived from indeterminate apical meristems are said to be monopodial. In sympodial stems, the cells of the apical meristem undergo terminal differentiation, commonly in a flower or inflorescence, and the uppermost axillary bud continues the basic body plan of the shoot to produce a linear array of reiterative sympodial units along the axis of growth.

Cotton (*Gossypium* spp.) is the world's most important textile fiber and a significant oilseed crop with a worldwide economic impact estimated at $500 billion annually [1]. *Gossypium* species are native to the arid and semi-arid regions of tropics and subtropics of both the old and new worlds, and includes approximately 45 diploid and 5 allotetraploid species [2]. They are long-lived perennials with architectural variation ranging from trailing and herbaceous to 15 m trees [2]. Most, if not all, are day-length sensitive and undergo repeated annual cycles of vegetative growth in long-day seasons with reproductive development triggered by short-day photoperiods, but cooler air temperature and dry seasons also commonly promote flowering [3].

Domestication of the two allotetraploids that comprise the majority of world-wide cultivations, *G. hirsutum* (upland or American cotton) and *G. barbadense* (Pima or extra-long staple cotton), initiated independently at least 5000 years ago [2], and the two have similar architectures. The main-stem apical meristem is monopodial and remains vegetative for the life of the plant. At each node, there is a leaf with stipules and two axillary buds: one of these generally remains dormant while the other may grow to form either a vegetative or reproductive axillary branch [4]. Vegetative branches are monopodial and reiterate the main stem while reproductive branches are sympodial and are called fruiting branches [5]. On fruiting branches, the apical meristem of each sympodial unit produces an internode, node, leaf with stipules (called the subtending leaf) and two axillary buds. The apical meristem then converts to a determinate floral meristem to produce a flower and ultimately a boll, and one of the axillary-bud meristems continues growth to form the next sympodial unit [4,5].

Varieties domesticated for temperate climates were bred for day-neutrality and are cultivated and harvested as annual row crops [5]. This management strategy is well suited to highly mechanized production practices but is at odds with the plant's inherent perennial nature. For example, vegetative growth continues after reproductive development initiates, and flower and fruit set are not synchronous but continue through the growing season. These competing sinks divert resources from fiber and seed production such that late-forming fiber is inferior to early-forming fiber and can discount crop value [5–7]. To control cotton's perennial growth habit, growth inhibitors are used during the growing season and defoliants are used at season's end to terminate the crop in preparation for mechanical harvest [6,7].

In addition to retaining perennial growth habits, modern cultivated cotton suffers from restricted genetic diversity attributed to multiple bottlenecks during domestication and the current focus on a limited number of elite breeding lines [8]. Ancestral accessions, however, are a rich but underutilized source of variation affecting fiber quality and yield, and resistance to biotic and abiotic stresses [8–11]. Introgressing this diversity has potential for crop improvement but differences in the onset of flowering limit breeding to annual cycles in greenhouses or tropical territories unless photoperiod is artificially shortened. Some accessions require additional environmental cues, such as temperature, to initiate reproductive growth [3] and the specific conditions required may be difficult to replicate.

The photoperiodic pathway to flowering has many conserved elements irrespective of whether flowering is promoted by long or short days (reviewed by [12–15]). It is now well-established that the flowering hormone florigen, proposed by Mikhail Chailakhyan in the 1930s (see [16]) is the protein encoded by *FLOWERING LOCUS T* (*FT*) in Arabidopsis and its orthologs in other plants (e.g., *SINGLE FLOWER TRUSS* [*SFT*] in tomato and *HEADING DATE 3a* [*Hd3a*] in rice) [17–19]. Furthermore, *FT* orthologs govern the seasonal reproductive cycles of perennials [20,21] and *FT* overexpression can overcome the extended juvenile stage of many flowering trees [22–25]. Florigen is not however only a flowering hormone but also promotes, in a dose-dependent fashion, a more determinate habit in all aerial organs including apical meristems, leaf primordia, and lateral meristems [26,27]. Florigen is therefore a general growth hormone that contributes to the overall vegetative and reproductive architecture, and artificial selection at *FT* loci and related family members has contributed to domestication of several crops from exotic progenitors [21,26,28,29].

Cotton presents a unique opportunity to analyze perennial and "annualized" plants from the same gene pool. With the long-term goal of characterizing the genetics of annual and perennial growth habits, we focused first on characterizing the conditions required for flowering in a perennial, ancestral, photoperiodic accession and in a domesticated, day-neutral variety of *G. hirsutum*. We demonstrate that day-length affects flowering, plant architecture, and leaf shape in a coordinated and reversible manner, leading us to hypothesize that these changes are florigen-dependent. To test the effects of florigen on cotton growth and to facilitate simple crosses between diverging lines, we developed a system for transient expression of the Arabidopsis *FT* gene from a disarmed *Cotton leaf crumple virus* (dCLCrV) vector in both domesticated and ancestral *G. hirsutum*. Ectopic expression of *FT* in both ancestral and domesticated cotton mimics the effects of day-length, suggesting that day-length functions mainly via FT in cotton. In addition, we show that "virus-induced flowering" (VIF) is a useful breeding tool to facilitate introgression of desirable traits from ancestral accessions into elite cotton cultivars. Together, our findings suggest that judicious manipulation of *FT* and related genes may enhance "annualization" and crop management by attenuating perennial characteristics.

Materials and Methods

1. Plasmid Construction

Plasmid constructions were by standard procedures [30,31] using *Escherichia coli* XL1-Blue (Stratagene, La Jolla, CA) as the host strain. Restriction endonucleases were from New England Biolabs (Beverly, MA), oligonucleotides were from Invitrogen (Carlsbad, CA), and *ExTaq* polymerase (TaKaRa, distributed by Fisher Scientific, Pittsburg, PA) was used for PCR amplification of plasmid components. DNA for cloning was purified with Wizard columns from Promega (Madison, WI). Clones incorporating a PCR product were sequenced to ensure accuracy (MWG, Huntsville, AL).

FT cDNA was reverse transcribed (SuperScript III, Invitrogen) and PCR amplified from Arabidopsis plants overexpressing *CONSTANS* [32] using oligonucleotides FT-fwd-*Spe*I (5′-ctcgtgactagtatgtctataaatataagagacc) and FT-rev nt559-*Nhe*I (5′-ctcgtggctagcaatatcaattggttataaagg), and sub-cloned in the *Spe*I and *Nhe*I sites (underlined) of pCLCrVJRT0008 (Genbank EU541443) [33] to construct an in-frame fusion with the start codon of the viral coat protein. Use of the *Spe*I restriction site resulted in the incorporation of Ala, Cys, Leu, Gln, Thr, Ser, and Met codons between Met-1 and the *FT* open reading frame. Plasmid DNA used for biolistic bombardment was prepared by standard alkaline lysis followed by column purification (Promega).

2. Plant Material and Growth Conditions

G. hirsutum Delta Pine 61 (DP61; GRIN PI 607174) and Texas 701 (TX701; GRIN PI 165329) seeds were a generous gift from Jack McCarty (USDA). To promote germination, seeds were delinted in concentrated sulfuric acid for 20 s, rinsed with running water for 10 min, neutralized with 100 mM sodium bicarbonate for 20 s, and rinsed with running water for 10 min. Delinted seeds were then heat-treated in an 80°C water bath for 1 min, plunged into ice water for 1 min, and allowed to air-dry before planting in MetroMix 366 potting medium (Sun Gro Horticultural, Bellevue, WA).

Biolistic bombardments used a PDS1000-He gene gun (Bio-Rad, Hercules, CA). 5 μg of each dCLCrV-A viral genome (dCLCrV::A, dCLCrV::αChlI and dCLCrV::FT) was pre-mixed with 5 μg of CLCrV-B DNA, precipitated onto gold particles as per the manufacturer's instructions (Bio-Rad), and stored overnight at −20°C. The abaxial surfaces of cotton seedling cotyledons aged 1–7 days post-germination (dpg), prior to emergence of the first true leaves, were bombarded using 900 psi and 1350 psi rupture disk pressures with the viruses adhered to 1.6 μm, 1.0 μm, and 0.6 μm gold particles. Four to six seedlings were bombarded with each treatment. 4 d after infection, bombarded and untreated seedlings were transferred to 4 gallon pots containing MetroMix 366 and grown under long-days (16 h light/8 h dark) in a greenhouse with supplemental light provided by a combination of metal halide and mercury lamps (up to 1300 μmol photons m^{-2} s^{-1} light intensity at leaf level) or short-day (10 h light/14 h dark) growth room conditions with metal halide, mercury, and T5 fluorescent lighting (200 μmol photons m^{-2} s^{-1} light intensity at leaf level). The efficiency of establishing an infection under each condition and with each vector is indicated in Fig. S1.

Cross-pollination using TX701 as the pollen donor and DP61 as the maternal line were conducted as suitably-aged flowers developed. DP61 flowers were emasculated at the "candle stage" (prior to anthesis when the petals remain tightly closed and extend past the sepals) [5], and the pistil was covered with a paper sleeve from a soda straw. Approximately 24 h later, the sleeve was removed and the pistils were pollinated using *FT*-induced flowers from TX701 plants. Self-pollination was allowed to occur in flowers not used for crosses. Seeds were removed from the mature bolls and were dried and stored at room temperature.

3. RNA and DNA Analyses

A leaf disk (100 mg) was removed from the unexpanded fifth true leaf of bombarded and untransfected cotton plants, homogenized with liquid nitrogen, and RNA extracted using the RNeasy Plant Mini Kit as per the manufacturer's protocol (Qiagen, Valencia, CA). The RNA obtained was treated with RNase-free DNase (New England Biolabs) and cleaned with Trizol (Invitrogen) as per the manufacturers' protocols. For first strand synthesis, 500 ng of RNA was used as template with a random hexamer primer and Superscript III reverse transcriptase as per the manufacturer's protocol. One microliter of cDNA was used for PCR to detect *FT* and the internal reference gene *glyceraldehyde 3-phosphate dehydrogenase* A subunit (*GAPDH*) [33] with oligonucleotides FT-fwd nt51 (5′-gacgttcttgatccgtttaatag), FT-rev nt461 (5′-ccgagattgtagatctcagc), GAPDH fwd (5′-gggcaccatgactaccac), and GAPDH rev (5′-cagttgaagtcgggacg), for 25 cycles at 60°C annealing temperature using Phire DNA polymerase (Finnzyme, distributed by New England Biolabs).

A young leaf was removed from TX701, DP61, and F1 progeny plants, homogenized in liquid nitrogen, and genomic DNA extracted using CTAB followed with RNase A digestion [34]. PCR was performed using 0.5 μL of genomic DNA as template with oligonucleotides CLCrV-Rep-fwd (5′-taatcgccctcctcttggc), CLCrV-Rep-rev (5′-gacgccaacgccgtcaag), Chl-fwd (5′-cggtgaccct-tataactcgg), and Chl-rev (5′-gattacctgagccgatgag), and RedTaq Genomic DNA polymerase (Sigma, MO). CLCrV *Rep* and *Chl1* were amplified in the same 12.5 μL reaction for 30 cycles at 55°C annealing temperature.

Results

1. Virus Induced Flowering Uncouples Determinate Growth from Photoperiod

Texas 701 (TX701) is a short-day *G. hirsutum* landrace that does not flower in the normal cotton growing season of the continental United States, but will produce fruit annually at the cotton winter nursery in Tecoman, Mexico [9]. In addition to being photoperiodic, TX701 has relatively long internodes and the monopodial vegetative branches grow close to vertical, giving plants a tall and columnar stature (Fig. 1A). Leaves are deeply lobed and are referred to as the okra-leaf shape. This is in striking contrast to the architecture of domesticated cotton, exemplified by DeltaPine 61 (DP61) used in this study. DP61 is day neutral and starts producing sympodial fruiting branches early in its life cycle. With the exception of the first, each sympodial unit (SU) of a fruiting branch grows from the axillary bud of the preceding SU. This gives fruiting branches a zig-zag appearance and contributes to greater horizontal spread relative to the vegetative branches that form only from the lowest buds. These features, along with large broad leaves and shorter internodes, contribute to a compact and bushy appearance (Fig. 1B).

Under short-day conditions in a growth room, TX701 plants developed fruiting branches from the main stem at node 20.6±3.1 ($n = 7$ plants; Table S1). This is the node of first fruiting branch (NFB), the first node above the cotyledons from which a fruiting branch develops, and this heritable measure reflects the "earliness" of the cultivar which is an important cropping characteristic [9]. Floral buds (*i.e.* squares) were first evident 92 days post-germination (dpg) and reached anthesis by 150 dpg. When grown under long–day conditions (16 hr light) in the greenhouse, TX701 did not flower by 146 dpg (when growth was terminated), but instead continued indeterminate growth to yield tall plants with numerous vegetative branches emerging from the main stem (Fig. 1A). In the same greenhouse conditions, DP61 NFB was at node 5.1±0.9 ($n = 10$) with squares evident at 33 dpg and anthesis at 64 dpg (Fig. 1B, Table S1).

Day-length affected flowering time and leaf shape in a coordinated manner. When TX701 plants were moved from long-day to short-day conditions to induce flowering and bulk seed stocks, newly emerging leaves transitioned from the okra shape with deep palmate lobes to lanceolate simple leaves. Leaf shape transitioned back to deeply lobed when plants were returned to long day conditions (Fig. 2). These changes in leaf shape correlated with the photoperiodic onset, and subsequent termination, of reproductive growth. Development of lanceolate leaves also correlated with the onset of flowering on plants grown outside with only natural sunlight: lanceolate leaves and floral squares were evident by October 1, after the vernal equinox, and the transition was greater further along the fruiting branches (Fig. S2).

Given that the transition to reproductive growth was concurrent with altered leaf shape, we hypothesized that these changes were florigen-dependent and predicted that overexpressing Arabidopsis *FT* in ancestral, short-day cotton would uncouple flowering and determinate leaf growth from photoperiod. Generating transgenic cotton to ectopically express *FT* in ancestral lines is not practical since transformation is only efficient in domesticated Coker 312 (GRIN PI 529278) and regeneration takes about one year [35–37]. Rather, we utilized a disarmed *Cotton leaf crumple virus* (dCLCrV) vector to transiently deliver sequences of interest [33]. In this geminivirus-based system, the sequence of interest replaces the coat protein gene on the A genome and when co-delivered with the B genome, both replicate autonomously and spread systemically throughout the host as un-encapsidated DNA. The virus is considered disarmed since the coat protein is essential for whitefly-mediated plant-to-plant transmission. *AtFT* cDNA was cloned downstream of the coat protein promoter generating dCLCrV::FT, and, along with dCLCrV, an empty-vector control construct, and dCLCrV::αChl1 containing antisense sequence to the *G. hirsutum* magnesium chelatase subunit 1 [33], was used to infect TX701 seedlings.

Under non-inductive, 16 hr-long days, TX701 plants infected with dCLCrV::FT were clearly reproductive and developed sympodial fruiting branches with productive flowers (Fig. 3, Fig. S1B, Table S1). Floral buds were evident by 33 dpg on node 5 of the main stem, but these earliest squares did not result in fertile flowers. The first flowers to reach anthesis did so at 71 dpg

Figure 1. Wild and domesticated cotton have different architectures and flower at different times. (**A**) TX701, a photoperiodic accession, has deeply lobed "okra" leaves and pronounced apical dominance. TX701 does not flower under long days and this plant is completely vegetative. (**B**) DP61, a day-neutral cultivar, has a bushy growth habit with normal leaves and flowers profusely. Arrows are pointing to floral buds ("squares"). Both plants were grown under long-day conditions (16/8 h day/night) in a greenhouse with supplemental lighting. Scale bars, 25 cm.

(Fig. 3C, Table S1). This timing compared favorably to DP61 plants which, under the same conditions, produced fruiting branches as early as node 5 and reached anthesis by 64 dpg. None of the untransfected TX701 plants, nor dCLCrV or dCLCrV::αChl1 infected TX701 flowered under these conditions. To correlate *FT* expression from dCLCrV::FT with flowering in TX701 plants, total RNA was isolated from the fifth leaf of the main stem at 29 dpg while it was an unexpanded sink leaf (*i.e.,* importing virus from infected mature leaves) and expression of *FT* determined by RT-PCR. *AtFT* expression was detected only in plants inoculated with dCLCrV::FT that subsequently flowered under non-inductive conditions (Fig. 3E).

Young VIF-treated TX701 plants had deeply lobed, palmate leaves on the main stem but more lanceolate leaves developed as the plants aged, and were more pronounced further along the fruiting branches (Fig. 2A). These changes in leaf shape were observed among 100% of flowering VIF-treated TX701 plants but not observed among dCLCrV-infected (Fig. 2B) and uninfected TX701 plants grown under long days, and imply that *FT* overexpression impacts leaf development. These observations that leaf marginal meristems in the okra-leaf background were suppressed by photoperiod (Fig. S2) and *FT* overexpression (Fig. 2) in an age-related fashion are consistent with models for plant development in which signals for determinate growth become stronger and/or more penetrant as the plant ages [27,38]. Together, ectopic expression of *FT* in TX701 yields phenotypes similar to the pleiotropic effects of short day-length.

2. High Florigen Promotes an "Annualized" Cotton Phenotype

dCLCrV::FT infection made both wild and domesticated cotton more determinate in all aerial organs, consistent with observations in tomato overexpressing *SFT* [17,27]. Relative to controls, DP61 plants infected with dCLCrV::FT had a more compact stature because internodes were shorter (data not shown) and fruiting branches had fewer sympodial units (Fig. 4A–D). These plants also had more synchronous flower and fruit set because fruit close to the main stem matured without continued formation of new flowers further along the fruiting branch (Fig. 4A–D). The extent of these phenotypes varied among infected plants, as expected given that dCLCrV does not infect every cell during systemic spread as evident from the chlorotic patches shown in a dCLCrV::αChl1-infected plant (Fig. S1A) and the severity of infection varies from plant to plant [33,38]. For this reason, quantifying *FT* in individual buds was not pursued since levels in one bud do not predict levels in another, and destructive extractions preclude the possibility of monitoring bud fate. Rather, to measure growth in relation to dCLCrV::FT infection, whole-plant architecture phenotypes were correlated with the severity of leaf crumpling, which is a valid and commonly-used measure for virus titer [39] (Fig. 2 and Fig. 4B). The most compact plant architectures correlated with more robust dCLCrV::FT viral infection. A more determinate growth habit was not observed among plants infected with dCLCrV, despite leaf crumpling symptoms equal to or exceeding the severity of dCLCrV:FT infected plants, indicating that *FT* was indeed the cause of the growth alterations.

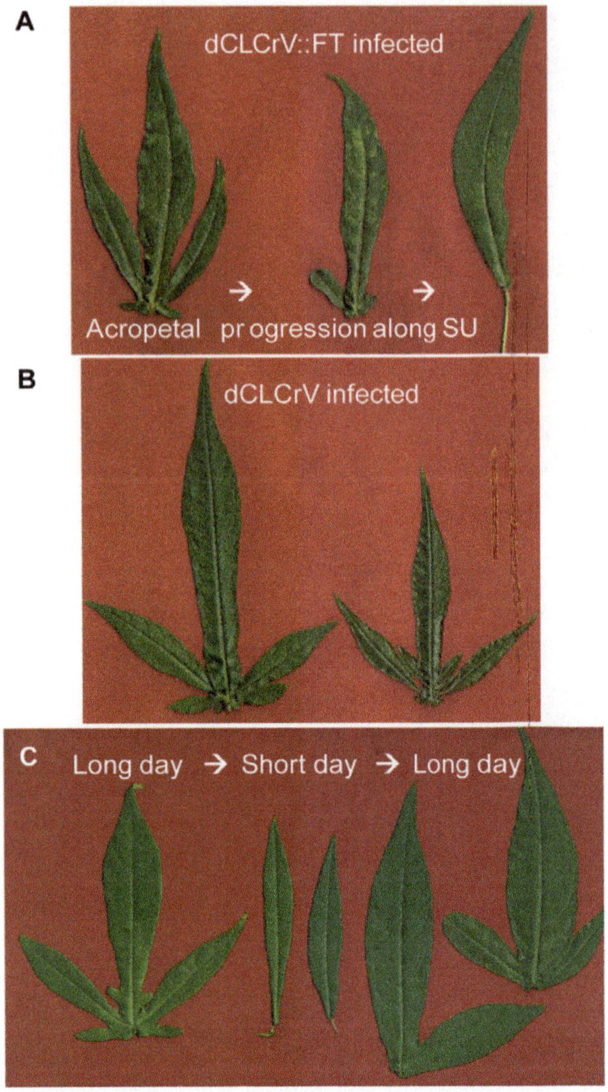

Figure 2. Florigen promotes determinate leaf growth. (**A**) Leaf shape along the fruiting branch (SU, sympodial units) of a dCLCrV::FT-infected TX701 plant transitions from highly lobed to lanceolate as the fruiting branch gets older. (**B**) dCLCrV-infected TX701 plants do not show the same transition. Note the leaf crumpling symptoms of dCLCrV-infected plants (**A**, **B**). (**C**) Changes in TX701 leaf correlate with changes in day length and reproductive *vs.* vegetative growth; these leaves are from the same plant that transitioned from long-day, to short day and back to long day growth conditions, with ~10 weeks in each condition. See also Fig. S2 for plants grown exclusively in natural sunlight.

In the ancestral accession TX701, the fruiting branches of dCLCrV::FT-infected plants occasionally terminated with a floral cluster (Fig. 4E, F), which was not observed in short-day-induced plants (not shown) or in any DP61 plants. This is consistent with both the apical bud and one or both of the axillary buds of the terminal SU converting to a reproductive fate (*i.e.*, becoming determinate) roughly simultaneously before forming the vegetative components of the next SU (*i.e.*, internode and subtending leaf with new axillary buds). In both dCLCrV::FT-infected DP61 and TX701, the apical buds of the main stem and vegetative branches did not form flowers: the monopodial branches remained indeterminate (Fig. 4C–E).

As described above, NFB for TX701 was 20.6±3.1 ($n=7$ plants) under short days and as early as 5 in long days with VIF (Table S1). This suggests that high florigen enhances earliness more effectively than photoperiod. Flowers and bolls also matured faster, but we attribute this to greater photosynthesis under long days rather than to dCLCrV::FT. NFB was also improved in day neutral DP61: 5.1±0.9 for untransfected plants ($n=10$), 5.0±1.7 for dCLCrV-infected plants ($n=3$), and 3.0±0.0 for dCLCrV::FT-infected plants ($n=3$). Therefore, ectopic *FT* expression results in earlier NFB and more synchronized fruiting, both of which are important for cotton cultivation.

3. *FT*-induced TX701 Flowers can be Crossed with Day-neutral DP61 Flowers to Produce Fertile F1 Plants

Because flowering time was comparable between VIF-treated TX701 and DP61, we recognized that VIF could be used to introgress desirable germplasm into elite cultivated cotton lines. To test if VIF is a viable tool for breeding programs, we used *FT*-induced TX701 flowers as pollen donors in crosses with DP61. Cross-pollinated flowers formed healthy bolls with good seed yields (Fig. 3D and Fig. S3). To demonstrate that these crosses were successful, the F_1 generation was scored for three traits: leaf shape, NFB, and petal spot. TX701 plants have an extreme okra-leaf phenotype with three to five deep and narrow lobes while DP61 plants have normal leaves with three broad lobes (Fig. 5A). The main-stem leaves from all DP61 × TX701 F_1 plants ($n=46$) exhibited an intermediate phenotype (Fig. 5A). Under 16 hr long-day conditions, NFB among DP61 plants was 5.1±0.9 ($n=10$ plants), TX701 plants did not produce fruiting branches by node 25 when results were scored ($n=8$ plants) and F_1 NFB was 14.7±2.2 ($n=46$; Fig. 5B), consistent with previous reports from conventional crosses between TX701 and DP61 [9]. TX701 flowers have creamy white petals with prominent burgundy petal spots (Fig. 3C, Fig. S3) while DP61 flowers are devoid of petal spots (Fig. 5C, top). All DP61 × TX701 F_1 plants had prominent petal spots like the male parent (Fig. 5C, bottom). Each of these three phenotypes demonstrates that the F_1 population was the result of the controlled crossing event between emasculated DP61 and *FT*-induced TX701 flowers.

In a previous study with plants infected with a dCLCrV::GFP construct, GFP fluorescence was sometimes observed in the outer boll wall, in the central column of the ovary, and in the ovule integuments [33]. Although it is well-established that geminiviruses are not ovule- or pollen-transmitted [40,41], we screened the F_1 generation by PCR for the presence of the viral *Rep* gene for confirmation. The dCLCrV *Rep* gene was absent from uninoculated DP61 and TX701 plants (Fig. 5D, lanes 1 and 2) and from all analyzed F_1 plants, as expected (Fig. 5D, lanes 3 through 11). *Rep* sequences were detected exclusively in a TX701 plant infected with dCLCrV::FT and in a plasmid control (Fig. 5D, lane 12 and 13). Detection of the endogenous *Chl1*, present in all plant samples, served as an internal control (Fig. 5D). Consequently, although the pollen donor was manipulated with a recombinant virus, progeny of VIF plants do not carry DNA originating from the dCLCrV vectors, and should not be considered "genetically modified organisms" (GMOs).

Discussion

FT is a key integrator of signaling pathways leading to floral induction. Extensive research demonstrates the conserved function of this gene product across diverse plant species irrespective of photoperiod. *FT* orthologs from a spectrum of eudicots have been expressed in Arabidopsis and induced early flowering, and *FT*

Figure 3. dCLCrV::FT infection uncouples flowering from photoperiod. (**A**) Arrows point to some of many squares on a representative dCLCrV::FT infected TX701 plant under long day conditions. Scale bar is 25 cm. (**B**) Close-up of sympodial growth on a fruiting branch of an FT-induced TX701 plant. (**C**) An open bloom from an FT-induced TX701 plant reveals characteristic dark red petal spots. (**D**) A boll on a DP61 plant resulting from cross-pollination with an FT-induced TX701 flower (See also Fig. S3). (**E**) RT-PCR demonstrates that *FT* expression is limited to plants inoculated with dCLCrV::FT; *GAPDH* expression serves as an internal control. Lanes are: 1 and 2, untransfected TX701; 3, no reverse transcriptase control; 4, dCLCrV::FT-infected plant shown in (C); 5, dCLCrV::FT-infected plant shown in (B); 6, dCLCrV::FT-infected plant shown in (A); 7 and 8, other dCLCrV::FT-infected TX701 plants that flowered (plants not shown); 9) dCLCrV::FT-bombarded TX701 plant which did not flower – note the absence of *FT*; 10) dCLCrV::FT plasmid template control; 11) no template control.

orthologs expressed in species other than Arabidopsis have similarly promoted early flowering. For example, rice is a short-day monocot, but overexpressing *HEADING DATE 3a* in Arabidopsis under non-inductive conditions is as effective at promoting early flowering as overexpressing *FT* [42–44]. However, the function of FT is now considered in a broader context wherein it advances determinate growth in all organs. In this model, which was developed in perennial tomato [17,27], local balances between *SFT* and *SELF-PRUNING* (*SP*), a potent *SFT*-dependent *SFT* inhibitor, control patterns of indeterminate and determinate growth: High *SFT/SP* expression ratios promote determinacy while low *SFT/SP* expression ratios promote indeterminacy. *SP* is a homolog of *Antirrhinum major CENTRORADIALIS* and Arabidopsis *TERMINAL FLOWER 1*. In this balance model, import of florigen (the product of *SFT*) into organs already expressing *SFT* communicates the status of distal tissues to modulate the *SFT/SP* ratio in new growth [17]. Importantly, high levels of florigen promote determinate growth in all organs and thus florigen functions as a general growth regulator rather than specifically a flowering hormone. In tomato, florigen confers different flowering responses in primary and secondary shoots, regulates the pattern of sympodial branches, accelerates the

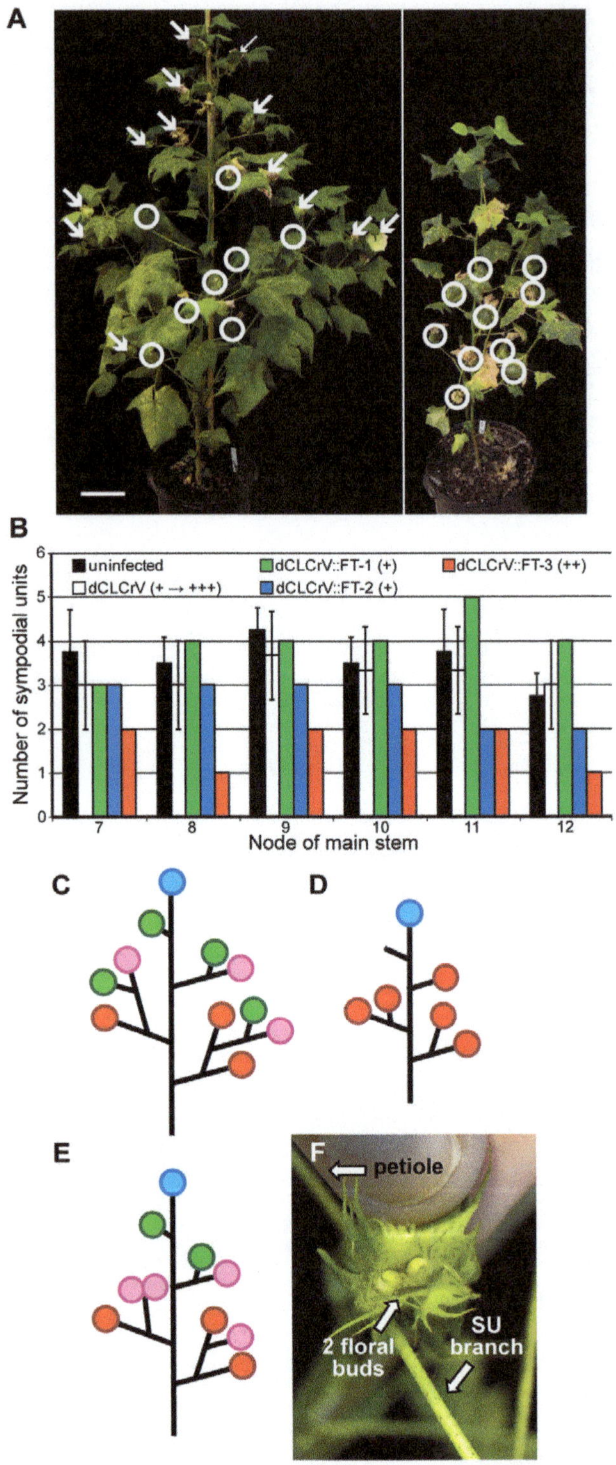

Figure 4. FT promotes determinate growth and synchronizes flowering. (**A**) dCLCrV::FT-infected DP61 plants exhibit a more compact growth habit. Shown are dCLCrV-infected (i.e., empty virus) and dCLCrV::FT-infected DP61 plants (left and right, respectively; both plants are 87 dpg and were grown at the same time in the same greenhouse). White circles highlight maturing bolls and arrows point to flowers before or in bloom. Note that the dCLCrV::FT-infected plant has only maturing bolls and no immature flowers. Scale bar is 25 cm. (**B**) dCLCrV::FT-infected plants demonstrate more determinate growth. Shown are the mean number of sympodial units along fruiting branches among untransfected plants ($n=4$, black bar), dCLCrV-infected plants ($n=3$, white bar), and three dCLCrV::FT-infected plants represented individually (green, blue and red bars) to show the range of variation; dCLCrV::FT-3 is the plant shown in (A). The severity of viral infection was scored as mild (+) or stronger (++) based on leaf crumpling. (**C, D, E**) Schematic representations of growth patterns observed among ~90 d-old plants: (**C**) Uninfected and dCLCrV-infected DP61; (**D**) dCLCrV::FT infected DP61, (**E**) dCLCrV::FT infected TX701. Red circles represent maturing bolls; magenta circles represent immature or blooming flowers; green circles represent active buds reiterating sympodial growth; blue circles represent the monopodial bud of the main stem; and branches without a circle represent buds that have terminated without a flower or a fruit. Leaves are not represented, and the number of branches and internode lengths are not to scale. (**F**) Representative floral cluster (two floral buds inside a common bract whorl) terminating a fruiting branch on a dCLCrV::FT-infected TX701 plant. "SU branch" is the internode of the terminal sympodial unit; "petiole" is the petiole of the leaf subtending the floral cluster. No other vegetative growth is evident.

maturation and complexity of compound leaves, and promotes abscission zone formation, among other effects [27]. It is now apparent that the ratios of SFT and SP (FT and CEN/TFL1), as well as other members of the *CENTRORADIALIS/TERMINAL FLOWER 1/SELF-PRUNING* (*CETS*) gene family, have influenced the balance of vegetative (indeterminate) and reproductive (determinate) growth in many annuals and perennials, and have played important roles in the transition of wild exotics to domesticated crops [20,21,28,29,45].

Ectopic expression of *FT* affected growth in both photoperiodic TX701 and day-neutral DP61 cotton in ways that are consistent with the balance model for indeterminate and determinate growth regulation [17,27], and thus contributes evidence that the balance model is broadly applicable. Using dCLCrV to deliver *FT* perturbed the existing balance to favor determinate growth, and the effects of high florigen included earlier flowering time uncoupled from photoperiod, the precocious transition to determinate plant architecture, and changes in leaf shape. Several TX701 fruiting branches terminated with a floral cluster rather than continuing sympodial reiterations (Fig. 4). We interpret this as roughly simultaneous conversion of the apical and axillary meristems to floral identity in the terminal sympodial unit, analogous to phenotypes observed in tomato with high SFT/SP ratios. Changes in leaf morphology, from deeply lobed palmate to simple lanceolate on the fruiting branches in response to both VIF and short-day induction, parallel the reduced complexity of compound leaves in tomato overexpressing *SFT* [27].

Flowers of VIF-treated TX701 frequently did not reach anthesis and many of those that did abscised before producing mature bolls (data not shown). Since *FT* promotes determinate growth and was expressed from a strong promoter on a multi-copy viral vector, floral maturation may have ceased because of premature termination. The accumulation of virus and *FT*-derived signal will vary from bud to bud as a function of viral replication and source/sink relationships, and we propose that those flowers that reached anthesis received a tolerable dose while those that terminated development received an excessive dose. It is worth noting that ectopic overexpression of the Arabidopsis *FT* in transgenic poplar [46] and from *Apple latent spherical virus* in apple [24] resulted in aberrant floral morphologies.

FT overexpression in DP61 resulted in fruiting branches developing at the third node, which was earlier than uninfected controls. However, since day-neutral cultivars already have NFB in the range of 5 to 7, this impact of increased FT levels on cotton agronomy will be modest. Of much greater significance is the role of *FT* overexpression on flower and fruit synchronization. Asynchronous flowering and fruit set throughout the growing season are indeterminate perennial characteristics that compro-

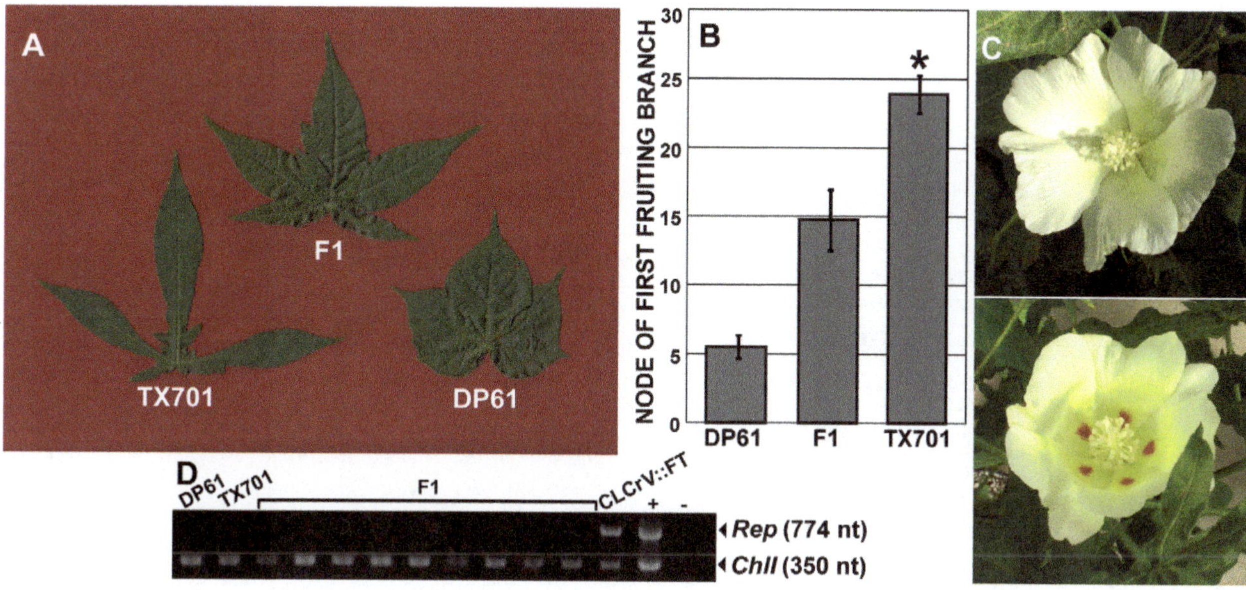

Figure 5. F_1 progeny of crosses with a VIF-treated parent have expected traits and are virus-free. (**A**) The shape of main-stem leaves harvested from nodes 8–9 of TX701, DP61, and the F_1, as labeled. TX701 leaves exhibit the okra leaf phenotype with five deep lobes and reduced lamina area. DP61 exhibits the normal leaf phenotype with three smaller lobes and a well-expanded lamina. The F_1 from the DP61 × *FT*-induced TX701 cross exhibit an intermediate phenotype with leaves having five lobes of intermediate length. (**B**) Node of first fruiting branch is intermediate in the F_1 population. Under non-inductive long days, the F_1 produce floral buds ($n = 46$) in contrast to the TX701 parent which does not demonstrate reproductive growth by node 25 ($n = 8$ plants). Day-neutral DP61 produce floral buds earlier in development ($n = 6$ plants). (**C**) The maternal DP61 flowers (top) do not have the flower spots characteristic of the paternal TX701 pollen donor (see Fig. 3c); F_1 progeny flowers (bottom) have the flower spot trait. (**D**) dCLCrV is not transmitted to the F_1 progeny. From left to right, the viral *Rep* gene is not detected in uninfected DP61 or TX701 (lanes 1 and 2), nor in 9 different F1 progeny plants (lanes 3–11, collectively labeled F1), but is readily detected by PCR from a dCLCrV::FT-infected plant (lane 12, labeled dCLCrV::FT; this is the same plant shown in Fig. 3A and E, lane 6) and from a plasmid template (lane 13, labeled +). Detection of the endogenous magnesium chelatase subunit I (*ChlI*) serves as an internal control (plasmid control in the case of lane 13, +). No sequences were amplified in the absence of DNA (lane 14, labeled -).

mise yields and complicate crop management when cotton is grown as an annual row crop [5,6,47,48]. *FT* overexpression reduced these perennial traits in DP61. Flowering and fruit maturation were more synchronized since the fruiting branches terminated rather than reiterating the pattern of sympodial units. This prevented continued production of new flowers while fruits closer to the main stem were maturing, and in combination with shorter internodes, resulted in more compact plants (Fig. 4). Unlike TX701, however, floral clusters were not observed at the termini of VIF-treated DP61 plants. The reason for this is not clear, but may relate to a higher balance of determinate growth factors in domesticated cultivars that result in cessation of bud growth before flower initiation. Flowers that did form on *FT*-overexpressing DP61 plants formed where expected (*i.e.*, at the ends of each sympodial unit) and unlike many of those on VIF-treated TX701, progressed through anthesis and formed mature bolls. This difference in DP61 and TX701 treated plants may also relate to different balances between determinacy and indeterminacy factors.

Our results presented here with cotton show that manipulating *FT* expression can alter growth habit in ways that would, with judicious control of expression levels, benefit yields and crop management: Earlier and more synchronized flowering and fruiting, a stockier growth habit, and a shorter growing season for focused agricultural inputs are all desired agronomic traits. In essence, achieving a more determinate and "annualized" growth habit is a prominent goal in modern cotton breeding. The fact that perennial traits of asynchronous flowering and continued vegetative growth remain, argues that, perhaps, traditional breeding has reached its limits, and that architectural manipulation through biotechnology will be required for further gains. Although yield per plant may be reduced, this could be compensated by denser planting, which in turn would have the benefits of faster field coverage to control weeds. Furthermore, the extent of vegetative growth and onset of reproductive growth could be customized to region-specific environments through manipulations of CETS. There already exists substantial support for these speculations. As noted, the CETS family is prominent in the domestication of several crops, and in tomato is directly linked to significant developments in the 20th century. Tomato, like cotton, is naturally an indeterminate perennial with continued flower and fruit set through the growing season. However, the *sp* mutant of tomato, which results in a high SFT(FT)/SP(TFL) ratio, has accelerated termination of sympodial growth, and results in a more compact, determinate plant with nearly homogeneous fruit set. Identifying the *sp* phenotype [49] "was the single most important genetic trait in the development of modern agrotechniques for this crop plant because the 'determinate' growth habit facilitates mechanical harvest" [50,51].

It was previously shown that expressing Arabidopsis *FT* from *Zucchini yellow mosaic virus* and *Apple latent spherical virus* promoted flowering under non-inductive photoperiods in melon [52] and soybean [53], respectively, and reduced the juvenile phase of apple trees [24]. In addition to promoting determinate cotton plant architecture, we and others [24] demonstrate that VIF is a promising breeding tool. Domestication of wild species often contributes to genetic homogeneity in crops, and loss of diversity is associated with loss of tolerance/resistance to abiotic and biotic stresses, and increased susceptibility to pathogens, pests and environmental change. The Irish potato famine, caused by the rapid spread of *Phytophthora infestans* among genetically-uniform cultivated potatoes [54] and over-exploitation of the Tcms cytoplasm in maize, which confers cytoplasmic male sterility but also susceptibility to *Helminthosporium maydis*, the causative agent of southern corn leaf blight [55], are classic examples of how reliance on inbred crops can have devastating agricultural and social impacts. Genetic diversity across the *Gossypium* genus has been estimated and there is consensus that domesticated lines are highly inbred while wild races contain extensive diversity [2,10,11,56]. The two species of highest economic importance, *G. hirsutum* and *G. barbadense*, are both derived from a common allotetraploid and this polyploidization event was the first major bottleneck [2]. Moreover, Upland cotton, which accounts for ~90% of USA production, derives from race 'latifolium' [10]. Spread of cotton boll weevil early in the twentieth century further limited diversity [10], and pedigrees of commercially successful modern lines indicate frequent use of relatively few parents [11,56]. Furthermore, most transgenic cultivars contain a linkage group of the transgene donor parent, Coker 312 [56]. Based on this history, it is apparent that hybridization with lines outside of 'latifolium' will be required to increase diversity in Upland cotton. Since these are photoperiodic, VIF will benefit breeding efforts. For that same reason, VIF may also accelerate the generation of recombinant-inbred lines for linkage-association mapping and nested-association mapping [57–59], and reduce germplasm maintenance costs by permitting self-fertilization of photoperiodic or long-seasoned accessions in temperate regions.

In summary, the *FT* gene product, florigen, is increasingly recognized as a general growth hormone that regulates shoot architecture by advancing organ-specific and age-related determinate growth via its local expression and systemic distribution (as reviewed in [60]). It is now clear that the ratio of florigen to other members of the same gene family evolved to influence patterns of vegetative and reproductive growth in annual [28], biennial [21], and perennial plants [20], and that changes in these ratios contributed to the domestication of many crops from wild progenitors [28,29]. Knowledge of these expanded roles beyond control of photoperiodic flowering opens new avenues for exploiting florigen with the applied goals of enhancing plant productivity and crop management. We demonstrate here that virus-induced flowering (VIF) promotes determinate growth: earlier and photoperiod-independent flowering, precocious transition to reproductive architecture, and lanceolate leaf shape were coordinated with ectopic expression of *FT* in ancestral cotton; and greater fruit synchronization and "annualized" growth were observed in domesticated cotton infected with dCLCrV::FT. Our results indicate that judicious manipulation of *FT* and related genes could benefit cotton architecture to improve crop management.

Supporting Information

Figure S1 Biolistic delivery of dCLCrV::αChlI and dCLCrV::FT promotes silencing and flowering, respectively, in TX701. (**A**) Cotton plants inoculated with dCLCrV::αChlI show systemic and sustained silencing of the magnesium chelatase I (ChlI) subunit, resulting in chlorotic sectors. Shown are 6-week old TX701 plants that were inoculated with dCLCrV (left) and dCLCrV::αChlI (right) at the seedling stage. (**B**) Biolistic rupture disk pressure and size of gold particles affect transfection of TX701 cotton. Shown is the percentage of plants exhibiting silencing when inoculated with dCLCrV::αChlI adhered to different sized gold particles and delivered at 900 and 1350 psi (for each treatment, $n = 4$). Also shown is the percentage of TX701 plants that flowered under non-inductive long days when inoculated with dCLCrV::FT using the same biolistic parameters ($n = 4$ for each treatment except with 0.6 μm gold and 1350 psi where $n = 6$).

Figure S2 Leaves from TX701 plants grown outside with only natural sunlight. (**A**) Main stem leaf that formed under long day conditions while plants had no reproductive growth. (**B**) A fruiting branch from the same plant that developed under short day conditions. Note the floral squares and the simpler, more lanceolate subtending leaves towards the tip of the branch. Scale bars, 5 cm.

Figure S3 Cross-pollinating emasculated DP61 flowers with *FT*-induced TX701 male parents. (**A**) Flowers from dCLCrV::FT-infected TX701 plants were used as pollen donors to (**B**) cross-pollinate emasculated DP61 flowers. (**C, D**) Healthy bolls formed with good seed set (**E**).

Table S1 VIF accelerates the time to flower irrespective of photoperiod in short-day TX701 and day-neutral DP61 cotton.

Acknowledgments

We are grateful to Elison Blancaflor and Alan Sparks at the Noble Foundation, in Ardmore, OK, for use of their PDS-1000 gene gun; Jack McCarty for seeds; Richard Tuttle for DNA constructs and protocols; Jane Dever and Kater Hake for helpful discussion; and Yuval Eshed for critically reviewing the manuscript.

Author Contributions

Conceived and designed the experiments: RCM BGA. Performed the experiments: RCM. Analyzed the data: RCM BGA. Contributed reagents/materials/analysis tools: RCM BGA. Wrote the paper: RCM BGA.

References

1. Chen ZJ, Scheffler BE, Dennis E, Triplett BA, Zhang T, et al. (2007) Toward sequencing cotton (*Gossypium*) genomes. Plant Physiol 145: 1303–1310.
2. Wendel JF, Brubaker C, Alvarez I, Cronn R, Stewart JM (2009) Evolution and natural history of the cotton genus. In: Paterson AH, ed. Genetics and Genomics of Cotton. New York: Springer Science and Business Media, LLC. pp 3–22.
3. Mauney JR, Phillips LL (1963) Influence of daylength and night temperature on flowering of *Gossypium*. Bot Gaz 124: 278–283.
4. Gore U (1935) Morphogenic studies of the inflorescence of cotton. Bot Gaz 97: 118–138.
5. Oosterhuis DM (1990) Growth and development of the cotton plant. In: Miley WN, Oosterhuis DM, eds. Nitrogen nutrition in cotton: Practical Issues Proc Southern Branch Workshop for Practicing Agronomists. Madison, WI: Publ. Amer. Soc. Agron. pp 1–24.
6. Jost P, Whitaker J, Brown SM, Bednarz C Use of plant growth regulators as a management tool in cotton. University of Georgia Cooperative Extension Service Bulletin 1305: http://pubs.caes.uga.edu/caespubs/pubcd/B1305.htm.
7. Shurley D, Bednarz CW, Anthony S, Brown SM (2004) Increasing cotton yield, fiber quality, and profit through improved defoliation and harvest timeliness University of Georgia Cooperative Extension Service Bulletin 1252 AGECON-04-94.
8. Paterson AH, Boman RK, Brown SM, Chee PW, Gannaway JR, et al. (2004) Reducing the genetic vulnerability of cotton. Crop Sci 44: 1900–1901.
9. Guo YF, McCarty JC, Jenkins JN, Saha S (2008) QTLs for node of first fruiting branch in a cross of an upland cotton, *Gossypium hirsutum* L., cultivar with primitive accession Texas 701. Euphytica 163: 113–122.

10. Lubbers EL, Chee PW (2009) The worldwide gene pool of *G. hirsutum* and its improvement. In: Paterson AH, ed. Genetics and Genomics of Cotton. New York: Springer Science and Business Media, LLC. pp 23–52.
11. Percy RG (2009) The worldwide gene pool of *Gossypium barbadense* L. and its improvement. In: Paterson AH, ed. Genetics and Genomics of Cotton. New York: Springer Science and Business Media, LLC. pp 53–68.
12. Ayre B (2010) The flowering hormone - Florigen: A protein hormone. In: Davies P, ed. Plant Hormones, Biosynthesis, Signal Transduction, Action! New York: Springer. pp 539–548.
13. Turck F, Fornara F, Coupland G (2008) Regulation and identity of florigen: *FLOWERING LOCUS T* moves center stage. Annu Rev Plant Biol 59: 573–594.
14. Hayama R, Yokoi S, Tamaki S, Yano M, Shimamoto K (2003) Adaptation of photoperiodic control pathways produces short-day flowering in rice. Nature 422: 719–722.
15. Tsuji H, Taoka K, Shimamoto K (2011) Regulation of flowering in rice: two florigen genes, a complex gene network, and natural variation. Curr Opin Plant Biol 14: 45–52.
16. Chailakhyan MK (1968) Internal factors of plant flowering. Annu Rev Plant Physiol 19: 1–37.
17. Lifschitz E, Eviatar T, Rozman A, Shalit A, Goldshmidt A, et al. (2006) The tomato *FT* ortholog triggers systemic signals that regulate growth and flowering and substitute for diverse environmental stimuli. Proc Natl Acad Sci USA 103: 6398–6403.
18. Corbesier L, Vincent C, Jang SH, Fornara F, Fan QZ, et al. (2007) FT protein movement contributes to long-distance signaling in floral induction of Arabidopsis. Science 316: 1030–1033.
19. Tamaki S, Matsuo S, Wong HL, Yokoi S, Shimamoto K (2007) Hd3a protein is a mobile flowering signal in rice. Science 316: 1033–1036.
20. Hsu C-Y, Adams JP, Kim H, No K, Ma C, et al. (2011) *FLOWERING LOCUS T* duplication coordinates reproductive and vegetative growth in perennial poplar. Proc Natl Acad Sci USA 108: 10756–10761.
21. Pin PA, Benlloch R, Bonnet D, Wremerth-Weich E, Kraft T, et al. (2010) An antagonistic pair of *FT* homologs mediates the control of flowering time in sugar beet. Science 330: 1397–1400.
22. Bohlenius H, Huang T, Charbonnel-Campaa L, Brunner AM, Jansson S, et al. (2006) *CO/FT* regulatory module controls timing of flowering and seasonal growth cessation in trees. Science 312: 1040–1043.
23. Hsu CY, Liu YX, Luthe DS, Yuceer C (2006) Poplar FT2 shortens the juvenile phase and promotes seasonal flowering. Plant Cell 18: 1846–1861.
24. Yamagishi N, Sasaki S, Yamagata K, Komori S, Nagase M, et al. (2011) Promotion of flowering and reduction of a generation time in apple seedlings by ectopical expression of the *Arabidopsis thaliana FT* gene using the *Apple latent spherical virus* vector. Plant Mol Biol 75: 193–204.
25. Endo T, Shimada T, Fujii H, Kobayashi Y, Araki T, et al. (2005) Ectopic expression of an *FT* homolog from *Citrus* confers an early flowering phenotype on trifoliate orange (*Poncirus trifoliata* L. Raf.). Transgenic Res 14: 703–712.
26. Danilevskaya ON, Meng X, McGonigle B, Muszynski MG (2011) Beyond flowering time: Pleiotropic function of the maize flowering hormone florigen. Plant Signal Behav 6: 1–4.
27. Shalit A, Rozman A, Goldshmidt A, Alvarez JP, Bowman JL, et al. (2009) The flowering hormone florigen functions as a general systemic regulator of growth and termination. Proc Natl Acad Sci USA 106: 8392–8397.
28. Blackman BK, Strasburg JL, Raduski AR, Michaels SD, Rieseberg LH (2010) The role of recently derived *FT* paralogs in sunflower domestication. Curr Biol 20: 629–635.
29. Tian Z, Wang X, Lee R, Li Y, Specht JE, et al. (2010) Artificial selection for determinate growth habit in soybean. Proc Natl Acad Sci USA 107: 8563–8568.
30. Ausubel FM, Brent R, Kingston RE, Moore DD, Seidman JG, et al. (2002) Short protocols in molecular biology: a compendium of methods from Current protocols in molecular biology. 5th ed. Hoboken, NJ: John Wiley & Sons, Inc.
31. Sambrook J, Russel DW (2001) Molecular cloning: a laboratory manual. Cold Spring Harbor: Cold Spring Harbor Laboratory Press.
32. Ayre BG, Turgeon R (2004) Graft transmission of a floral stimulant derived from *CONSTANS*. Plant Physiol 135: 2271–2278.
33. Tuttle JR, Idris AM, Brown JK, Haigler CH, Robertson D (2008) Geminivirus-mediated gene silencing from *Cotton leaf crumple virus* is enhanced by low temperature in cotton. Plant Physiol 148: 41–50.
34. Doyle JJ, Doyle JL (1990) Isolation of plant DNA from fresh tissue. Focus 12: 13–15.
35. John ME, Stewart JM (2010) Genetic engineering applications in crop improvement. In: Stewart JM, Oosterhuis DM, Heitholt JJ, Mauney JR, eds. Physiology of Cotton. New York: Springer Science+Business Media. pp 394–403.
36. Trolinder NL, Goodin JR (1987) Somatic embryogenesis and plant regeneration in cotton (*Gossypium hirsutum* L.). Plant Cell Rep 6: 231–234.
37. Wilkins T, Mishra R, Trolinder N (2004) *Agrobacterium*-mediated transformation and regeneration of cotton. Food, Ag and Environ 2: 179–197.
38. Prusinkiewicz P, Erasmus Y, Lane B, Harder LD, Coen E (2007) Evolution and development of inflorescence architectures. Science 316: 1452–1456.
39. Idris AM, Tuttle JR, Robertson D, Haigler CH, Brown JK (2010) Differential *Cotton leaf crumple virus*-VIGS-mediated gene silencing and viral genome localization in different *Gossypium hirsutum* genetic backgrounds. Physiol Mol Plant Pathol 75: 13–22.
40. Mink GI (1993) Pollen and seed-transmitted viruses and viroids. Annu Rev Phytopathol 31: 375–402.
41. Sudarshana MR, Wang HL, Lucas WJ, Gilbertson RL (1998) Dynamics of bean dwarf mosaic geminivirus cell-to-cell and long-distance movement in *Phaseolus vulgaris* revealed, using the green fluorescent protein. Mol Plant Microbe Interact 11: 277–291.
42. Kojima S, Takahashi Y, Kobayashi Y, Monna L, Sasaki T, et al. (2002) *Hd3a*, a rice ortholog of the *Arabidopsis FT* gene, promotes transition to flowering downstream of *Hd1* under short-day conditions. Plant Cell Physiol 43: 1096–1105.
43. Komiya R, Ikegami A, Tamaki S, Yokoi S, Shimamoto K (2008) *Hd3a* and *RFT1* are essential for flowering in rice. Development 135: 767–774.
44. Takahashi Y, Teshima KM, Yokoi S, Innan H, Shimamoto K (2009) Variations in Hd1 proteins, *Hd3a* promoters, and *Ehd1* expression levels contribute to diversity of flowering time in cultivated rice. Proc Natl Acad Sci USA 106: 4555–4560.
45. Navarro C, Abelenda JA, Cruz-Oro E, Cuellar CA, Tamaki S, et al. (2011) Control of flowering and storage organ formation in potato by FLOWERING LOCUS T. Nature 478: 119–122.
46. Zhang HL, Harry DE, Ma C, Yuceer C, Hsu CY, et al. (2010) Precocious flowering in trees: the *FLOWERING LOCUS T* gene as a research and breeding tool in *Populus*. J Exp Bot 61: 2549–2560.
47. Cothren J, Oosterhuis D (2010) Use of growth regulators in cotton production. In: Stewart J, Oosterhuis D, Heitholt J, Mauney J, eds. Physiology of Cotton. New York: Springer. pp 289–303.
48. Kerby T, Bourland F, Hake K (2010) Physiological rationales in plant monitoring and mapping. In: Stewart J, Oosterhuis D, Heitholt J, Mauney J, eds. Physiology of Cotton. New York: Springer. pp 304–317.
49. Yeager AF (1927) Determinate growth in tomato. J Hered 18: 263–265.
50. Pnueli L, Carmel-Goren L, Hareven D, Gutfinger T, Alvarez J, et al. (1998) The *SELF-PRUNING* gene of tomato regulates vegetative to reproductive switching of sympodial meristems and is the ortholog of *CEN* and *TFL1*. Development 125: 1979–1989.
51. Rick CM (1978) The tomato. Sci Am 239: 76–87.
52. Lin MK, Belanger H, Lee YJ, Varkonyi-Gasic E, Taoka KI, et al. (2007) FLOWERING LOCUS T protein may act as the long-distance florigenic signal in the cucurbits. Plant Cell 19: 1488–1506.
53. Yamagishi N, Yoshikawa N (2010) Expression of *FLOWERING LOCUS T* from *Arabidopsis thaliana* induces precocious flowering in soybean irrespective of maturity group and stem growth habit. Planta 233: 561–568.
54. Fry WE, Goodwin SB (1997) Resurgence of the Irish Potato Famine fungus. Bioscience 47: 363–371.
55. Ullstrup AJ (1972) The impacts of the southern corn leaf blight epidemics of 1970–1971. Annu Rev Phytopathol 10: 37–50.
56. Wallace TP, Stelly DM, Smith W, Robinson F, Stewart JM, et al. (2009) Status of the USA cotton germplasm collection and crop vulnerability. Genet Resour Crop Evol 56: 507–532.
57. Bergelson J, Roux F (2010) Towards identifying genes underlying ecologically relevant traits in Arabidopsis thaliana. Nat Rev Genet 11: 867–879.
58. Kump KL, Bradbury PJ, Wisser RJ, Buckler ES, Belcher AR, et al. (2011) Genome-wide association study of quantitative resistance to southern leaf blight in the maize nested association mapping population. Nat Genet 43: 163-U120.
59. Yu JM, Holland JB, McMullen MD, Buckler ES (2008) Genetic design and statistical power of nested association mapping in maize. Genetics 178: 539–551.
60. McGarry RC, Ayre BG (2012) Manipulating plant architecture with members of the *CETS* gene family. Plant Sci 188: 71–81.

12

Development and Evaluation of a Genome-Wide 6K SNP Array for Diploid Sweet Cherry and Tetraploid Sour Cherry

Cameron Peace[1]*, Nahla Bassil[2], Dorrie Main[1], Stephen Ficklin[1], Umesh R. Rosyara[3], Travis Stegmeir[3], Audrey Sebolt[3], Barbara Gilmore[2], Cindy Lawley[4], Todd C. Mockler[5], Douglas W. Bryant[5,6], Larry Wilhelm[7], Amy Iezzoni[3]

1 Department of Horticulture, Washington State University, Pullman, Washington, United States of America, **2** USDA-ARS, National Clonal Germplasm Repository, Corvallis, Oregon, United States of America, **3** Department of Horticulture, Michigan State University, East Lansing, Michigan, United States of America, **4** Illumina Inc., Hayward, California, United States of America, **5** The Donald Danforth Plant Science Center, St. Louis, Missouri, United States of America, **6** Intuitive Genomics, Inc., St. Louis, Missouri, United States of America, **7** Oregon Health Sciences University, Portland, Oregon, United States of America

Abstract

High-throughput genome scans are important tools for genetic studies and breeding applications. Here, a 6K SNP array for use with the Illumina Infinium® system was developed for diploid sweet cherry (*Prunus avium*) and allotetraploid sour cherry (*P. cerasus*). This effort was led by RosBREED, a community initiative to enable marker-assisted breeding for rosaceous crops. Next-generation sequencing in diverse breeding germplasm provided 25 billion basepairs (Gb) of cherry DNA sequence from which were identified genome-wide SNPs for sweet cherry and for the two sour cherry subgenomes derived from sweet cherry (*avium* subgenome) and *P. fruticosa* (*fruticosa* subgenome). Anchoring to the peach genome sequence, recently released by the International Peach Genome Initiative, predicted relative physical locations of the 1.9 million putative SNPs detected, preliminarily filtered to 368,943 SNPs. Further filtering was guided by results of a 144-SNP subset examined with the Illumina GoldenGate® assay on 160 accessions. A 6K Infinium® II array was designed with SNPs evenly spaced genetically across the sweet and sour cherry genomes. SNPs were developed for each sour cherry subgenome by using minor allele frequency in the sour cherry detection panel to enrich for subgenome-specific SNPs followed by targeting to either subgenome according to alleles observed in sweet cherry. The array was evaluated using panels of sweet (n = 269) and sour (n = 330) cherry breeding germplasm. Approximately one third of array SNPs were informative for each crop. A total of 1825 polymorphic SNPs were verified in sweet cherry, 13% of these originally developed for sour cherry. Allele dosage was resolved for 2058 polymorphic SNPs in sour cherry, one third of these being originally developed for sweet cherry. This publicly available genomics resource represents a significant advance in cherry genome-scanning capability that will accelerate marker-locus-trait association discovery, genome structure investigation, and genetic diversity assessment in this diploid-tetraploid crop group.

Editor: Rongling Wu, Pennsylvania State University, United States of America

Funding: This work was partially funded by the USDA's National Institute of Food and Agriculture – Specialty Crop Research Initiative project, "RosBREED: Enabling marker-assisted breeding in Rosaceae" (2009-51181-05808). The funders had no role in study design, data collection and analysis, decision to publish, or preparation of the manuscript. No additional external funding was received for this study.

* E-mail: cpeace@wsu.edu

Introduction

Within *Prunus* (Rosaceae), two cherry species, sweet (*P. avium*) and sour cherry (*P. cerasus*), are highly valued for their excellent quality fruit. These two species represent a natural diploid-tetraploid series with the tetraploid sour cherry ($2n = 4x = 32$) arising through natural hybridization between sweet cherry ($2n = 2x = 16$) and the wild tetraploid ground cherry (*P. fruticosa*) [1,2]. In cherry, linkage maps constructed for the genetically less complex sweet cherry are primarily based on simple sequence repeat (SSR) markers [3,4,5,6]. The application of these linkage maps for other studies such as quantitative trait locus (QTL) discovery is limited by the low-throughput and, in many cases, low density and low levels of polymorphism in cultivated sweet cherry germplasm for the SSR markers.

High-throughput and low-cost next generation sequencing (NGS) is a powerful approach to identify single nucleotide polymorphisms (SNPs) which can be used as markers for the development of high-density genome scans. This approach has been successfully used to develop genome scan platforms for diploid crops such as apple [7], maize [8], rice [9] and polyploid crops such as potato [10,11], and wheat [12]. While low diversity in a crop still reduces the proportion of observed polymorphic markers, the sheer number of markers that can be efficiently generated by the NGS approach overcomes this historical limitation. A 9K publicly available SNP array for peach, recently developed by an international consortium, is widely being used for advanced genetic studies of peach [13]. The development of this array was facilitated by release of the peach (*Prunus persica*) reference genome sequence by the International Peach Genome Initiative [14].The high quality of the peach sequence was reinforced by high congruence between genetic and physical map positions of peach array SNPs (I. Verde, pers. comm.).

Extensive synteny conserved among the diploid *Prunus* species [3] suggests that the peach genome sequence can be used as a template to develop a genome-wide set of markers for *Prunus* crops for which a high quality genome sequence is not available. For example, comparisons of available sweet cherry genetic maps with the high-density peach×almond "T×E" *Prunus* reference genetic map [3]identified extensive co-linearity [4], yet complete co-linearity could not be rigorously tested as the largest sweet cherry population used consisted of just 118 individuals [6]. Dirlewanger et al. [15] have estimated the genome size of sweet cherry to be 338 Mb.

Herein we use a cherry-peach comparative genomics strategy to develop a moderate-density cherry SNP array relevant for sweet and sour cherry breeding germplasm based on SNPs discovered using next generation sequencing platforms. This effort was led by RosBREED, a community initiative to enable marker-assisted breeding for rosaceous crops [16] which led the recent equivalent developments of SNP arrays for apple [7] and peach [13].

Materials and Methods

The workflow and design parameters described below are summarized in Figure 1.

Whole genome re-sequencing of cherry breeding accessions

A SNP detection panel of 16 sweet and 8 sour cherry accessions was chosen for whole genome, low-coverage resequencing (Table 1). The accessions were founders, intermediate ancestors, or important parents used in U.S breeding programs. For each accession, paired-end libraries were prepared as recommended by manufacturer protocols (Illumina Inc., San Diego, CA, USA). For sweet cherry, equimolar amounts of four libraries were pooled (to create four pools in total) while for sour cherry, with twice the genome size of sweet cherry, equimolar volumes of two libraries were pooled (to create another four pools in total) (Table 1). Each library pool was sequenced in one lane of Illumina GA II with 80 cycles per read at the Center for Genome Research and Biocomputing (CGRB; Oregon State University, Corvallis, OR, USA). The raw sequence data was retrieved and kept separate for each cherry accession and then aligned to the reference genome of 'Lovell' peach [14] using *SOAP* [17] with parameters of $M = 4$ (find best hits for each seed, ≤2 mismatches allowed), $r = 1$ (repeats aligned 1 time randomly), and $v = 2$ (≤2 mismatches allowed).

Detection and Stage 1 filtering of SNPs

SNPs from resequenced accessions were detected using *SOAPsnp* (http://soap.genomics.org.cn/soapsnp.html) as recommended by Li et al. [17] The detected SNPs were filtered ("Stage 1" filtering) and kept if: (1) *SOAPsnp*'s "quality score" metric of the consensus genotype was greater than 30; (2) the sequencing depth at the putative SNP position was at least 8; and (3) the sequencing depth at the putative SNP position was no greater than 1254, corresponding to the average read depth of all SNPs plus three standard deviations. For each gene, the exonic, intronic, and intergenic space was defined using the Peach v1.0 'dhLovell' genome annotation [14]. This filtration yielded "Stage 1 SNPs".

SNP validation with the GoldenGate® assay

A set of 144 SNPs were chosen from among Stage 1 SNPs to validate the efficiency of SNP detection and adjust subsequent filtering parameters (Figure 1). The initial choice comprised 80 Stage 1 SNPs evenly spaced over the sweet cherry genome and 40 SNPs evenly spaced over the sour cherry genome. In the absence of the availability of a whole genome sequence for cherry, the whole genome sequence of peach, i.e., the eight pseudomolecules of the haploid chromosomes and linkage groups (LGs) of the Peach v1.0 'dhLovell' genome assembly [14], was used as the proxy cherry genome. One sweet cherry SNP was chosen to be located within 100 thousand basepairs (kb) from each end of each LG. An even number of sweet cherry SNPs chosen between these ends were then evenly physically spaced along each LG according to LG genetic distances of [3], corresponding to one SNP every 2–4 million basepairs (Mb). The spacing of the 80 sweet cherry SNPs across all LGs averaged 3.00 Mb (standard deviation of 0.52 Mb), with a minimum average of 2.40 Mb (±0.11 Mb) for LG8 and a maximum average of 3.8 Mb (±0.09 Mb) for LG2. A total of 40 sour cherry SNPs (many of which were also SNPs for sweet cherry) were chosen equidistant between pairs of sweet cherry SNPs (first and second, third and fourth, etc.) such that the average physical distance between sour cherry SNPs was twice that of sweet cherry. For both crop sources of the 120 SNPs, 40% were chosen to be located in exons (CDS) of annotated genes, 20% in introns, 20% in 5′ or 3′ untranslated regions (UTR, outside genes but within 2 kb of start or stop codons), and the final 20% in intergenic regions. Sixteen further sweet cherry SNPs and eight further sour cherry SNPs spanned an 863 kb region on LG2 between the simple sequence repeat markers CPSCT038 and BPPCT034, the location of a major trait locus associated with fruit size [18]. While these further 24 trait locus-targeted SNPs were chosen for variation in genic regions where possible, preference was given to achieving uniform physical spacing in the designated windows. Approximately 20% of the 144 validation SNPs were planned to be accession-specific, i.e., their minor allele would be detected in only one re-sequenced accession of the detection panel. Twenty-two of the sweet cherry SNPs and nine of the sour cherry SNPs met this criterion within accessions of their respective crops, and a further eight sweet cherry SNPs were accession-specific within sour cherry accessions of the detection panel. Accession-specific SNPs were from 12 of the 16 sweet cherry accessions and five of the eight sour cherry accessions. The 144 SNPs also deliberately included a wide range of minor allele frequencies (MAFs).

To examine filtering parameters for associations with genotyping efficiency, validation panels of 79 sweet cherry (Table S1) and 81 sour cherry (Table S2) accessions were genotyped for the 144 SNPs described above. Individuals in the validation panel were founders, intermediate ancestors, and important breeding parents of modern cherry cultivars and included the 24 accessions of the SNP detection panel. Genomic DNA was purified from each accession using the E-Z 96 Tissue DNA Kit (Omega Bio-Tek, Inc., Norcross, GA, USA). DNA was quantitated with the Quant-iT™ PicoGreen® Assay (Invitrogen, Carlsbad, CA, USA), using the Victor multiplate reader (Perkin Elmer Inc., San Jose, CA, USA). Concentrations were adjusted to a minimum of 50 ng/µl in 5 µl aliquots and were submitted to the Research Technology Support Facility at Michigan State University (East Lansing, MI, USA) where the GoldenGate® assay was performed following the manufacturer's protocol (Illumina Inc.). After amplification, PCR products were hybridized to VeraCode microbeads via the address sequence for detection on a BeadXpress Reader. SNP genotypes were scored with the Genotyping Module of GenomeStudio Data Analysis software v2010.3 [19].

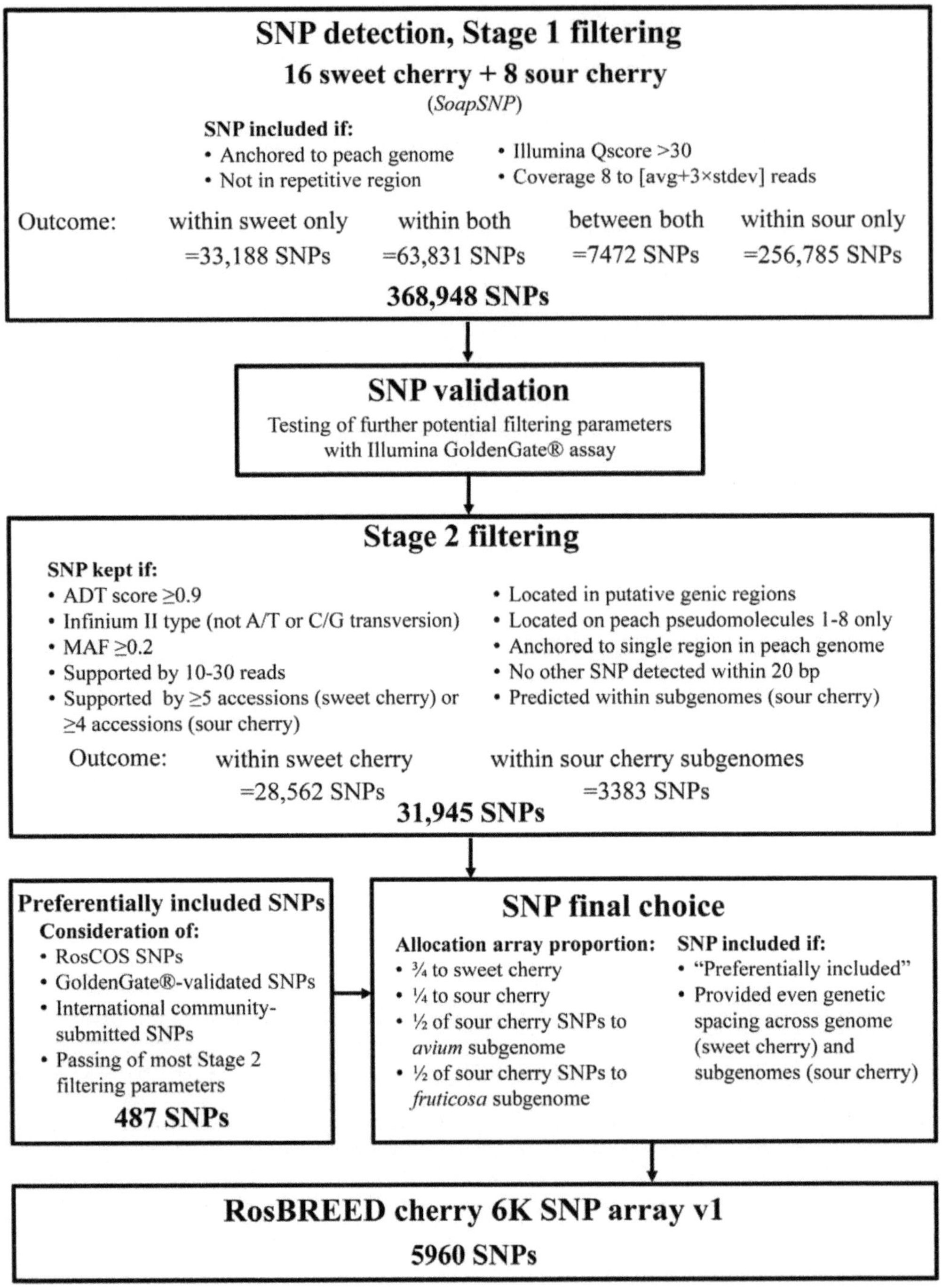

Figure 1. Workflow for SNP detection, validation, and final choice in development of the RosBREED 6K cherry SNP array v1. Stage 1 filtered 1.9 million cherry SNPs anchored to the peach genome to almost 40K SNPs. More stringent filtering criteria in Stage 2, guided by a prior validation step with a small SNP subset examined for a range of potential filters, putatively enriched the quality of the remaining 32K SNP pool. Finally, the 6K array SNPs were chosen from among stage 2 SNPs by attempting to achieve even genetic spacing over species genomes and subgenomes with pre-determined proportional allocations, after preferential inclusion of certain SNPs. ADT = Illumina's Assay Design Tool. MAF = minor allele frequency.

SNP final choice for RosBREED 6K array

Two further rounds of filtering were used, considering validation results. Stage 1 SNPs were converted to Illumina Assay Design Tool (ADT) format with custom scripts and their ADT scores calculated; only SNPs with a score ≥0.90 were retained. A/T and C/G transversions were removed from further consideration, thus retaining only SNPs of normalization bin "C" (for Infinium® II compatibility). SNPs with MAF <0.2 were discarded, as well as those not supported by between 10 and 30 reads and at least five detection panel accessions for sweet cherry or four accessions for sour cherry. Non-intragenic SNPs were removed for sweet cherry, i.e., only exonic and intronic SNPs were considered

Table 1. Cherry accessions used for low-coverage re-sequencing and subsequent SNP detection.

Cherry type	Accession	Adaptors	Read count (million)	No. aligned reads (million)
Sweet	PMR-1	GGGT	10.9	1.5
Sweet	New York 54	TCGT	10.7	1.5
Sweet	Ambrunes	CCAT	8.8	1.2
Sweet	Schmidt	AGCT	11.3	1.6
Sweet	Regina	GGGT	10.1	1.3
Sweet	Van	TCGT	9.9	1.4
Sweet	Lambert	CCAT	9.0	1.3
Sweet	Gil Peck	AGCT	10.3	1.3
Sweet	Emperor Francis	GGGT	16.5	2.0
Sweet	Attika	TCGT	16.0	2.0
Sweet	Napoleon	CCAT	9.0	1.2
Sweet	Windsor	AGCT	20.8	2.7
Sweet	Cashmere	GTGT	16.5	2.4
Sweet	Krupnoplodnaya	TCGT	13.7	1.8
Sweet	Black Republican	AGCT	1.6	0.2
Sweet	Early Burlat	CCAT	20.5	2.9
Sour	Újfehértói Fürtös	GGGT	5.8	0.7
Sour	Montmorency	TCCT	5.7	0.8
Sour	Rheinische Schattenmorelle	CCAT	16.0	2.2
Sour	23 23 (13)	AGCT	5.1	0.8
Sour	R1 (1)	GTGT	26.1	3.9
Sour	Nana	AACT	22.9	3.6
Sour	Englaise Timpurii	TGGT	13.7	2.4
Sour	Pitic de Iasi	CCCT	17.6	2.8
Total			308.6	43.5

The adaptors (used to retrieve accession-specific sequences from pools), numbers of Illumina Genome Analyzer II reads, and numbers of reads aligned to the peach 'Lovell' genome are indicated.

for the array. Non-intragenic SNPs were included in the final array for sour cherry, rather than intragenic, due to a mistake in filtering. SNPs not assigned to the first eight pseudomolecules representing peach's eight chromosomes were excluded. Non-uniquely anchored SNPs (those SNPs for which at least 90% of the 60 bp flanking sequences were anchored to more than one peach genome location) were also removed, and there could be no other detected SNP within 20 bp of the targeted SNP. For sour cherry, SNPs predicted to be between the *avium* and *fruticosa* subgenomes (i.e., AABB) or polymorphic in both subgenomes (i.e., ABAB) were removed (strategy described below). This filtering process yielded "Stage 2 SNPs" (Figure 1).

Subgenome specificity of sour cherry SNPs was deduced by the following strategy. MAF within the sour cherry detection panel was used as a proxy for allele dosage. SNPs close to a 1:1 ratio for two alternative SNP alleles were assumed to represent either SNPs between the *avium* and *fruticosa* subgenomes (i.e., AABB), and therefore not expected to segregate, or SNPs segregating in both subgenomes (i.e., ABAB), as neither were desirable for the final array. Sour cherry SNPs were examined for divergence from a 1:1 ratio by Chi-squared analysis, with those significantly different ($p<0.05$) assumed to represent subgenome-specific SNPs (i.e., ABAA or AAAB). Finally, to determine whether such SNPs were polymorphic within the *avium* or *fruticosa* subgenome, sequence information from sweet cherry detection panel accessions was consulted: presence of the rare sour cherry allele in sweet cherry was assumed to indicate that the SNP identified polymorphism within the *avium* subgenome; absence of the allele in sweet cherry indicated a *fruticosa* subgenome SNP.

Members of the international cherry genomics community requested inclusion of 487 SNPs in the final array. These "preferentially included SNPs" were: 304 sweet cherry RosCOS SNPs [6]; 150 Stage 1 SNPs spanning 3 Mb at a LG2 fruit size locus [18] and passing Stage 2 filtering criteria except for inclusion of: intergenic SNPs, any MAF, and sour cherry SNPs significantly different from 1:1 dosage at $p<0.10$; 21 of the GoldenGate®-validated SNPs with high MAF; and 12 pre-validated SNPs from other research programs.

Choosing SNPs for the 6K SNP array considered the final number designated to each crop (approximately 75% to sweet cherry and 25% to sour cherry), their estimated genetic location, the subgenome targeted for sour cherry, and the two sources of available SNPs (i.e., Stage 2 SNPs and preferentially included SNPs). Preferentially included SNPs were automatically included, leaving 5513 SNPs to be chosen. For the ~75% SNPs allocated to sweet cherry and ~25% to sour cherry, SNPs were evenly spaced genetically across each crop's genome. For sour cherry, approximately half of the chosen SNPs were targeted to evenly genetically spanning the *avium* subgenome and half to the *fruticosa* subgenome. Genetic location in the cherry genome of each Stage 2 SNP was estimated by physically anchoring 80 genetically mapped markers (RosCOS, SSR, and CAPS from [6]) to the peach whole genome sequence and then calibrating the genetic location of SNPs between each pair of unambiguously genetically mapped markers according to physical locations. Genetic locations of Stage 2 SNPs in the T×E *Prunus* reference genetic map [3] were also determined using the same approach, anchored by 153 SSR and RFLP markers.

SNP array evaluation

The RosBREED cherry 6K SNP array v1 was evaluated with panels of sweet cherry (n = 269) and sour cherry (n = 330) germplasm that included a diversity of cultivars, ancestors, founders, and progeny individuals forming a complex pedigree structure linking North American cultivated cherry germplasm for these crops (Tables S3 and S4). The array, employing exclusively Illumina Infinium® II design probes and dual color channel assays (Infinium HD Assay Ultra, Illumina), was used for genotyping following manufacturer recommendations. SNP genotypes were determined using GenomeStudio Genotyping Module Version v2010.3 [19]. All DNA samples were above the GenCall Score threshold of 0.15 and were therefore used in further analyses following the protocols and instructions provided [20]. For sweet cherry, after clustering with the GenomeStudio built-in clustering algorithm, Gentrain2 [21], all SNPs were visually examined for an expected maximum of three clusters (AA, AB, and BB) and then classified as failed, monomorphic, or polymorphic. The AB scores were converted into base pair calls by referencing the Top strand [22]. In contrast to diploid sweet cherry, five genotypes were possible in tetraploid sour cherry for each SNP: AAAA, AAAB, AABB, ABBB, and BBBB. Some sweet cherry accessions (n = 105) were included in the sour cherry GenomeStudio file to help discern the homozygous classes ("AAAA" and "BBBB") and the balanced heterozygous class ("AABB"). Manual editing was done to check and adjust clusters to the expected genotypic classes following the GenomeStudio polyploid protocol [23]. This recent version of GenomeStudio (Version 2010.3) allows more than three clusters (five in the case of sour cherry) to be manually defined. SNP informativeness in sour cherry was classified using the same criteria as sweet cherry (i.e., failed, monomorphic, or polymorphic) except that a fourth class, termed "unresolved polymorphic", was used for polymorphic markers that exhibited ambiguous clusters. For each polymorphic SNP (excluding unresolved), MAF among all cultivar and advanced selection panel accessions were determined with GenomeStudio for sweet cherry and manually for sour cherry.

Results

SNP detection

A total of 15.7 Gb of sweet cherry and 9 Gb of sour cherry DNA reads were obtained from 308.6 million 80-base reads generated for the 16 sweet cherry and eight sour cherry accessions (Table 1). Excluding 'Black Republican' that generated a relatively low number of reads (1.6 million), the total number of sweet cherry reads generated using the Illumina platform averaged 3.1× coverage per accession and ranged from ~8.8 million reads in 'Ambrunes' to 20.8 million reads in 'Windsor'. The total number of sour cherry reads obtained averaged 1.9× coverage per accession (assuming a genome size of 599 Mb) and ranged from 5.1 million reads in selection 23 23 (13) to 26.1 million reads in selection R1 (1). Approximately 14.1% of the cherry reads were aligned to the Peach v1.0 'dhLovell' whole genome sequence [14]. A total of 1,900,695 SNPs anchored to the peach genome were identified, including polymorphism between peach and cherry, approximately a third of which were within-cherry. Passing the filtering criteria for Stage 1 detection were 1,005,660 SNPs (52.9%), of which 368,948 were within-cherry SNPs – 97,019 within sweet cherry (i.e., polymorphic among sweet cherry detection panel accessions), 320,816 within sour cherry, 63,831 within both crops (included in previous two figures), and 7472 between but not within sweet and sour cherry only (i.e., homozygous in each crop for different alleles) (Figure 1).

SNP validation

SNP performance via the GoldenGate® assay with a subset of SNPs screened over 160 sweet and sour cherry accessions depended on SNP source, crop screened, and various parameter scores (Table 2). Sweet cherry SNPs screened with sweet cherry accessions performed similarly to sour cherry SNPs screened with sour cherry accessions, with approximately a third of attempted SNPs polymorphic in both cases (Table 2). Evenly spaced and trait locus-targeted SNPs performed similarly for sweet cherry SNPs/accessions. However, for sour cherry the trait locus SNPs were more often polymorphic (50% of SNPs compared to the average of 35%) and failed less often (13% compared to the average of 33%). Accession-specific SNPs performed relatively poorly for both cherry types, as did SNPs in UTRs and intergenic regions. Exonic regions provided the best performing SNPs (42% and 47% polymorphism for sweet and sour cherry, respectively) while intronic SNPs performed only ~80% as well as exonic SNPs. SNPs with ADT scores of ≥0.90 were the most often polymorphic compared to SNPs with lower ADT scores. Similarly, SNPs with intermediate MAFs (11–40% for sweet cherry and 21–30% for sour cherry) were more often polymorphic than SNPs with extreme MAFs. Sweet cherry SNPs were not as successful on sour cherry accessions (20% polymorphism) while none of the 48 sour cherry SNPs were polymorphic among sweet cherry accessions (Table 2).

SNP final choice

Stage 2 filtering resulted in 31,945 SNPs suitable for the final array. Of these, 28,562 were derived from polymorphism within sweet cherry while the other 3383 were expected to be within sour cherry subgenomes, 1730 indicative of polymorphism within the *avium* subgenome and 1653 within the *fruticosa* subgenome. Partial Stage 2 filtering of the 487 preferentially included SNPs resulted in 413 suitable SNPs.

SNP choices for the 6K array consisted of 4408 sweet cherry SNPs (227 of those being RosCOS SNPs) and 1552 sour cherry SNPs (788 *avium* and 764 *fruticosa*). After a small degree of loss due to technical issues in array manufacturing by Illumina Inc., 5696 SNPs were included on the final array, with 4214 (74%) targeting the sweet cherry genome (221 RosCOS) and 1482 (26%) targeting the sour cherry genome that consisted of 752 for the *avium* subgenome and 730 for *fruticosa* (Table 3). Sweet cherry SNPs targeting each chromosome averaged 527 and ranged from 392 (chromosome 8) to 902 (chromosome 1), which depended on the known genetic length of each chromosome (Table 3). Sour cherry chromosomes of the *avium* subgenome were targeted with an average of 94 SNPs ranging from 66 (chromosome 5) to 164 (chromosome 1) SNPs, while the *fruticosa* chromosomes were targeted with an average of 91 SNPs and ranging from 73 (chromosome 4) to 161 (chromosome 1) SNPs (Table 3). Included RosCOS SNPs were well distributed across the genome, accounting for 3.8 to 6.7% of the sweet cherry SNPs for any given chromosome (Table 3).

Table 2. Outcomes of a GoldenGate® validation assay for a set of 144 SNPs screened with 160 cherry accessions to test potential filtering parameters.

		Proportion of SNPs[a]					
		In sweet cherry accessions			In sour cherry accessions		
SNP type	Total[a]	Failed	Monomorphic	Polymorphic	Failed	Monomorphic	Polymorphic
Evenly spaced	80/40	**0.41**/0.28	**0.25**/0.73	**0.34**/0.00	0.64/**0.38**	0.18/**0.30**	0.19/**0.33**
Trait locus	16/8	**0.38**/0.25	**0.31**/0.75	**0.31**/0.00	0.50/**0.13**	0.25/**0.38**	0.25/**0.50**
Accession-specific	22/9	**0.45**/0.22	**0.41**/0.78	**0.14**/0.00	0.50/**0.22**	0.41/**0.67**	0.09/**0.11**
Genic location:							
exonic	38/17	**0.37**/0.29	**0.21**/0.71	**0.42**/0.00	0.47/**0.41**	0.16/**0.12**	0.37/**0.47**
intronic	18/13	**0.33**/0.23	**0.33**/0.77	**0.33**/0.00	0.61/**0.31**	0.22/**0.31**	0.17/**0.38**
UTR	17/9	**0.47**/0.33	**0.29**/0.67	**0.24**/0.00	0.65/**0.22**	0.29/**0.44**	0.06/**0.33**
intergenic	23/9	**0.48**/0.22	**0.26**/0.78	**0.26**/0.00	0.83/**0.33**	0.13/**0.56**	0.04/**0.11**
ADT score:							
<0.8	19/13	**0.47**/0.23	**0.26**/0.77	**0.26**/0.00	0.63/**0.23**	0.21/**0.38**	0.16/**0.38**
0.8–0.9	24/15	**0.42**/0.33	**0.33**/0.67	**0.25**/0.00	0.67/**0.40**	0.25/**0.33**	0.08/**0.27**
≥0.9	53/20	**0.38**/0.25	**0.23**/0.75	**0.40**/0.00	0.58/**0.35**	0.15/**0.25**	0.26/**0.40**
MAF (detection panel)							
1–10%	24/6	**0.42**/0.67	**0.38**/0.33	**0.21**/0.00	0.50/**0.50**	0.33/**0.50**	0.17/**0.00**
11–20%	19/23	**0.32**/0.17	**0.32**/0.83	**0.37**/0.00	0.58/**0.26**	0.21/**0.43**	0.21/**0.30**
21–30%	18/15	**0.39**/0.20	**0.17**/0.80	**0.44**/0.00	0.67/**0.27**	0.06/**0.13**	0.28/**0.60**
31–40%	18/4	**0.33**/0.50	**0.17**/0.50	**0.50**/0.00	0.72/**0.75**	0.06/**0.00**	0.22/**0.25**
41–50%	17/0	**0.59**/na	**0.24**/na	**0.18**/na	0.65/**na**	0.24/**na**	0.12/**na**
Total	96/48	**0.41**/0.27	**0.26**/0.73	**0.33**/0.00	0.61/**0.33**	0.19/**0.31**	0.20/**0.35**

Figures shown in **bold** are for those cases where the crop for which the SNPs were developed matches the crop on which the SNPs were screened.
[a]Of the 96/48 SNPs examined that targeted sweet cherry/sour cherry.

SNP array evaluation

Approximately a third of the SNPs on the RosBREED cherry 6K SNP array v1 were observed to be polymorphic in the sweet and sour cherry evaluation panels, and some of the SNPs developed for one crop were successful for the other. Informative SNPs were randomly distributed over the genome such that genetic variation was successfully sampled at medium density in any region of the genome (Figure 2).

A total of 1825 SNPs were polymorphic across the sweet cherry evaluation panel of breeding germplasm, representing 32% of SNPs present on the array (Table 3). Of the remainder, the vast majority (63%) were monomorphic in sweet cherry while 4.7% failed (Table 3). Sweet cherry RosCOS SNPs were highly polymorphic (92% of those tested), dramatically higher than for non-RosCOS sweet cherry SNPs (38%). Of the 4214 SNPs specifically chosen to target the sweet cherry genome, 1589 (38%) were polymorphic for that crop's evaluation panel, while the remaining 236 SNPs polymorphic for sweet cherry were from sour cherry – 221 *avium* (29% of those SNPs) and only 15 *fruticosa* (2.9% of SNPs targeted to that subgenome). A large proportion of sweet cherry SNPs (2432, 58%) were monomorphic in that crop while few failed (193, 5%).

The 1825 SNPs polymorphic in the sweet cherry evaluation panel provided an average physical spacing of approximately 120 kb between SNPs across the sweet cherry genome, considering the peach genome as the proxy for cherry (Table 4). Some gaps were closed by the 236 sour cherry SNPs polymorphic in sweet cherry. Chromosomes 5 and 8 had the smallest average gap length between polymorphic SNPs, each just below 100 kb (Table 4). The largest single gaps between polymorphic SNPs in the sweet cherry genome were on chromosomes 1 and 2 at just under 1.8 Mb each. These largest physical gaps represented estimated genetic gaps of only 0.73 cM (T×E reference map; [3]) or 1.64 cM (sweet cherry RosCOS map; [6]) for chromosome 1 and only 1.28 cM (T×E reference map) or 1.97 cM (cherry RosCOS map) for chromosome 2. The actual largest estimated genetic gap between polymorphic SNPs occurred elsewhere on chromosome 1, at 5.1 cM, the only estimated gap that was >5 cM (Table 4). Average genetic coverage achieved by polymorphic SNPs for the sweet cherry genome was one SNP per 0.29 cM (Table 4).

Screening of the array with the sour cherry evaluation panel revealed 2058 SNPs (36% of the array) to be polymorphic with their genotypes resolvable (Table 5) – more than for sweet cherry despite only a third as many SNPs being specifically targeted to sour cherry and the accidental use of intergenic SNPs that were expected to be only a quarter as polymorphic as intragenic (Table 2). The 677 polymorphic sour cherry SNPs in this total represented 46% of those targeted to this crop's genome, with slightly more than half being *avium* subgenome SNPs and the rest *fruticosa* (Table 5). Another 1743 SNPs (31%) were polymorphic but could not be resolved into clusters representing the range of expected SNP dosage, while 32% of the array's SNPs were monomorphic for sour cherry and 1.5% failed (Table 5). Of those SNPs targeted to the sour cherry *avium* subgenome, 48% were polymorphic, 41% were unresolved polymorphic, 9% monomorphic, and 1.3% failed for the sour cherry evaluation panel. For *fruticosa* SNPs, these numbers were 43% polymorphic, 39% unresolved polymorphic, 17% monomorphic, and 0.6% failed

Table 3. SNP informativeness in sweet cherry for the eight sets of chromosomes based on whether the SNP was derived from polymorphism in sweet cherry or in one of the two sour cherry subgenomes (i.e., *avium* or *fruticosa*).

		Number of SNPs									
Chromosome	SNP source[a]	SNPs chosen[b]	Failed	Monomorphic	Polymorphic	MAF <0.05	MAF 0.05–0.1	MAF 0.1–0.2	MAF 0.2–0.3	MAF 0.3–0.4	MAF 0.4–0.5
1	Sweet	902 (50)	29 (0)	561 (8)	312 (42)	47 (6)	43(3)	72 (4)	62 (9)	33 (7)	55 (13)
	Sour	164/161	6/5	112/155	46/1	10/0	8/1	11/0	8/0	1/0	8/0
2	Sweet	557 (21)	32 (0)	345 (4)	180 (17)	26 (1)	27(0)	30 (6)	53 (4)	20 (0)	23 (6)
	Sour	92/83	9/2	60/80	23/1	5/1	5/0	6/0	2/1	3/0	2/0
3	Sweet	434 (18)	28 (1)	230 (3)	176 (14)	40 (1)	16(4)	45 (2)	39 (6)	18 (0)	18 (1)
	Sour	87/74	5/4	57/68	25/2	7/2	4/0	1/0	5/0	5/0	3/0
4	Sweet	479 (26)	26 (0)	256 (0)	197 (26)	23 (6)	14(0)	37 (2)	42 (6)	44 (7)	37 (5)
	Sour	89/73	8/0	57/72	24/1	4/0	2/0	11/0	4/0	1/0	2/1
5	Sweet	489 (33)	14 (0)	304 (0)	171 (33)	25 (4)	23(5)	23 (4)	28 (3)	42 (7)	30 (10)
	Sour	66/84	6/1	47/83	13/0	4/0	1/0	3/0	1/0	2/0	2/0
6	Sweet	508 (32)	24 (0)	292 (2)	192 (30)	28 (4)	28(1)	34 (5)	33 (8)	38 (8)	31 (4)
	Sour	108/100	1/0	77/94	30/6	10/3	7/2	6/0	0/0	2/1	5/0
7	Sweet	453 (22)	16 (0)	261 (0)	176 (22)	60 (5)	18(1)	37 (2)	23 (5)	22 (6)	16 (3)
	Sour	71/75	2/0	49/72	20/3	7/2	1/0	4/0	4/1	3/0	1/0
8	Sweet	392 (19)	24 (0)	183 (0)	185 (19)	65 (5)	17(1)	23 (0)	25 (2)	25 (5)	30 (6)
	Sour	75/80	6/1	35/72	34/7	7/5	4/1	9/0	6/1	0/0	8/0
Total Sweet		4214 (221)	193 (1)	2432 (17)	1589 (203)	314 (32)	186 (15)	301 (25)	305 (43)	242 (40)	240 (48)
Total Sour		752/730	43/13	494/696	215/21	54/13	32/4	51/0	30/3	17/1	31/1
Grand Total		5696	249	3622	1825	381	222	352	338	260	272

The 269 sweet cherry accessions evaluated are listed in Table S2.

[a]Numbers of SNPs for the subgenomes of sour cherry are split (/) between *avium* and *fruticosa*.

[b]Numbers in parentheses are totals for RosCOS SNPs derived from sweet cherry and are included in the first number.

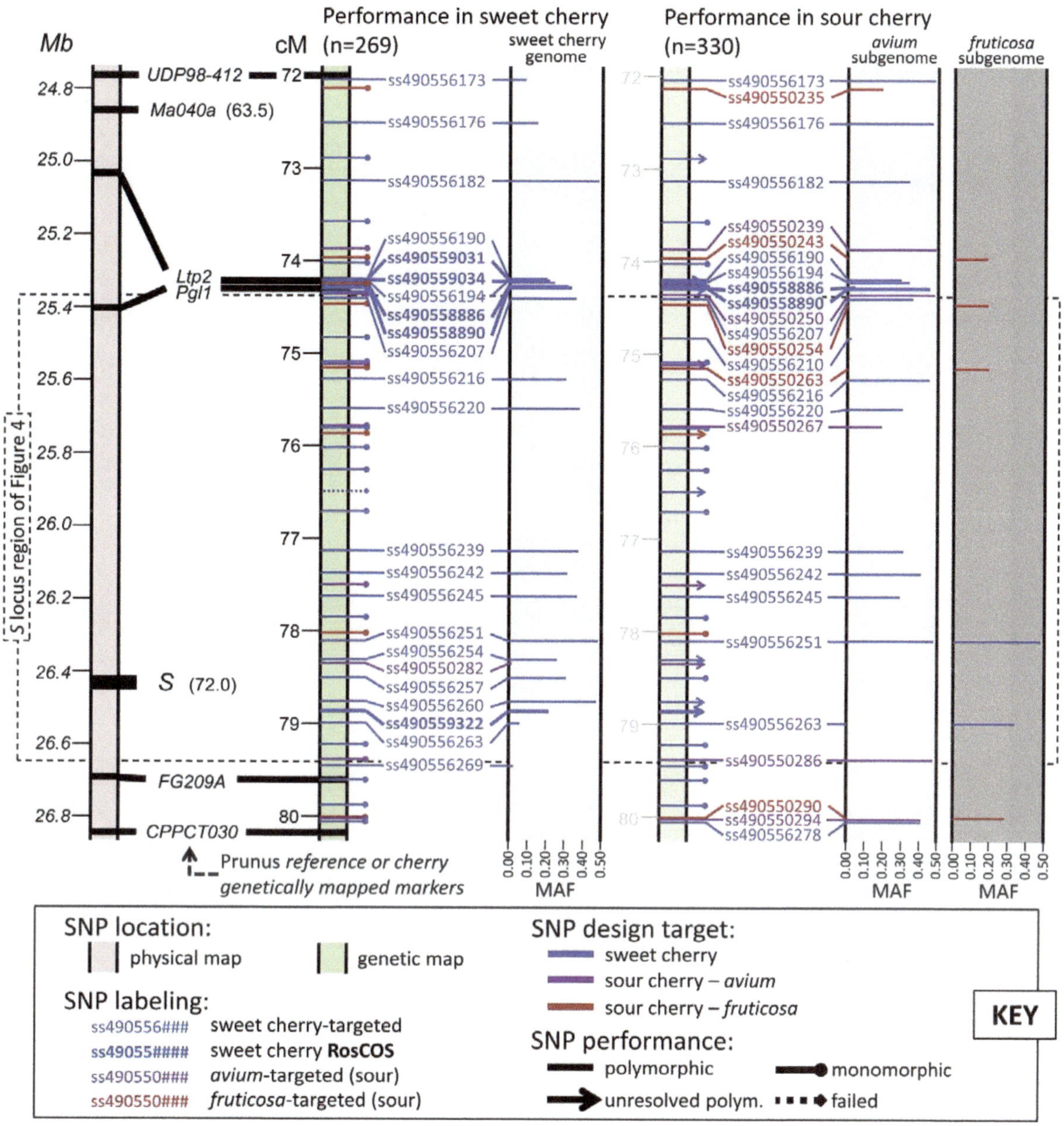

Figure 2. Performance of the RosBREED cherry 6K SNP array v1 for a sample region of the cherry genome. SNPs are shown in a 8.2 cM/ 2.1 Mb region around the *S* locus at the distal end of *Prunus* LG/chromosome/scaffold 6. Positions of SSR and RFLP markers positioned in the *Prunus* reference map [3] and physical map [14] are indicated. SNP positions are scaled genetically to the 8.2 cM window between reference-mapped markers UDP98-412 and CPPCT030. The *S* locus and Ma040a are also physically positioned, with their genetic locations in the sweet cherry genetic map [4] indicated in parentheses. SNPs designed for sweet cherry that were observed to be polymorphic in sour cherry are assumed here to be polymorphic within the *avium* subgenome of sour cherry except where haplotypes determined for Figure 3 identified polymorphism in both subgenomes (ss490556251) or in *fruticosa* (ss490556263).

(Table 5). Sweet cherry RosCOS SNPs were not as successful (57% polymorphism) among sour cherry accessions as they were for sweet cherry accessions but performed better than did other sweet cherry SNPs on sour cherry accessions (33% polymorphism).

The 2058 SNPs polymorphic in the sour cherry evaluation panel provided an average physical spacing of 124 kb between SNPs across the sour cherry *avium* subgenome and 666 kb across the sour cherry *fruticosa* subgenome (Table 6). As sweet cherry SNPs polymorphic in sour cherry were assumed in these calculations to be within the *avium* subgenome, this subgenome therefore had almost five times greater density of polymorphic SNPs than the *fruticosa* subgenome. Sour cherry's *avium* chromosome 7 had the largest single gap between polymorphic SNPs of 1.9 Mb (Table 6), corresponding to an estimated 1.50 cM (T×E reference map) or 5.92 cM (cherry RosCOS map) in this region. For the *fruticosa* subgenome, the largest physical gap, of 8.3 Mb, was on chromosome 2 (Table 6). Such a gap in this region corresponded to an estimated 9.55 cM (T×E reference map) or 6.62 cM (sweet cherry RosCOS map). However, the largest estimated genetic gaps for each subgenome were elsewhere in the

Table 4. Estimated physical and genetic distribution of SNPs of the RosBREED cherry 6K SNP array v1 across sweet cherry chromosomes evaluated for 269 sweet cherry accessions.

Chromosome	Physical			Genetic		
	Average gap (kb)	Largest gap (kb)	Number of gaps >150 kb	Average gap (cM)	Largest gap (cM)	Number of gaps >5 cM
1	38.1[a] (130.6[b])	881.3 (1773)	34 (89)	0.09[a] (0.23[b])	3.3 (5.1)	0 (1)
2	36.1 (129.3)	705.4 (1740)	28 (55)	0.10 (0.27)	1.1 (2.8)	0 (0)
3	36.7 (107.1)	622.7 (1248)	14 (36)	0.11 (0.25)	1.1 (2.3)	0 (0)
4	46.8 (133.8)	1107.7 (1174)	24 (62)	0.13 (0.27)	4.0 (4.2)	0 (0)
5	28.7 (99.7)	898.0 (1330)	11 (35)	0.11 (0.29)	2.8 (5.0)	0 (0)
6	40.2 (125.4)	396.4 (966.0)	18 (63)	0.17 (0.39)	2.9 (5.0)	0 (0)
7	37.2 (111.8)	387.2 (1065)	19 (46)	0.15 (0.35)	1.1 (3.6)	0 (0)
8	39.7 (96.5)	457.5 (738.0)	20 (49)	0.15 (0.27)	1.6 (1.9)	0 (0)
Genome-wide	37.9 (118.2)	1107.7 (1773)	168 (435)	0.12 (0.29)	4.0 (5.1)	0 (1)

[a]Based on the 4214 attempted sweet cherry SNPs.
[b]Based on all 1825 polymorphic SNPs.

genome, on chromosome 6 for *avium* (7.3 cM) and chromosome 7 for *fruticosa* (11.2 cM) (Table 7). Although the SNPs specifically targeting the sour cherry *avium* and *fruticosa* subgenomes resulted in average densities of one polymorphic SNP per 1.3 and 1.5 cM (T×E reference map), respectively, sweet cherry SNPs polymorphic in this other cherry crop improved genetic resolution for *avium* to one SNP per 0.29 cM, assuming all such SNPs were indeed within this subgenome (Table 7). Only two gaps between polymorphic SNP markers for the *avium* subgenome and 13 gaps for *fruticosa* were >5 cM (Table 7).

Half of the polymorphic SNPs in sweet cherry had a MAF >0.2, while a third were <0.10; in contrast, half of the polymorphic SNPs in sour cherry were >0.4 MAF and only 6% were <0.10 (Figure 3). Of the 1036 SNPs polymorphic in both crops, there was little correlation between MAFs (r = −0.13). This observation held considering just the sweet cherry SNPs polymorphic in both crops (r = −0.11, n = 919), the sour cherry *avium* SNPs (r = −0.22, n = 110), or the sour cherry *fruticosa* SNPs (r = not applicable, n = 7). Average heterozygosity in the 16 sweet cherry accessions of the detection panel for sweet cherry-targeted SNPs was 0.118 (s.d. 0.013) and was similar for all other non-seedling accessions of the evaluation panel (average 0.110, s.d. 0.012). For sweet cherry accessions such as the MIM series and 'Walpurgis' with no known pedigree connections to detection panel accessions, heterozygosity was similar (results not shown). For all SNPs on the array, heterozygosity of sweet cherry accessions averaged 0.103 and ranged from 0.075 to 0.124.

In a ~1.2 Mb region spanning the self-incompatibility *S* locus on *Prunus* LG6 for the sweet cherry cross 'New York 54' × 'Emperor Francis', offspring representing the four possible non-recombinant haplotypes were identified (Figure 4A). One individual, 3 (56), was identified to have resulted from a recombination of 'New York 54' haplotypes, with the recombination site localized to a ~400 kb, ~1.5 cM interval (T×E reference map). For this sweet cherry cross, all sour cherry-derived SNPs were homozygous in this region. SNP haplotypes constructed for the same *S* locus region could also be followed from parents to offspring of the sour cherry cross 'Ujfehertoi Furtos' × 'Surefire' (Figure 4B). The two sour cherry cultivars shared two of their four *S*-alleles (S_4 and S_{35}) for which their SNP haplotypes were identical across the examined region; the S_4 haplotype was also shared with the sweet cherry cultivar Emperor Francis (Figure 4A and 4B). A recombination between the S_4 and $S_{13'}$ haplotypes of 'Surefire' was identified in progeny individual 27-03-29, localized to a ~100 kb, ~0.5 cM interval. Nine of the 11 sweet cherry-derived SNPs were informative in this cross as the SNPs were heterozygous within at least one of the subgenomes. In this sour cherry cross, three of four sour cherry-targeted SNPs were heterozygous only in the *fruticosa* subgenome while polymorphism observed for the fourth, ss490550286, was between but not within the *avium* and *fruticosa* subgenomes (Figure 4B). One sweet cherry-targeted SNP, ss490556254, was polymorphic in the sour cherry cross; however, dosage could not be resolved. Another sweet cherry-targeted SNP, ss490556251, was polymorphic in both the *avium* and *fruticosa* subgenomes (Figure 4B, also indicated in Figure 2).

Public availability of SNP information

All 5696 SNPs on the RosBREED cherry 6K SNP array v1 (4214 sweet cherry and 1482 sour cherry; Tables S5, S6, S7) were deposited in NCBI's dbSNP repository [24] available at www.ncbi.nlm.nih.gov/projects/SNP and also at the Genome Database for Rosaceae (GDR; [25]) at www.rosaceae.org. GDR provides a downloadable Excel file on the peach genome project page containing the genomic locations, flanking sequence, and web links to a GBrowse viewer for these cherry SNPs on the peach genome.

Discussion

The RosBREED cherry 6K SNP array v1was determined to be useful for various genetics studies of cultivated cherry, despite the use of a relatively low number of detection panel accessions each sequenced at low depth. The number of SNP detection panel accessions, 16 for sweet cherry and eight for sour cherry, achieved a total depth of genome coverage (46.3× for sweet cherry and 15× for sour cherry) that was less than that of recent SNP detection and array development for peach and apple, two of cherry's rosaceous relatives. The recently developed International Peach SNP Consortium peach 9K SNP array v1 used 56 accessions that achieved 118× total genome coverage and for which 84.3% of included SNPs were informative when evaluated on diverse breeding germplasm [13]. The 27 accessions used for the International RosBREED SNP Consortium apple 8K SNP array

Table 5. SNP informativeness in sour cherry for the eight sets of chromosomes based on whether the SNP was derived from polymorphism in sweet cherry or in one of the two sour cherry subgenomes (i.e., *avium* or *fruticosa*).

			Number of SNPs									
Chromosome	SNP source[a]	SNPs chosen[b]	Failed	Monomorphic	Unresolved Polymorphic	Polymorphic	MAF <0.05	MAF 0.05–0.1	MAF 0.1–0.2	MAF 0.2–0.3	MAF 0.3–0.4	MAF 0.4–0.5
1	Sweet	902 (50)	8 (0)	364 (11)	211 (8)	319 (31)	14 (2)	7 (3)	43 (5)	28 (6)	88 (6)	139 (9)
	Sour	164/161	1/2	14/18	68/62	81/79	1/0	0/1	3/44	10/13	13/9	54/12
2	Sweet	557 (21)	10 (0)	199 (5)	166 (5)	182 (11)	11 (3)	3 (0)	23 (2)	32 (1)	23 (1)	90 (4)
	Sour	92/83	2/0	12/20	40/26	38/37	3/0	0/0	2/8	2/21	2/3	29/5
3	Sweet	434 (18)	4 (0)	165 (4)	107 (6)	158 (8)	5 (2)	7 (2)	10 (3)	28 (1)	20 (0)	88 (0)
	Sour	87/74	1/2	13/12	33/35	40/25	0/0	0/4	1/7	4/9	4/0	31/5
4	Sweet	479 (26)	9 (0)	176 (0)	141 (9)	153 (17)	7 (4)	5 (1)	11 (3)	38 (5)	20 (2)	72 (2)
	Sour	89/73	1/0	8/15	45/28	35/30	0/0	0/2	2/6	9/17	4/1	20/4
5	Sweet	489 (33)	4 (0)	208 (8)	128 (7)	149 (18)	11 (7)	4 (1)	8 (3)	34 (3)	14 (1)	78 (3)
	Sour	66/84	0/0	4/16	30/36	32/32	0/0	0/1	1/0	1/18	2/3	28/10
6	Sweet	508 (32)	13 (0)	188 (3)	145 (11)	162 (18)	10 (2)	7 (6)	12 (1)	12 (3)	36 (3)	85 (3)
	Sour	108/100	0/0	8/14	43/31	57/55	0/1	0/0	3/25	3/14	3/5	48/10
7	Sweet	453 (22)	6 (0)	184 (5)	113 (3)	150 (14)	12 (5)	2 (0)	10 (2)	21 (2)	20 (3)	85 (2)
	Sour	71/75	2/0	2/16	22/29	45/30	0/0	0/0	3/4	6/18	3/0	33/8
8	Sweet	392 (19)	14 (0)	130 (4)	140 (6)	108 (9)	7 (0)	4 (1)	12 (2)	13 (3)	16 (1)	56 (2)
	Sour	75/80	3/1	9/17	28/36	35/26	3/0	0/1	2/4	2/8	10/4	18/9
Total Sweet		4214 (221)	68 (0)	1614 (40)	1151 (55)	1381 (126)	77 (25)	39 (14)	129 (21)	206 (24)	237 (17)	693 (25)
Total Sour		752/730	10/5	70/128	309/283	363/314	7/1	0/9	17/98	37/118	41/25	261/63
Grand Total		5696	83	1812	1743	2058	85	48	244	361	303	1017

The 330 sour cherry accessions evaluated are listed in Table S3.
[a]Numbers of SNPs for the subgenomes of sour cherry are split (/) between *avium* and *fruticosa*.
[b]Numbers in parentheses are totals for RosCOS SNPs derived from sweet cherry and are included in the first number.

Table 6. Estimated physical distribution of SNPs of the RosBREED cherry 6K SNP array v1 across sour cherry chromosomes evaluated for 330 sour cherry accessions.

Chromosome	*avium* subgenome			*fruticosa* subgenome		
	Average gap (kb)	Largest gap (kb)	Number of gaps >150 kb	Average gap (kb)	Largest gap (kb)	Number of gaps >150 kb
1	282.0[a] (116.8)[b]	2259 (1384)	111 (92)	288.6 (588.2)	2812 (4166)	107 (66)
2	286.3 (120.1)	1472 (1548)	53 (48)	310.7 (696.9)	2442 (8302)	49 (28)
3	249.2 (109.8)	1073 (1207)	54 (37)	291.5 (862.9)	1545 (3428)	48 (21)
4	333.6 (158.5)	1649 (1281)	73 (64)	401.2 (884.4)	3077 (8057)	56 (27)
5	276.1 (101.3)	2438 (973.5)	26 (30)	215.9 (565.3)	1755 (2701)	43 (22)
6	262.4 (131.3)	1611 (1267)	72 (61)	282.0 (509.4)	2215 (2990)	67 (46)
7	308.9 (114.1)	1106 (1900)	51 (43)	290.3 (714.5)	2113 (4204)	47 (26)
8	289.3 (151.9)	1642 (925.6)	47 (43)	266.8 (810.2)	2024 (4350)	51 (25)
Genome-wide	284.8 (123.7)	2438 (1900)	487 (418)	291.2 (665.5)	3077 (8302)	468 (261)

[a]Based on the 1482 attempted sour cherry SNPs (752 for the *avium* subgenome and 730 for *fruticosa*).
[b]Based on all 2058 polymorphic SNPs; for the *avium* subgenome, this calculation included all sweet cherry SNPs polymorphic for sour cherry.

v1 achieved 89× total genome coverage and 70.6% of SNPs were informative [7]. However, sequencing depth per accession was similar among all these efforts, averaging 3.1× genome coverage per accession for sweet cherry and 2.0× for sour cherry (Table 1), 2.2× for peach [13], and 3.3× for apple [7]. Sour cherry coverage was comparable because the sequencing was performed at twice the depth of sweet cherry. As the proportion of informative cherry SNPs was only half that of peach and apple (only 38% and 46% of those targeted to sweet cherry and sour cherry breeding germplasm, respectively), the relative limitation of the cherry SNP detection panel was either the number or relevance of detection panel accessions. The number of accessions was a compromise based on the available budget and intended allocation of array attention of 75% to sweet cherry and 25% to sour cherry, utilizing state-of-the-art NGS technology (Illumina GA II). Because of the intended use of the array in characterizing genetic variation in cultivated cherry, the few accessions of the detection panel were carefully chosen to efficiently represent cherry breeding germplasm. For both sweet and sour cherry, choices were based on prior knowledge of geographic origin, pedigree, SSR and RosCOS SNP diversity (sweet cherry, [6]) or isozyme diversity (sour cherry, [26]), and the potential to confer disease resistance [27,28]. The detection panel choice appears to have been suitable because the array performed as well on the original detection panel accessions as on other breeding germplasm, including on material unconnected by pedigree. With the large number of SNPs on the array, the observed levels of polymorphism and heterozygosity translate to an unprecedented resolution for assaying genetic variation in cherry. In sweet cherry, for example, the array is expected to reveal 1500–2000 polymorphic genome-wide SNPs for any given set of cultivars, and 400–700 SNPs heterozygous within any given cultivar, from our observations of polymorphism and heterozygosity in the sweet cherry evaluation panel. This estimate is supported by [29] who reported 515 to 634 SNPs heterozygous for each of four cultivars when this 6K array was used as the basis for the two highest density genetic maps to date of a *Prunus* species.

Table 7. Estimated genetic distribution of SNPs of the RosBREED cherry 6K SNP array v1 across sour cherry chromosomes evaluated for 330 sour cherry accessions.

Chromosome	*avium* subgenome			*fruticosa* subgenome		
	Average gap (cM)	Largest gap (cM)	Number of gaps >5 cM	Average gap (cM)	Largest gap (cM)	Number of gaps >5 cM
1	0.48[a] (0.20)[b]	6.1 (5.8)	1 (1)	0.49 (0.99)	4.7 (5.6)	0 (2)
2	0.52 (0.24)	2.7 (1.5)	0 (0)	0.56 (1.26)	2.8 (9.6)	0 (1)
3	0.54 (0.25)	3.4 (2.0)	0 (0)	0.62 (1.84)	3.3 (7.4)	0 (2)
4	0.60 (0.30)	4.5 (4.0)	0 (0)	0.73 (1.74)	4.2 (4.2)	0 (0)
5	0.74 (0.28)	5.8 (5.0)	1 (1)	0.59 (1.52)	5.8 (6.6)	1 (2)
6	0.76 (0.40)	3.6 (7.3)	0 (0)	0.81 (1.48)	5.2 (8.3)	1 (3)
7	0.92 (0.35)	3.2 (4.1)	0 (0)	0.87 (2.13)	4.8 (11.2)	0 (2)
8	0.82 (0.43)	4.3 (2.7)	0 (0)	0.72 (2.10)	2.5 (5.8)	0 (1)
Genome-wide	0.64 (0.29)	6.1 (7.3)	2 (2)	0.65 (1.50)	5.8 (11.2)	2 (13)

[a]Based on the 1482 attempted sour cherry SNPs (752 for the *avium* subgenome and 730 for *fruticosa*).
[b]Based on all 2058 polymorphic SNPs; for the *avium* subgenome, this calculation included all sweet cherry SNPs polymorphic for sour cherry.

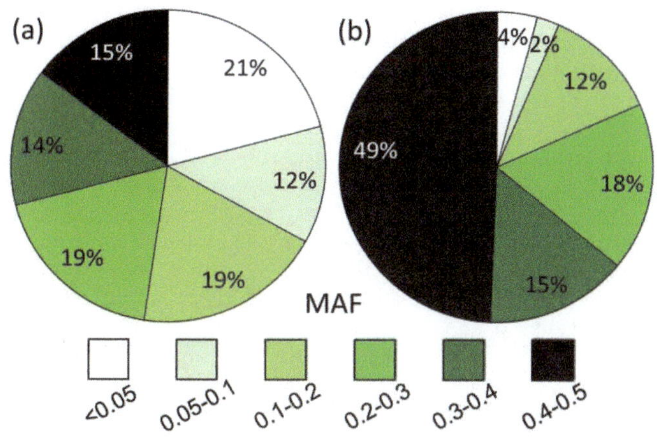

Figure 3. Minor allele frequency (MAF) of polymorphic cherry SNPs. MAF was determined in sweet cherry (n = 50) and sour cherry (n = 37) evaluation panels for 1825 and 2058 polymorphic SNPs, respectively.

Therefore, the set of SNPs on the array developed from a small but diverse set of breeding-relevant accessions is expected to be generally informative for cherry germplasm in cultivation and in breeding programs.

The proportion of polymorphic markers verified in the sour cherry germplasm was nearly double that found in the sweet cherry set. However, dosage could be resolved for only approximately half these SNP markers. Therefore, despite having twice the number of chromosomes, and using intergenic rather than intragenic SNPs, the final number of resolvable polymorphic sour cherry SNPs was just slightly more than that for sweet cherry, 2058 versus 1825. Given that only 314 of these markers were chosen to putatively target the *fruticosa* subgenome, it is suspected that the sour cherry *avium* subgenome will have significantly denser marker coverage than the *fruticosa* subgenome. However, it is probable that some of the sweet cherry SNPs that were polymorphic in sour cherry were within the *fruticosa* subgenome, given that 215 of the 236 sour cherry SNPs polymorphic in sweet cherry were *avium*-targeted.

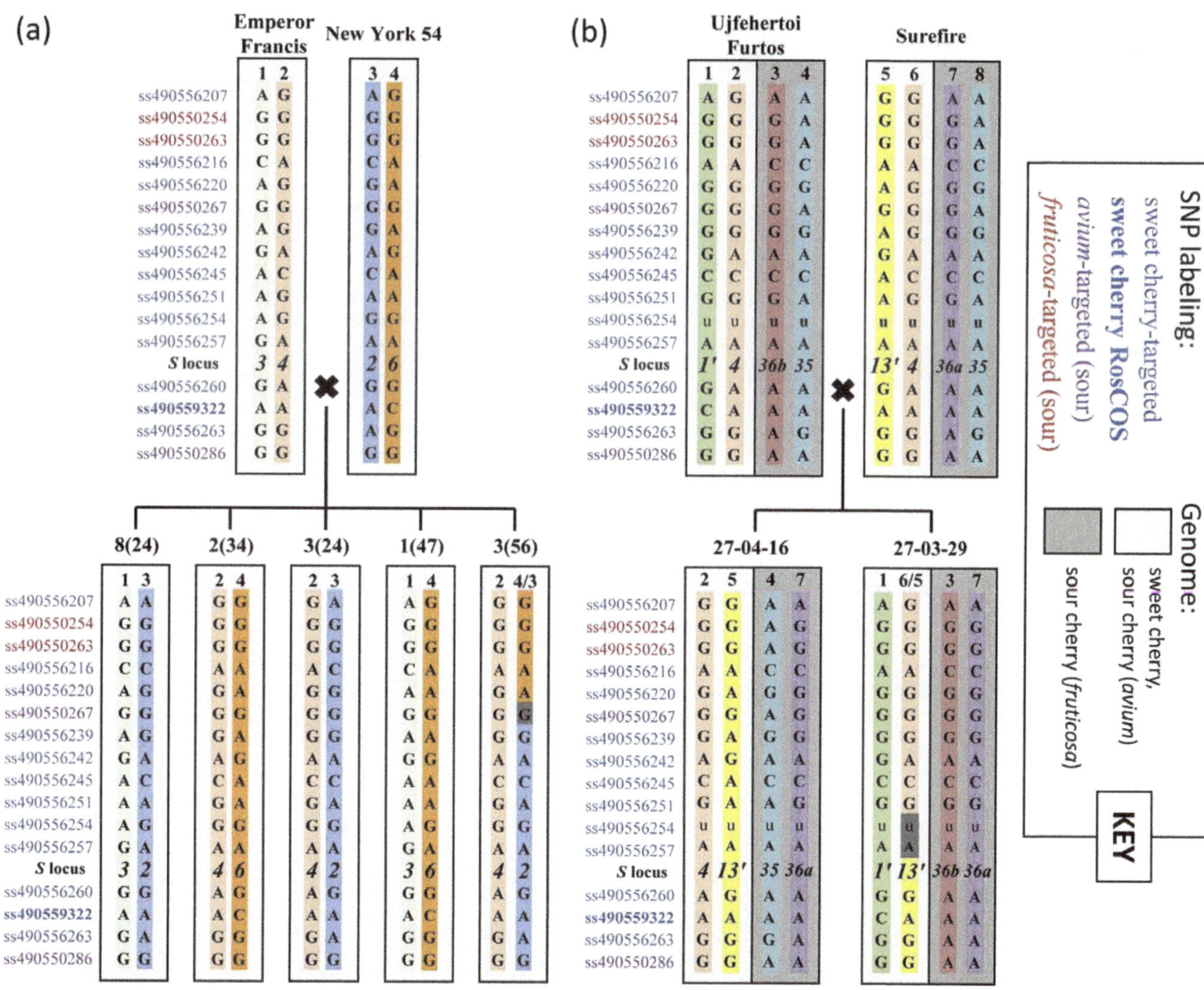

Figure 4. Reconstruction of a ~1.2 Mb region spanning the self-incompatibility *S* locus and its inheritance in cherry. (a) Sweet cherry, with four parental haplotypes (1–4). (b) Sour cherry, with eight parental haplotypes (1–8). Identical haplotypes have the same background colors. Haplotypes are shown for five sweet cherry and two sour cherry seedlings. Monomorphic SNPs within cross-over regions are highlighted in grey. Genotypes indicated as "u" are for an unresolved polymorphic SNP in sour cherry.

Here, a comparative genomics approach relied heavily on the peach reference genome sequence for SNP detection (aligning sequence reads) and for final choice of SNPs for the array (evenly spanning the "cherry" genome rather than random positions). Cherry and peach belong to the same genus, with the same chromosome number and apparent genetic co-linearity [3,4,6]. However, the two crops are in divergent subgenera [30] and not cross-compatible; therefore, there may be significant genomic differences at the micro-syntenic and DNA sequence levels. The latter may account for the low proportion (14.1%) of raw cherry reads that could be aligned to the peach genome given the strict alignment criteria used here, such that only sequences from the most conserved regions between cherry and peach genomes were interrogated for SNPs. If the 86% of unaligned cherry reads were due to substantial sequence-level divergence in localized regions (spanning a few cM or more), the array will be non-informative for such regions of the cherry genome. Similarly, large-scale deletions in the peach genome compared to cherry would result in non-sampled regions. If such deletions occurred at chromosome ends, even the successful genetic mapping of these cherry SNPs would not detect the missing regions. Translocations between the two subgenera occurring between previously genetically mapped markers and on the scale of up to millions of base pairs or several cM are not expected to affect the array's genome coverage. As genetic locations used to evenly space SNPs across the "cherry" genome were based on RosCOS SNPs previously genetically mapped in the sweet cherry genome, SNP spacing should not be affected by any differences in recombination rates between peach and cherry.

The high degree of observed monomorphism appears to be due to total sequencing depth of the SNP detection panel. Of the sweet cherry SNPs that did align to the peach genome and passed the various filters to be included on the array, the large proportion (58%) of monomorphism and low failure rate (5%) in sweet cherry accessions indicates that sequences were accurately aligned to unique locations in the peach genome but that detected SNPs were false positives. While a low failure rate (1%) for sour cherry SNPs in sour cherry accessions also indicates successful sequence alignment to the peach genome, the relatively low rate of monomorphism (13%) suggests that most of the originally detected sour cherry SNPs were true SNPs for that crop. However, if unresolved polymorphic sour cherry SNPs were actually monomorphic because their variation in genotype clustering (apparent polymorphism) was due to sequence variation in flanking sequences, then such an adjusted monomorphism rate for sour cherry (53%) would be similar to that of sweet cherry. For peach and apple, monomorphism and failure rates were lower, totaling 16% for peach [13] and 28% for apple [7], probably due the greater total depth of detection panel sequencing for those crops as discussed above. In fact, there is a strong linear relationship between polymorphism rate (P) and total sequencing depth (D) among the three diploid crops of apple, peach, and sweet cherry ($P = 0.75D$, $R^2 = 0.96$), suggesting that, using similar filtering parameters among these crops, false positives due to sequencing errors and failure due to undetected polymorphism in SNP-flanking sequences could be effectively avoided with a total sequencing depth of at least 133×. This prediction is an expected consequence of the pseudo-random genomic sampling that underlies shotgun sequencing approaches; detection of individual SNPs requires deep sequencing at individual bases and detection of SNPs genome-wide requires broad sequencing over the genome space. Therefore sufficiently deep sequencing is required to identify not only individual SNPs that are potentially of value for downstream assays but to also discover nearby SNP-flanking polymorphisms that can cause SNP assay failures.

The above considerations suggest some alternative array development strategies. The number of polymorphic SNPs in the cherry array could have been doubled by focusing available resources for SNP detection on one crop or the other (sweet or sour cherry) to double the sequencing depth for that crop. Due to a shared genomic background, such a single cherry crop array would have been useful for the other crop to a certain extent. Following simple extrapolations from our observed cross-species results (Tables 3 and 5), a 6K sweet cherry array based on a detection panel of only sweet cherry accessions (achieving an estimated 100× total genome coverage and 75% polymorphism rate) would be expected to provide polymorphic SNPs of ~4300 for sweet cherry and ~1900 for sour cherry highly skewed to the *avium* subgenome. Focusing on sour cherry SNP detection instead, a 6K sour cherry array would be expected to provide polymorphic SNPs of ~4300 for sour cherry (equally distributed over both subgenomes) but only ~900 for sweet cherry. Considering that SNP polymorphism was desired for both crops, both hypothetical scenarios are inferior to the actual dual-crop SNP detection strategy employed that achieved polymorphism of 1825 SNPs in sweet cherry and 2058 SNPs in sour cherry. However, the 2058 SNPs polymorphic in sour cherry are likely to be skewed somewhat toward the *avium* subgenome because most were developed for sweet cherry (*P. avium*). A better strategy for the dual-crop array would have been to strongly bias the choice of sour cherry SNPs to the *fruticosa* subgenome; in hindsight, targeting all sour cherry SNPs to the *fruticosa* subgenome would have better balanced subgenome coverage.

The RosBREED cherry 6K SNP array v1 appears to be informative beyond just cherry breeding and cultivated germplasm. The *avium* subgenome of sour cherry is believed to represent a subset of *P. avium* species diversity that pre-dates modern cultivated sweet cherry [2]. Revealingly, 29% of *avium*-subgenome sour cherry SNPs (215) were polymorphic for sweet cherry and 33% of sweet cherry SNPs (1380) were polymorphic for sour cherry. Therefore, ~30% of SNPs on the array are expected to be polymorphic for any diverse set of sweet cherry germplasm, including wild populations. A similar transferability rate may extend to wild *P. cerasus* populations for the 1482 sour cherry SNPs. Transferability of SNP polymorphism between the distinct *P. avium* and *P. fruticosa* species (effectively targeted with 4966 and 730 SNPs on the array, respectively) is estimated to be only a tenth of the transferability within *P. avium* given that 10% of the sour cherry SNPs polymorphic for sweet cherry accessions were originally developed for *fruticosa*. The ~200 RosCOS SNPs performed exceptionally well because of their previous validation in cherry germplasm [6]. These RosCOS markers provide a valuable tool for comparative genomics in Rosaceae such as comparing functional genetic variation across genera. More than 100 RosCOS SNPs are expected to be polymorphic in apple from the 128 such markers included on the apple 8K array [7]. While only 14 RosCOS SNPs were polymorphic for peach [13], hundreds of other markers polymorphic in both peach and cherry provide sufficient anchors for comparative genomics within the *Prunus* genus [3,4].

Close examination of haplotype segregation for a 16-SNP region at the *S* locus (Figure 4) indicated that use of the array should enable monitoring of all recombination events in cherry germplasm for any two subsequent generations examined at a time. Unique haplotypes and recombinations between them were successfully detected and localized for both sweet and sour cherry around the *S* locus. The unprecedented resolution of genetic variation in cherry germplasm revealed by genome-scanning with the array opens the door to fine-scale QTL dissection and other linkage-based analyses, identification of incorrect pedigree records, and deduction of identity by descent for chromosomal segments across the genome. In the region examined, common haplotypes, indicating strong evidence for common ancestry, were identified both within a crop (S_4 and S_{35} in sour cherry) and between the two crops (S_4). The $S_{1'}$, S_4, and $S_{13'}$ alleles of sour cherry are known to be *avium*-derived as these *S* alleles are present in sweet cherry germplasm [31,32,33]. For example, the S_4 allele of the sweet cherry parent 'Emperor Francis' was observed to be identical to the S_4 haplotype present in both sour cherry parents. In contrast, the S_{35}, S_{36a}, and S_{36b} alleles are considered to be *fruticosa*-derived as they have not been observed in sweet cherry germplasm [34]. Reconstruction of the two sour cherry subgenomes that is ongoing using the SNP data and a linkage mapping approach is predicted to be complicated by the lack of balanced subgenomes within sour cherry germplasm as sour cherry is known to be a segmental allotetraploid [34,35]. Initial insight into reconstruction of sour cherry subgenomes was illustrated with the identification of seven sour cherry *S*-allele haplotypes. However, as the *S* locus region is known to be subject to recombinational suppression and as a result, a site for the accumulation of polymorphism, this level of subgenome-balanced polymorphism should not be extrapolated genome-wide.

Conclusion

The cherry SNP array described here will foster genetics studies in the Rosaceae and help bridge the gap between genomics and breeding in cherry because breeding germplasm was the basis of detected SNPs and SNP choices of the final array. The RosBREED cherry 6K SNP array v1 is commercially available from Illumina and we expect that it will be used worldwide for genetic studies in cherry and related species. The SNP markers included in the cherry 6K Illumina arrays are available for download in Excel format and viewable in GBrowse at the Genome Database for Rosaceae (GDR; http://www.rosaceae.org).

Supporting Information

Table S1 Sweet cherry validation panel of 79 sweet cherry accessions, used in the GoldenGate® assay for validation of detected SNPs.

Table S2 Sour cherry validation panel of 81 sour cherry accessions, used for validation of detected SNPs.

Table S3 Sweet cherry evaluation panel of 269 sweet cherry accessions, used in the Infinium® II assay for SNP array evaluation.

Table S4 Sour cherry evaluation panel of 330 sour cherry accessions, used for SNP array evaluation.

Table S5 All SNPs of the RosBREED cherry 6K SNP array v1. Information includes physical and genetic positions, targeted genome, and performance outcomes for sweet cherry (n = 269) and sour cherry (n = 330) evaluation panels.

Table S6 Markers of the *Prunus* reference T×E genetic map [3] used for estimating genetic positions of cherry 6K array SNPs.

Table S7 Markers of the sweet cherry genetic map [6] used for estimating genetic positions of cherry 6K array SNPs.

Acknowledgments

We thank all international researchers who participated in the Rosaceae SNP Summit, Stellenbosch, South Africa, 11–12 November 2010, for their contribution to international needs assessment and design of SNP filtering criteria. The sweet cherry breeder Nnadozie Oraguzie (Washington State University, WA, USA) helpfully provided leaf material for sweet cherry accessions.

Author Contributions

Conceived and designed the experiments: CP NB DM AI. Performed the experiments: NB AS BG CL. Analyzed the data: CP SF URR TS CL TCM DWB LW AI. Contributed reagents/materials/analysis tools: NB CL TCM DWB. Wrote the paper: CP NB URR TS CL TCM AI.

References

1. Olden EJ, Nybom N (1968) On the origin of *Prunus cerasus* L. Hereditas 59:327–345.
2. Tavaud M, Zanetto A, David JL, Laigret F, Dirlewanger E (2004) Genetic relationships between diploid and allotetraploid cherry species (*Prunus avium*, *Prunus* × *gondouinii* and *Prunus cerasus*). Heredity 93: 631–638.
3. Dirlewanger E, Graziano E, Joobeur T, Garriga-Caldere F, Cosson P, et al. (2004) Comparative mapping and marker-assisted selection in Rosaceae fruit crops. Proc Natl Acad Sci 101: 9891–9896.
4. Olmstead JW, Sebolt AM, Cabrera A, Sooriyapathirana SS, Hammar S, et al. (2008) Construction of an intra-specific sweet cherry (*Prunus avium* L.) genetic linkage map and synteny analysis with the *Prunus* reference map. Tree Genet Genomes 4: 897–910.
5. Clarke J, Sargent D, Bošković R, Belaj A, Tobutt K (2009) A cherry map from the inter-specific cross *Prunus avium* 'Napoleon' × *P. nipponica* based on microsatellite, gene-specific and isoenzyme markers. Tree Genet Genomes 5: 41–51.
6. Cabrera A, Rosyara UR, de Franceschi P, Sebolt A, Sooriyapathirana SS, et al. (2012) Rosaceae conserved orthologous sequences marker polymorphism in sweet cherry germplasm and construction of a SNP-based map. Tree Genet Genomes 8: 237–247.
7. Chagné D, Crowhurst RN, Troggio M, Davey MW, Gilmore B, et al. (2012) Genome-wide SNP detection, validation, and development of an 8K SNP array for apple. PLoS ONE 7(2): e31745.
8. Ganal MW, Durstewitz G, Polley A, Bérard A, Buckler ES, et al. (2011) A large maize (*Zea mays* L.) SNP genotyping array: Development and germplasm genotyping, and genetic mapping to compare with the B73 reference genome. PLoS ONE 6(12): e28334.
9. McNally KL, Childs KL, Bohnert R, Davidson RM, Zhao K, et al. (2009) Genome wide SNP variation reveals relationships among landraces and modern varieties of rice. Proc Natl Acad Sci U S A 106: 12273–12278.
10. Hamilton JP, Hansey CN, Whitty BR, Stoffel K, Massa AN, et al. (2011) Single Nucleotide Polymorphism discovery in elite North American potato germplasm. BMC Genomics 12: 302.
11. Felcher K, Coombs J, Massa AN, Hansey C, Hamilton JP, et al. (2012) Integration of two diploid potato linkage maps with the potato physical map. PLoS ONE, in press.
12. Akhunov E, Nicolet C, Dvorak J (2009) Single nucleotide polymorphism genotyping in polyploidy wheat with the Illumina GoldenGate assay. Theor Appl Genet 119:507–517.
13. Verde I, Bassil N, Scalabrin S, Gilmore B, Lawley CT, et al. (2012) Development and evaluation of a 9K SNP array for peach by internationally coordinated SNP detection and validation in breeding germplasm. PLoS ONE 7(4): e35668.
14. International Peach Genome Initiative, Peach Genome, Genome Database for Rosaceae website. Available: www.rosaceae.org/peach/genome. Accessed 2012 Jul 3.

15. Dirlewanger E, Claverie J, Wünsch A, Iezzoni AF (2007) Cherry. In: Kole, C, editor. Genome Mapping and Molecular Breeding in Plants. Fruits and Nuts. Springer-Verlag Berlin Heidelberg. pp. 103–118.
16. Iezzoni A, Weebadde C, Luby J, Yue CY, Peace CP, et al. (2010) RosBREED: Enabling marker-assisted breeding in Rosaceae. Acta Hortic 859: 389–394.
17. Li R, Li Y, Kristiansen K, Wang J (2008) SOAP: short oligonucleotide alignment program. Bioinformatics 24: 713–714.
18. Zhang G, Sebolt AM, Sooriyapathirana SS, Wang D, Bink MCAM, et al. (2010) Fruit size QTL analysis in an F_1 population derived from a cross between a domesticated sweet cherry cultivar and a wild forest sweet cherry. Tree Genet Genomes 6: 25–36.
19. Illumina Inc. (2010a) GenomeStudio Genotyping module v1.0, User Guide. Illumina Inc., Towne Centre Drive, San Diego, CA, USA.
20. Illumina Inc. (2010b) Infinium® Genotyping Data Analysis: A guide for analyzing Infinium® genotyping data using the Illumina® GenomeStudio Genotyping Module. Illumina Inc., Towne Centre Drive, San Diego, CA, USA.
21. Illumina Inc. (2010c) Improved cluster generation with Gentrain2, Tech Note. Illumina Inc., Towne Centre Drive, San Diego, CA, USA.
22. Illumina Inc. (2006) "TOP/BOT" strand and "A/B" allele, Tech Note. Illumina Inc., 5200 Research Way, San Diego, CA, USA.
23. Illumina Inc. (2011) Calling polyploid genotypes with GenomeStudio software v2010.3/v1.8. Tech Note. Illumina Inc., Towne Centre Drive, San Diego, CA, USA.
24. Sherry ST, Ward MH, Kholodov M, Baker J, Phan L, et al. (2001) dbSNP: the NCBI database of genetic variation. Nucl Acids Res 29: 308–311.
25. Jung S, Staton M, Lee T, Blenda A, Svancara R, et al. (2008) GDR (Genome Database for Rosaceae): integrated web-database for Rosaceae genomics and genetics data. Nucleic Acids Res 36: D1034–1040.
26. Beaver JA, Iezzoni AF, Ramm CW (1995) Isozyme diversity in sour, sweet, and ground cherry. Theor Appl Genet 90: 847–852.
27. Olmstead JW, Lang GA, Grove GG (2001) Inheritance of powdery mildew resistance in sweet cherry. HortScience 36: 337–340.
28. Wharton PS, Iezzoni AF, Jones AL (2003) Screening cherry germplasm for resistance to cherry leaf spot. Plant Disease 87: 471–477.
29. Klagges C, Campoy JA, Quero-García J, Guzmán A, Mansur L, et al. (2012) Construction and comparative analyses of highly dense linkage maps of two sweet cherry intra-specific progenies of commercial cultivars. PLoS ONE (submitted).
30. Wen J, Berggren ST, Lee CH, Ickert-Bond S, Yi TS, et al. (2008) Phylogenetic inferences in *Prunus* (Rosaceae) using chloroplast *ndh*F and nuclear ribosomal ITS sequences. J Syst Evol 46: 322–332.
31. Bošković R, Tobutt KR (2001) Genotyping cherry cultivars assigned to incompatibility groups by analysing stylar ribonucleases. Theor Appl Genet 103: 475–485.
32. Hauck NR, Ikeda K, Tao R, Iezzoni AF (2006) The mutated S_1-haplotype in sour cherry has an altered *S*-haplotype-specific F-Box protein gene. J Heredity 97: 514–520.
33. Tsukamoto T, Hauck NR, Tao R, Jiang N, Iezzoni AF (2006) Molecular characterization of three non-functional *S*-haplotypes in sour cherry (*Prunus cerasus*). Plant Mol Biol 62: 371–383.
34. Tsukamoto T, Hauck NR, Tao R, Jiang J, Iezzoni AF (2010) Molecular and genetic analysis of four non-functional *S* haplotype variants derived from a common ancestral *S* haplotype identified n sour cherry (*Prunus cerasus* L.). Genetics 184: 411–427.
35. Beaver J, Iezzoni AF (1993) Allozyme inheritance in tetraploid sour cherry (*Prunus cerasus* L.). J Amer Soc Hort Sci 118: 873–877.

Genetic Patterns of Domestication in Pigeonpea (*Cajanus cajan* (L.) Millsp.) and Wild *Cajanus* Relatives

Mulualem T. Kassa[1], R. Varma Penmetsa[1], Noelia Carrasquilla-Garcia[1], Birinchi K. Sarma[1,2], Subhojit Datta[1,3], Hari D. Upadhyaya[4], Rajeev K. Varshney[4], Eric J. B. von Wettberg[5,6], Douglas R. Cook[1]*

1 Department of Plant Pathology, University of California Davis, Davis, California, United States of America, **2** Department of Mycology and Plant Pathology, Banaras Hindu University, Varanasi, India, **3** Indian Institute of Pulses Research, Uttar Pradesh, India, **4** International Crops Research Institute for the Semi-Arid Tropics, Patancheru, Andhra Pradesh, India, **5** Department of Biological Sciences, Florida International University, Miami, Florida, United States of America, **6** Center for Tropical Plant Conservation, Fairchild Tropical Botanic Garden, Coral Gables, Florida, United States of America

Abstract

Pigeonpea (*Cajanus cajan*) is an annual or short-lived perennial food legume of acute regional importance, providing significant protein to the human diet in less developed regions of Asia and Africa. Due to its narrow genetic base, pigeonpea improvement is increasingly reliant on introgression of valuable traits from wild forms, a practice that would benefit from knowledge of its domestication history and relationships to wild species. Here we use 752 single nucleotide polymorphisms (SNPs) derived from 670 low copy orthologous genes to clarify the evolutionary history of pigeonpea (79 accessions) and its wild relatives (31 accessions). We identified three well-supported lineages that are geographically clustered and congruent with previous nuclear and plastid sequence-based phylogenies. Among all species analyzed *Cajanus cajanifolius* is the most probable progenitor of cultivated pigeonpea. Multiple lines of evidence suggest recent gene flow between cultivated and non-cultivated forms, as well as historical gene flow between diverged but sympatric species. Evidence supports that primary domestication occurred in India, with a second and more recent nested population bottleneck focused in tropical regions that is the likely consequence of pigeonpea breeding. We find abundant allelic variation and genetic diversity among the wild relatives, with the exception of wild species from Australia for which we report a third bottleneck unrelated to domestication within India. Domesticated *C. cajan* possess 75% less allelic diversity than the progenitor clade of wild Indian species, indicating a severe "domestication bottleneck" during pigeonpea domestication.

Editor: Shin-Han Shiu, Michigan State University, United States of America

Funding: This work was supported by the Indian Council of Agricultural Research under the Indo-US Agricultural Knowledge Initiative. The funders had no role in study design, data collection and analysis, decision to publish, or preparation of the manuscript.

Competing Interests: The authors have declared that no competing interests exist.

* E-mail: drcook@ucdavis.edu

Introduction

One common feature of domesticated organisms is reduced genetic diversity compared to their wild relatives. Two major forces that cause the reduction in genetic diversity are small population sizes ("founder effect") that occur during the initial formation of a domesticated lineage, and selective sweeps and/or directional selection for genes associated with domestication traits [1]. Intensive breeding, which is a recent phenomenon relative to domestication, typically causes further reductions to genetic diversity [2] and understanding such shifts at the molecular genetic level can inform crop improvement programs. Although the impact of such processes on genetic diversity are reasonably well described for major crops such as maize, wheat, soybean and rice [3–7], for many minor crops, which are often of significant regional importance, the circumstances of domestication are poorly described.

As much as domestication is a human-driven process, it can also be influenced by random gene flow from wild relatives. Many crops, particularly minor crops of regional importance, are still grown alongside their wild relatives, increasing the opportunity for gene flow between cultivated and non-cultivated populations. Although such gene flow reduces our ability to characterize domestication-related processes, its occurrence over protracted periods can allow for the contribution of novel traits from locally-adapted wild populations of related species into domesticated forms [8]. Geographical and/or environmental factors can also constrain genetic change during domestication. For example, in cases where recent (i.e., post-Columbian) expansion of minor crops has taken cultivated genotypes beyond areas of their historical domestication, analyses of genetic diversity may reveal bottlenecks and nested patterns of domestication that reflect new populations adapting to new environments or regional human preferences.

Pigeonpea (*Cajanus cajan* (L.) Millsp.) is a widely adapted, drought tolerant food legume crop cultivated throughout the semi arid tropics and subtropics. Though considered a minor crop, pigeonpea is of considerable importance in areas of South Asia (mainly on the Indian-subcontinent), Africa, the Caribbean and Latin America, where it is a prominent source of protein nitrogen in the human diet, as well as wood for fuel and light duty structural applications such as thatch for roofing. Grown on 4.63 million hectares, pigeonpea ranks 6th among grain legumes in production

[9]. The genus *Cajanus* is composed of 34 species [10], among which pigeonpea is the only cultivated member, with the remaining wild relatives assigned to the secondary or tertiary gene pools according to the gene pool concept of Harlan and de Wet [11]. Hybridization is widespread in the genus and many wild species can be crossed to cultivated *C. cajan*, a feature that has enabled the use of inter-specific crosses in breeding programs [12–13]. The majority of *Cajanus* species are endemic and confined either to Southern/South-Eastern Asia or Australia [14,15]. Given this substantial overlap in geographic distribution and the high degree of cross-compatibility among species, it seems probable that many *Cajanus* species are parts of species complexes that arose through current or recent natural gene flow.

Morphological evidence suggests that *C. cajanifolius*, which is native to the Indian subcontinent, is the progenitor of pigeonpea [16,17]. Nevertheless, historians and scientists have debated the center of origin for pigeonpea, with arguments in favor of either an African [18,19] or Indian [16,17,20] origin. Proponents of the African center of origin typically cite the presence of a single endemic wild species (*C. kerstingii*) in Africa [18,19]. The bulk of evidence, however, favors an Indian origin, with some authors postulating a Northern Indian origin no earlier than 3,500 years ago [16,17,20]. Evidence in favor of an Indian origin includes the presence, in India, of several wild, morphologically diverse species including the putative wild progenitor (*C. cajanifolius*), as well as archaeological remains, linguistic evidence, and a variety of uses in the daily cuisine within India [17]. Archeological records reveal pigeonpea seeds in Maharashtra, a State in India, from the 2nd century BC to the 3rd century BC [21]. The proposed route of dispersion of pigeonpea from India is to Malaysia and East Africa, on to West Africa and finally to the West Indies where it was named pigeonpea in 1962. Pigeonpea arrived in the New World through the slave trade from Africa [22].

Pigeonpea germplasm represents a diverse set of landraces and heterogeneous feral forms that are adapted to various agro-ecological settings [23]. Despite extensive phenotypic diversity, molecular evidence from Simple Sequence Repeats (SSRs) [24] and Diversity Array Technology (DArT) [25] suggests very low genetic diversity within cultivated pigeonpea when compared to its wild relatives. The only means to broaden the genetic base of domesticated *C. cajan* is to introgress genetic diversity from the wild gene pool [23], and thus understanding how diversity is assorted among pigeonpea and its wild relatives has practical implications.

With the objective of understanding genetic diversity among *Cajanus* species and inferring patterns of domestication, we examined allelic variation in domesticated pigeonpea and its wild relatives using a set of gene-based single nucleotide polymorphisms. The genetic signatures of domestication that we identify suggest a primary bottleneck within subtropical India, the likely center of domestication, and a nested bottleneck associated with pigeonpea that is cultivated in disperse tropical regions, which we speculate is the consequence of breeding for adaptation to a new environment. Moreover, we provide evidence of both modern and archaic gene flow between pigeonpea and wild relatives, including a third genetic bottleneck in Australian *Cajanus* species that is unrelated to the India-centric domestication of modern pigeonpea.

Results

Single nucleotide polymorphisms (SNPs) were assayed in a total of 110 accessions representing cultivated *C. cajan* (79 accessions) and its wild relative relatives (31 accessions) (Table S1), all of which belong to the genus *Cajanus*. The wild accessions represent 13 of 34 known species, while the cultivated group includes modern varieties, pre-breeding material, land races, as well as perennial pigeonpea accessions obtained from non-agricultural settings that are presumed to be feral forms. These genotypes originate from widespread geographical regions, spanning the known distribution of *Cajanus* species and represent both tropical and subtropical environments of Africa, Asia, Latin America, the Caribbean, the Indian sub-continent and Australia.

Individual SNPs were identified based on comparisons of sequences of *C. cajan* accession ICP 28 and *C. scarabaeoides* accession ICPW 94 in a set of low-copy orthologous genes. These two species span an evolutionary distance that is wider than the proposed domestication gradient from cultivated *C. cajan* to its presumed progenitor *C. cajanifolius*. Excluding failed sequencing reactions and monomorphic loci, single nucleotide polymorphisms were identified in a total of 670 unique genes from which 752 SNPs were used to design a GoldenGate genotyping assay (Table S2). Within this set of nucleotide variants, a minimum of 16.6% represent shared ancestral variation, while 36.8% and 68.2% of loci were polymorphic within *C. scarabaeoides* and/or the remaining *Cajanus* species group, respectively.

Phylogenetic Relationships between Wild and Domesticated Groups

The combined data set was used to calculate pair-wise distances between all genotypes and thereby generate a dissimilarity matrix from which a weighted Neighbor Joining tree was calculated. Previous analysis of nuclear *ITS* and chloroplast *trnL-F* spacer sequences suggested that *C. scarabaeoides* is the most basal member of the *Cajanus* clade (MT. Kassa, PhD dissertation) and thus *C. scarabaeoides* was used to root the Neighbor Joining tree. Similarly, a parsimony phylogenetic analysis was performed using all wild accessions and two representative samples of the domesticated (*C. cajan*) accessions. Both parsimony and Neighbor Joining trees resolved congruent topology with overall similarity on major clades (Figures 1 and 2). Three well-resolved clades were evident from this analysis, including a basal set of *C. scarabaeoides* of Indian origin, and two sister clades representing wild species of Australian origin and a more diverse but well-supported clade containing the remaining wild *Cajanus* species that were exclusively of Indian origin (Figures 1 and 2).

The tree topology in Figures 1 and 2 reflects both the distinctiveness of species and the geographical origin of species' groups. Thus, domesticated accessions formed a monophyletic group that was internal to, and significantly less diverse than, the group of non-*scarabaeoides* wild species of Indian origin. *C. cajanifolius* has been nominated as the progenitor of domesticated pigeonpea based largely on morphometric and alpha-taxonomic criteria [17,20]. Indeed, two accessions of *C. cajanifolius* (C7847 and ICP 15632) are sister to the large group of domesticated genotypes in maximum parsimony analysis with strong bootstrap support (91%), validating *C. cajanifolius* as the most recent progenitor for pigeonpea. These *C. cajanifolius* accessions, which we speculate are true wild representatives, are closely related to the wild non-*scarabaeoides* species of Indian origin with the expected affinities to coherent sets of *C. platycarpus* and *C. sericeus* genotypes, as well as to individual representatives of *C. albicans* and *C. crassus*.

Phylogenetic tools are not suited to analysis of individual genotypes with strongly reticulate histories, as is the case with genotypes from a single species or breeding pool. Nevertheless, Neighbor Joining analysis does identify similarity among sets of genotypes (see Figure 1) and many of these similarities are congruent with the known history of individual accessions and

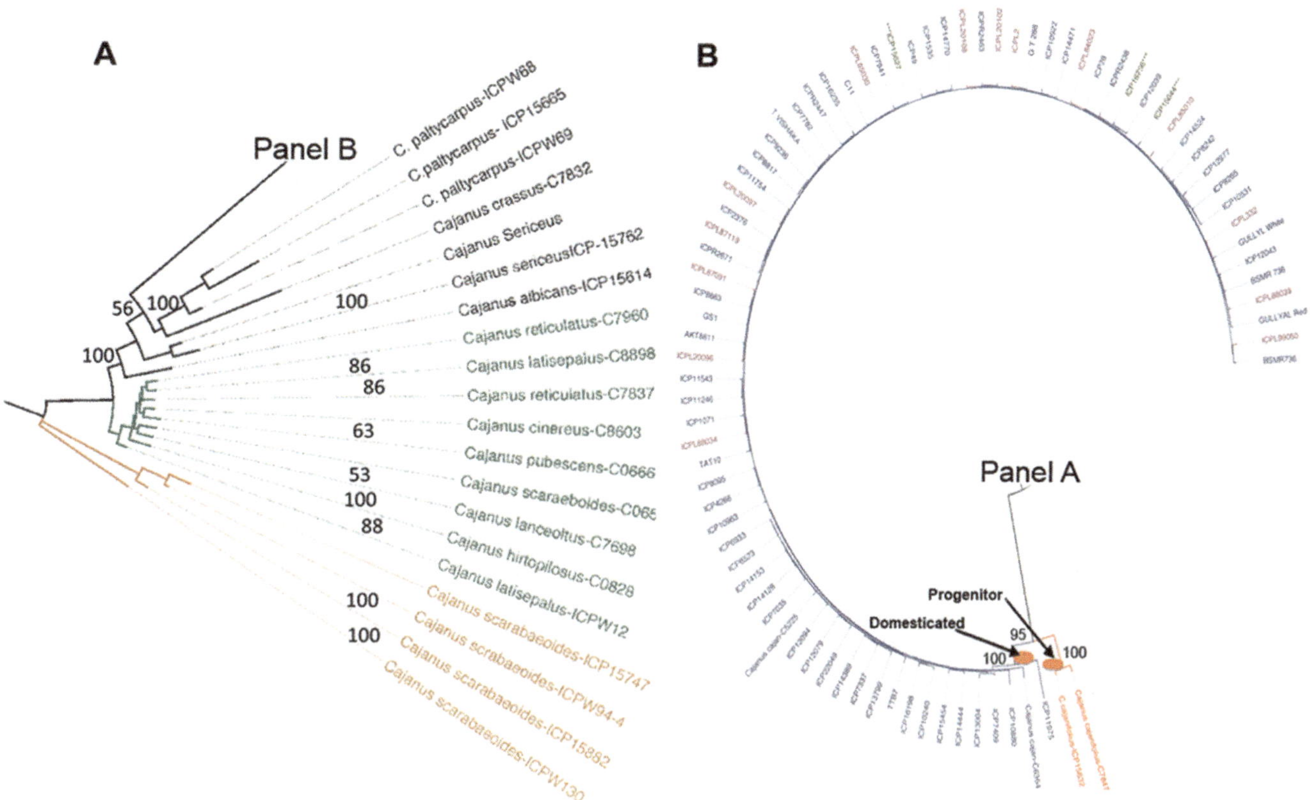

Figure 1. Neighbor Joining phylograms depicting wild and domesticated pigeonpea. Panel **A**, relationships among wild Cajanus species. Species groups are designated by color: orange, Wild-*Scarabaeoides*; green, Wild-Australia; black, Wild-India. Panel **B**, expansion of domesticated lineages from Panel A. The nature of accessions is reflected in their colors: red, wild progenitor; blue, cultivars and genebank accessions; pink, landraces. *** indicates mislabeled accessions annotated as wild species. The two Neighbor-Joining trees are linked to each other at Panel B and Panel A respectively. Bootstrap values of ≥50% are shown above their respective branches.

supported by subsequent population genetic analyses (see below). Thus, *C. cajan* C6364 and ICP 11975 were the most basal genotypes among the domesticated accessions. Interestingly, *C. cajan* C6364 is annotated as a naturally occurring, semi-domesticated and rarely found Australian woody herbaceous pigeonpea, while ICP 11975 is a genotype from the Philippines. The distant relationship of *C. cajan* C6364 to other *Cajanus* spp. of Australian origin, and its close affinity with domesticated *C. cajan*, is consistent with the origin of *C. cajan* C6364 as a feral genotype, and in fact both *C. cajan* C6364 and ICP 11975 show evidence of genetic admixture (see below). The data also suggest that at least three accessions are misclassified, as they are annotated as wild non-*cajan* species (i.e., ICP 15627, *C. albicans*; ICP 15756, *C. scarabaeoides*, and ICP 15644, *C. lineatus*) but were well integrated into the domesticated group.

To further assess relationships among accessions we conducted a Principal Coordinate Analysis (PCoA) using GenAlEx v.6.3 [26]. This multivariate approach was chosen to complement phylogenetic analysis because phylogenetic analyses are more sensitive to relationships between related individuals whereas PCoA is more informative regarding distances among major groups [27]. Principal Coordinate Analysis (PCoA) distinguished three groups of individuals (I, II and III) along discriminate axes 1 and 2, which accounted for 85.81% and 8.02% of the genetic variation, respectively (Figure 3). Along the first axis, wild accessions were resolved from domesticated accessions, while the second axis resolved the Indian *C. scarabaeoides* group (group I) from the remaining wild accessions of both Australian and Indian origin (group II). Within group II the Australian set forms a homogenous subgroup and the Indian genotypes form a more diverse assemblage, consistent with the previous phylogenetic analysis. Group III contained the domesticated *C. cajan* cluster. The low level of variation of the domesticated cluster is reflected in the tight clustering of most genotypes. Interestingly, ICP 11975 is an outlier from the main domestication group in Figure 3, supporting its basal affiliation to the domestication lineage predicted by Neighbor Joining analysis (Figure 1). A single accession of *C. cajan* from the Philippines (ICP 12765), as well as *C. cajanifolius* accessions ICPW 29 and ICP 15629, and *C. lineatus* ICPW 46, were also outliers in the PCoA analysis (Figure 3). Analysis of allele frequencies (see below) suggested a high proportion of genetic admixture for these genotypes and we suggest that these accessions originated as hybrids between wild and cultivated forms.

Genetic Structure of Wild and Domesticated Pigeonpea

To investigate genetic relationships among accessions and to search for evidence of genetic admixture between cultivated and wild genotypes, we utilized the Bayesian algorithm STRUCTURE [28,29]. STRUCTURE uses allele frequencies to derive subsets from a set of sampled individuals that approximate Hardy-Weinberg equilibrium, and thus represent subpopulations in the genetic sense. In the current study the taxonomic divisions are species level distinctions and, with the exception of *C. cajan*, sampling of multiple accessions within a species was limited. Thus,

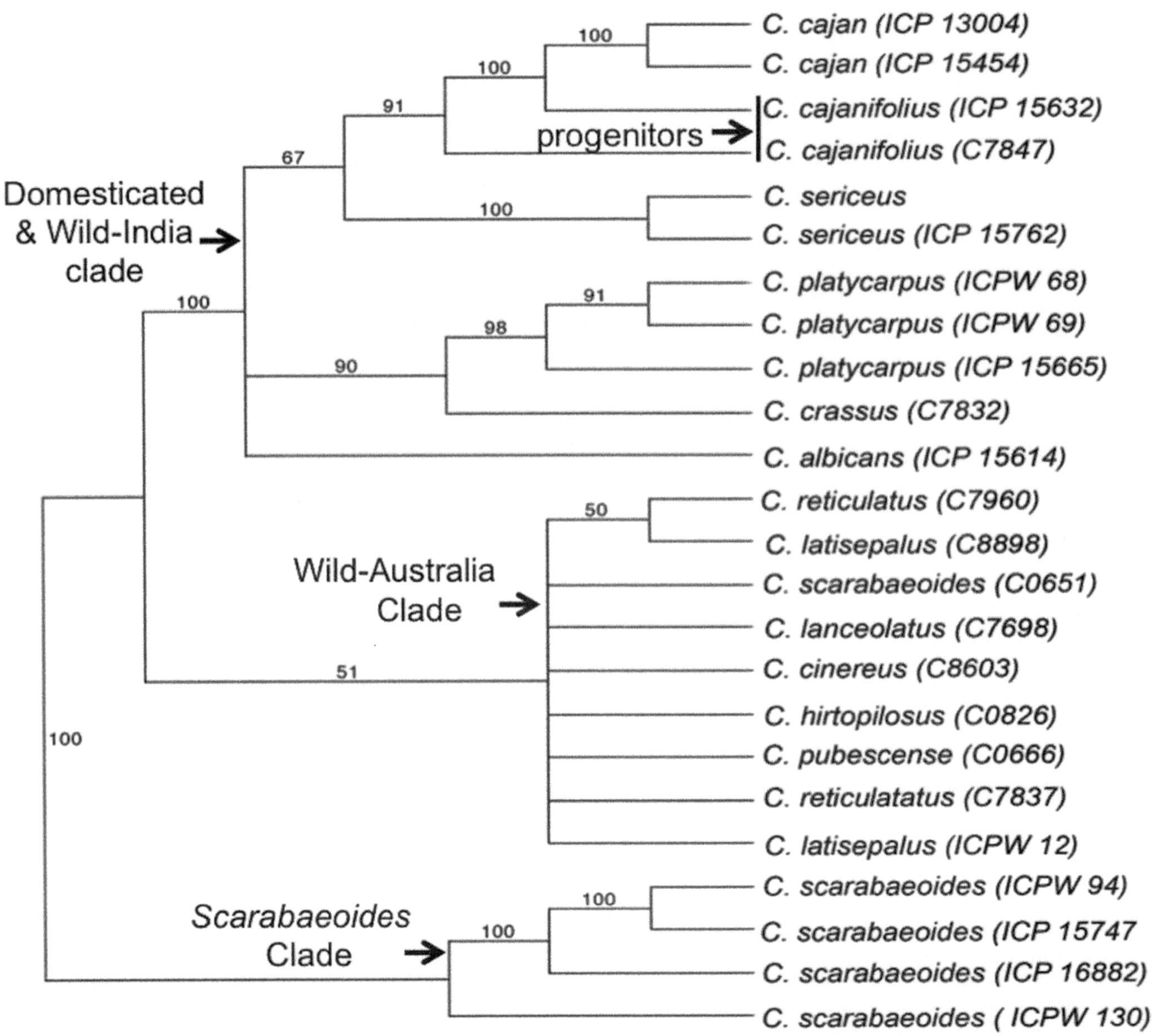

Figure 2. Phylogeny of *Cajanus* species depicted as a 50% majority rule consensus tree. Tree topology was inferred with maximum parsimony via heuristic searches among 1000 trees. Numbers above branches indicate bootstrap support (>50%). The vertical bar indicates the putative projentitor species (two *C. Cajanifolius* accessions). Tree length = 2145, consistency index (CI) = 0.638, and retention index (RI) = 0.837. The Bootstrap support for the sister relationship between wild-Australia and Wild-India clades (in the absence of the two *C. cajan* accessions) is 100% (data not shown).

the population genetic processes that STRUCTURE is sensitive to occur within the cultivated group of accessions, but not the more basal groups of diverged species. Nevertheless, in the combined analysis STRUCTURE served to delimit the primary subdivisions circumscribed by phylogenetic analysis and provided the basis to investigate the possibility of gene flow between groups.

When one has knowledge of the biology and history of a set of accessions, analyzing the partitioning of accessions into "K" subgroups can be informative (Figure 4 and Table S3). At optimal K = 3 domesticated accessions were resolved from a second group of wild accessions from India and Australia, and from a third group of *C. scarabaeoides* accessions, congruent with previous phylogenetic and Principal Coordinate Analysis (PCoA) (Figures 1, 2 and 3). At K = 2 STRUCTURE distinguished the wild and domesticated groups, mirroring axis one of the PCoA analysis (Figure 3), and only at K = 5 were the major phylogenetic lineages shown in Figures 1 and 2 well resolved. At K values of 2, 3 and 4 the wild species of Indian origin consistently shared partial membership (12–17%) with the domesticated group, although their primary membership was with wild Australia. This shared membership is not unexpected given the hypothesis that domesticated *C. cajan* is derived from the wild India group, and indeed among 209 loci reporting shared membership 36% of SNP were common to a majority (>7 of 9) of wild India accessions. By contrast, 12% of these 209 SNPs were shared between domesticated accessions and only one of the wild India species; for example, half of this set (6% total) was associated only with *C. crassus*, the most basal of wild-India species. This observation is

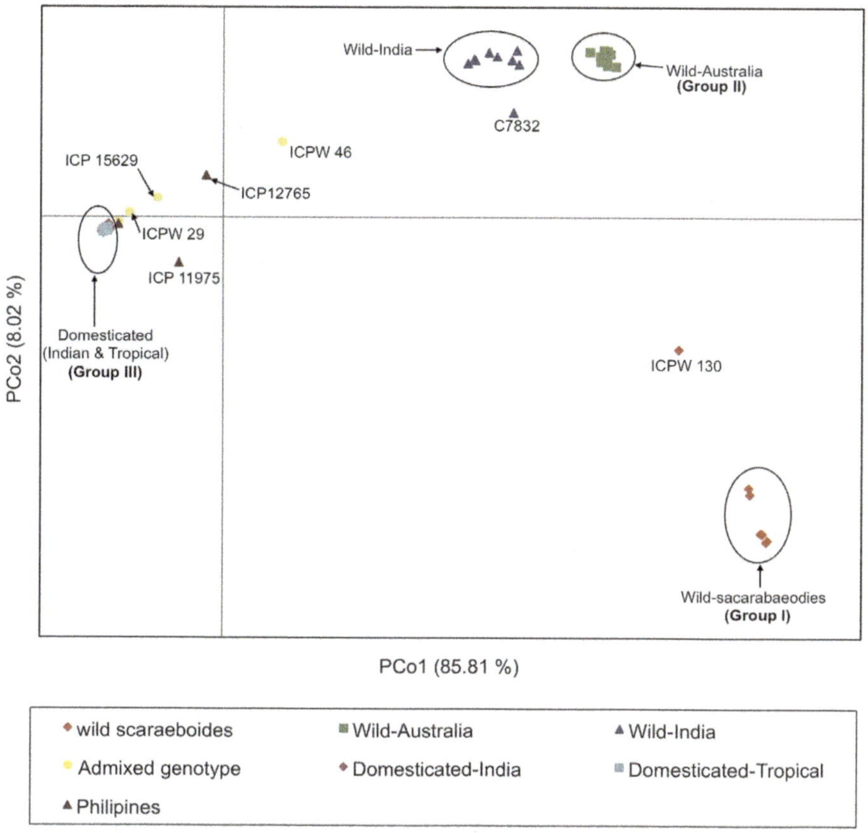

Figure 3. Principal coordinates analysis of domesticated pigeonpea and wild relatives. Red diamonds, wild *C. scarabaeoides*; green squares, wild-Australia *Cajanus* spp; dark blue triangles, wild-India *Cajanus* spp; light blue squares and pink diamonds represent domesticated *C. cajan* of tropical and Indian origin, respectively. Admixed genotypes between domesticated and wild species are labeled with yellow circles, or in the case of admixed Philippines accessions with dark red triangles. Accession numbers were added for accessions mentioned in the text.

consistent with admixture between domesticated accessions and *C. crassus* and raises the possibility of significant rates of gene flow between domesticated lineages and other species within the wild-India clade.

We analyzed the cultivated accessions by themselves to determine if genetic structure could be detected without the confounding effect of far greater differences among the wild taxa. As shown in Figures 5 and Table S4, the results indicate a predominant genetic subdivision that mirrors the eco-geographic history of the associated genotypes, with one group containing all accessions from tropical regions in Africa, the Caribbean, and Latin America, and the second group containing genotypes exclusively of Indian origin. Thirty percent of "domesticated" accessions showed evidence of genetic admixture between Indian and tropical accessions (Figure 5B). Analysis of admixed SNPs revealed examples of simple admixture between the tropical and Indian groups and complex patterns of admixture involving alleles from the wild-India lineage. When considered together, genotypes with complex patterns of admixture contained 165 segregating SNPs, of which 76% occurred only in one admixed genotype. These divergent SNP patterns are consistent with the origin of these genotypes through independent hybridization events rather than common ancestry. Genotypes with complex patterns of admixture include three genotypes from the Philippines, including those previously inferred as being admixed based on Neighbor Joining and PCoA analyses, and five additional *C. cajan* accessions of Indian origin (ICP 12977, TAT10, ICPL 85010, ICPL 99050 and ICP 13004). Fifty-eight SNPs (29 loci) among 15 cultivated accessions were implicated in the simple admixture. Figure 6 depicts a proposed network of relationships among these individuals that is suggestive of breeding history, though we stress that deeper genotyping is necessary to fully validate these inferences.

Only in the case of *C. cajan* ICP 11975 was there evidence of genetic contribution from the *C. scarabaeoides* gene pool to a domesticated genetic background. This is interesting given the status of ICP 11975 as a Philippines "*C. cajan*" accession. ICP 11975 was also unusual as an outlier in the PCoA analysis (Figure 3), reflecting its unique genetic constitution. *C. cajanifolius* ICPW 29 and ICP 15629 had >95% membership in the domesticated cluster, with less than 5% contribution from the wild background, and thus would be more appropriately referred to as admixed accessions of *C. cajan*, rather than as accessions of *C. cajanifolius*. *C. lineatus* (ICPW 46) was admixed in nearly equal proportions from both wild and domesticated backgrounds, and had been previously proposed as an admixed genotype based on morphology by van der Maesen (personal communication).

Genetic Variation among Wild and Domesticated Accessions

To find groups with reduced genetic variation, we used Analysis of Molecular Variance (AMOVA) to partition variance among hierarchical sets of individual genotypes (three groups and six populations) that were circumscribed by a combination of phylogenetic analysis, geographical origin, breeding history, and

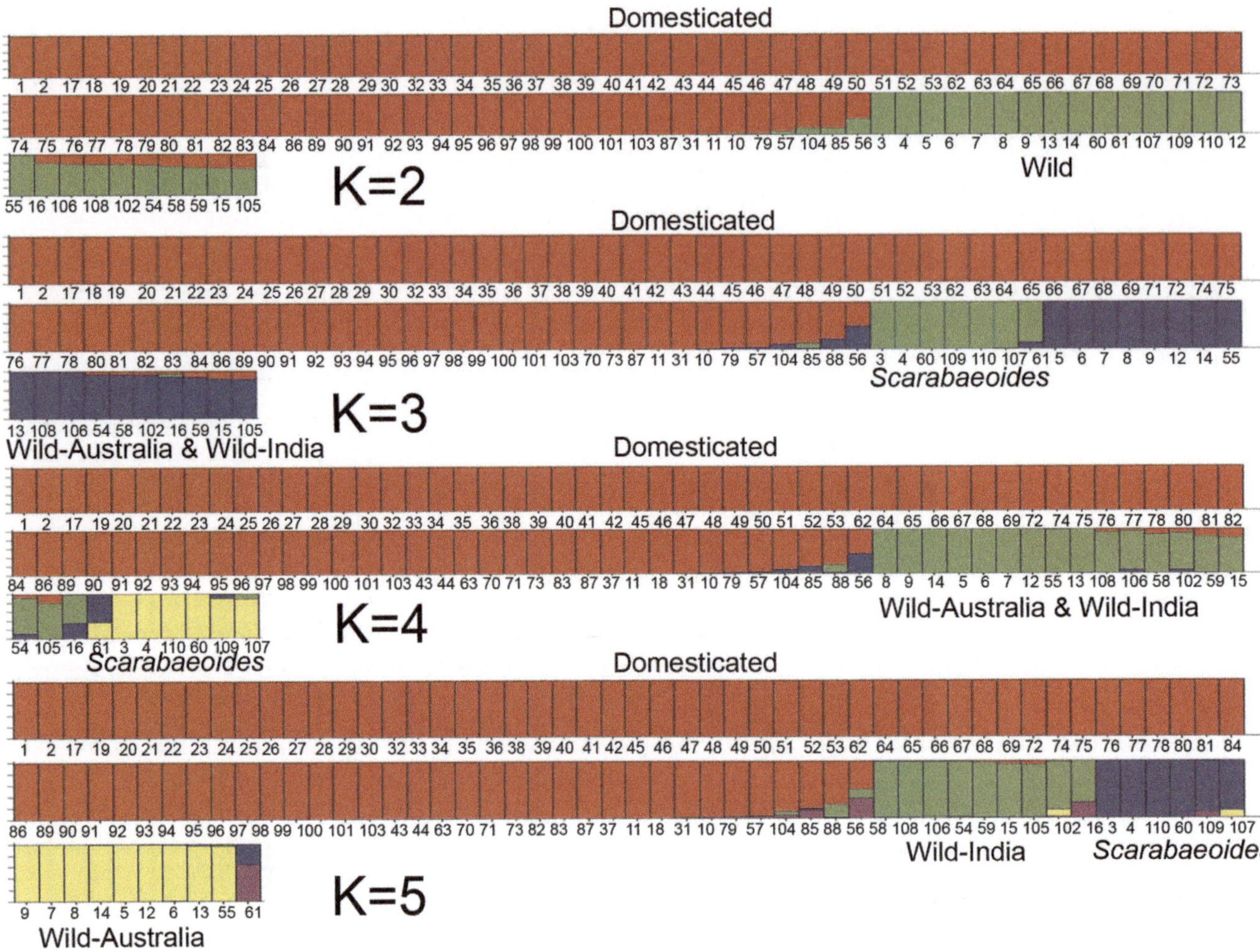

Figure 4. Population structure among the wild and domesticated genotypes of *Cajanus*. Output of the population genetic program STRUCTURE at increasing K values of 2 to 5. The primary divisions at increasing K values mirror phylogenetic groupings shown in Figures 1 and 2. Memberships of individual genotypes to specific subgroupings are indicated by colored bars. Genetic admixture is evidenced by fractional membership in multiple subgroups. The correspondence between numbers below each genotype and specific accessions is given in Table S1. Likelihood values for each value of K are given in Table S3.

the outputs of PCoA and STRUCTURE: Group I, wild-*scarabaeoides*; Group II, wild-Australia and wild-India; Group III, domesticated-India, domesticated-tropical, and domesticated Philippines. As shown in Table 1, most genetic variation was attributed to differences among the three groups (89%), although genetic differentiation was evident at all levels of analysis. These broad patterns of genetic differentiation reflect patterns established in previous phylogenetic, PCoA and STRUCTURE analyses.

Genetic polymorphism was highest within wild groups of Indian origin (both wild-India and wild-*scarabaeoides*), with rates ~3-fold higher than within the wild-Australian group (~37% polymorphism compared to ~12%) (Table 2). In contrast, the lowest rates of polymorphism were documented for the domesticated populations, with the domesticated-tropical population having ~60% the polymorphism of the domesticated-India population and no private alleles relative to the domesticated-Indian group. As expected, there was little genetic differentiation between domesticated-India and domesticated-Tropical populations (Fst = 0.05). In contrast, the Philippines accessions showed relatively high genetic differentiation when compared to these domesticated groups (Fst = 0.179) and possessed overall high levels of polymorphism, consistent with our previous suggestion of independent admixture events leading to these individual accessions.

Domesticated populations were considerably less differentiated from the wild-Indian population relative to differentiation from either wild-Australia or wild-*scarabaeoides* (Table 3). Even lower genetic differentiation was observed between wild-India and wild-Australia (Fst = 0.290), which is consistent with phylogenetic and allele frequency analyses that establish these populations as sister groups. In contrast to the low genetic distance (Table 4) with wild-India, wild-Australia was strongly differentiated from the domesticated groups, suggesting that the divergence between wild-India and wild-Australia was archaic relative to domestication of *C. cajanifolius* from wild-India, or that domestication occurred from an isolated subpopulation within the wild-India group.

Discussion

Here we have investigated the genetic diversity and population structure of domesticated pigeonpea and its wild relatives in the genus *Cajanus*. Because the genotypes we studied represent both shallow sampling of widely diverged species as well as relatively deep sampling of the single cultivated species, we combined both

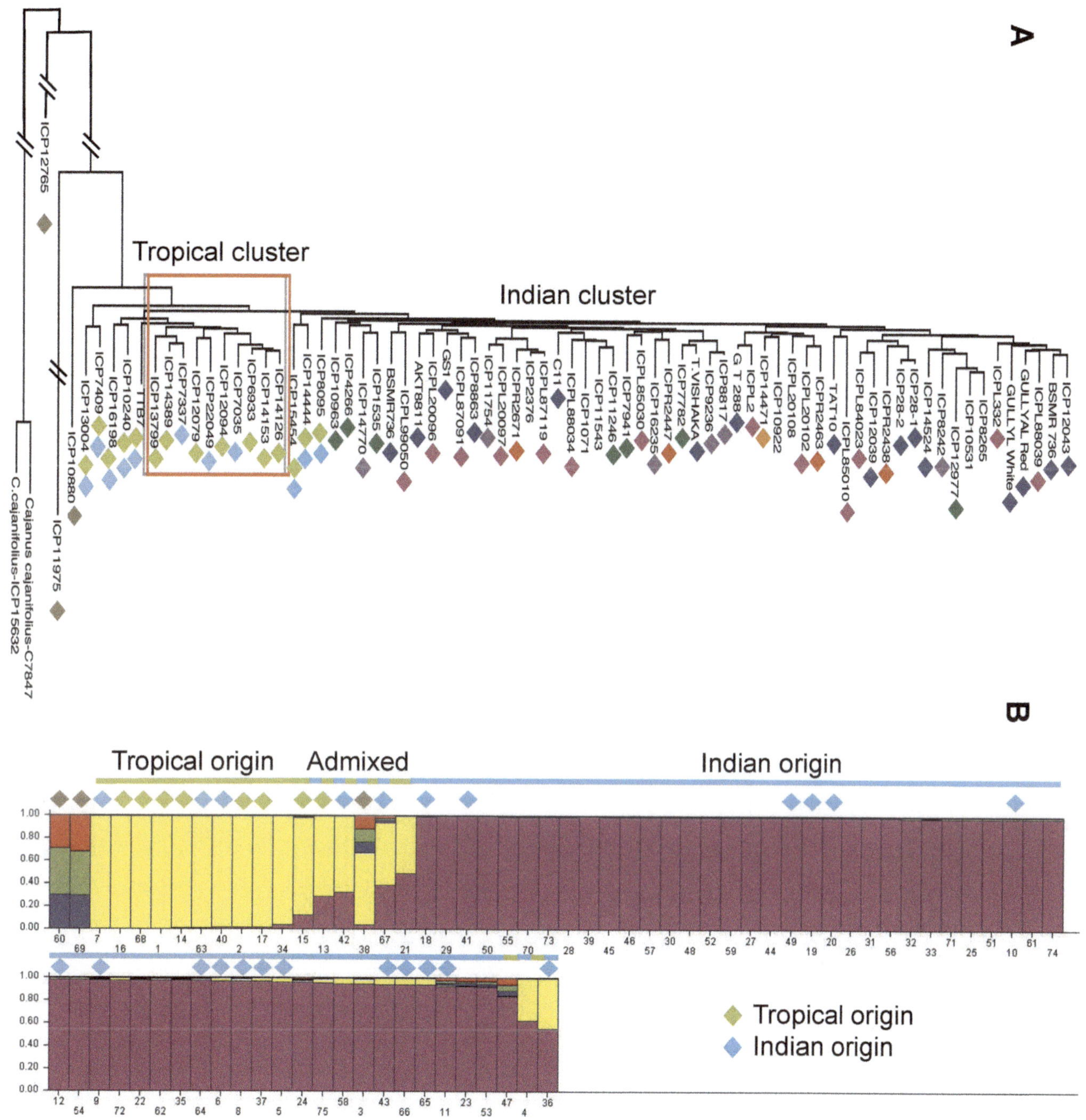

Figure 5. Population structure of cultivated *Cajanus cajan*. Panel **A**, Weighted Neighbor-Joining tree depicting pairwise relationships between accessions. Colors denote the nature of individual accessions: Blue diamonds, cultivars and elite varieties; Pink diamonds, landraces; purple diamonds, ICRISAT reference material; Green diamonds, Core collection; Red diamonds, R-line; Orange diamonds, Minicore; Light green diamonds, Tropical; light blue, Indian; Light brown diamonds, Philipines. Genotypes with admixture between Indian and Tropical subgroups are designated by both light green and light blue diamonds. Panel **B**, population subdivisions with cultivated genotypes revealed by STRUCTURE. Green diamonds, genotypes with tropical distribution; blue diamonds, genotypes with an Indian sub-tropical distribution. Admixed genotypes are those with fractional membership in multiple groups. Likelihood values for each value of K are given in Table S4.

phylogenetic and population genetic analyses. The data were sufficient to derive relationships that were simultaneously congruent with, and more detailed than, previous plastid and nuclear gene phylogenies (MT. Kassa, PhD Dissertation). Moreover, the results permit assignment of *C. cajanifolius* as the most probable progenitor species, and they allow us to infer the origin of modern cultivated pigeonpea from nested population bottlenecks, with an initial domestication in India and subsequent spread of cultivation to tropical regions beyond India.

Crop domestication is accompanied by genome-wide reduction in genetic diversity [1]. This reduction derives from a population bottleneck imposed during the founding of a new crop lineage [30] and subsequently due to selection on specific loci that confer agronomically important traits [31]. Bottleneck severity varies

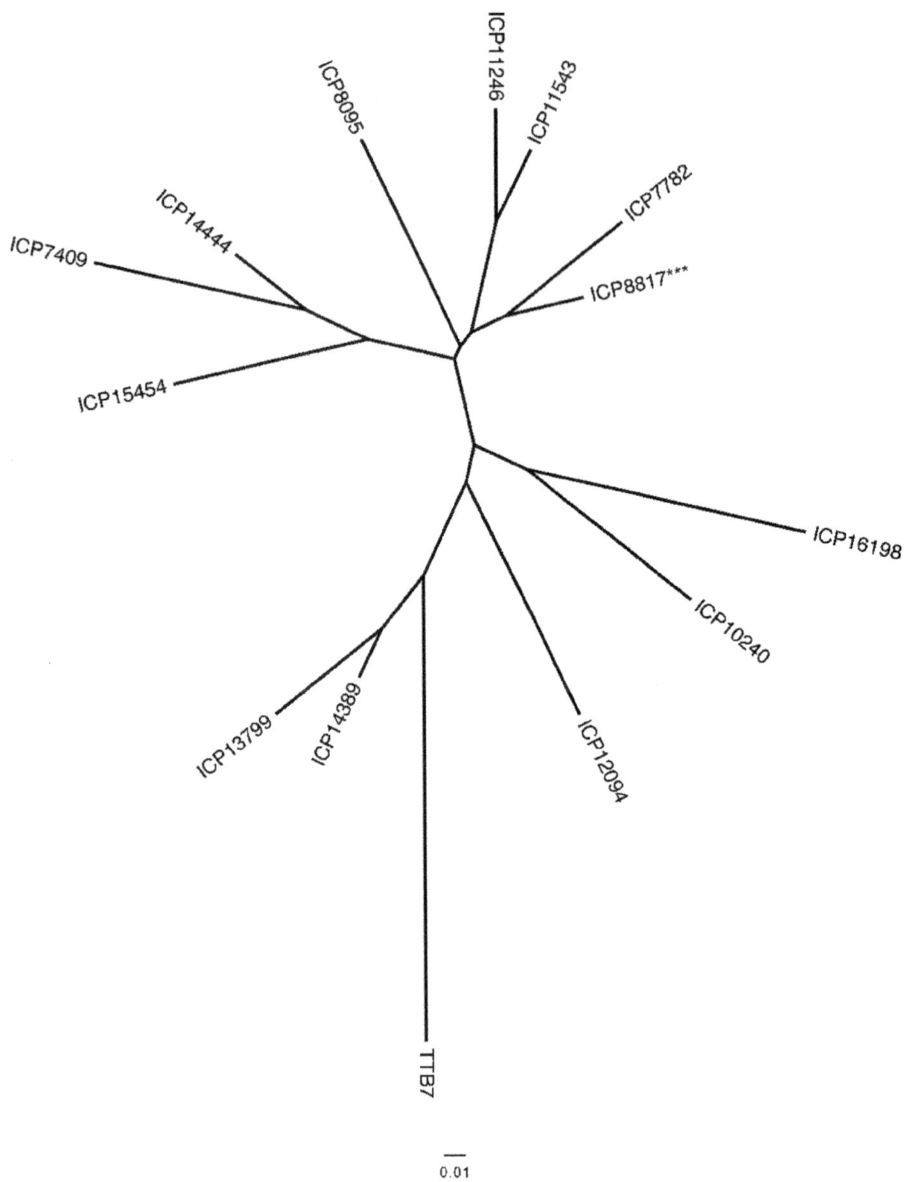

Figure 6. Inferred relationships among admixed *Cajanus cajan* accessions. An un-rooted Neighbor Joining tree depicting pairwise similarity between genotypes with simple admixture between domesticated-India and domesticated-tropical accessions. Admixed genotypes were identified based on mixed-group membership defined in Figure 5B. Accession ICP 8817 has an identical haplogype to accession ICP 9236, and thus only ICP 8817 is shown.

Table 1. Summary results of AMOVA analyses within and among populations of 95 accessions of the domesticated and wild groups.

Source	df	SS	MS	Est. Var.	% Variation	Statistics	Value	P
Among groups	2	18135.809	9067.905	492.714	**89%**	**Fct**	**0.895**	**0.010**
Among pops within groups	3	1107.072	369.024	30.265	**5%**	**Fsc**	**0.521**	**0.010**
Among Individuals within pops	89	2476.214	27.823	27.823	**5%**	**Fst**	**0.949**	**0.010**
Total	94	21719.095		550.801	100%			

d.f.: Degrees of freedom; SS, sum of squared observations; MS, mean of squared observations; Est. var., estimated variance % Var., percentage of total variance; Fct, proportion of the total genetic variance between groups; FSc, proportion of the total genetic variance among populations within a group; Fst, proportion of the total genetic variance within populations.

Table 2. Percentage of Polymorphic loci in wild and domesticated groups.

Sub-groups	Number of accessions	Genetic Status	Polymorphic loci (%)
Wild *scarabaeoides*	4	Wild	36.7%
Wild Australian	9	Wild	11.84%
Wild Indian	9	Wild	37.37%
Domesticated Indian	58	Domesticated	8.64%
Domesticated Tropical	12	Domesticated	5.45%
Philippines	3	Domesticated	23.94%
Mean			**20.66%**
SE			**5.78%**

among crop species depending on the duration of domestication and number of domestication events. For example, several grasses have about two-thirds of the genetic diversity found in their wild relatives [32], and simulations reveal a more severe bottleneck for rice than maize [31,33]. Previous studies using SSR [24] and DArT [25] markers detected a reduction in levels of genetic diversity in domesticated pigeonpea compared to wild relatives though the degree of a bottleneck effect was not quantified.

Here we quantify the reduction in genetic diversity, estimating that domesticated pigeonpea contains only ?25% of polymorphic loci present in the progenitor wild-India group. Only 62 markers detected variation among the domesticated *C. cajan* group (excluding the Philippines accessions) in comparison to 283 SNP markers that were polymorphic among the progenitor wild-India accessions. It is noteworthy that landraces (primitive cultivars) and improved (elite) cultivars that comprise the domesticated portion of our genotype panel (Table 3) contained similar levels of polymorphic SNPs, indicating that much of the diversity that survived through the incipient stages of domestication was retained in current day cultivars and breeding lines. Despite the genetically narrow base of pigeonpea, the cultigen is noted for high levels of morphological diversity. Thus, different genotypes are adapted for acceptable agronomic yield in both tropical and semi-arid regions of the world, as reflected in the eco-geographical variation in collection sites for accessions used in this study. Similar genetic bottleneck effects have also been observed in other crop species such as soybean [34,35], sunflower [36], and lima beans [37].

Although there was no clear distinction between landraces and modern cultivars, domesticated genotypes were resolved into two sections based primarily on the results of Neighbor Joining and STRUCTURE analyses (Figures 5A and 5B). The subdivision reflects the geographical origin of the respective genotypes, further supporting the validity of the groups, with one lineage of Indian origin containing approximately twice the genetic diversity of a second lineage of tropical origin. Both of these populations are depauperate of genetic diversity, with low genetic differentiation and low genetic distance between them. Taken together, these results suggest that primary domestication occurred in India, with a more recent nested bottleneck associated with genotypes grown in tropical regions. We suggest that the genetic distinctiveness of the tropical and Indian subgroups within *C. cajan* likely derives from breeding for the geographically-wide but agro-climatically similar tropical regions versus semi-arid environments.

Although limited within-taxon sampling reduces our ability to assess genetic diversity in the wild species, we can still make preliminary assessments of diversity in the *Cajanus* species that are important members of the secondary gene pool. In particular, we note low diversity in the wild accessions collected from Australia. This situation is curious, because the wild-Australia group contains seven distinct taxonomic species, yet possesses less than one-third the polymorphism found in the taxonomically homogeneous *C. scarabaeoides* lineage of Indian origin. A majority of these Australian species are endemic to Australia and possess similar morphological characters (e.g. leaf shape, leaf and flower color and the growth habit) [25]. Australia has been designated as an important center of species diversity for *Cajanus* [38], but our results argue against this conclusion because genetic diversity was quite low among the seven species used in this analysis.

Several lines of evidence indicate that the Australian lineage is closely related to the lineage of *non-C. scarabaeoides* wild-India species, including the sister relationship of these two lineages in Parsimony analysis (see legend to Figure 2 for clarification) and their low genetic differentiation (Fst = 0.290) relative to other among group comparisons (Table 3). As noted above, the Australian lineage is genetically homogeneous, with polymorphism rates less than twice that observed in the domesticated-

Table 3. Pairwise estimates of FST among wild and domesticated groups.

	Wild *scarabaeoides*	Wild Australian	Wild Indian	Domesticated Indian	Domesticated Tropical	Philippines
Wild *scarabaeoides*						
Wild Australian	0.533					
Wild Indian	0.496	0.290				
Domesticated Indian	0.808	0.812	0.565			
Domesticated Tropical	0.812	0.829	0.573	0.050		
Philippines	0.712	0.667	0.448	0.165	0.179	

Table 4. Pairwise estimates of Nei's genetic distance (lower diagonal) and gene flow (Nm) values (upper diagonal) among wild and domesticated groups.

	Wild *scarabaeoide*	Wild Australian	Wild Indian	Domesticated Indian	Domesticated Tropical	Philippines
Wild *scarabaeoides*		0.219	0.254	0.059	0.058	0.101
Wild Australian	0.349		0.611	0.058	0.052	0.125
Wild Indian	0.476	0.135		0.192	0.186	0.309
Domesticated Indian	1.578	0.661	0.414		4.705	1.264
Domesticated Tropical	1.574	0.662	0.416	0.004		1.148
Philippines	1.254	0.549	0.333	0.028	0.029	

India pigeonpea. These results are consistent with recent introduction of *Cajanus* into Australia from India, with a corresponding genetic bottleneck. Could human migration have been a factor? It is likely that migrating humans carried seed for nutrition, if not planting. If so, then genetic drift and new climates would have had pronounced effects on the characteristics of even casually collected seed stocks; for example, Australia's climate is highly varied and differing moisture regimes would likely drive rapid divergence in adaptive leaf traits without a requirement for corresponding genome-wide diversification. Such morphological diversification could explain the proliferation of species assignments based on morphometric criteria, despite correspondingly low genetic diversity.

Interestingly, genetic evidence suggests that humans may have colonized Australia by migration from the Indian subcontinent. A proposed but controversial early migration route includes movement from the Indian sub-continent to Australia in the late Pleistocene, i.e., >10,000 years ago [39], while a proposed more recent event corresponds to changes to the anthropological record in Australia around 5,000 to 3,000 years ago [40]. Although we have no evidence that humans either cultivated or carried pigeonpea along this migration route, the apparent origin of related *Cajanus* spp in India and the presence of a narrow genetic base of derived *Cajanus* species in Australia are consistent with this possibility. If true, then this "Australian-focused *Cajanus* bottleneck" was entirely independent of the recognized Indian-domestication, because the modern cultivated pigeonpea is genetically distinct from its Australian relatives.

Gene flow subsequent to initial domestication is also likely to have contributed to the character of cultivated pigeonpea, and the haplotypes of several accessions provided evidence of recent genetic admixture between wild and cultivated gene pools. The potential for gene flow is significant, as insect-aided natural out-crossing for pigeonpea may range up to 70% [41–45] and recently up to 17% natural out-crossing has been observed for wild species [44], with the highest out-crossing rate recorded for *C. lineatus*. Van der Maesen (personal communication) notes that multiplication of *C. lineatus* in experimental gardens is common practice in India and therefore the occurrence of spontaneous hybrids involving *C. lineatus* and nearby cultivated *C. cajan* should be expected and has been observed. Here we identified one highly admixed genome described as *C. lineatus*.

In addition to the above referenced *C. lineatus* accession, the highest rates of admixture from the wild-India population were observed for two Philippines accessions (ICP 12765 and ICP 10880), ICRISAT reference set accession (ICP 11975), and two accessions of *C. cajanifolius* (C7847 and ICPW 32) collected from the field. Of particular interest are two accessions of *C. cajan* (ICP 15629 and ICPW 29) that possess genomes that are predominantly domesticated, but with 5–10% membership of the wild-Indian group. *C. cajan* ICP 15629 and ICPW 29 served as parental lines to develop a stable cytoplasmic male sterility (CMS) system in pigeonpea. The CMS accession, ICP 2039A, was derived from an inter-specific hybrid of ICPW 29 and cultivar ICP 11501 [46]; we speculate that intentional hybridization and repeated backcrossing may have contaminated the genome of these parental accessions.

Conclusions

This molecular diversity study corroborates the long held alpha-taxonomic hypothesis that *C. cajanifolius* is the most recent progenitor of cultivated pigeonpea and supports India as the most likely center of pigeonpea domestication. However, crop domestication is a progressive process that may involve both independent derivations within the range of the ancestral species (in the sense of Allaby et al. [47]) and hierarchical selection events that together span thousands of years and serve to adapt germplasm to diverse eco-geographical conditions. Moreover, when crops are grown in the vicinity of locally adapted wild species, there is great potential for both intentional and accidental genetic admixture, which would further impact allele content in the cultivated gene pool. Cultivated pigeonpea would be particularly prone to such admixture, given its significant out-crossing rates and documented cross-compatibility with local wild species. In the current analysis, we used population structure analysis and AMOVA to reveal genetic admixture between wild and cultivated genomes, suggesting the involvement of gene flow between wild and domesticated species. The wild gene pool of *Cajanus* contains not only high genetic diversity but also unique and rare alleles for agronomically important traits (e.g. trichomes of *C. scarabaeoides* for pod borer resistance; *C. platycarpus* has shown to be the only source of resistance to the P3 race of *Phytophthora* blight disease). Thus pigeonpea breeding and improvement programs would benefit from the continued and expanded use of this bounty of genetic diversity prevailing in the wild gene pool.

Materials and Methods

Plant Materials

As listed in Table S1, the 110 accessions of *C. cajan* and allied species used in this study derive from diverse environments in Asia, Australia, Africa and the Caribbean. With the exception of 12 wild accessions, which were acquired from the Western Australia Herbarium and were originally collected from the field in Australia, all other genotypes were obtained from the gene bank of the International Crops Research Institute for the Semi-Arid Tropics (ICRISAT) or from germplasm resources at Panjabrao Deshmukh Agricultural University (PDAU), University of Agricultural Sciences (UAS)-Bangalore, or UAS-Dharwad. Within the cultivated genotypes several accessions derive from core and mini-core accessions circumscribed by ICRISAT based on various morphological descriptors, and agro-morphology traits and SSR diversity data [48–50].

Molecular Methods

Genotyping was based on a set of 752 SNPs discovered in a comparison of 1440 sequenced amplicons between *C. cajan* accession ICP 28 and *C. scarabaeoides* accession ICPW 94 (Table S2 and available at www.comparative-legumes.org/). Polymorphisms were identified by amplicon re-sequencing and sequence alignment involving *C. cajan* (ICP 28) and *C. scarabaeoides* (ICPW 94). The target sequences were a set of primarily single copy orthologous genes, whose orthology was inferred initially from legume EST data (i.e., the transcriptomes of *Medicago truncatula*, *Lotus japonicus* and *Glycine max*) and subsequently based on conserved genome location in a multi-species comparative genetic analysis (Penmetsa and Cook, unpublished data). Individual SNPs meeting assay design criteria, determined by Illumina Inc. using their proprietary Assay Design Tool, were converted to a 768 Illumina GoldenGate genotyping assay. Together these SNP assays survey biallelic states at 670 distinct genes. For purposes of genotyping, DNAs were extracted from the 110 *Cajanus* accessions using the Qiagen DNeAsy protocol using a Retch mixer mill, according to manufacturer's instruction, and delivered to the UC Davis Genome Center DNA Technologies Core for analysis (http://dnatech.genomecenter.ucdavis.edu/). Allele calls were curated using the Illumina Beadstudio software package (Illumina, San Diego, CA, USA). To minimize the confounding effects of technical error, all SNP calls with more than 10% missing data were excluded from the analysis. Details on sequence context of SNPs, and the curated genotyping data (locus x genotype) are provided in Table S2.

Data Analysis

As our genotypes included both diverged species and samples from the breeding pool where interbreeding and/or admixture occurs or has occurred often in the recent past, we took both phylogenetic and population genetic approaches to the data. We first used phylogenetic methods to distinguish the major groups, having a greater number of informative markers than previous studies. We then took the least well-resolved group of domesticated germplasm and most closely related wild species and analyzed it with population genetic approaches. Phylogenetic analyses were conducted using both the character based phylogenetic analysis of maximum parsimony and the distance-based analysis of Neighbor Joining. The maximum parsimony analysis was based on the complete data set of the concatenated SNPs at 670 loci and the analysis was performed using PAUP* 4.0b10 [51]. A full heuristic search was performed with 1000 random addition sequence replicates using Tree Bisection Reconnection (TBR) branch swapping. Clade support was evaluated through bootstrapping with 500 replicates using TBR branch swapping and the results were used to generate a consensus parsimony tree. To further deduce overall similarity among accessions and resolve relationships between individual genotypes, pairwise dissimilarity was calculated by simple matching according to the method of Saito and Nei [52]. The resulting dissimilarity matrix was used to derive a weighted Neighbor Joining tree [52] with 1000 bootstraps. This weighted Neighbor Joining analysis employs a likelihood-based criterion that models distance between genotypes as random variables that obey a Gaussian distribution [53]. The analysis was carried out with DARwin5 software [54].

Genetic structure was analyzed using the program STRUCTURE 2.1 [28,29]. STRUCTURE assigns individual genotypes to a specified number of groups "K" based on membership coefficients calculated from the genotype data. The analysis was run from $K=1$ to 10 using a burn-in period of 50,000 steps followed by 500,000 MCMC (Monte Carlo Markov Chain) replicates with 2 iterations, assuming admixture and correlated allele frequencies. Optimal K, which is adopted as the number of sub-populations from which the analyzed accessions derive, was determined using an *ad hoc* static ΔK based on the rate of change in the natural log probability of data between successive K values as described by Evanno and colleagues [55]. At optimal K, individual sub-populations were extracted and analyzed separately using STRUCTURE 2.1 to resolve additional genetic relationships.

Based on the output of STRUCTURE and principal coordinate analysis (PCoA) (see below), combined with phylogenetic inferences and common knowledge of geography and breeding history, we circumscribed 6 subgroups for further analysis. Analysis of Molecular Variance (AMOVA) was conducted, based on the hierarchical model and permutational procedures of Excoffier and colleagues [56], to assess the level of variation among these wild and domesticated groups. To avoid the potential confounding effects of admixture (revealed by STRUCTURE), we removed all admixed genotypes (except the Philippines accessions) and AMOVA was performed on 95 genotypes. Genetic variation within groups (Fct), variation within populations (Fst) and variation between populations within a group (Fsc), population polymorphism, and Nei's Genetic distance and gene flow (Nm) were analyzed using GenAlEx v.6.3 [26] and Arlequin [57]. A principal coordinate analysis (PCoA) using GenAlEx v.6.3 was conducted to complement the output of the phylogenetic analyses, with the former being most sensitive to differences among groups and the later more sensitive to differences between closely related individuals [27].

Supporting Information

Table S1 Details of individual *Canajus* spp accessions.

Table S2 SNP assay design and genoytpe data.

Table S3 Ln of the probability and its variance for K from 1 to 10, provided as supporting information for Figure 4.

Table S4 Ln of the probability and its variance for K from 1 to 5, provided as supporting information for Figure 5.

Acknowledgments

The authors acknowledge BOYSCAST fellowships from the Indian Department of Science and Technology to BKS and SD. This paper is contribution number 231 from the Tropical Biology Program of Florida International University, associated with EJBvW. We are grateful to Professor Nigel Barker of Rhodes University Professor Jos van der Maesen of Wageningen University The Netherlands, and Associate Professor David Remington of the University of North Carolina at Greensboro for their helpful comments during preparation of the manuscript, and to

Dr Reyazul Rouf Mir and Dr Rachit K Saxena for their valuable assistance with germplasm selection. We also acknowledge the important contributions of certain DNA samples used in this study from Dr Pawan Kulwal of Dr. Panjabrao Deshmukh Agricultural University (PDAU), Akola (India), Dr B Fakrudin University of Agricultural Sciences (UAS)-Bangalore (India), and Dr M Byregowda of UAS-Dharwad (India).

Author Contributions

Conceived and designed the experiments: MTK RVP EJBvW DRC. Performed the experiments: MTK RVP NCG BKS SD. Analyzed the data: MTK EJBvW DRC. Contributed reagents/materials/analysis tools: RVP HDU RKV DRC. Wrote the paper: MTK EJBvW DRC.

References

1. Tanksley SD, McCouch SR (1997) Seed banks and molecular maps: unlocking genetic potential from the wild. Science 277: 1063–1666.
2. Kilian B, Ozkan H, Kohl J, von Haeseler A, Barale F, et al. (2006) Haplotype structure at seven barley genes: relevance to gene pool bottlenecks, phylogeny of ear type and site of barley domestication. Mol Gen Genomics 276: 230–241.
3. Matsuoka Y, Vigouroux Y, Goodman MM, Sanchez JG, Buckler E, et al. (2002) A single domestication for maize shown by multilocus microsatellite genotyping. Proc Natl Acad Sci USA 99: 6080–6084.
4. Dubcovsky J, Dvorak J (2007) Genome plasticity a key factor in the success of polyploidy wheat under domestication. Science 316: 1862–1866.
5. Morrell PL, Clegg MT (2007) Genetic evidence for a second domestication of barley (*Hordeum vulgare*) east of the Fertile Crescent. Proc Natl Acad Sci USA 104: 3289–3294.
6. Li YH, Li W, Zhang C, Yang L, Chang RZ, et al. (2010) Genetic diversity in domesticated soybean (*Glycine max*) and its wild progenitor (*Glycine soja*) for simple sequence repeat and single-nucleotide polymorphism loci. New Phytologist 188: 242–253.
7. He Z, Zhai W, Wen H, Tang T, Wang Y, et al. (2011) Two Evolutionary Histories in the Genome of Rice: the Roles of Domestication Genes. PLoS Genetics 7: e1002100.
8. Bagavathiannan MV, Van Acker RV (2008) Crop ferality: Implications for novel trait confinement. Agriculture, Ecosystems and Environment 127: 1–6.
9. FAOStat (2008) Online agricultural statistics. Available: http://www.faostat.org. Accessed 2012 Jun 4.
10. Lewis G, Schrire B, MacKinder B, Lock M (2005) Legumes of the world. UK: Royal Botanic Gardens, Kew publishing. 592 p.
11. Harlan JR, de Wet JMJ (1971) Toward a rational classification of cultivated plants. Taxon 20: 509–517.
12. Mallikarjuna N, Saxena KB, Jadhav DR (2011) *Cajanus*. In: Cole C, ed. Wild Crop Relatives: Genomic and breeding resources of Legume Crops and Forages. Germany: Springer. pp 21–33.
13. Bohra A, Mallikarjuna N, Saxena K, Upadhyaya H, Vales I, et al. (2010) Harnessing the potential of crop wild relatives through genomics tools for pigeonpea improvement. Jour Appl Biol 37: 83–98.
14. Fortunato RH (2000) Systematic relationships in *Rhynchosia* (Cajaninae Phaseoleae–Papilionoideae–Fabaceae) from the neotropics. In: Herendeen PS, Bruneau A, eds. Advances in legume systematics series No 9. The Royal Botanic Gardens, Kew, UK. 339–354.
15. Van der Maesen LJG (2003) Cajaninae of Australia (Leguminosae: Papilionoideae). Australian Systematic Botany 16: 219–227.
16. Van der Maesen LJG (1980) India is the native home of the pigeonpea. Liber gratulatorius in honorem HCD de Wit. Misc. Paper 19. Wageningen, the Netherlands. pp 257–262.
17. Van der Maesen LJG (1990) Pigeonpea origin, history, evolution, and taxonomy. In: Nene YL, Halls D, Sheila VK, eds. The Pigeonpea. UK: CAB International, pp 15–46.
18. Purseglove JW (1968) Tropical Crops: Dicotyledons, UK: Longman group. 719 p.
19. Tindall HD (1988) Vegetables in the Tropics. UK: Macmillan International edition. 533 p.
20. De DN (1974) Pigeonpea. In: Hutchinson JB, ed. Evolutionary studies on world crops. Cambridge University Press, Cambridge. pp 79–87.
21. Kajale MD (1974) Plant economy at Bhokardan. In: Dev SB, Gupta RS, eds. Excavations at Bhokardan (Bhogavardana). Nagpur University and Maharashtra Marathwada University, Nagpur, India. pp 7–224.
22. Van der Maesen LJG (1986) *Cajanus* DC. and *Atylosia* W. & A. (Leguminosae). Wageningen papers. The Netherlands: Agricultural University, Wageningen. 225 p.
23. Saxena KB (2008) Genetic Improvement of Pigeonpea- A Review. Tropical Plant Biology 1: 159–178.
24. Odeny DA, Jayashree B, Ferguson M, Hoisington D, Cry LJ, et al. (2007) Development, characterization and utilization of microsatellite markers in pigeonpea. Plant Breeding 126: 130–136.
25. Yang S, Ash G, Harper J, Varling J, Wenzl P, et al. (2006) Low level of genetic diversity in cultivated pigeonpea compared to its wild relatives is revealed by diversity arrays technology. Theoretical and Applied Genetics 113: 585–595.
26. Peakall R, Smouse PE (2006) GENALEX 6: Genetic analysis in Excel. Population genetic software for teaching and research. Molecular Ecology Notes 6: 288–295.
27. Hauser LA, Crovello TJ (1982) Numerical analysis of genetic relationships in Thelypodieae (Brassicaceae). Systematic Botany 7: 249–268.
28. Pritchard JK, Stephens P, Donnelly P (2000) Inference of population structure using multilocus genotype data. Genetics 155: 945–959.
29. Falush D, Stephens M, Pritchard JK (2003) Inference of population structure using multilocus genotype data: linked loci and correlated allele frequencies. Genetics 164: 1567–1587.
30. Eyre-Walker A, Gaut RL, Hilton H, Feldman DL, Gaut BS (1998) Investigation of the bottleneck leading to the domestication of maize. Proc Natl Acad Sci USA 95: 4441–4446.
31. Tenaillon MI, U'Ren J, Tenaillon O, Gaut BS (2004) Selection versus demography: a multilocus investigation of the domestication process in maize. Molecular Biology and Evolution 21: 1214–1225.
32. Buckler ES, Thornsberry JM, Kresovich S (2001) Molecular diversity, structure and domestication of grasses. Genetic Research 77: 213–218.
33. Zhu Q, Zheng X, Luo J, Gaut BS, Ge S (2007) Multilocus analysis of nucleotide variation of *Oryza sativa* and its wild relatives: Severe bottleneck during domestication of rice. Molecular Biology and Evolution 24: 875–882.
34. Guo J, Wang Y, Song C, Zhou J, Qiu L, et al. (2010) A single origin and moderate bottleneck during domestication of soybean (*Glycine max*): implications from microsatellites and nucleotide sequences. Annals of Botany 106: 505–514.
35. Hyten DL, Song Q, Zhu Y, Choi IY, Nelson RL, et al. (2006) Impacts of genetic bottlenecks on soybean genome diversity. Proc Natl Acad Sci USA 103: 16666–16671.
36. Liu A, Burke JM (2006) Patterns of nucleotide diversity in wild and cultivated sunflower. Genetics 173: 321–330.
37. Motta-Aldana JR, Serrano ML, Hernández-Torres J, Castillo-Villamizar G, Debouck DG, et al. (2010) Multiple origins of lima bean landraces in the Americas: Evidence from chloroplast and nuclerar DNA polymorphisms. Crop Science 50: 1773–1787.
38. Nene YL, Sheila VK (1990) Pigeonpea: geography and importance. In: Nene YL, Hall SH, Sheila VK, eds. The pigeonpea. CAB International, Wellingford, UK. pp 1–14.
39. Macaulay V, Hill C, Achilli A, Rengo C, Clarke D, et al. (2005) Single, rapid coastal settlement of Asia revealed by analysis of complete mitochondrial genomes. Science 308: 1034–1036.
40. Redd AJ, Roberts-Thomson J, Karafet T, Bamshad M, Jorde LB, et al. (2002) Gene Flow from the Indian Subcontinent to Australia: Evidence from the Y chromosome. Current Biology 12: 673–677.
41. Bhatia GK, Gupta SC, Green JM, Sharma D (1981) Estimates of natural cross-pollination in *Cajanus cajan* (L.) Millsp. Several experimental approaches. In: Kumble V, ed. Proceedings of International Workshop on Pigeonpeas, ICRISAT, Patancheru, India. pp 129–136.
42. Ratnaparkhe MB, Gupta VS, Venmurthy MR, Ranjekar PK (1995) Genetic fingerprinting of pigeonpea (*Cajanus cajan* (L.) Millsp.) and its wild relatives using random amplified polymorphic DNA markers. Theoretical and Applied Genetics 91: 893–898.
43. Souframanien J, Manjaya JG, Krishna TG, Pawar ME (2003) Random amplified polymorphic DNA analyses of cytoplasmic male sterile and male fertile pigeon pea (*Cajanus cajan* (L.) Millsp. Euphytica 129: 293–299.
44. Saxena KB, Kumar RV (2010) Insect-aided natural out-crossing in four wild relatives of pigeonpea. Euphytica 173: 329–335.
45. Saxena KB, Kumar RV (2003) Development of a cytoplasmic- nuclear male-sterility system in pigeonpea using *C. scarabaeoides* (L.) Thouars. Indian Journal of Genetics 63: 225–229.
46. Saxena KB, Kumar RV, Srivastava N, Bao S (2005) A cytoplasmic-nuclear male-sterility system derived from a cross between *Cajanus cajanifolius* and *Cajanus cajan*. Euphytica 145: 289–294.
47. Allaby RG, Fuller DQ, Brown TA (2008) The genetic expectations of a protracted model of the origins of domesticated crops. Proc Natl Acad Sci USA 105: 13982–13986.
48. Reddy LJ, Upadhyaya HD, Gowda CLL, Singh S (2005) Development of core collection in pigeonpea [*Cajanus cajan* (L.) Millsp.] Using geographic and qualitative morphological descriptors. Genetic Resources and Crop Evolution 52: 1049–1056.
49. Upadhyaya HD, Reddy LJ, Gowda CLL, Reddy KN, Singh S (2006) Development of a Mini-Core Subset for Enhanced and Diversified Utilization of Pigeonpea Germplasm Resources. Crop Science 46: 2127–2132.
50. Upadhyaya HD, Reddy KN, Shivali S, Varshney RK, Bhattacharjee R, et al. (2011). Pigeonpea composite collection for enhanced utilization of germplasm in crop improvement programs. Plant Genet Resou Charact Utiliz 9: 97–108.
51. Swofford D.L (2002) PAUP*. Phylogenetic Analysis Using Parsimony (*and Other Methods). Version 4. Sinauer Associates, Sunderland, Mass.
52. Saitou N, Nei M (1987) The Neighbor-Joining method: A new method for reconstructing phylogenetic trees. Molecular Biology and Evolution 4: 406–425.
53. Bruno WJ, Socci ND, Halpern AL (2000) Weighted Neighbor Joining: A Likelihood-Based Approach to Distance-Based Phylogeny Reconstruction. Molecular Biology and Evolution 17: 189–197.

54. Perrier X, Jacquemoud-Collet JP (2006) DARwin. Available: http://darwin.cirad.fr/darwin. Accessed 2012 Jun 4.
55. Evanno G, Regnaut S, Goudet J (2005) Detecting the number of clusters of individuals using the software STRUCTURE: a simulation study. Molecular Ecology 14: 2611–2620.
56. Excoffier L, Smouse PE, Quattro JM (1992) Analysis of molecular variance inferred from metric distances among DNA haplotypes: application to human mitochondrial DNA restriction data. Genetics 131: 479–491.
57. Excoffier L, Laval G, Schneider S (2005) Arlequin (version 3.0): An integrated software package for population genetics data analysis. Evolutionary Bioinformatics Online 1: 47–50.

14

Recent and Projected Increases in Atmospheric CO_2 Concentration Can Enhance Gene Flow between Wild and Genetically Altered Rice (*Oryza sativa*)

Lewis H. Ziska[1]*, David R. Gealy[2], Martha B. Tomecek[1], Aaron K. Jackson[2], Howard L. Black[2]

1 Crop Systems and Global Change Laboratory, United States Department of Agriculture, Agricultural Research Service, Beltsville, Maryland, United States of America,
2 Dale Bumpers National Rice Research Center, United States Department of Agriculture, Agricultural Research Service, Stuttgart, Arkansas, United States of America

Abstract

Although recent and projected increases in atmospheric carbon dioxide can alter plant phenological development, these changes have not been quantified in terms of floral outcrossing rates or gene transfer. Could differential phenological development in response to rising CO_2 between genetically modified crops and wild, weedy relatives increase the spread of novel genes, potentially altering evolutionary fitness? Here we show that increasing CO_2 from an early 20th century concentration (300 μmol mol^{-1}) to current (400 μmol mol^{-1}) and projected, mid-21st century (600 μmol mol^{-1}) values, enhanced the flow of genes from wild, weedy rice to the genetically altered, herbicide resistant, cultivated population, with outcrossing increasing from 0.22% to 0.71% from 300 to 600 μmol mol^{-1}. The increase in outcrossing and gene transfer was associated with differential increases in plant height, as well as greater tiller and panicle production in the wild, relative to the cultivated population. In addition, increasing CO_2 also resulted in a greater synchronicity in flowering times between the two populations. The observed changes reported here resulted in a subsequent increase in rice dedomestication and a greater number of weedy, herbicide-resistant hybrid progeny. Overall, these data suggest that differential phenological responses to rising atmospheric CO_2 could result in enhanced flow of novel genes and greater success of feral plant species in agroecosystems.

Editor: Dorian Q. Fuller, University College London, United Kingdom

Funding: Research was performed as an employee of the U.S. Government (USDA-ARS) and no financial conflict of interest exists. There are no current external funding sources for this study.

Competing Interests: The authors have declared that no competing interests exist.

* E-mail: l.ziska@ars.usda.gov

Introduction

Gene transfer between organisms is acknowledged as a significant influence on evolutionary change [1]. Consequently, a scientific understanding of biotic and abiotic factors that promote gene flow is of considerable interest. Some biotic conditions are already well-recognized. For example, in plant populations cross pollination must occur with synchronicity in flowering, spatial proximity and genetic compatibility [2]. Naturally occurring abiotic or physical phenomenon are also understood; for example, the role of wind in pollen transfer for anemophilous plants, such as ragweed.

However, in addition to these natural drivers, there is increasing recognition that gene flow can be altered by anthropogenic activities [3,4]. One prominent feature of anthropogenic disturbance that has been well quantified globally is the increase in atmospheric carbon dioxide, CO_2. Since 1959, concentrations of atmospheric CO_2 have increased from 318 to 392 μmol mol^{-1}; and, depending on anthropogenic emission rates, may exceed 1000 μmol mol^{-1} by the end of the century [5].

Carbon dioxide is highly relevant to biology as it represents the sole source of carbon available in plant metabolism. Increases in CO_2 concentration *per se* have been shown in numerous studies to increase carbon fixation with subsequent effects on plant growth, development, morphology and reproduction [6]. It is also evident that inter and intraspecific variation to rising CO_2 occurs [7,8], and such variation could lead to differences in reproductive phenology and gene flow. These latter differences have never been quantified. Their occurrence, however, given the inclusive nature of the CO_2 increase, has extensive biological consequences since it suggests a specific human-disturbance capable of altering gene transfer for plant communities globally.

One plant community where gene flow is of explicit interest is the agro-ecosystem. The introduction of novel genes in domesticated plant taxa has led to concerns that gene flow between genetically transformed and uncultivated relatives could alter the fecundity and management of feral plant species [9] with subsequent impacts on their evolutionary fitness [2,10,11]. For agro-ecosystems, increased gene exchange either from the crop to the related wild/weedy species–or vice-versa–is consequential, since the end result will be an introduction of non desired traits (e.g. herbicide resistance) and the formation of an unwanted biotype of the cultivated crop.

Historically, wild or weedy rice (henceforth referred to as wild rice) has been controlled through utilization of a transplant flooded culture system. With modern advances in mechanization and selective herbicides, and declining water availability globally, direct seeding systems have been increasingly emphasized. As this trend has continued, wild rice has emerged as a pervasive and serious impediment to the productivity of cultivated rice globally [12,13]. Often called red rice due to the coloration of its pericarp, wild rice is the same species as the cultivated crop (*Oryza sativa*; not to be confused with *Oryza rufipogon* or *Zizania latifolia* Griseb.). It can be very difficult to control, because seasonal shattering allows seed bank retention, while undispersed weed seeds may be inadvertently collected and replanted in subsequent crops.

Wild and cultivated rice are primarily self-pollinated [14]. However, each type can outcross at significant levels depending on the environment and their genetic background [15,16]. Overall, rice is recognized globally as one of the five most-widely grown crops prone to gene flow [17].

An emerging strategy to manage wild rice is to apply herbicides in conjunction with the implementation of herbicide-resistant cultivated rice. One such herbicide resistant rice cultivar is "Clearfield™ 161" (CL 161). Derived from chemical mutagenesis (non-transgenic), this and related cultivars are resistant to acetolactate synthase-inhibiting imidazolinone herbicides such as imazethapyr [14], and are now widely grown in the southern U.S. where wild rice infestation occurs. However, if potential transfer of herbicide resistance between wild and cultivated rice populations were to occur in response to rising CO_2, it would accelerate the introduction of herbicide resistant hybrids into rice systems with subsequent effects on evolutionary fitness, diversity, spread and management of wild rice.

For this study we apply the hypothesis that recent 20th century CO_2 increases (300 to 400 μmol mol^{-1}), and those projected for the 21 century (up to 600 μmol mol^{-1}), could differentially affect the spatial and temporal development of reproductive traits in wild and cultivated rice with modifications in phenological development, outcrossing rates and gene transfer between populations. The data presented here confirm this hypothesis and indicate that rising CO_2 may be an additional factor in the spread of novel genes between wild and cultivated rice populations.

Results

Specific developmental patterns of interest that would be associated mechanistically with directional gene flow from reproductive structures would include changes in height at anthesis, changes in tiller and panicle formation, and synchronicity of flowering times between populations. An examination of these parameters included a test of both populations at planting frequencies associated with cultivation; i.e. approximately 2 wild rice plants per m^2 of cultivated rice, utilizing environmental chambers at constant macro-environmental conditions in order to maintain consistent CO_2 values between early 20th century and mid-21st century levels. It is clear that as CO_2 increased, plant height, tiller production and panicle number all increased to a greater extent in the wild vs. the cultivated populations (Figure 1). Plant height is thought to be an important determinant of pollen outcrossing and gene flow, given the limited range of pollen release in rice [14]; here CO_2-induced height differences between the cultivated and wild populations increased from 8 to 26 cm (Figure 1). Tiller and panicle numbers, which provide the structural basis for floral initiation, also rose to a greater extent for the wild relative to the cultivated populations doubling in response to the CO_2 increase from 300 to 600 μmol mol^{-1} (Figure 1).

In rice, gene transfer escalates when flowering times (anthesis) are in sync between wild and cultivated populations. Interestingly, anthesis for the cultivated population was not affected, whereas in contrast, flowering times were initiated earlier (~8 days) in response to rising CO_2 for the wild population (Figure 2).

Field evaluation of seed collected from these populations indicated that outcrossing from the cultivated to the wild population was low, and did not change as a function of CO_2. In contrast, outcrosses per plot, and the percent outcrossing significantly increased from the wild population relative to the cultivated line as a function of CO_2 concentration (Figure 3). This disproportionate increase in outcrossing and gene transfer from the wild population was associated with the morphological and phenological changes arising from the increase in atmospheric CO_2, namely, increased height, greater floral (panicle) production and enhanced temporal initiation of anthesis.

Discussion

Biotic or abiotic factors that differentially impact outcrossing and gene flow between species will influence population size, genetic diversity, metapopulation dynamics and, potentially, the rate and trajectory of evolution [18].

There are compelling arguments that human activity, from habitat destruction to global warming, is beginning to change reproductive biology and outcrossing in plant communities (3). To date, much of this work is focused on how anthropogenic disturbances are likely to change species range, demography [19], pollinator populations [20], and plant phenology [21], with little empirical evidence regarding outcrossing rates and gene transfer.

Among anthropogenic factors, CO_2 has a fundamental influence on plant metabolism, independent of any effect on climate. For example, with respect to reproduction, increasing CO_2 has been shown to alter flowering times [22], increase pollen amounts [23], increase flower numbers and size [24] and enhance nectar production [25]. Differential reproductive responses between and within plant species have also been widely documented in response to atmospheric CO_2 increases above pre-industrial concentrations [26]. However, a clear association between differential species responses to CO_2 and quantification of gene flow has been unavailable.

Here the disparate responses to recent and projected CO_2 between related rice lines resulted in a directional increase in gene flow from feral to cultivated populations. Such gene shifts have been noted previously in rice systems [14], emphasizing that gene flow can occur in both directions. This may seem counter-intuitive given the relative population ratios of cultivated to wild rice (i.e., ca 7:1); however increases in height, tiller and panicle production, and synchronicity of flowering observed here for wild rice are consistent with facilitating gene flow from wild to cultivated rice plants [27].

We emphasize that for agro-ecosystems, increased gene flow in either direction is of consequence since the trait of interest (whether herbicide resistance or any other gene introduced via mutation or genetic engineering) will now be associated with an undesirable biotype of the preferred crop. That is, Clearfield plants that receive wild rice genes will produce hybrid seed whose phenotypic characteristics, including herbicide resistance, shattering, nutritional composition, etc., are undesirable relative to the quantitative and qualitative traits of the cultivated rice [27].

Therefore, the CO_2 induced changes in gene flow observed here enhanced the creation, occurrence and perpetuation of 'weedy', herbicide-resistant hybrid progeny. Such progeny, in turn, will have negative consequences with respect to rice yields (e.g., enhanced competition), quality (e.g. nutrition) and weed management (e.g. herbicide resistance).

The results reported here for the differential response to rising CO_2 are consistent with previous results for multiple red rice

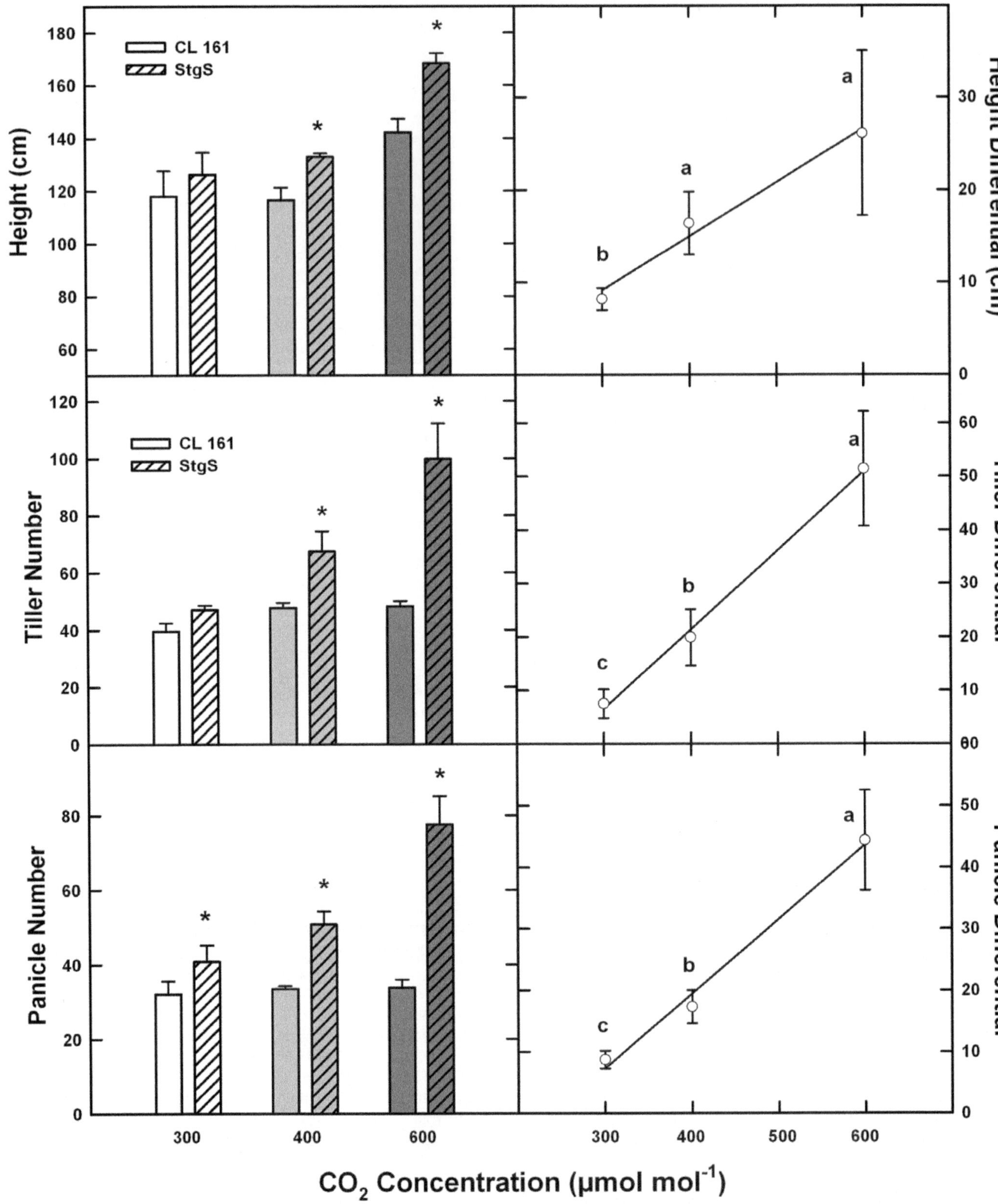

Figure 1. Relative change in morphological parameters (per plant) that influence reproductive onset and outcrossing as a function of recent and projected increases in CO_2 concentration (μmol mol^{-1}) for a wild and cultivated rice population. * indicates a significant difference at the $P<0.05$ level as a function of a given CO_2 concentration between populations; different letters indicate a significant difference in the degree of CO_2 enhancement for the measured parameter (e.g. height). Bars are ±SE.

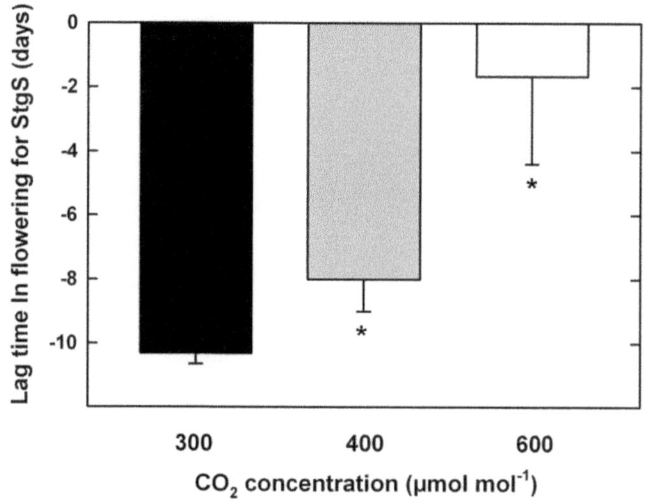

Figure 2. Lag time in days following flowering in the cultivated rice population (CL 161) that wild rice (StgS) flowered as a function of CO_2 concentration. * indicates a significant difference relative to the 300 µmol mol^{-1} baseline; Bars are±SE.

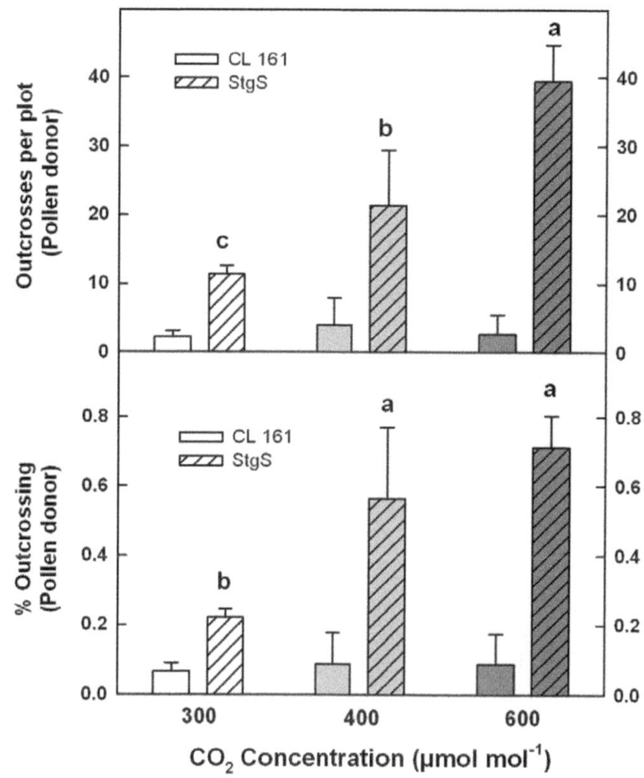

Figure 3. Percent outcrossing and outcrosses per plot for the pollen donor as detected in cultivated and wild rice populations in the field plots. Outcrosses were always significantly higher for the StgS relative to the CL 161 population regardless of CO_2 concentrations (not shown). Different letters indicate a significant population by CO_2 concentration interaction for Ithe StgS relative to the CL 161 population. Bars are ±SE

biotypes vs. cultivated lines when grown at intervals of 100 µmol mol^{-1} [28] above the pre-industrial CO_2 baseline. It has been suggested that the greater degree of genetic and phenotypic plasticity among wild or weedy rice biotypes may allow for greater exploitation of abiotic resources (e.g., CO_2) as compared to commercial rice cultivars that have been developed to be phenotypically uniform, early in maturity, and short in stature [28,29].

Because of selection pressures associated with weed management, many of the plant species present in agroecosystems are genetically related to the desired crop species [30]; consequently it is recognized that outcrossing can, and does occur (e.g., oat and wild oat, sorghum and shattercane, nightshade/bittersweet and potato, cultivated and wild sunflower *inter alia*) [31,32,33,34]. Based on the results shown here for rice, studies on outcrossing and gene flow should also be conducted to ascertain whether the ongoing increase in atmospheric CO_2 has, or will, increase the spread of novel genes via hybridization for these other agroecosystems. The greater introduction and spread of such genes (e.g. herbicide resistance) will have significant effects on pest management and food security.

Materials and Methods

Controlled environment chambers (EGC Corp., Chagrin Falls, OH) were used because, at present, no methodology is available that can expose plants to sub-ambient carbon dioxide concentrations [CO_2], for 24 h day^{-1} [35]. Temperature for each chamber was varied in a diurnal mode from an overnight low of 22°C, to a maximum afternoon value of 32°C, with an average daily (24 h) value of 24.4°C. Light as photosynthetically active radiation (PAR) varied diurnally in conjunction with temperature, with the highest PAR (~900 µmol m^2 s^{-1}) occurring during the afternoon. The daily light period was 14 h, supplied by a mixture of high-pressure sodium and metal halide lamps, and averaged 20.9 mol m^{-2} day^{-1} for all chambers. Injection of either CO_2 or CO_2-free air was determined using a TC-2 controller that monitored [CO_2] measured from an absolute infrared gas analyzer (WMA-2, PP Systems, Haverhill, MA, USA). Carbon dioxide concentrations were set at 300, 400 and 600 µmol mol^{-1} 24 h day^{-1}. These concentrations approximated the carbon dioxide values present during the end of the 19th century, the current value, and that projected by the end of the 21st century [5]. Mean 24 h day^{-1} [CO_2] values (±SD) throughout the experiment were 312±17.3, 413±18.6, and 624±31.5 µmol mol^{-1}. Temperature, humidity and [CO_2] were recorded every 15 min and averages determined on a daily basis for all experimental runs.

An awnless, strawhull red rice type from Arkansas, StgS (PI653423; henceforth referred to as StgS wild rice), typical of the most widespread biotypes in Arkansas/southern U.S.A. [36], was selected as the outcrossing partner for cultivated rice. Bulk seed were obtained from a collection maintained by USDA-ARS. CL 161 rice, which contains resistance to imazethapyr herbicide (HR) was selected as the outcrossing partner for wild rice. Seed are available from a commercial source (Horizon-Ag, Memphis, TN). Seed of these *O. sativa* biotypes were planted at commercial densities: CL 161 rice at ~25 plants per m^2 and StgS at 2 plants m^2 in each of three chambers, with each chamber programmed to one of the CO_2 treatment concentration. All plants were watered to the drip point daily with a complete nutrient solution. Non destructive observations of phenology, including emergence, tillering, height, heading and flowering were determined for each biotype and CO_2 treatment. Plants were grown until maturity (final harvest) which was determined when >50% of the panicles for both biotypes had senesced. Because seed from wild rice shatters, three panicle subsamples from individual StgS plants for each experimental treatment were wrapped in cheesecloth, and an allometric relationship between seed and panicle weight was used to estimate seed yield. At maturity individual plants were cut at soil level, tillers and panicles counted before separation into tillers, panicles, seed and leaf laminae. Because of the difficulty in separating roots of cultivated and wild rice, no attempt was made to determine root biomass. Seed yield was determined for CL 161 rice by gently shaking panicles following cutting. All plant material was dried at 65°C or until dry weight was constant. Seeds were separated by biotype, air dried and corrected to 10% moisture.

At the end of a given experiment, the CO_2 treatments were randomly reassigned, and the entire experiment repeated with runs over time considered as replications (blocks). This occurred over a two year period from 2009 through 2010, with three runs for the 300 and 600 µmol mol^{-1} and two runs for the 400 µmol mol^{-1} CO_2 concentrations. PAR, humidity and temperature were quantified prior to, and at the end of each run in order to determine within chamber and among chamber variability. Values for temperature, PAR and humidity were consistent among chambers and runs. All measured and calculated growth and reproductive parameters were analyzed using analyses of variance including CO_2 concentration, run and biotype (StgS vs. CL 161) as fixed effects (Statview software, Cary, NC, USA). Seed from each run and biotype was collected and shipped to the USDA-ARS Dale Bumpers Rice National Rice Research Center in Stuttgart, Arkansas.

Before planting, the weights of two groups of 100 randomly selected seeds of both rice and StgS from each growth chamber harvest were averaged so that a constant number of seeds could be weighed out and planted in each screening plot. Seeds harvested from the growth chambers in August 2009 to February 2010, and in December 2010 to May 2011 were planted in the field on May 5, 2010, and May 9, 2011, respectively, at the University of Arkansas Rice Research and Extension Center located near Stuttgart, AR (34.49° N, 91.55° W). Soil was a DeWitt silt loam (fine smectitic, thermic, Typic Albaqualfs) with 1.2% organic matter and pH 5.8. Cultivated and wild rice were drill-seeded at a depth of approximately 2 cm into separate plots 12.2-m-long with nine rows spaced 18 cm apart. Seeding density was 370 seeds m^{-2}. This was equivalent to 7200 seeds per 12.2 m×1.62 m plot. Plots were firmed with a roller to facilitate seed-soil contact and assist seed germination.

Herbicides were applied to plots to control barnyardgrass and other weeds as necessary. Clomazone + quinclorac (0.26 kg active ingredient (ai) ha^{-1}+0.21 kg ai ha^{-1}) were applied after planting (pre-emergence to late pre-emergence). In order to ensure complete kill of all non herbicide-resistant wild rice plants (i.e. plants that did not contain the HR gene), imazethapyr + CS-7 surfactant (0.07 kg ai ha^{-1}+0.25% v/v; recommended rate), was applied 15 to 18 days after crop emergence when rice was at the 3 to 4 leaf stage and 15 cm tall, and again at 1 and 2 weeks after the initial application. Surviving plants in plots planted with StgS rice seeds were considered to be putative F_1 hybrids that had acquired the HR gene from CL 161 pollen [15,16] as a result of natural outcrossing in the growth chambers. In plots planted with CL 161 seeds, tall, late-flowering plants with pubescent leaves were considered to be putative F_1 hybrids [27] that had acquired StgS genes from natural outcrossing in the growth chambers. Urea at 34 kg nitrogen ha^{-1} was broadcast over all plots before establishment of the permanent flood on June 18, 2010, and June 27, 2011.

Healthy, green leaf tissue sampled from putative F_1 hybrid plants in field plots were collected in September of each year and kept frozen until DNA extraction. DNA was extracted and polymerase chain reaction (PCR) was performed following procedures described previously [37,38]. Samples were separated on an ABI Prism 3730 DNA Analyzer (Applied Biosystems, Foster City, CA) and the sizes of SSR fragments were determined and alleles binned using GeneMapper version 3.7 software (Applied Biosystems, Foster City, CA).

Six genetic markers (Rid12, RM5, RM232, RM234, and RM253, and RM 277) previously shown to differentiate between CL 161 rice and StgS or closely related wild rice types [36,39] were selected to determine the genotypes of putative hybrid plants. True F_1 hybrids between CL 161 and StgS were confirmed as being heterozygous for the six selected loci. In true hybrid plants, the expected allele size pairings (rice/wild rice) for genetic markers were: Rid 12 (151 bp/165 bp), RM232 (156 or 154 bp/148 bp), RM234 (135 bp/158 or 154 bp), RM253 (131 bp/139 or 141 bp), and RM 277 (114 bp/118 bp). Outcrossing rates were calculated as: 100×(number of true F_1 hybrids in screening plot)/(number of seedlings in screening plot). Data on outcrossing rates and gene flow were tabulated for each year, biotype and CO_2 treatment for all field plots. Data were analysed using analyses of variance with chamber run, CO_2 concentration, year, plot and biotype (StgS vs. CL 161) as fixed effects (Statview software, Cary, NC, USA).

Acknowledgments

We thank Ms. Maria Valerio and Mr. Robert Erdman for expert technical assistance.

Author Contributions

Conceived and designed the experiments: LHZ DRG. Performed the experiments: LHZ DRG MBT AKJ HLB. Analyzed the data: LHZ DRG MBT AKJ HLB. Contributed reagents/materials/analysis tools: LHZ DRG MBT AKJ HLB. Wrote the paper: LHZ DRG.

References

1. Slatkin M (1987) Gene flow and the geographic structure of natural populations. Science 236: 787–792.
2. Ellstrand NC, Prentice HC, Hancock JF (1999) Gene flow and introgression from domesticated plants into their wild relatives. Ann Rev. Ecol. Syst. 30: 539–563.
3. Eckert CG, Kalisz S, Geber MA, Sargent MA, Elle E, et al. (2009) Plant mating systems in a changing world. Trends Ecol Evol 25: 35–44.
4. Aguilar R, Ashworth L, Galleto L, Adrian Aizen M (2006) Plant reproductive susceptibility to habitat fragmentation: review and synthesis through a meta-analysis. Ecol Lett 9: 968–980.
5. Intergovernmental Panel on Climate Change (2007) Summary for policymakers. In: Climate Change 2007: The Physical Science Basis. Contribution of Working Group I to the Fourth Assessment Report of the Intergovernmental Panel on Climate Change (Solomon, S., et al., eds), Cambridge University Press.
6. Bazzaz FA (1990) The response of natural ecosystems to the rising global CO_2 levels. Ann Rev Ecol Syst 21: 167–196.
7. Curtis PS, Snow AA, Miller AS (1994) Genotype-specific effects of elevated CO_2 on fecundity in wild radish (*Raphanus raphanistrum*). Oecol 97: 100–105.
8. Kimball BA (1993) Effects of increasing atmospheric CO_2 on vegetation. Vegetatio 104/ 105: 65–83.
9. Snow AA (2002) Transgenic crops-why gene flow matters. Nature Biotech. 20: 542.
10. Ellstrand NC (2003) Dangerous Liasons? When Cultivated Plants Mate with Their Wild Relatives. Johns Hopkins University Press, 264 pages.
11. Lu B, Snow AA (2005) Gene flow from genetically modified rice and its environmental consequences. Biosci 55: 669–678.
12. Delouche JC, Labrada R (2007) Weedy rices – origin, biology, ecology and control. FAO Plant Production and Protection Paper # 188, Rome, Italy, 141 pages.
13. Estorninos LH Jr., Gealy DR, Talbert RE, Gbur EE (2005) Rice and red rice interference. I. Response of red rice (*Oryza sativa*) to sowing rates of tropical *japonica* and *indica* rice cultivars. Weed Sci 53: 676–682.

14. Gealy DR, Mitten DH, Rutger JN (2003) Gene flow between red rice (*Oryza sativa*) and Herbicide resistant rice (*O. sativa*): Implications for weed management. Weed Technol 17: 627–645.
15. Shivrain VK, Burgos N, Anders M, Raiguru S, Moore J, et al. (2007) Gene flow between Clearfield™ rice and red rice. Crop Prot. 26: 349–356.
16. Shivrain VK, Burgos NR, Sales MA, Mauromoustakos A, Gealy DR, et al. (2009) Factors affecting the outcrossing rate between Clearfield™ rice and red rice (*Oryza sativa*). Weed Sci 57: 394–403.
17. Council for Agricultural Science and Technology (CAST) (2007) Implications of Gene Flow in the Scale-up and Commercial Use of Biotechnology-derived Crops: Economic and Policy Considerations. Issue Paper 37. CAST, Ames, Iowa. 132 p.
18. Takebayashi N, Morrell P (2001) Is self-fertilization a dead end? Revisiting an old hypothesis with genetic theories and a macroevolutionary approach. Am J Bot 88: 1143–1150.
19. McDonald AJ, Riha S, Ditomasso A, DeGaetano A (2009) Climate change and the geography of weed damage: Analysis of US maize systems suggests the potential for significant range transformations. Ag Ecosys Environ 130: 131–140.
20. Hegland SJ, Nielsen A, Lazaro A, Bjerknes A-L, Totland O (2009) How does global warming affect plant-pollinator interactions? Ecol Lett 12: 184–195.
21. Cleland EE, Chuine I, Meuzel A, Mooney HA, Schwartz MD (2007) Shifting plant phenology in response to global change. Trends Ecol. Evol 22: 357–365.
22. Johnson SL, Lincoln DE (2000) Allocation responses to CO_2 enrichment and defoliation by a native annual plant *Heterotheca subaxillaris*. Global Change Biol 6: 767–778.
23. Wayne PM, Foster S, Connolly J, Bazzaz F, Epstein P (2002) Production of allergenic pollen by ragweed (*Ambrosia artemisiifolia* L.) is increased in CO_2-enriched atmospheres. Ann Allergy Asthma Immunol 80: 669–679.
24. Lake JC, Hughes L (1999) Nectar production and floral characteristics of *Tropaeolum majus* L. grown in ambient and elevated carbon dioxide. Ann of Bot 84: 535–541.
25. Dag A, Eisikowitch D (2000) The effect of carbon dioxide enrichment on nectar production in melons under greenhouse conditions. J Apicultural Res 39: 88–89.
26. Springer CJ, Ward JK (2007) Flowering time and elevated atmospheric CO2. New Phytol 176: 243–255.
27. Gealy DR, Yan W, Rutger JN (2006) Red rice (*Oryza sativa*) plant types affect growth, coloration and flowering characteristics of first and second generation crosses with rice. Weed Technol 20: 839–852.
28. Ziska LH, McClung (2008) Differential response of cultivated and weedy (red) rice to recent and projected increases in atmospheric carbon dioxide. Agron J. 100: 1259–1263.
29. Treharne K (1989) The implications of the 'greenhouse effect' for fertilizers and agrochemicals. p 67–78. *In*: R.C. Bennet (ed.) Ministry of Agriculture, Fisheries and Food, London, UK.
30. Andersson MS, de Vicente MC (2010) Gene Flow Between Crops and Their Wild Relatives. Johns Hopkins University Press, Baltimore, MD 564 pages.
31. Murray BG, Morrison IN, Friesen LF (2002) Pollen mediated gene flow in wild oat. Weed Sci 50: 321–325.
32. Paterson A, Schertz K, Lin Y, Liu S, Chang Y (1995) The weediness of wild plants: molecular analysis of genes influencing dispersal and persistence of johnsongrass, *Sorghum halepense* (L.) Pers. Proc. Natl. Acad. Sci. USA 92: 6127–31.
33. Eijlander R, Stiekema WJ (1994) Biological containment of potato (*Solanum tuberosum*): outcrossing to the related wild species black nightshade (*Solanum nigrum*) and bittersweet (*Solanum dulcamara*). Sex Plant Reprod. 7: 29–40.
34. Arias DM, Rieseberg LH (1994) Gene flow between cultivated and wild sunflowers. Theor Appl Genet 89: 655–660.
35. Mayeux HS, Johnson HB, Polley HW, Durnesnil MJ, Spanel GA (2003) A controlled environment chamber for growing plants across a sub-ambient CO_2 gradient. Funct. Ecol. 7: 125–133.
36. Gealy DR, Agrama H, Jia MH (2012) Genetic analysis of atypical U.S red rice phenotypes: Indications of prior gene flow in rice fields? Weed Sci (In Press, DOI: 10.1614/WS-D-11-00159.1).
37. Xin, Z, Velten JP, Oliver MJ, Burke JJ (2003) High-throughput DNA extraction method suitable for PCR. BioTechniques 34: 820–826.
38. Bryant R, Proctor A, Hawkridge M, Jackson A, Yeater K, et al. (2012) Genetic variation and association mapping of silica concentration in rice hulls using a germplasm collection. Genetica (In Press, DOI: 10.1007/s10709-012-9637-x).
39. Gealy DR, Agrama HA, Eizenga GC (2009) Exploring genetic and spatial structure of U.S. weedy red rice (*Oryza sativa*) in relation to rice relatives worldwide. Weed Sci 57: 627–643.

15

Diversity and Relationships of Eggplants from Three Geographically Distant Secondary Centers of Diversity

Maria Hurtado[1], Santiago Vilanova[1], Mariola Plazas[1], Pietro Gramazio[1], H. Hemal Fonseka[2], Ramya Fonseka[3], Jaime Prohens[1]*

1 Instituto de Conservación y Mejora de la Agrodiversidad Valenciana, Universitat Politècnica de València, Valencia, Spain, **2** Horticultural Crop Research and Development Institute, Gannoruwa, Sri Lanka, **3** Department of Crop Science, Faculty of Agriculture, University of Peradeniya, Peradeniya, Sri Lanka

Abstract

Eggplant (*Solanum melongena* L.) was domesticated in the Indo-Birmanian region, which is also the primary center of diversity for this crop. From there eggplant spread to other regions, and diversity accumulated in several secondary centers of diversity. We have assessed the diversity and relationships of 52 accessions of eggplant from three geographically distant secondary centers of diversity (China, Spain, and Sri Lanka) using 28 morphological descriptors and 12 highly polymorphic genomic SSRs. A wide variation was found for most morphological traits, and significant differences among the three centers of diversity were detected for 22 of these traits. The PCA analysis showed that eggplants from the three origins were morphologically differentiated, and accessions from each of the three secondary centers of diversity presented a typical combination of morphological characteristics. In this respect, discriminant analysis showed that accessions could be correctly classified to their origin using only six traits. The SSR characterization identified 110 alleles and allowed obtaining a unique genetic fingerprint for each accession. Many alleles were found to be private to each origin, but no universal alleles were found for any of the origins. The PCA analysis showed that the genetic differentiation among origins was less clear than for morphological traits, although the analysis of the population structure shows that accessions mostly group according to the origin, but also provides evidence of migration among the three secondary centers of diversity. The genetic diversity (H_T) within each origin was high, ranging between $H_T = 0.5400$ (Sri Lanka) and $H_T = 0.4943$ (China), while the standardized genetic differentiation (G'_{ST}) among origins was moderate ($G'_{ST} = 0.2657$). The correlation between morphological and SSR distances was non-significant ($r = 0.044$), indicating that both data are complementary for the conservation of germplasm and breeding of eggplant. These results are relevant for the management of genetic resources, breeding programmes, and evolutionary studies of eggplant.

Editor: Randall P. Niedz, United States Department of Agriculture, United States of America

Funding: This work was financed by the Ministerio de Ciencia e innovación (grants AGL2009-07257 and RF-2008-00008-00-00) (http://www.micinn.es). The funders had no role in study design, data collection and analysis, decision to publish, or preparation of the manuscript.

Competing Interests: The authors have declared that no competing interests exist.

* E-mail: jprohens@btc.upv.es

Introduction

Eggplant (*Solanum melongena* L.) is the most important *Solanum* crop native to the Old World, and it ranks as one of the most important vegetable crops in the world with an annual worldwide production of more than $41 \cdot 10^6$ t (2010; FAO data). This results in an average production of 6.1 kg/person/year or 16.7 g/person/day. Eggplant is one of the vegetables with highest antioxidant capacity [1], which is a consequence of its high content in phenolics [2]. Despite the economic and nutritional importance of eggplant, breeding efforts in this crop have been limited [3]. The use of exotic germplasm in breeding programmes can be of great relevance for the improvement of the crop and for addressing future breeding challenges. In this respect, Muñoz-Falcón et al. [4] found that the genetic diversity of modern cultivars of black eggplants was reduced, and that incorporation of black fruit materials from different origins could increase the genetic base of this cultivar type and contribute to better exploitation of the heterosis resulting from the crosses of genetically distant materials [5].

Assessment of the diversity and relationships of the cultivated species facilitates the establishment of conservation strategies, the use of genetic resources in breeding programmes, and the study of the crop evolution. The domestication and evolution of eggplant has been the subject of a number of studies in which historical [6,7], morphological [8], and molecular [9,10] data have been used. There is general consensus that eggplant was domesticated in South-East Asia from the wild relative *Solanum incanum* L, and morphological [8], and molecular [9–14] data, as well as the high fertility of F1 hybrids with *S. melongena* [3,8] support this hypothesis. It is unknown how *S. incanum*, which is naturally distributed in Africa and the Middle East, reached the Indo-Birmanian region, although it has been speculated that it could have arrived there unintentionally as a weed or dragged by ocean currents from Africa to India, or intentionally because of the value of its berries for tanning hides [8,15]. In any case, as occurred with tomato [16], the domestication of eggplant outside of the area of natural distribution of its wild ancestor resulted in an important genetic bottleneck [3,8].

Diffusion of the crop from its primary center of diversity, situated in the Indo-Birmanian region [17,18], where primitive cultivars and weedy forms exist [3,8,10], to other areas resulted in the diversification of the crop due to micro-evolutive forces like mutation, selection (natural and artificial), genetic drift, or gene flow, and to recombination, and led to the accumulation of diversity in several secondary centers of diversity [10,19]. For example, the cultivation of eggplant in China has been documented since more than 2000 years ago [7]; its introduction into Europe, through Spain, was brought about by the Arabs before the 10th century [19]. The evolution of the crop in geographically distant secondary centers of diversity led to the differentiation of cultivar types specific to different regions of the world [3]. In this respect, two large groups of eggplant cultivars are considered by breeders: "Occidental" or "Western" eggplants (Middle East, Africa, Europe, and America) and "Oriental" or "Asian" eggplants (Eastern and Southeastern Asia) [3,6,10]. Within each of these large groups, several cultivar types are distinguished [3].

Although a number of studies have been devoted to the study of the diversity of eggplant materials from specific countries or regions, like India [18], Spain [19], China [20,21], Turkey [22], or Asia [10,23], up to now no comprehensive assessment has been performed on the comparative diversity and regional differentiation of eggplant materials from different regions of the world. The study of the diversity of geographically distant centers of diversity, where eggplant was introduced through different routes, can provide information of interest for understanding the structure of variation in eggplant, as well as for the conservation of genetic resources and breeding of this crop. Similar studies have been undertaken in other crops, such as lentil (*Lens culinaris* Med.) [24], coffee (*Coffea arabica* L.) [25], maize (*Zea mays* L.) [26] and sorghum (*Sorghum bicolor* (L.)) [27].

The characterization of eggplant with both morphological descriptors and molecular markers has proved useful for the study of the diversity and relationships of different varietal groups of eggplant, as they sample different levels of diversity [4,19,28,29]. Availability of characterization data for traits of agronomic interest is essential for breeding programs. Morphological descriptors for the characterization of eggplant are available as a result of the European Eggplant Genetic Resources Network (EGGNET) [30]. These descriptors have been used and validated in a number of characterizations of eggplant genetic resources and breeding materials [4,19,28,29,31]. Also, molecular markers are useful to study the genetic diversity of eggplant [4,10–12,14,20–23,28,29,32–35]. SSRs have revealed as the best presently available markers to study the relationships of different groups of cultivars of eggplant, as they have wide genome coverage, are highly variable, have a highly repeatability, are easy to use, and are amenable to high throughput [36,37]. For example, Muñoz-Falcón et al. [28,29,38] have found that SSRs are much better than AFLPs at resolving the relationships between and within cultivar groups as well as for assigning correctly accessions to their cultivar groups. Also, Demir et al. [22], when comparing SSRs with RAPDs, found that the former were more adequate than the latter to study the diversity and relationships of local landraces. Several hundred SSR markers are available in eggplant [14,23,32–35,39], and among them genomic SSRs have proved to be more polymorphic than EST-SSRs [38].

Here we assess the morphological and the molecular (SSR) diversity of a collection of eggplants from three geographically distant secondary centers of diversity: China, Spain, and Sri Lanka. Our objective is to obtain information on the diversity, relationships, and differentiation among accessions of these three different origins. These results will be of interest for the conservation of genetic resources, breeding, and study of the evolution of the cultivated eggplant.

Results

Morphological Characterization

The collection of eggplants studied, as well as each of the three groups of accessions originating from China, Spain, or Sri Lanka, displayed a wide variation for most of the morphological traits studied (Table 1). Although for the 28 morphological traits considered there is overlap in the range of variation of individual accessions among the three groups, significant differences among means of the three origins are found for 22 of the morphological traits considered (Table 1). In this respect, the number of significant differences among origin means for the traits studied has been of 12 for China vs. Spain accessions, 20 for China vs. Sri Lanka accessions, and 15 for Spain vs. Sri Lanka accessions.

Chinese accessions are less vigorous than those of Spain or Sri Lanka, so that the Chinese accessions have lower plant height (P-Height) and smaller leaves (L-Pedicel, L-Length, L-Breadth) than either the Spanish or Sri Lankan eggplants (Table 1). Regarding anthocyanic pigmentation of the vegetative parts of the plant (P-Anthocyans, L-Anthocyans, Fr-CAnthocyans), the highest pigmentation is found in the Chinese eggplants, the lowest in the Sri Lankan ones, while in the Spanish accessions it is intermediate. While the plant prickliness (P-Prickles) has been in general low, Sri Lankan accessions have presented, as a mean, a lower prickliness (P-Prickles) than either the Chinese or Spanish eggplants. Leaves of Sri Lankan accessions had a smaller leaf blade breadth/width ratio (L-Ratio) than the Chinese of Spanish eggplants, and the Chinese accessions had a lower leaf blade lobing (L-Lobing) than the Spanish or Sri Lankan eggplants (Table 1).

For flower traits, Chinese eggplants had, in general, a smaller number of flowers per inflorescence (Fl-Number) than either the Spanish or Sri Lankan eggplants (Table 1). Spanish eggplants had more petals per flower (Fl-Petals) than that of the Chinese or Sri Lankan accessions.

The fruits of Sri Lankan accessions, in general, have higher fruit length (Fr-Length) and lower fruit breadth (Fr-Breadth) and, in consequence, are more elongated (Fr-Ratio) than those of China or Spain (Table 1). Spanish eggplants have a larger part of the fruit covered by the calyx (Fr-CLength) than those of China or Sri Lanka, while the intensity of the fruit skin covered by the calyx (Fr-UnderC) is higher in Sri Lankan accessions, followed by the Spanish ones, and finally by the Chinese accessions,. Contrary to what occurred in a number of Chinese and Spanish accessions, the Sri Lankan eggplants studied did not present chlorophylls in the fruit skin (Fr-Chlorophyll) and therefore presented lower mean values than Chinese or Spanish accessions for this trait (Table 1). Regarding the skin colour, Sri Lankan materials had, on average higher L* (Fr-L*) values (indicating closer to white) than Chinese or Spanish accessions. Fruits of Sri Lankan accessions were, on average, more glossy (Fr-Gloss) than those of the Spanish accessions, which, in turn were also more glossy than the Chinese accessions. Finally, Sri Lankan accessions had a significantly higher flesh browning (Fr-Browning) than Chinese accessions (Table 1).

According to the scree plot method, a total of five principal components were found to be relevant in the morphological PCA study. These five principal components account for 63.4% of the total variation. However, given that the first and second components account, respectively, for 22.9% and 16.2% of the total variation (Figure 1), and that in the graphical analyses no

Table 1. Mean and range for the traits measured in the eggplant accessions originating from China, Spain, and Sri Lanka used for the present study, and probability of the *F*-statistic, obtained from ANOVA analyses, for differences among means.

Trait	Code	China		Spain		Sri Lanka		Prob. *F*
		Mean[a]	Range	Mean[a]	Range	Mean[a]	Range	
No. of accessions		20		14		18		
Plant growth habit	P-Habit	2.1a	1.0–4.2	1.6a	0.7–5.0	2.1a	1.0–5.0	0.394
Plant height (cm)	P-Height	81.2a	48.3–108.2	105.4b	45.0–149.5	95.2b	61.2–120.3	0.006
Shoot tips anthocyanins intensity	P-Anthocyans	6.2b	1.0–9.0	5.1ab	3.0–9.0	4.0a	1.0–7.0	0.010
Stem prickles	P-Prickles	2.3b	0.0–3.0	2.6b	0.0–3.0	0.7a	0.0–3.0	<0.001
Leaf pedicel length (cm)	L-Pedicel	4.9a	2.9–6.2	7.2c	5.1–9.7	5.6b	4.3–6.7	<0.001
Leaf blade length (cm)	L-Length	15.9a	11.6–20.2	20.7c	17.6–28.2	19.2b	17.3–21.9	<0.001
Leaf blade breadth (cm)	L-Breadth	9.4a	5.8–13.3	12.2b	9.4–15.1	12.3b	9.8–14.6	<0.001
Leaf blade length/breadth ratio	L-Ratio	1.71b	1.43–2.01	1.70b	1.53–1.88	1.58a	1.35–1.86	0.005
Leaf blade lobing	L-Lobing	3.5a	1.0–7.0	4.9b	3.0–7.0	4.8b	3.0–7.0	0.009
Leaf anthocyanins intensity	L-Anthocyans	6.9c	0.0–9.0	4.6b	0.0–9.0	2.4a	0.0–9.0	<0.001
Flowers per inflorescence	Fl-Number	1.4a	1.0–4.0	2.5b	1.0–6.0	2.8b	1.5–4.0	0.001
Flower diameter (cm)	Fl-Diameter	3.9a	3.0–5.2	4.1a	2.5–5.9	3.5a	2.5–5.7	0.162
Petals per flower	Fl-Petals	5.5a	5.0–7.0	6.1b	5.7–7.0	5.3a	5.0–6.0	<0.001
Corolla colour	Fl-Corolla	3.4a	1.0–5.0	2.9a	0.0–5.0	3.2a	1.0–7.0	0.606
Fruit longitudinal perimeter (cm)	Fr-Perimeter	30.2a	22.9–46.4	26.8a	17.6–38.7	29.7a	20.6–36.4	0.253
Fruit length (cm)	Fr-Length	13.3a	6.9–22.7	12.5a	6.4–18.9	16.2b	7.6–20.7	0.019
Fruit breadth (cm)	Fr-Breadth	7.6b	3.0–13.5	7.6b	5.5–10.0	5.2a	3.5–8.6	0.002
Fruit length/breadth ratio	Fr-Ratio	2.26a	0.58–5.00	1.70a	0.85–2.99	3.42b	0.89–4.73	<0.001
Fruit weight (g)	Fr-Weight	426b	139–982	332ab	61–685	209a	97–392	0.030
Relative fruit calyx length	Fr-CLength	2.6a	1.0–5.0	5.7b	1.0–9.0	3.1a	1.0–9.0	<0.001
Fruit calyx anthocyanins intensity	Fr-CAnthocyans	7.6c	0.0–9.0	3.6b	0.0–9.0	0.6a	0.0–9.0	<0.001
Fruit colour intensity under calyx	Fr-UnderC	0.7a	0.0–5.0	2.1b	0.0–5.0	4.2c	3.0–5.0	<0.001
Fruit skin chlorophyll	Fr-Chlorophyll	4.3b	0.0–9.0	4.4b	0.0–9.0	0.0a	0.0–0.0	<0.001
Fruit skin L*primary colour	Fr-L*	25.8a	19.6–58.5	34.1a	21.3–54.7	52.5b	22.3–74.8	<0.001
Fruit skin a*primary colour	Fr-a*	8.4a	−17.9–19.2	6.7a	−2.2–16.0	8.7a	−0.8–16.7	0.649
Fruit skin b*primary colour	Fr-b*	2.6a	−2.1–25.14	4.4a	0.3–14.1	4.8a	−1.08–14.01	0.335
Fruit skin gloss (gloss units)	Fr-Gloss	7.1a	1.9–11.9	10.6b	5.6–17.9	13.7c	8.9–18.5	<0.001
Fruit flesh browning	Fr-Browning	4.0a	1.1–8.4	4.9ab	1.6–9.4	6.3b	0.9–13.2	0.035

[a]Means separated by different letters within a row are significantly different according to the Student-Newman-Keuls multiple range test at P≤0.05.

relevant changes are introduced by including the third, fourth or fifth principal components, we just present and discuss the data referring to the first and second components. The first component was positively correlated to plant vigour traits (L-Breadth, L-Length, L-Pedicel, Fl-Number, and P-Height) as well as to traits related to elongated fruits (Fr-Length, Fr-Ratio, and Fr-Perimeter), and to those associated to the light skin colour, like a high luminosity (Fr-L*), high gloss (Fr-Gloss), and also to a high intensity of the color intensity under calyx (Fr-UnderC) (Figure 1). Negative correlations for this first component were found for traits associated to broad fruits (Fr-Breadth) as well as to traits related to the dark pigmentation (due to content in anthocyanins) of the different plant parts (L-Anthocyanins, Fr-CAnthocyanins, P-Anthocyanins, and Fl-Corolla), or of the fruit caused by the presence of chlorophylls (Fr-Chlorophyll) (Figure 1). Negative correlations of this first component were also found with the leaf breadth/width ratio (L-Ratio), presence of prickles in the plant (P-Prickles), or fruit flesh browning (Fr-Browning). The second component was positively correlated to fruit and flower size (Fr-Weight, Fl-Petals, and Fl-Diameter), to plant vigour traits (L-Breadth, L-Length, L-Pedicel, Fl-Number, and P-Height), to fruit breadth (Fr-Breadth), as well as to the content of chlorophyll in the fruit skin (Fr-Chlorophyll) (Figure 1). This second component was negatively correlated to traits related to elongated fruits (Fr-Ratio, Fr-Length, and Fr-Perimeter) and to prostrate plant habit (P-Habit) (Figure 1).

The projections of the accessions on the PCA graph show that with the exception of two accessions (C02 and L11) the accessions of China, Spain and Sri Lanka plot in different parts of the PCA graph (Figure 2). The Chinese accessions are, in general, characterized by negative values of the first component, with low values of traits related to plant vigour (L-Breadth, L-Length, L-Pedicel, Fl-Number, and P-Height), dark fruits (low Fr-L* and Fr-Gloss, and high Fr-Chlorophyll), and high pigmentation of the plant (high L-Anthocyanins, Fr-CAnthocyanins, P-Anthocyanins, and Fl-Corolla), as well as for a wide range of values for the second component. Spanish accessions present intermediate-high values for both the first and second components, and therefore are

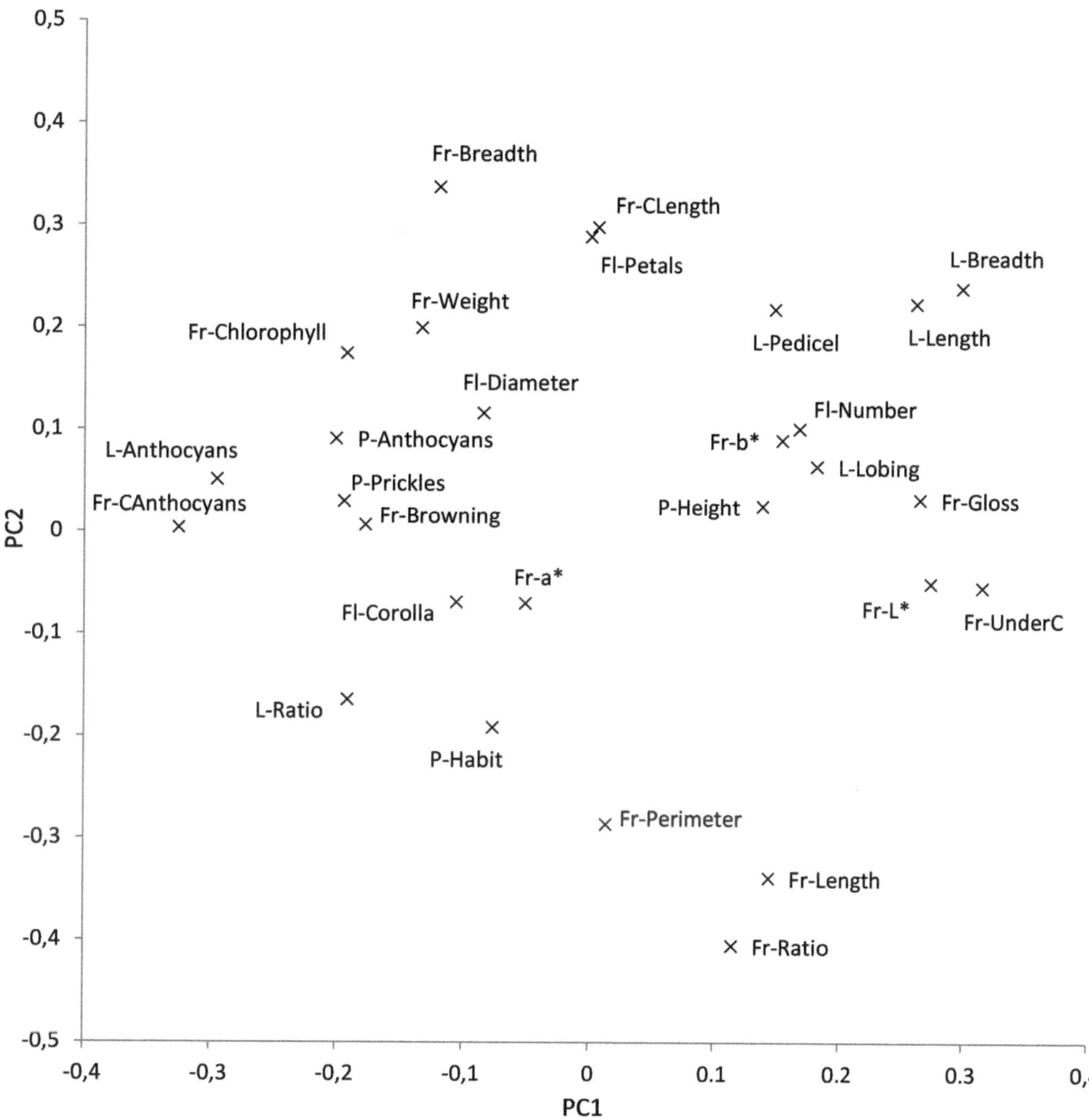

Figure 1. PCA relationships between morphological traits. The two first components (PC1 and PC2) of the principal components analysis account for 22.9% and 16.2% of the total variation, respectively. Results were obtained after the characterization of 52 *S. melongena* accessions from China, Spain, and Sri Lanka using 28 morphological descriptors (see text and Table 2).

associated to high plant vigour (high L-Breadth, L-Length, L-Pedicel, Fl-Number, and P-Height), large flower and fruit size (high Fl-Petals, Fl-Diameter, Fr-Breadth, Fr-Weight), and negatively to elongated fruits (low Fr-Ratio, Fr-Length, and Fr-Perimeter). Sri Lankan accessions present positive values of the first component and relatively low values of the second component, and therefore are associated to elongated fruits and small flowers and fruits (high Fr-Ratio, Fr-Length, and Fr-Perimeter, and low Fl-Petals, Fl-Diameter, Fr-Breadth, and Fr-Weight) with light skin fruit (high Fr-L* and Fr-Gloss and low Fr-Chlorophyll) and low plant pigmentation (L-Anthocyanins, Fr-CAnthocyanins, P-Anthocyanins, and Fl-Corolla) (Figure 2). The outlier Chinese accession C02 is, together with C01, the only accession from this origin that does not have anthocyanins in the fruit (high Fr-L* and Fr-UnderC values), and is also characterized by large leaves (high L-Pedicel, L-Length, L-Breadth) and high fruit flesh browning (Fr-Browning), which results in high values of the first component. Also, the outlier Sri Lankan accession L11 is the only one in this group that does not present elongated fruits (high L-Breadth and low Fr-Length and Fr-Ratio) and therefore has high values for the second component of the PCA and is intermingled with the Spanish accessions (Figure 2).

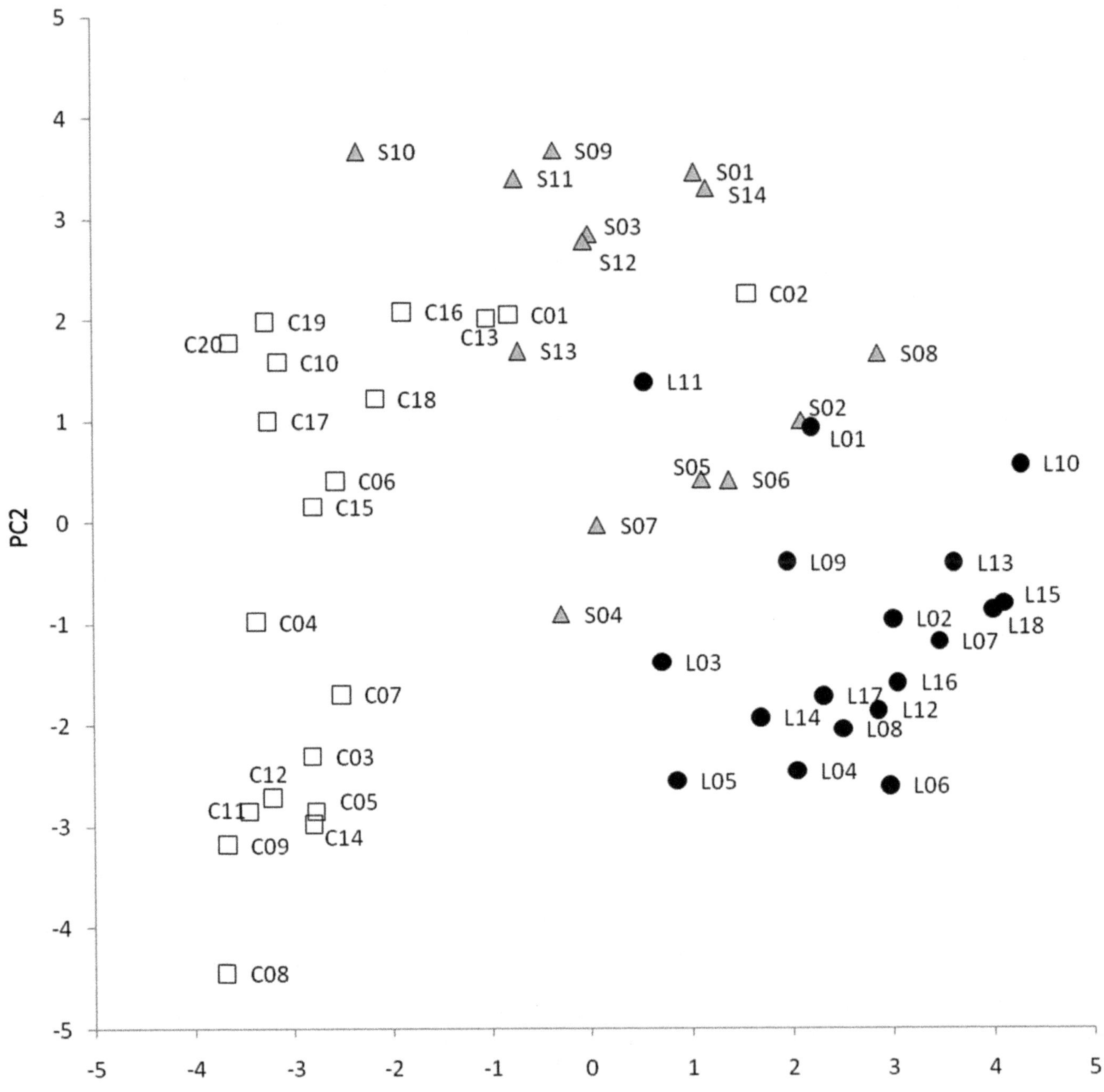

Figure 2. PCA morphology-based relationships between accessions. A total of 52 *S. melongena* accessions from China (white squares), Spain (grey triangles), and Sri Lanka (black circles) (Table 1) were morphologically evaluated with 28 morphological traits (see text and Table 2) and were represented on the two first components (PC1 and PC2) of the principal components analysis (22.9% and 16.2% of the total variation explained by the first and second component, respectively).

The discriminant analysis of the 28 morphological traits revealed that 100% of the accessions were correctly classified to their country of origin. The forward stepwise discriminant analysis showed that a correct classification of all the accessions could be achieved using a minimum set of six traits: L-Length, Fr-L*, Fr-UnderC, Fr-b*, P-Prickles, and Fr-Weight. The two discriminating functions for this model were highly significant ($P<0.001$). For all these traits, with the exception of Fr-b*, important and highly significant ($P<0.001$) differences were found among origins (Table 1).

SSR Characterization

The twelve SSR loci evaluated were polymorphic in the materials evaluated and amplified 110 alleles (average of 9.2 alleles per locus) (Table 2). The number of alleles per locus ranged between five for CSM54 and CSM73 and 16 for CSM30. PIC values ranged between 0.251 for CSM43 and 0.722 for CSM30,

Table 2. SSR markers, number of alleles per locus for all the samples and for each origin (China, Spain, and Sri Lanka), number of private alleles (i.e., present in one or more accessions of each origin), and PIC value for each SSR locus.

SSR locus	All samples	China		Spain		Sri Lanka		PIC
	Alleles (n)	Alleles (n)	Private alleles (n)	Alleles (n)	Private alleles (n)	Alleles (n)	Private alleles (n)	
CSM7	6	4	1	3	0	4	1	0.489
CSM27	11	4	1	4	2	8	4	0.569
CSM30	16	9	4	7	3	6	4	0.722
CSM31	12	7	4	4	2	6	3	0.626
CSM32	12	5	2	6	2	6	4	0.598
CSM36	6	3	1	2	0	5	3	0.351
CSM43	6	1	0	3	2	4	3	0.251
CSM44	11	3	1	5	2	7	5	0.606
CSM52	6	3	0	4	1	4	1	0.550
CSM54	5	5	3	2	0	2	0	0.422
CSM73	5	1	0	4	2	4	1	0.331
CSM78	14	7	3	6	2	7	5	0.720
Mean	*9.2*	*4.4*	*1.7*	*4.2*	*1.5*	*5.25*	*2.8*	*0.574*
Total	110	52	20	50	18	63	34	---

with an average value of 0.574. A unique genetic fingerprint was obtained for each of the individual accessions.

When considering the number of alleles present in accessions from different origins, the Sri Lanka materials presented the highest number of alleles (63), followed by Chinese (52), and Spanish (50) accessions, which corresponds, respectively, to 57.3%, 47.3%, and 45.5% of the total number of alleles detected (Table 2). However, no private and universal SSR alleles were found for any of the three origins. For the materials of Sri Lanka, a total of 34 alleles were private (i.e., unique to one or more accessions of this group), while for China and Spain the number of private alleles was of 20 and 18, respectively (Table 2). Most of these private alleles specific to each origin were present in low frequencies, so that when considering each of the origins separately the average value of the frequency of the private alleles was of 0.077, with a maximum frequency value of 0.308. Eighteen alleles were present in all the groups, while 9 were exclusive of China and Spain, 5 of China and Sri Lanka, and 5 to Spain and Sri Lanka (Figure 3).

The first and second components of the PCA analysis performed with SSR data account for 22.0% and 19.1% of the total variation, respectively. Although in this analysis accessions of the different origins are intermingled (Figure 4), accessions from China and Sri Lanka are mostly distributed in different areas of the plot. With the exception of L05 and L11, accessions from Sri Lanka have a combination of low and moderate values for the first and second components; on the other hand, with the exception of C01, C11, C12, and C14, accessions from China present positive values of either the first or second components (Figure 4). Accessions from Spain are scattered all over the graph with the exception of the bottom part of the plot (i.e., the area with values <-0.2 for the second component). Inclusion of a third component in the analysis (17.9% of the total variation explained by this component) shows that no accessions from Spain present low values for the third component (all accessions have values >-0.15), while no accessions from Sri Lanka present high values for this component (all accessions have values <0.15). On the other hand, accessions for China present a wide range of values for the third component (Figure 4).

The Evanno's test indicated that the most informative number of populations (K) was 3. The inferred population structure for $K=3$ obtained with the STRUCTURE software showed that most of the accessions of each origin are assumed to belong to the same population (Figure 5). Therefore, we have named these populations as CH, SP, and SL, corresponding to populations containing mostly Chinese, Spanish, and Sri Lankan accessions, respectively. In this respect, when the highest value for the membership coefficient (q_i) is used to assign one accession to a population, for the Chinese accessions, 14 out of the 20 accessions are assigned to the CH population; for the rest of Chinese accessions, C14, C16, C17, and C18 are assigned to the SP population, and C01 and C11 to the SL population. In the case of the Spanish accessions, 10

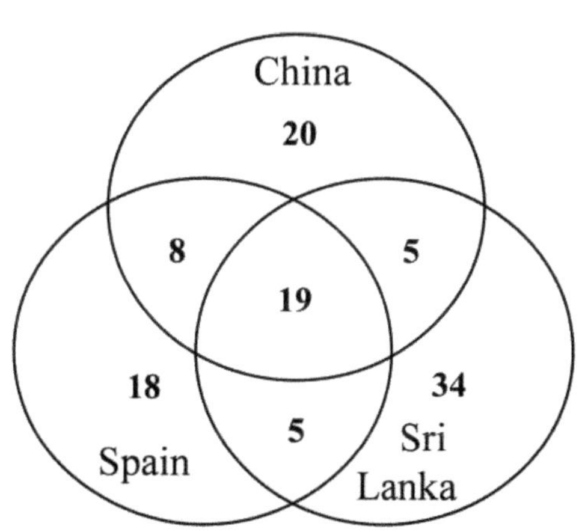

Figure 3. SSR alleles in the different origins. The figures indicate the number of SSR alleles private to each of the secondary centers of diversity, as well as the number of alleles shared by accessions from two or the three different origins.

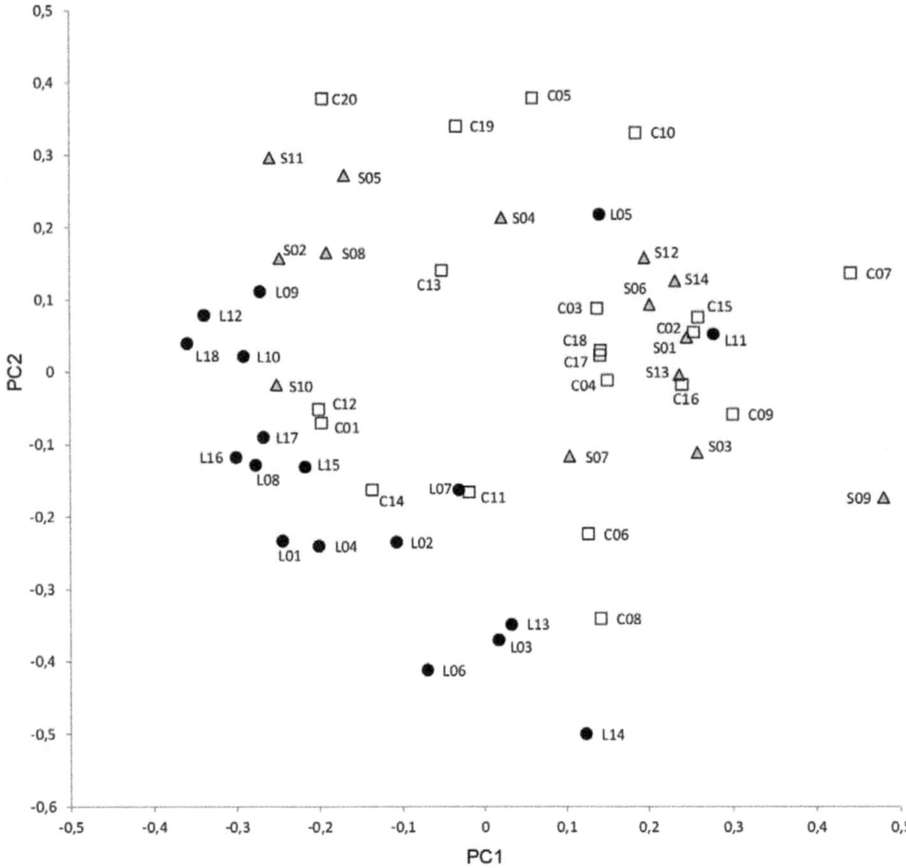
0,5
0,4
0,3
0,2
0,1
0
-0,1
-0,2
-0,3
-0,4
-0,5
-0,6
PC2
-0,5
-0,4
-0,3
-0,2
-0,1
0
0,1
0,2
0,3
0,4
0,5
PC1
C20
C05
C19
C10
S11
S05
S04
L05
S02
S08
S12
C13
S14
C07
L09
S06
C03
C15
L12
C02
L11
C18
S01
L18
L10
C17
S13
C04
C16
S10
C12
C09
C01
L17
S03
L16
L15
S07
L08
L07
C11
C14
S09
C06
L01
L04
L02
L13
C08
L03
L06
L14

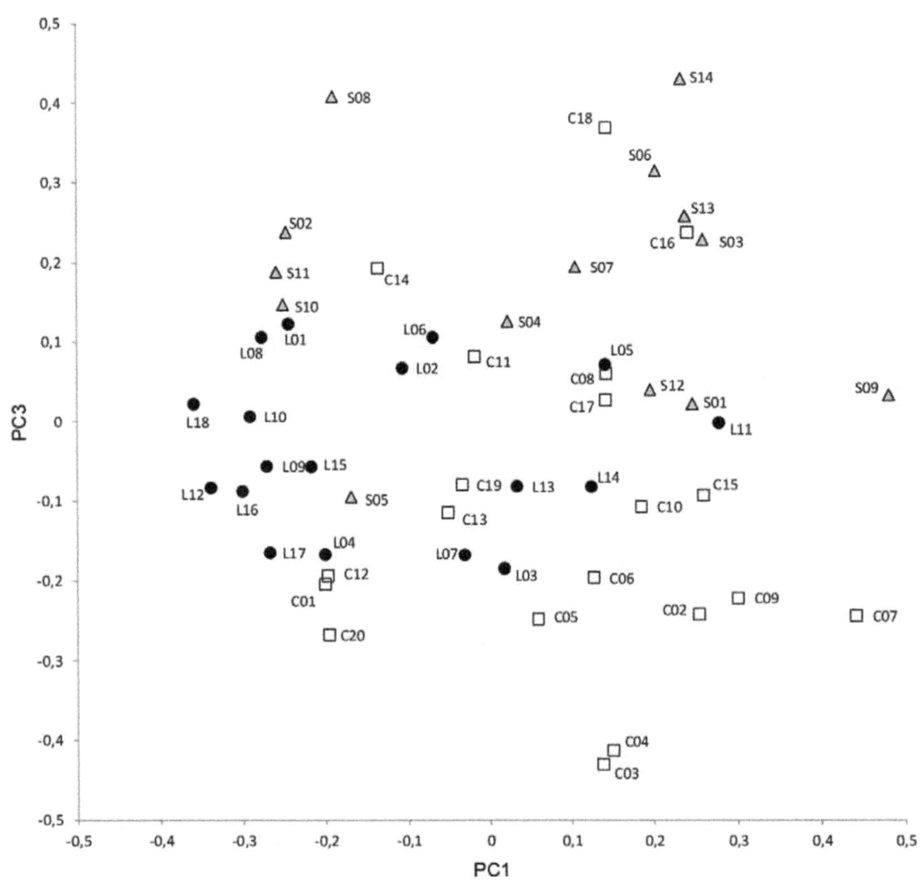
0,5
0,4
0,3
0,2
0,1
0
-0,1
-0,2
-0,3
-0,4
-0,5
PC3
-0,5
-0,4
-0,3
-0,2
-0,1
0
0,1
0,2
0,3
0,4
0,5
PC1
S14
S08
C18
S06
S13
S02
C16
S03
S11
C14
S07
S10
S04
L06
L08
L01
C11
L05
L02
C08
S09
S12
C17
S01
L18
L10
L11
L09
L15
L14
C19
L13
C15
L12
L16
S05
C10
C13
L04
L17
L07
C12
L03
C06
C01
C09
C02
C07
C05
C20
C04
C03

Figure 4. PCA SSR-based relationships between accessions. A total of 52 *S. melongena* accessions from China (white squares), Spain (grey triangles), and Sri Lanka (black circles) (see Table 1) were evaluated using 12 polymorphic SSRs (see text and Tables 3 and 5) and were represented on the three first components (PC1, PC2 and PC3) of the principal components analysis (22.0%, 19.1%, and 17.9% of the total variation explained by the first, second, and third principal components, respectively). Scatterplots show the projections of the accessions on the first and second principal components (above) and on the first and third principal components (below).

out of the 14 accessions are assigned to the SP population; three of the rest of Spanish accessions (S01, S09, and S12) to the CH population, while the remaining one (S10) to the SL population. Finally, the 15 out of the 18 Sri Lankan accessions are assigned to the SL population; two out of three of the rest of accessions (L05 and L11) to the CH population, and the last one (L02) to the SP population (Figure 5). For most of the accessions (69.2%), the membership coefficient q_i to one of the populations was higher than 0.8, while the rest (30.8%) could be considered as admixed ($q_i \leq 0.8$). Seven of the admixed accessions were from China (C01, C08, C12, C10, C14, C15, and C19), five from Spain (S03, S05, S09, S10, and S12), and four from Sri Lanka (L02, L03, L07, and L10) (Figure 5).

The three origins present a high level of total genetic diversity (H_T), which ranges from $H_T = 0.4943$ for Chinese accessions to $H_T = 0.5400$ for Sri Lankan accessions (÷). The partition of the total diversity shows that most of the genetic diversity is found within each of the groups ($D_{ST} = 0.0481$ and $H_S = 0.4909$), which results in a moderate value of the relative magnitude of genetic differentiation ($G_{ST} = 0.0892$). However, the standardized G_{ST} (G'_{ST}) reached a much higher value ($G'_{ST} = 0.2657$) (Table 3). The values of heterozigosity for individual accessions were low, with an average value of 0.052, with non-significant differences among origins ($P > 0.05$).

The lowest genetic distance (and consequently highest genetic identity) values are found between the Chinese and Sri Lankan origins, with a value of 0.1204 (Table 4). The genetic distance values of Spanish and Chinese origins and of Spanish and Sri Lankan eggplants are similar, with values of 0.1613 and 0.1604, respectively (Table 4). Correlations obtained from the Mantel test between the morphological and SSR distances were low ($r = 0.044$) and non-significant ($t = 0.5729$; $P = 0.7166$).

Discussion

Morphological and molecular data are not only complementary for the study of the diversity and relationships of eggplants from these distant centers of diversity, but also synergistic, as information of interest becomes conspicuous when comparing both type of data. Several studies are available on the relationships between morphological and molecular data in eggplant, but these studies have been restricted to specific cultivar groups [4,28–29], to study Spanish accessions [19], or to compare both type of data for predicting the performance of hybrids [5]. In this respect, our study is the first one to study simultaneously the morphological and molecular diversity of a significant number of eggplant accessions from geographically distant centers of diversity of eggplant.

Morphological Diversity

The morphological data show that within each of the origins, a considerable diversity exists, confirming what had already been observed within Spanish eggplants [19]. For most morphological traits significant differences among origins are found for most traits, and eggplants from the different origins mostly plot in different areas of the PCA graph, showing that accessions from each of the three secondary centers of diversity are morphologically differentiated and present a typical combination of traits [28,29]. In this way, a combination of just six morphological traits is enough to correctly assign any of the accessions to its origin. Amazingly, in this study we have found that the number of

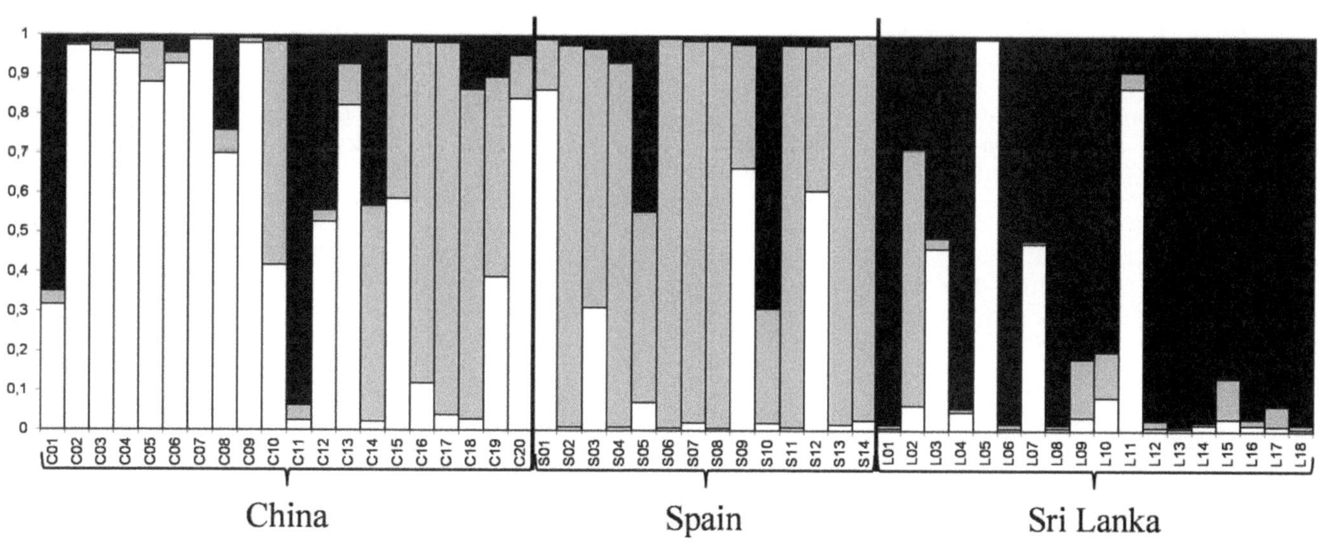

Figure 5. Assignment tests of accessions to populations. The estimated number of populations (parameter *K*) was set to 3 in the STRUCTURE software (Pritchard et al., 2000). Accessions are organized according to their origin (China, Spain, and Sri Lanka). The three populations are named according to the origin of the majority of the accessions constituting each population (CH for China, SP for Spain, and SL for Sri Lanka). Each accession is represented by a horizontal line, which is partitioned into colored segments (white for the CH population, grey for the SP population, and black for the SL population).

Table 3. Total genetic diversity (H_T), among groups genetic diversity (D_{ST}), within groups genetic diversity (H_S), relative magnitude of genetic differentiation (G_{ST}) and standarized G_{ST} (G'_{ST}), estimated from SSR data for the eggplant accessions according to their origin.

Group	Sample size	H_T	D_{ST}	H_S	G_{ST}	G'_{ST}
Total	52	0.5390	0.0481	0.4909	0.0892	0.2697
Origins						
China	20	0.4943				
Spain	14	0.5286				
Sri Lanka	18	0.5400				

differences between eggplants from China and Sri Lanka, has been much greater than those between any of the former origins and Spanish eggplants. These results indicate that although cultivated eggplants have often been classified into "Western" or "Occidental" and "Asian" or "Oriental" cultivar groups [3,6,10,35], the real situation, in particular of the "Asian" eggplants may be more complex, with several groups with important morphological differences. For example, in general, the Chinese accessions used in our study have more anthocyanin pigmentation, are more prickly, have darker fruits, with more chlorophyll in the skin, and less vigorous, with smaller leaves, and less elongated and glossy fruits than Sri Lankan accessions. The few outlier accessions that in the morphological PCA group with accessions from other origins (C02 and L11) present a few morphological characteristics that differ greatly from the rest of accessions of their origin (e.g. lack of anthocyanin pigmentation, large leaves, and high fruit flesh browning in C02, and round fruits in L11). However, for the rest of traits present values similar to those of the Chinese (C02) and Sri Lankan (L11) accessions and therefore, when using the discriminant traits are correctly classified to their origins.

The wide range found for most morphological traits within each of the origins suggests that genetic drift may not have been an important force in the morphological differentiation of eggplant within each of the three secondary centers of diversity [40]. On the other hand, the morphological differentiation found among origins may indicate that natural selection and/or, more likely, artificial selection may have been responsible for the combination of morphological differences characteristic of accessions from different secondary centers of diversity [41]. Some of the traits for which significant differences exists among origins (e.g., prickliness, anthocyanin pigmentation, fruit chlorophyll, presence of pigmentation under the calyx) are monogenic or oligogenic and present a high degree of penetration and expression [3,42]; therefore, the

Table 4. SSR-based estimates of genetic distances (above the diagonal) and genetic identities (below the diagonal) between different eggplant origins.

Origin	China	Spain	Sri Lanka
China		0.1613	0.1204
Spain	0.8510		0.1604
Sri Lanka	0.8813	0.8514	

frequencies of phenotypes change quickly in response to selection. In consequence, the fact that for these traits the means are different, even though a wide range of variation is present in each origin, suggests that artificial selection, very likely linked to different uses or to different management practices, may have been responsible for the morphological differentiation among origins. This suggests that ethnobotanical aspects may have had an important role in the differentiation of eggplant landraces from separate regions.

The important differences among origins for morphological traits may be relevant for eggplant breeding, and indicates that sources of variation of interest for several relevant agronomic traits of materials from a center of diversity may be found in exotic materials from other centers of diversity. For example, when considering some traits of interest for breeding modern cultivars [3], in general Chinese accessions seem to be a good source of variation for large fruits and low fruit flesh browning, Spanish ones for vigorous (e.g., high with large leaves) plants, and Sri Lankan accessions for low prickliness and high glossiness.

Molecular Diversity

The high morphological variation in the materials studied is matched by high levels of molecular diversity. In this sense, we have found that the 12 genomic SSR loci were polymorphic and allowed the detection of a total of 110 SSR alleles (9.2 alleles/polymorphic locus). These 12 SSR loci were developed by Vilanova et al. [35] and were selected for the present study for its high polymorphism. However, the level of polymorphism found here has been higher than that found in the original work of Vilanova et al. [35], in which a total of 92 alleles (7.7 alleles/polymorphic locus) were detected in 22 eggplant materials from different cultivar types and origins. In other studies, the level of SSR polymorphism has been much lower. For example, Muñoz-Falcón et al. [38] in a study of 42 eggplant accessions from different origins and cultivar groups used 49 SSR loci (17 corresponding to genomic SSRs and 32 to EST-SSRs) of which only 21 were polymorphic and which allowed the detection of 85 alleles (4.1 alleles/polymorphic locus). In a similar study, Muñoz-Falcón et al. [4] found polymorphism for 11 out of 14 SSR loci (2 genomic SSRs and 12 EST-SSRs), giving a total of 61 alleles (4.7 alleles/polymorphic locus) in a collection of 44 eggplants. Stàgel et al. [32] found that only 11 out of 38 EST-SSR loci were polymorphic in a collection of 38 eggplant cultivars of different origins and allowed the detection of 39 alleles (3.5 alleles/polymorphic locus). Ge et al. [23] in a study of 88 EST-SSRs of 42 Asian accessions found polymorphism for 79 of them, with a total of 323 polymorphic alleles (4.1 alleles/polymorphic locus). Also, when we consider each of the three origins, a high diversity is found, and the number of alleles per polymorphic locus is of 4.3, 4.2, and 5.3 for accessions from China, Spain, and Sri Lanka, respectively, which is a confirmation that these three countries can be considered as secondary centers of diversity [19–21]. The fact that a high number of private alleles are found in each of the origins we have studied, although most of them are present at low frequencies, has important implications for the conservation of the genetic diversity. For example, it suggests that the establishment of core collections for this crop should include accessions from different origins so that most of the allelic diversity is represented [43].

The multivariate analysis of SSR data shows a less clear picture of the differentiation among centers of diversity than the analysis of morphological data, with a higher degree of admixture among accessions of different origins in the SSR PCA plots than in the morphological PCA graph. In any case, the combination of the first and second components separate most of the Sri Lankan accessions from most of the Chinese accessions, while the third component differentiates mostly the Spanish and Sri Lankan accessions. Therefore, despite the fact that some of the accessions are intermingled, differentiation is observed among the three origins in the PCA analysis. The Bayesian-based analysis without a priori assignment of accessions to populations resulted in the creation of three populations, each of which was formed mostly by accessions from a single origin, which suggests an important differentiation among origins [25,26]. The results of the PCA analysis show a good agreement with the population structure analysis. In this respect, accessions from one origin that in the PCA analysis plotted closer to most of the accessions of another origin, were assigned to the population formed mostly by accessions of this latter origin in the population structure analysis. More than two thirds of the accessions were clearly assigned to one of the populations, as they presented a membership coefficient q_i to one of the populations higher than 0.8 [25,44]; the rest of accessions could be considered as admixed, although in most cases were basically an admixture of only two populations. Accessions from one origin (e.g., China) assigned in the population structure analysis to populations mostly composed by accessions from another origin (e.g. Spain or Sri Lanka) were mostly not admixed accessions, which may indicate that introduction of these materials have taken place in relatively recent times. Also, the fact that eggplant is fundamentally autogamous [45] may have helped to maintain the genetic integrity of the materials introduced from another center of diversity.

The analysis of the genetic diversity with the total diversity (H_T) parameter [46] shows that the total diversity present in each of the origins is high and similar to the total diversity present in the whole collection, which is an indication of the high diversity present in each origin. This result is somewhat surprising, as the diversity present in a small country, like Sri Lanka, is even higher than the diversity found in a large country, like China. Given that Sri Lanka is close to India, which is part of the primary center of diversity [3,9,10], may have allowed the accumulation of a high diversity in Sri Lanka.

Despite the fact that the genetic diversity among groups (D_{ST}) is considerably lower than the genetic diversity within groups (H_S), the standardized genetic differentiation (G'_{ST}), which is more adequate than G_{ST} for SSR data [47] is greater than 0.25, which is a moderate value and shows that there is a considerable differentiation among the three origins [25,47,48]. When considering the genetic relationships among the three origins, the lowest genetic distance is found between Chinese and Sri Lankan accessions, while the genetic distance between Spanish and Chinese or Sri Lankan accessions are similar. Given that China and Sri Lanka are geographically closer than to Spain probably has contributed to a greater exchange of materials and genetic flux among these two countries, resulting in lower genetic distances.

The heterozygosity values observed were low, and similar to the values found by others in eggplant landraces of the black type [4]. This result was expected, given the high degree of autogamy of eggplant [3,45], which results in the fixation of alleles in homozygosis in populations in which there is no artificial control of pollination.

Comparison of Morphological and Molecular Diversity

The correlations between morphological and SSR molecular data have not been significant, which is contrast to the moderate value ($r = 0.38$) found by Muñoz-Falcón et al. [4] when comparing morphological and SSR data in a study of black eggplants which included modern F1 hybrids, old non hybrid cultivars, Spanish landraces, and Non-Spanish landraces. Significant correlations between morphological data and AFLPs were also found in the former study, and also by Prohens et al. [19] in a study of the diversity of Spanish eggplants. The correlations between morphological and molecular data are usually very variable and dependent, among others, on the plant material, morphological descriptors, and molecular markers used. In this respect, in other solanaceous crops, like pepper [49] or potato [50] no significant correlations were found among both types of data.

The lack of correlations in our collection suggests that both types of data sample different levels of diversity and, therefore, both of them should be considered for the management and conservation of germplasm [51]. In this respect, morphological markers in eggplant sample traits for which the variation is usually controlled by a few genes or QTLs [3,42], while genomic SSRs detect differences in the length of short tandem repeated sequences mostly situated in non-coding regions of the genome [36,37]. Wendel and Doyle [52] suggested that the lack of correlation between morphological and molecular data may be caused by the fact that both types of markers follow different evolutionary paths.

Conclusions

The results show that eggplant accessions from the three geographically distant secondary centers of diversity studied (China, Spain, and Sri Lanka) present a wide morphological and molecular diversity. At the morphological level, a clear differentiation exists among the different origins, indicating that different selection criteria have been applied in each secondary center of diversity, leading to a typical syndrome of traits for each origin. The molecular data also show that considerable differentiation exists at the molecular level among the three origins, although in this case there is evidence of migration between the different origins, as revealed by the study of the population structure. The lack of correlation between morphological and molecular diversity shows that both types of data provide complementary information and that both of them should be taken into account in the management of germplasm and formation on core collections [43,51]. The results also have importance for eggplant breeding, as it shows that sources of variation of interest can be found in the materials evaluated, and also suggests that crossing among individuals assigned to different populations may result in heterotic hybrids [5]. Finally, these results are also important for understanding the evolution and domestication of eggplant [9,10].

Materials and Methods

Permissions

No specific permits were required for the described field studies, which took place in an experimental field plot at the Universitat Politècnica de Valencia. This field plot is used by the authors of this paper affiliated to the aforementioned institution (MH, SV, MP, PG, and JP) for field trials for characterization of germplasm of cultivated species.

Plant Materials

Fifty-two accessions of *S. melongena*, of which 20 originated in China (C codes), 14 in Spain (S codes), and 18 in Sri Lanka (L codes) were used for the present study (Table 5). The materials correspond to local landraces or to selections within the local landraces and were chosen trying to represent the diversity of local materials of *S. melongena* of each country. The criteria used for choosing the accessions was based on previous available morphological and genetic data (for the Spanish accessions) [4,19,28,35], on distribution across geographic range, and also on availability of

Table 5. Plant materials used for the study of morphological and molecular (SSR) variation of a collection of eggplants from China, Spain, and Sri Lanka.

Accession name	Code	Place of origin
China		
Qi Xian Hei You Guan	C01	Qi
Tuo Cheng Qing Qie	C02	Tuocheng
Zao Xian Qie	C03	Acheng
Qie Zie	C04	Aihui
Zi Chang Qie	C05	Hubei
Tuan Qie	C06	Zhuix
Wu Xian Qie	C07	Anxiang
He Xian Qie	C08	Chenzhou
Yong Ji Xian Qie	C09	Yongji
Zi Yuan Qie	C10	Taoan
Niu Jiao Qie	C11	Changzhou
Niu Jiao Qie	C12	Jiangyin
Hong He Bao Qie	C13	Yichun
Yu Jiang Chang Xian Qie	C14	Yujiang
Bao Tou Niu Xin Qie	C15	Baotou
Tian Jin Da Min Qie	C16	Yinchuan
Tian Jin Er Min Qie	C17	Yinchuan
Cao Xian Yuan Zie Qie	C18	Cao
De Zhou Duan Ba Hong Qie	C19	Dehou
ASI-S-1	C20	Beijing
Spain		
AN-S-24	S01	Jaén
V-S-10	S02	Alicante
AN-S-3	S03	Córdoba
MU-S-7	S04	Murcia
B-S-3	S05	Baleares
MU-S-4	S06	Murcia
V-S-3	S07	Valencia
V-S-2	S08	Gandía
V-S-9	S09	Alicante
MU-S-8	S10	Murcia
MU-S-6	S11	Murcia
NC056471	S12	Murcia
H15	S13	Almagro
AN-S-23	S14	Jaén
Sri Lanka		
BW11	L01	Bombuwela
SM 164	L02	Gannoruwa
Thinnevelly Purple	L03	Jaffna
8890	L04	Matara
SA7MTE2	L05	Unknown
5124	L06	Gampaha
7517	L07	Colombo
1624	L08	Galle
558	L09	Nuwara Eliya
799	L10	Matale
8891	L11	Matara
1139	L12	Puttlam
2287	L13	Unknown
Ridiyagama	L14	Hambantota
Farmer Lenairi	L15	Anuradhapura
Kaluthavelly	L16	Batticaloa
Ampara	L17	Ampara
Welimada	L18	Welimada

seeds. The plant material used is part of the collection of the Instituto de Conservación y Mejora de la Agrodiversidad Valenciana (Valencia, Spain).

For each accession five plants raised from seed were grown in an open-air field plot (39°28'55" N; 0°20'11" W) in Valencia (Spain) following a completely randomized experimental design. Plants were spaced 1 m between the rows and 0.8 m apart in the row. The standard horticultural practices for eggplant production in the area of Valencia were followed.

Morphological Characterization

Individual plants were characterized using 23 primary descriptors, most of which were developed by EGGNET [4,19,30]. These descriptors include plant (P), leaf (L), flower (Fl), and fruit (Fr) characteristics. Thirteen traits corresponding to these primary descriptors are quantitative: plant height (cm; P-Height), leaf pedicel length (cm; L-Pedicel), leaf blade length (cm; L-Length), leaf blade breadth (cm; L-Breadth), leaf blade length/breadth ratio (L-Ratio), flowers per inflorescence (Fl-Number), flower diameter (cm; Fl-Diameter), petals per flower (Fl-Petals), fruit longitudinal section perimeter (cm; Fr-Perimeter), fruit length (cm, Fr-Length), fruit breadth (Fr-Breadth), fruit length/breadth ratio (Fr-Ratio), fruit weight (g; F-Weight). The other ten traits are measured in a scale with predetermined values (Table 6), and correspond to EGGNET descriptors [4,19,30]. Apart from these primary descriptors, the skin colour and brightness, as well as the fruit flesh browning were also measured objectively in at least three fruits per plant. For the skin colour the L* (Fr-L*), a* (Fr-a*), and b* (Fr-b*) Hunter colour coordinates of the predominant (primary) colour of the skin were assessed with a Minolta CR300 (Minolta Co., Osaka, Japan) chromameter. The skin gloss (gloss units; Fr-Gloss) was measured with a Novo-Curve Elcometer 400 (Elcometer, Manchester, UK) glossmeter. Finally, for the fruit flesh browning (Fr-Browning), the flesh colour in the central part of a transversal section of the fruit was measured with the CR300 chromameter at 0 min and 10 min after being cut, and the flesh browning was measured as the difference between the Hunter colour coordinate L* (luminosity) at 0 and 10 min [53], so that the higher the value of the difference the greater the browning.

Molecular Characterization

Genomic DNA was extracted from a mixture of young leaf from the five plants morphologically evaluated according to the CTAB method procedure [54]. The quality of DNA was checked on 1% agarose gels and the DNA concentrations estimated using a Nanodrop ND-1000 (Nanodrop Technologies, Wilmington, Delaware, USA) spectrophotometer.

Table 6. Morphological traits measured in a scale with pre-determined values of the descriptor states, and description of the scale used for the study of morphological variation in the eggplant accessions studied [4,19,30].

Traits	Codes	Range (scale)
Plant growth habit	P-Habit	1–9 (1 = upright; 9 = prostrate)
Shoot tip anthocyanins intensity	P-Anthocyans	0–9 (0 = absent; 9 = very strong)
Stem prickles	P-Prickles	0–9 (0 = absent; 9 = very many)
Leaf blade lobing	L-Lobing	1–9 (1 = very weak; 9 = very strong)
Leaf anthocyanins intensity	L-Anthocyans	0–9 (0 = absent; 9 = very strong)
Corolla colour	Fl-Colour	0–7 (0 = White; 7 = blue)
Relative fruit calyx length	Fr-CLength	1–9 (1 = very short (<10%); 9 = very long (>10%))
Fruit calyx anthocyanins intensity	Fr-CAnthocyans	0–9 (0 = absent; 9 = very strong)
Fruit color intensity under calyx	Fr-UnderC	0–9 (0 = none; 9 = very strong)
Fruit skin chlorophyll	Fr-Clorophyll	0–9 (0 = none; 9 = very strong intensity)

For the present study we used twelve simple sequence repeat (SSR) markers (Table 7) that proved to be highly polymorphic and that presented high PIC values in eggplant [35]. The map position of nine out of these 12 SSRs is known (Table 7). SSRs were amplified following the M13-tail method described by Schuelke [55] to facilitate the incorporation of a dye label during PCR. Amplifications were performed in a total volume of 12 μl, with 10 ng DNA, 1 mM $MgCl_2$, 0.05 μM of forward primer, 0.25 μM

Table 7. Primer sequences, expected size, annealing temperature, linkage group, and map position [35,39] of the twelve SSR markers used for molecular characterization of the materials studied.

SSR locus	Repeat	Primer sequence (5′-3′)	Annealing temperature	Linkage group
CSM7	CT	F- CGACGATCACCTTGATAACG	58.6	Unknown
		R- CCTAAATGCAGAGTTTCCAAAG		
CSM27	GA	F- TGTTTGGAGGTGAGGGAAAG	60.0	3[a]
		R- TCCAACTCACCGGAAAAATC		
CSM30	CT	F- CACTGTTCCTGGTTGCTGTG	60.1	9[b]
		R- TTTAGCTTTAGCCCATCTACCG		
CSM31	AG	F- CAACCGATATGCTCAGATGC	59.8	1[c]
		R- GCCCTATGGTCATGTTTTGC		
CSM32	AG	F- TCGAAAGTACAGCGGAGAAAG	59.6	4
		R- GGGGGTTTGATTTTCATTTTC		
CSM36	GA	F- CCTCAATGGCAGTAGGTCAGA	60.1	9[b]
		R- GTTCTTTGAGCCTCCAGTGC		
CSM43	AG	F- ATTTTAACCCCGGGAAAATG	59.6	1[c]
		R- ACCGCTTCTAGGTTTTGCAC		
CSM44	AG	F- CGTCGTTGTAACCCATCATC	58.7	3[a]
		R- TTGCCAAATTCCTTGTGTTC		
CSM52	TC	F- CTTGGGTCACAAAAGGTTCC	59.7	Unknown
		R- TCACCGAAAAAGATCCAACC		
CSM54	GA	F- ATGTGCCTCCATTCTGCAAG	61.1	9[b]
		R- TGGGTGGGATGCTGAGTAAG		
CSM73	CT	F- TTCAACATAGCCTGGACCATT	60.0	Unknown
		R- AATGCAGGGTTTGGACTTCA		
CSM78	CT	F- AGGGAGGAGCTCTCGTGTG	60.2	10
		R- CAATAACGTAGCTTAATTACTCCCAAG		

[a]Markers CSM27 and CSM44 are positioned at 10.8 cM and 88.7 cM, respectively, in linkage group 3.
[b]Markers CSM30, CSM36 and CSM54 are positioned at 29.3 cM, 71.2 cM, and 0.0 cM, respectively, in linkage group 9.
[c]Markers CSM31 and CSM43 are positioned at 182.0 cM and 161.8 cM, respectively, in linkage group 1.

of reverse primer, 0.2 µM fluorescent-labelled M-13 primer, 0.2 mM dNTPs and 1 unit of *Taq* polymerase in 1X PCR buffer. Conditions of the PCR amplification were as follows: 1 cycle for 2 minutes at 94°C, 35 cycles of 15 s at 94°C, 30 s at appropriated annealing temperature (Table 7), 45 s at 72°C, followed by 10 min extension at 72°C. SSR alleles were resolved on an ABI PRISM 3100 DNA (Applied Biosystems, Carlsbad, California, USA) genetic analyzer using GeneScan 3.7 (Applied Biosystems) software and precisely sized using GeneScan 500 LIZ molecular size standards with Genotyper 3.7 (Applied Biosystems) software.

Data Analysis

The range and mean values for the morphological traits for each of the groups of Chinese, Spanish, and Sri Lankan eggplants were calculated from the means of each accession. Analyses of variance were used to detect differences for the traits studied among the three groups of accessions. Kolmogorov-Smirnov and Bartlett tests were performed to test, respectively, the normality of data and homogeneity of variances among different origins. Principal components analysis (PCA) were performed for standardised morphological and agronomic data using pairwise Euclidean distances among accessions. Eigenvalues were calculated for each of the principal components, and relevant components were identified using the spree plot method [56]. Discriminant analysis was used to study the percentage of cases correctly classified, and the forward stepwise procedure was used for selecting the minimum subset of variables for discriminating among the three groups.

For the SSR data, pairwise genetic similarities were estimated using the Dice (Sorensen) similarity coefficient. Principal components analyses (PCA) were performed using the pairwise genetic similarities. Possible population structure associated to origin and likelihood of assignment of each accession to population was estimated using the Bayesian-based model procedure implemented in the software STRUCTURE v2.3.3 [57]. The analysis was carried out using a burning period of 10000 iterations. We tested a continuous series of *K*, from 1 to 10, in 10 independent runs [58]. No prior knowledge about the population of origin was introduced. The most informative *K* was identified using the statistic ΔK [59]. Subsequently, population structure was inferred for $K=3$ and using 50000 iterations. Genetic diversity was estimated with the total diversity (H_T) [46]. Total diversity was partitioned into diversity among origins (D_{ST}), and within origins (H_S). The relative magnitude of genetic differentiation among origins (G_{ST}) was calculated as the ratio D_{ST}/H_T [46]. The standarized G_{ST} (G'_{ST}), which standardizes the observed G_{ST} value with the maximum possible value that G_{ST} could obtain given the amount of observed diversity, was calculated as $(G_{ST}(1+H_T))/(1-H_S)$ [47]. For each accession, heterozygosity values for SSRs were calculated as $1-\sum(p_i^2)$, where p_i is the frequency of the ith allele [60]. Genetic distances and identities among groups were calculated according to Nei [61]. Correlations between morphological and SSR distance matrices were investigated using a Mantel [62] test.

Author Contributions

Conceived and designed the experiments: SV HHF RF JP. Performed the experiments: MH SV MP PG. Analyzed the data: MH SV HHF JP. Contributed reagents/materials/analysis tools: HHF RF JP. Wrote the paper: MH HHF RF JP.

References

1. Cao GH, Sofic E, Prior RL (1996) Antioxidant capacity of tea and common vegetables. J Agric Food Chem 44: 3426–3431.
2. Singh AP, Luthria D, Wilson T, Vorsa N, Singh V, et al. (2009) Polyphenols content and antioxidant capacity of eggplant pulp. Food Chem 114: 955–961.
3. Daunay MC (2008) Eggplant. In Prohens J, Nuez F, editors. Handbook of Plant Breeding - Vegetables II. New York: Springer. 163–220.
4. Muñoz-Falcón JE, Prohens J, Vilanova S, Nuez F (2009). Diversity in commercial varieties of black eggplants and implications for broadening the breeders' gene pool. Ann Appl Biol 154: 453–465.
5. Rodríguez-Burruezo A, Prohens J, Nuez F (2008) Performance of hybrids between local varieties of eggplant (*Solanum melongena*) and its relation to the mean of parents and to morphological and genetic distances among parents. Eur J Hort Sci 73: 76–83.
6. Daunay MC, Janick J (2007). History and iconography of eggplant. Chronica Hort 47(3): 16–21.
7. Wang JX, Gao TG, Knapp S (2008) Ancient Chinese literature reveals pathways of eggplant domestication. Ann Bot 102: 891–897.
8. Lester RN, Hasan SMZ (1991) Origin and domestication of the brinjal eggplant, *Solanum melongena*, from *S. incanum*, in Africa and Asia. In: Hawkes JG, Lester RN, Nee M, Estrada N, editors. Solanaceae III: taxonomy, chemistry, evolution. London: The Linnean Society of London. 369–387.
9. Weese TL, Bohs L (2010) Eggplant origins: out of Africa, into the Orient. Taxon 59: 49–56.
10. Meyer RS, Karol KG, Little DP, Nee MH, Litt A (2012) Phylogeographic relationships among Asian eggplants and new perspectives on eggplant domestication. Mol Phylogenet Evol: doi: 10.1016/j.y,pev.2012.02.006.
11. Isshiki S, Okubo H, Fujieda K (1994) Phylogeny of eggplant and related *Solanum* species constructed by allozyme variation. Sci Hort 59: 171–176.
12. Mace ES, Lester RN, Gebhardt CG (1999) AFLP analysis of genetic relationships among the cultivated eggplant, *Solanum melongena* L., and wild relatives (Solanaceae). Theor Appl Genet 99: 626–633.
13. Furini A., Wunder J (2004) Analysis of eggplant (*Solanum melongena*)-related germplasm: morphological and AFLP data contribute to phylogenetic interpretations and germplasm utilization. Theor Appl Genet 108: 197–208.
14. Tümbilen Y, Frary A, Mutlu S, Doganlar S (2011) Genetic diversity in Turkish eggplant (*Solanum melongena*) varieties as determined by morphological and molecular analyses. Int Res J Biotechnol 2: 16–25.
15. D'Arcy WG, Pickett KK (1993) Salt water flotation of *Solanum* fruits and possible dispersal of eggplant. Solanaceae Nwsl 3: 3–11.
16. Bai Y, Lindhout P (2007) Domestication and breeding of tomatoes: what have we gained and what can we gain the future? Ann Bot 100: 1085–1094.
17. Karihaloo JL, Rai M (1995) Significance of morphological variability in *Solanum insanum* L. (*sensu lato*). Plant Genet Res Nwsl 103: 24–26.
18. Behera TK, Sharma P, Singh BK, Kumar G, Kumar R, et al. (2006) Assessment of genetic diversity and species relationships in eggplant (*Solanum melongena* L.) using STMS markers. Sci Hort 107: 352–357.
19. Prohens J, Blanca JM, Nuez F (2005) Morphological and molecular variation in a collection of eggplant from a secondary center of diversity: implications for conservation and breeding. J Amer Soc Hort Sci 130: 54–63.
20. Li H, Chen H, Zhuang T, Chen J (2010) Analysis of genetic variation in eggplant and related *Solanum* species using sequence-related amplified polymorphism markers. Sci Hort 125: 19–24.
21. Ali Z, Xu ZL, Zhang DY, He XL, Bahadur S, et al. (2010) Molecular diversity analysis of eggplant (*Solanum melongena*) genetic resources. Genet Mol Res 10: 1141–1155.
22. Demir K, Bakir M, Sarikamiş G, Acunalp S (2010) Genetic diversity of eggplant (*Solanum melongena*) germplasm from Turkey assessed by SSR and RAPD markers. Genet Mol Res 9: 1568–1576.
23. Ge H, Li H, Liu Y, Li X, Chen H (2011) Characterization of novel developed expressed sequence tag (*EST*)-derived simple sequence repeat (*SSR*) markers and their application in diversity analysis of eggplant. Afr J Biotechnol 10: 9023–9031.
24. Erskine W, Muehlbauer FJ (1991) Allozyme and morphological variability, outcrossing rate and core collection formation in lentil germplasm. Theor Appl Genet 83: 119–125.
25. Silvestrini M, Junqueira M, Favarin AC, Guerreiro-Filho O, Maluf MP, et al. (2007) Genetic diversity and structure of Ethiopian, Yemen and Brazilian *Coffea arabica* L. accessions using microsatellites markers. Genet Res Crop Evol 54: 1367–1379.
26. Lu Y, Yan J, Guimarães CT, Taba S, Hao Z, et al. (2009). Molecular characterization of global maize breeding germplasm based on genome-wide single nucleotide polymorphisms. Theor Appl Genet 120: 93–115.
27. Strelchenko P, Okuizumi H, Shehzad T, Malinovskaya E, Kawase M, et al. (2010) Genetic relationships of sorghum germplasm in Asia and Africa revealed by rice cDNA-STS and indel markers. Japan Agric Resh Quarterly 44: 259–268.
28. Muñoz-Falcón JE, Prohens J, Vilanova S, Nuez F (2008) Characterization, diversity, and relationships of the Spanish striped (*Listada*) eggplants: a model for the enhancement and protection of local heirlooms. Euphytica 164: 405–419.
29. Muñoz-Falcón JE, Prohens J, Vilanova S, Ribas F, Castro A, et al. (2009) Distinguishing a protected geographical indication vegetable (*Almagro* eggplant) from closely related varieties with selected morphological traits and molecular markers. J Sci Food Agric 89: 320–328.

30. van der Weerden GM, Barendse GWM (2007) A web-searchable database developed for the EGGNET Project and applied to the Radboud University Solanaceae database. Acta Hort 745: 503–506.
31. Polignano G, Uggenti P, Bisignano V, Gatta C (2010) Genetic divergence analysis in eggplant (*Solanum melongena* L.) and allied species. Genet Res Crop Evol 57: 111–181.
32. Stàgel A, Portis E, Toppino L, Rotino GL, Lanteri S (2008) Gene-based microsatellite development for mapping and phylogeny studies in eggplant. BMC Genomics 9: 357.
33. Nunome T, Negoro S, Kono I, Kanamori H, Miyatake K, et al. (2009) Development of SSR markers derived from SSR-enriched genomic library of eggplant (*Solanum melongena* L.). Theor Appl Genet 119: 1143–1153.
34. Barchi L, Lanteri S, Portis E, Acquadro A, Valè G, et al. (2011) Identification of SNP and SSR markers in eggplant using RAD tag sequencing. BMC Genomics 12: 304.
35. Vilanova S, Manzur JP, Prohens J (2012) Development and characterization of genomic simple sequence repeat markers in eggplant and their application to the study of diversity and relationships in a collection of different cultivar types and origins. Mol Breed: doi 10.1007/s11032-011-9650-2.
36. Varshney RK, Graner A, Sorrells ME (2005) Genetic microsatellite markers in plants: features and applications. Trends Biotechnol 23: 48–55.
37. Kalia RK, Mai MK, Kalia S, Singh R, Dhawan AK (2011) Microsatellite markers: an overview of the recent progress in plants. Euphytica 177: 309–334.
38. Muñoz-Falcón JE, Vilanova S, Plazas M, Prohens J (2011) Diversity, relationships, and genetic fingerprinting of the *Listada de Gandía* eggplant landrace using genomic SSRs and EST-SSRs. Sci Hort 129: 238–246.
39. Vilanova S, Blasco M, Hurtado M, Muñoz-Falcón JE, Prohens J, et al. (2010) Development of linkage map of eggplant based on a *S. incanum* × *S. melongena* backcross generation. In: Prohens J, Rodríguez-Burruezo A, editors. Advances in genetics and breeding of Capsicum and eggplant. Valencia, Spain: Editorial Universitat Politècnica de València. 435–439.
40. Leinonen T, O'Hara RB, Cano JM, Merilä J (2008) Comparative studies of quantitative trait and neutral marker divergence: a meta-analysis. Journal Evol Biol 21: 1–17.
41. Allaby RG, Fuller DG, Brown TA (2008) The genetic expectations of a protracted model for the origins of domesticated crops. Proc Natl Acad Sci U S A 105: 13982–13986.
42. Doganlar S, Frary A, Daunay MC, Lester RN, Tanksley SD (2002) Conservation of gene function in the Solanaceae as revealed by comparative mapping of domestication traits in eggplant. Genetics 161: 1713–1726.
43. van Hintum TJL, Brown AHD, Spillane C, Hodgkin T (2000) Core collections of plant genetic resources. Rome, Italy: International Plant Genetic Resources Institute.
44. Burle ML, Fonseca JR, Kami JA, Gepts P (2010) Microsatellite diversity and genetic structure among bean (*Phaseolus vulgaris* L.) landraces in Brazil, a secondary center of diversity. Theor Appl Genet 121: 801–813.
45. Pessarakli MM, Dris R (2004) Pollination and breeding of eggplants. J Food Agric Environ 2: 218–219.
46. Nei M (1973) Analysis of genetic diversity in subdivided populations. Proc Natl Acad Sci U S A 70: 3321–3323.
47. Hedrick PW (2005) A standarized genetic differentiation measure. Evolution 59: 1633–1638.
48. Ægisdóttir HH, Kuss P, Stöcklin J (2009) Isolated populations of a rare alpine plant show high genetic diversity and considerable population differentiation. Ann Bot 104: 1313–1322.
49. Geleta LF, Labuschagne MT, Viljoen CD (2005) Genetic variability in pepper (*Capsicum annuum* L.) estimated by morphological data and amplified fragment length polymorphism markers. Biodivers Conserv 14: 2361–2375.
50. Veteläinen M, Gammelgard E, Valkonen JPT (2005) Diversity of Nordic landrace potatoes (*Solanum tuberosum* L.) revealed by AFLPs and morphological characters. Genet Res Crop Evol 52: 999–1010.
51. Bretting PK, Widrlechner MP (1995) Genetic markers and plant genetic resources management. Plant Breed Rev 13: 11–86.
52. Wendel JF, Doyle JJ (1998) Phylogenetic incongruence: window into genome history and molecular evolution. In: Soltis D, Soltis P, Doyle J, editors. Molecular systematics of plants II: DNA sequencing. Norwell, MA: Kluwer Academic Publishers. 265–296.
53. Massolo JF, Concellón A, Chaves AR, Vicente AR (2011) 1-Methylcyclopropene (1-MCP) delays senescence, maintains quality and reduces browning of non-climacteric eggplant (*Solanum melongena* L.) fruit. Postharvest Biol Technol 59: 10–15.
54. Doyle JJ, Doyle JL (1987) A rapid DNA isolation procedure for small quantities of fresh leaf tissue. Phytochem Bull 19: 11–15.
55. Schuelke M (2000) An economic method for the fluorescent labeling of PCR fragments. Nature Biotechnol 18: 233–234.
56. Jackson DA (1993) Stopping rules in principal components analysis: a comparison of heuristical and statistical approaches. Evolution 74: 2204–2214.
57. Pritchard JK, Stephens M, Donnelly P (2000) Inference of population structure using multilocus genotype data. Genetics 155: 945–959.
58. Falush D, Stephens M, Pritchard JK (2003) Inference of population structure using multilocus genotype data: linked loci and correlated allele frequencies. Genetics 164: 1567–1587.
59. Evanno G, Regnaut S, Goudet J (2005) detecting the number of clusters of individuals using the software STRUCTURE: a simulation study. Mol Ecol 14: 2611–2620.
60. Nei M (1987) Genetic variation within species. In: Nei M, editor. Molecular evolutionary genetics. New York, NY: Columbia University Press. 176–207.
61. Nei M (1972) Genetic distance among populations. Amer Naturalist 106: 283–292.
62. Mantel NA (1967) The detection of disease clustering and a generalised regression approach. Cancer Research 27: 209–220.

A Novel 5-Enolpyruvylshikimate-3-Phosphate Synthase Shows High Glyphosate Tolerance in *Escherichia coli* and Tobacco Plants

Gaoyi Cao[1,2]◈, Yunjun Liu[2]◈, Shengxue Zhang[2], Xuewen Yang[2], Rongrong Chen[2], Yuwen Zhang[1], Wei Lu[3], Yan Liu[2], Jianhua Wang[1], Min Lin[3], Guoying Wang[2]*

1 College of Agriculture and Biotechnology, China Agricultural University, Beijing, China, **2** Institute of Crop Sciences, Chinese Academy of Agricultural Sciences, Beijing, China, **3** Biotechnology Research Institute, Chinese Academy of Agricultural Sciences, Beijing, China

Abstract

A key enzyme in the shikimate pathway, 5-enolpyruvylshikimate-3-phosphate synthase (EPSPS) is the primary target of the broad-spectrum herbicide glyphosate. Identification of new *aroA* genes coding for EPSPS with a high level of glyphosate tolerance is essential for the development of glyphosate-tolerant crops. In the present study, the glyphosate tolerance of five bacterial *aroA* genes was evaluated in the *E. coli aroA*-defective strain ER2799 and in transgenic tobacco plants. All five *aroA* genes could complement the *aroA*-defective strain ER2799, and *AM79 aroA* showed the highest glyphosate tolerance. Although glyphosate treatment inhibited the growth of both WT and transgenic tobacco plants, transgenic plants expressing *AM79 aroA* tolerated higher concentration of glyphosate and had a higher fresh weight and survival rate than plants expressing other *aroA* genes. When treated with high concentration of glyphosate, lower shikimate content was detected in the leaves of transgenic plants expressing *AM79 aroA* than transgenic plants expressing other *aroA* genes. These results suggest that *AM79 aroA* could be a good candidate for the development of transgenic glyphosate-tolerant crops.

Editor: Alejandra Bravo, Universidad Nacional Autonoma de Mexico, Instituto de Biotecnologia, Mexico

Funding: This work was supported by the National Major Project for Transgenic Organism Breeding (2008ZX08003-001) and by the Scientific Research Foundation for Returned Overseas Chinese Scholars of the Ministry of Education of China. The funders had no role in study design, data collection and analysis, decision to publish, or preparation of the manuscript.

Competing Interests: The authors have declared that no competing interests exist.

* E-mail: gywang604@gmail.com

◈ These authors contributed equally to this work.

Introduction

Glyphosate (N-phosphonomethyl glycine), an important and potent herbicide, is widely used to control weeds in agricultural fields. Glyphosate inhibits the enzyme 5-enolpyruvylshikimate-3-phosphate synthase (EPSPS;EC 2.5.1.19), which converts phosphoenolpyruvate (PEP) and shikimate-3-phosphate (S3P) to 5-enolpyruvylshikimate-3-phosphate (EPSP) and shuts down the shikimate pathway, leading to plant death. Two types of EPSP synthases have been classified [1]. Type I EPSP synthases have been identified mainly in plants and bacteria, and type II EPSP synthases have been identified in some forms of bacteria. Type I EPSP synthases are naturally sensitive to glyphosate, whereas type II EPSP synthases are tolerant of glyphosate [2–4]. After the primary target of glyphosate was identified as EPSPS in the 1980s [5], EPSPS became the top choice for the development of transgenic glyphosate-tolerant crops. However, over-expression of most wild EPSPS in transgenic plants can not confer plants with glyphosate tolerance [6].

Mutagenesis of EPSPS is one way to obtain glyphosate-tolerant EPSPS. It was first reported that the expression of an altered EPSPS confers resistance to the herbicide glyphosate [7]. Expression of the *Salmonella typhimurium* EPSPS mutant (Pro101 to Ser) in transgenic tobacco plants confers tolerance to glyphosate [8]. Mutagenesis and structure analysis have revealed the mechanism of EPSPS function [4,9–12]. A single Ala residue at position 100 leads to the CP4 EPSPS (obtained from *Agrobacterium* sp. strain CP4) becoming insensitive to glyphosate, while natural plant and bacterial enzymes share a highly conserved Gly residue at this position [4]. Double mutation (T97I, P101S) of type I EPSPS cause the shift of the Gly residue at position 96 toward the glyphosate binding site, leading to glyphosate tolerance [12]. A P106L mutant of rice EPSPS was selected based on the directed evolution strategy and conferred high glyphosate tolerance in *E. coli* and in transgenic tobacco plants [13]. The change of two codons provides the maize EPSPS with glyphosate tolerance [14]. A proline-to-serine substitution at position 106 in the goosegrass's predicted EPSPS mature protein coding region produces a five fold higher glyphosate tolerance capability than the sensitive biotype [15].

In *E. coli*, EPSPS is encoded by *aroA*. Glyphosate-tolerant EPSPS can also be obtained by screening the bacteria grown in a glyphosate-contamination environment. Organisms can survive in chemically stressed environments through physiological adaptation resulting from modifications of gene expression or through adaptive mutation to relieve selective pressure [16–18]. In recent years, a number of glyphosate-resistant *aroA* genes have been cloned from bacteria [19–21]. However, only one of the type II EPSPS genes from *Agrobacterium* sp. CP4 and the point mutations of class I EPSPS from *E. coli* or maize have been successfully used in commercial transgenic crops [4].

In our previous work, several bacterial *aroA* genes (*G2 aroA*, *HTG7 aroA*, *A1501 aroA*, *RD aroA* and *AM79 aroA*), were cloned from *Pseudomonas fluorescens* [22], *Halomonas variabilis* [23], *Pseudomonas stutzeri* [24], and uncultured soil bacteria [25–26], respectively. Here, we evaluated the glyphosate tolerance of these five *aroA* genes in *E. coli* and in transgenic tobacco plants. Our results showed that *AM79 aroA* is potentially a better candidate for development of transgenic glyphosate-tolerant crops than other *aroA* genes.

Results

Evaluation of five *aroA* genes in the *aroA*-deficient *E. coli* mutant

Five *aroA* genes, named as *G2 aroA*, *HTG7 aroA*, *A1501 aroA*, *RD aroA* and *AM79 aroA*, were cloned from the bacteria grown in soil heavily contaminated with a high concentration of glyphosate [22–25]. An alignment analysis of these five EPSPS proteins with some other known EPSPS proteins showed that G2 EPSPS and AM79 EPSPS belong to class I, while HTG7 EPSPS, A1501 EPSPS and RD EPSPS belong to class II EPSPS (Figure 1).

To evaluate the glyphosate tolerance, these five genes were cloned into the *Eco*RI site of the low-copy plasmid pACYC184 and transformed into the *E. coli* strain ER2799, which is a stable *aroA*-defective mutant that cannot grow in minimal medium. The ER2799 containing different *aroA* genes could grow in the minimal M9 liquid medium with different concentrations of glyphosate. In the M9 medium containing 20 or 100 mM glyphosate, all strains expressing *aroA* genes grew well. When the glyphosate concentration was increased to 150 mM and 200 mM, strains expressing *RD aroA* and *HTG7 aroA* grew slower than the bacteria transformed with the other three *aroA* genes, indicating a lower glyphosate tolerance of *RD aroA* and *HTG7 aroA*. Although similar transcript levels were detected in strains expressing these *aroA* (Figure S1), strains expressing *AM79 aroA* and *A1501 aroA* showed significantly higher glyphosate tolerance than others (Figure 2), suggesting that the difference in glyphosate tolerance among strains expressing these *aroA* was due to the enzyme itself.

Comparson of enzyme kinetic parameters among the five bacterial EPSPS

Five *aroA* genes were cloned into pET-28a expression vector and the expressed proteins in *E. coli* were purified for enzyme assay. AM79 EPSPS and RD EPSPS had lower K_m values (14.59 and 7.34),

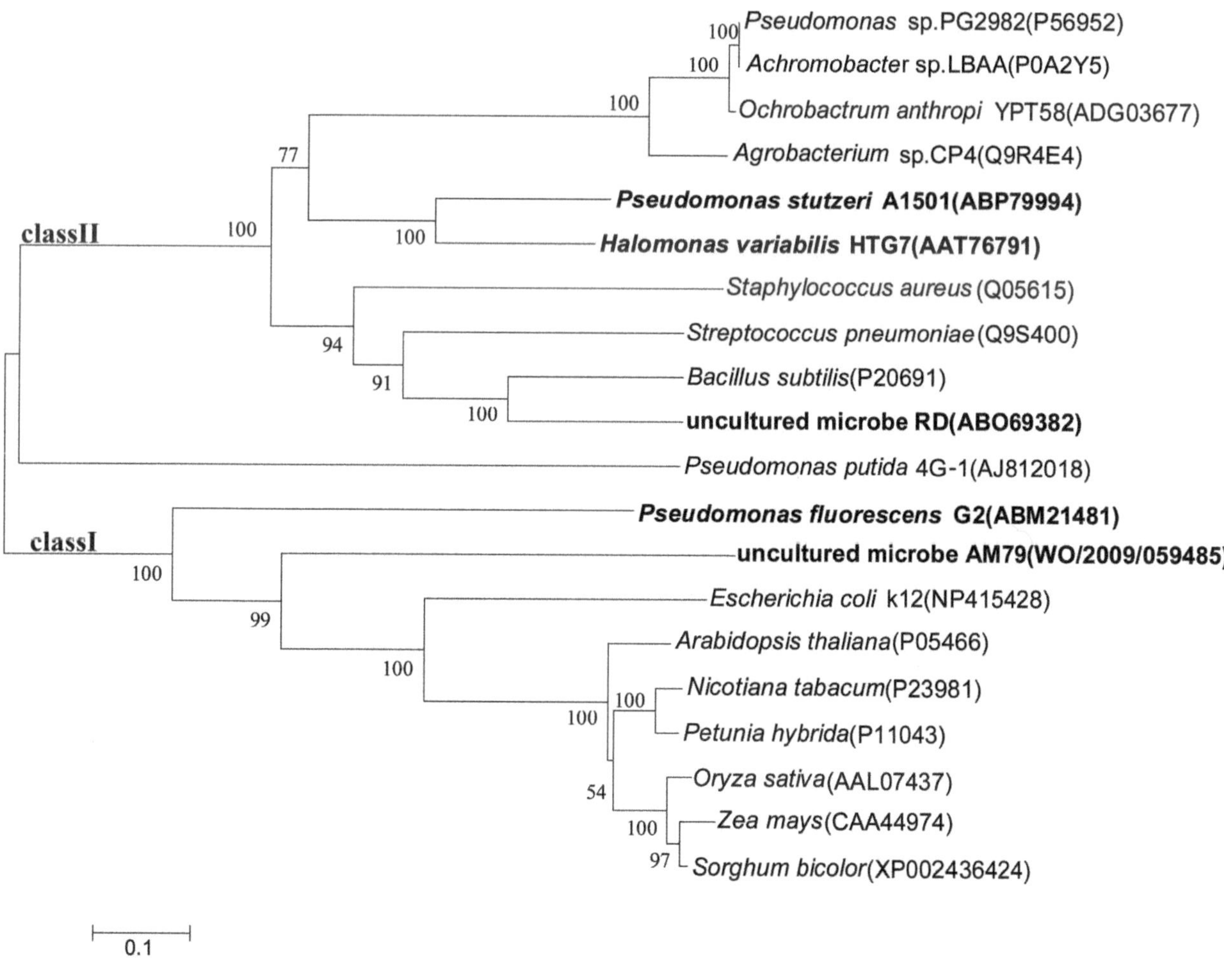

Figure 1. Phylogenetic analysis of the five EPSPS and related proteins. The phylogenetic tree was based on homologous sequences of the EPSPS proteins and the neighbor-joining methods (MEGA4.0). The percentage of the tree from 1000 bootstrap resamples supporting the topology is indicated when above 50. Accession numbers or international patent publication numbers are shown in parentheses. The scale bar represents 0.1 substitutions per position.

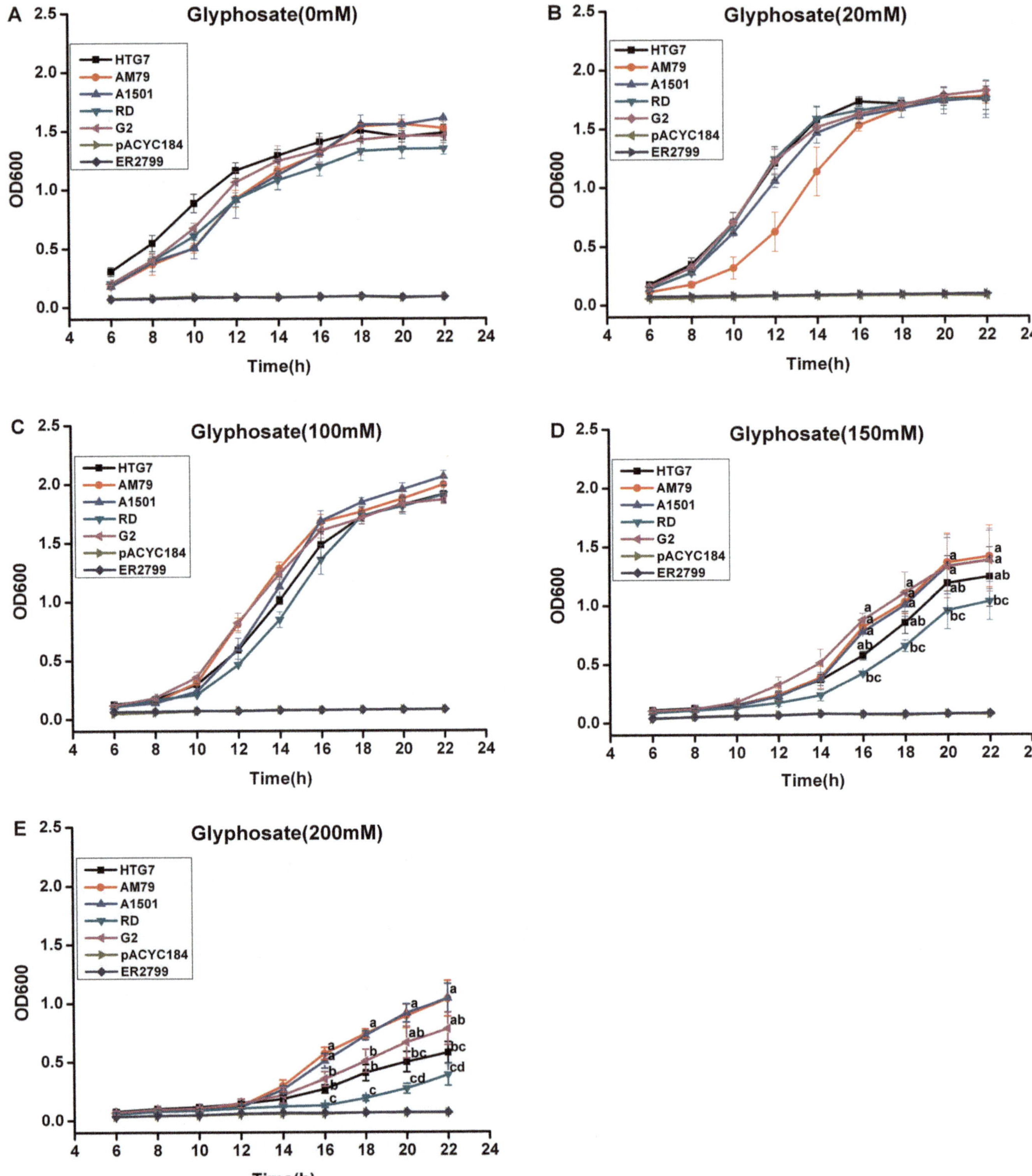

Figure 2. Glyphosate tolerance of *E. coli* containing five *aroA* genes. The plasmids pACYC184, pACYC-HTG7, pACYC-AM79, pACYC-A1501, pACYC-RD and pACYC-G2 were transformed into *E. coli* ER2799 competent cells for growth curve measurement. M9 liquid medium was supplemented with different concentrations of glyphosate. OD_{600} was recorded every two hours starting 6 h after treatment. Data are shown as the average ± S.E. of three independent experiments. Experimental data was tested by ANOVA analysis and different letter means significant difference at $P<0.05$ level. (A) Growth curve of ER2799 and the strain harboring different plasmids under 0 mM glyphosate. (B) Growth curve of ER2799 and the strain harboring different plasmids under 20 mM glyphosate. (C) Growth curve of ER2799 and the strain harboring different plasmids under 100 mM glyphosate. (D) Growth curve of ER2799 and the strain harboring different plasmids under 150 mM glyphosate. (E) Growth curve of ER2799 and the strain harboring different plasmids under 200 mM glyphosate.

indicating they had higher PEP affinities than other EPSPS proteins. A1501 EPSPS and AM79 EPSPS had higher K_i values than other three EPSPS proteins, meaning that they can tolerate high glyphosate concentrations (Table 1; Figure S2). Since glyphosate competes with PEP for binding in the active site of EPSPS, high K_i/K_m is critical for EPSPS to maintain enzymatic activity in the presence of glyphosate inhibitor. AM79 EPSPS had the highest K_i/K_m value (10.6) among the five EPSPS (Table 1), indicating *AM79 aroA* is a good candidate gene for transgenic crops.

Transgenic tobacco seedlings over-expressing AM79 aroA showed higher glyphosate tolerance

To investigate whether *AM79 aroA* is preferable to the development of glyphosate-resistant transgenic crops, we assessed the glyphosate tolerance conferred by the five *aroA* genes in transgenic tobacco plants. The signal peptide sequence of the pea rib-1,5-bisphospate carboxylase (*rbcS*) small subunit was fused in front of the *aroA* gene with the correct reading frame to direct the EPSPS to the plant chloroplast. The signal peptide and the EPSPS were controlled by the CaMV 35S promoter and cloned into the plasmid pCAMBIA3301 to construct the vectors for plant transformation (Figure 3). The vectors were introduced into tobacco plants (*Nicotiana tabacum* var. *Samsum*) via *Agrobacterium*-mediated transformation.

Tobacco T1 progeny seeds were germinated on MS medium without glyphosate using 10 mg L^{-1} phosphinothricin as a selective reagent to eliminate non-transgenic plant. The transgenic plants were confirmed by PCR amplification (Figure S3) and Southern blot analysis, which showed that most transgenic lines have one or two copy of transgenes (Figure 4). Our results also showed that the signal peptide did improve the glyphosate tolerance conferred by the *G2 aroA* in transgenic tobacco (Figure S4).

Seven to ten independent transgenic lines for each gene were chosen to assess glyphosate tolerance. Transgenic tobacco seedlings were grown vertically on the MS medium containing 1 mM glyphosate for fifteen days. On the medium without glyphosate, all of the transgenic plants grew similarly to the WT plants (Figure 5F). Treatment with 1 mM glyphosate killed the WT plants and inhibited the growth of the transgenic plants over-expressing the *aroA* gene (Figure 5). The transgenic plants expressing different *aroA* genes showed a different response to the 1 mM glyphosate treatment. Plants over-expressing *AM79 aroA* grew more quickly than the plants expressing other *aroA* genes, and their leaves remained green 15 days after treatment with 1 mM glyphosate.

The fresh weight of seven to ten independent transgenic lines for each gene was measured to assess the inhibition of plant growth by glyphosate. The significant difference was observed among transgenic plants expressing various *aroA*, and transgenic plants expressing *AM79 aroA* had the highest fresh weight (Figure 5). These results clearly show that the over-expression of *AM79 aroA* in plants offers higher glyphosate tolerance than other genes. We further chose five transgenic lines for each construct to investigate the transcription of *aroA* genes using quantitative real-time RT-PCR. Relative lower transcription level occurred for *A1501 aroA* and *G2 aroA* compared with other three *aroA* genes, might due to their higher GC content. Different transcription levels were observed among the transgenic lines expressing the same *aroA* (Figure S5), and the transcription level was correlated with the glyphosate tolerance of transgenic lines (data not shown).

The glyphosate tolerance of WT tobacco and transgenic seedlings were also assessed at various concentrations (Figure 6). The results showed that WT plants were only able to tolerate 0.1 mM glyphosate or less. However, transgenic tobacco seedlings expressing different *aroA* genes could tolerate a much higher concentrations of glyphosate. Compared to transgenic lines expressing other *aroA*, tobacco plants over-expressing *AM79 aroA* were more tolerant and could keep green leaves on the medium containing 10 mM glyphosate (Figure 6).

Table 1. Kinetic parameters of five EPSPS enzymes

	V_{max}(U mg^{-1})	K_m (PEP; μM)	K_i (glyphosate; μM)	K_i/K_m
HTG7	55.80±3.04	39.06±2.69	74.11±10.89	1.9
AM79	48.62±0.36	14.59±0.11	154.39±4.06	10.6
A1501	16.67±1.58	92.42±22.07	467.35±30.21	5.1
RD	94.08±4.67	7.34±1.97	28.83±4.50	3.9
G2	38.16±2.58	49.35±2.68	44.71±5.20	0.9

Data was shown as mean ± S.D. of three independent experiments.

Transgenic tobacco plants over-expressing AM79 aroA grown in a greenhouse showed higher glyphosate tolerance

To further confirm the glyphosate tolerance of transgenic plants, six-to-eight-leaf stage transgenic tobacco plants grown in a greenhouse were sprayed with Roundup® at an equal dose of 6 L ha^{-1}. Fifteen days after treatment, all WT tobacco plants died, whereas some transgenic plants remained living (Figure 7A). A survival rate of only approximately 10% was observed for transgenic plants expressing *A1501 aroA*, *RD aroA* and *G2 aroA*. Transgenic plants expressing *HTG7 aroA* or *AM79 aroA* had survival rates of 35.01% and 88.75%, respectively. ANOVA analysis showed significant difference among transgenic lines (Figure 7B). These results confirm that the over-expression of *AM79 aroA* can provide plants with a high glyphosate tolerance.

Transgenic plants over-expressing AM79 aroA had lower shikimate accumulation after glyphosate treatment

Glyphosate shuts down the shikimate pathway by inhibiting the EPSPS enzymes that convert PEP and S3P to EPSP. Glyphosate treatment increases the shikimate content and is thus an indicator of plant tolerance to glyphosate [27]. Tobacco plants were sprayed with 0.5, 1, or 2 L ha^{-1} Roundup®, and the shikimate content in the leaves were measured after 0, 1, 3, 5 and 7 d of treatment. Before treatment, WT and transgenic plants had similar shikimate contents. Treatments with different amounts of glyphosate all increased the shikimate contents in the leaves of WT and transgenic plants, reaching a maximum on the fifth day. However, significantly more shikimate contents occurred in WT leaves than that in transgenic plant leaves (Figure 8). When treated with high glyphosate concentration (2 L ha^{-1}), the lowest shikimate accumulation occurred in the transgenic plants expressing *AM79 aroA*, compared with transgenic plants expressing other *aroA* genes, indicating AM79 EPSPS can confer transgenic plants with high glyphosate tolerance.

Discussion

EPSPS were chosen as the first choice to develop transgenic glyphosate-tolerant crops because they are the primary target of glyphosate [5]. Since the 1980s, researchers have begun to isolate glyphosate-insensitive EPSPS from bacteria or plants, and numerous promising enzymes have been identified by microbial

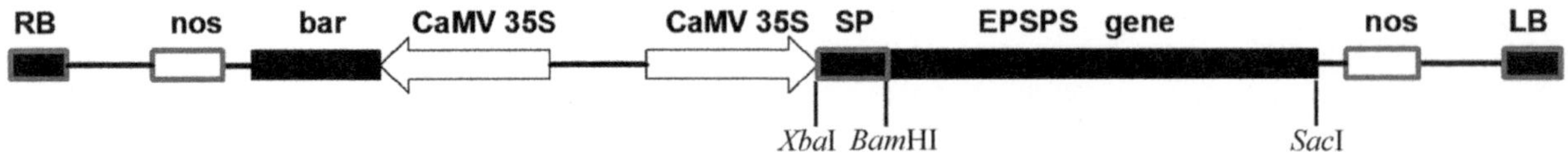

Figure 3. T-DNA cassette containing *aroA* gene for plant transformation. LB and RB, left and right border of T-DNA; bar, phosphinothricin acetyltransferase gene; nos, *NOS* terminator; CaMV 35S, cauliflower mosaic virus 35S promoter; SP, coding sequence of signal peptide of pea rib-1,5-bisphospate carboxylase (*rbcS*) small subunit; EPSPS gene, five microorganism *aroA* gene.

screening and selective evolution [4,13,28,29]. The *Agrobacterium* sp. strain CP4 is a naturally occurring, glyphosate-tolerant microbe in environments contaminated with high concentration of glyphosate, and the CP4 EPSPS has been commercially used in genetically modified crops [30]. Some other types of EPSPS were identified from other bacteria species such as *Streptococcus pneumonia* [31], *Ochrobactrum anthropi* [21], *Pseudomonas* sp. PG2982 [32] and *Staphylococcus aureus* [20].

Five *aroA* genes (*G2 aroA*, *HTG7 aroA*, *A1501 aroA*, *RD aroA* and *AM79 aroA*) have been isolated from highly glyphosate-tolerant bacterial strains [22–26]. Because the glyphosate tolerance of the native bacteria cannot mirror the tolerance of EPSPS itself, these five genes were cloned into the *Eco*RI site of the low-copy plasmid pACYC184 and transformed into the *aroA* mutant *E. coli* strain ER2799 to evaluate their glyphosate tolerance. Under low concentrations of glyphosate, all bacteria transformed with *aroA* genes showed tolerance to glyphosate, and their tolerant levels were similar. When the glyphosate concentration increased to 150 mM and 200 mM, *E. coli* strains containing *RD aroA* and *HTG7 aroA* grew slower than strains containing other three *aroA* genes, indicating a lower glyphosate tolerance of *RD aroA* and *HTG7 aroA* (Figure 2D, E). Similar transcript level of *aroA* were detected in different strains, suggesting that the different degree of glyphosate tolerance of individual EPSPS was due to the enzyme itself but not due to the different transcript level. At 200 mM glyphosate concentration, *AM79 aroA* and *A1501 aroA* showed

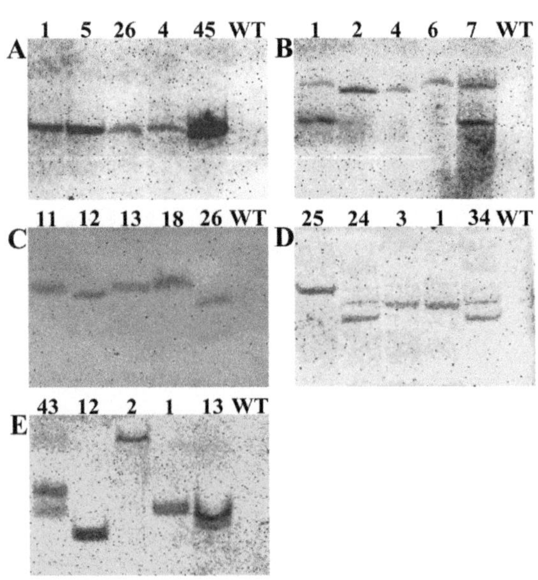

Figure 4. Southern blot analysis of transgenic tobacco plants. (A) HTG7; (B) AM79; (C) A1501; (D) RD; (E) G2. Five transgenic lines each construct were analyzed. 100 μg genomic DNA was digested with *Hin*dIII which has only one site in the plasmid, so the band numbers are equal to the copy number of transgene.

more glyphosate tolerance than other *aroA* genes. We further compared AM79 EPSPS with the commercially used CP4 EPSPS from *Agrobacterium* sp. strain CP4. Both genes were cloned into the *Bam*HI/*Sal*I site of the low-copy plasmid pACYC184 and transformed into the *E. coli* strain ER2799. The results showed that AM79 EPSPS could provide glyphosate tolerance as effective as CP4 EPSPS (Figure S6).

The native bacterial strains tolerated high concentrations of glyphosate. The *Halomonas variabilis* strain HTG7 grew quickly on MOPS agar containing 300 mM glyphosate [23]; however, the *E. coli* strain ER2799 containing *HTG7 aroA* did not grow well under the 200 mM glyphosate condition. In the original microbial strain, other genes may also be involved in glyphosate tolerance by hydrolysis, acetylation, and oxidative cleavage of glyphosate. The glyphosate N-acetyltransferase (*GAT*) gene and the glyphosate oxidoreductase (*GOX*) gene could both take part in these pathways [33,34]. Therefore, evaluation of the tolerance level in the same system (e.g. in the *E. coli* strain ER2799) is a better means to evaluate the glyphosate tolerance of individual *aroA* genes from different strains.

K_i value of EPSPS enzyme reflects the affinity of EPSPS binding glyphosate, and high K_i/K_m is critical for EPSPS to maintain enzymatic activity in the presence of glyphosate inhibitor. For the *E. coli aroA* with T97I/P101S double mutation, which has been used to develop glyphosate-tolerant crops, its K_i/K_m value is 24 [12]. CP4 EPSPS, which has also been used in glyphosate-tolerant maize and soybean, also has high K_i/K_m value which is 32 [4]. AM79 EPSPS had the highest K_i/K_m value (10.6) among the five EPSPS enzymes (Table 1), indicating *AM79 aroA* is a good candidate gene to develop transgenic crops. A1501 EPSPS had the highest K_i value (467.35) among the five EPSPS enzymes, indicating that it can tolerate high glyphosate concentrations. However, A1501 EPSPS had lower K_i/K_m value (5.1) than AM79 EPSPS, which might be the reason that the glyphosate tolerance of transgenic plants over-expressing *A1501 aroA* was not as good as the transgenic plants over-expressing *AM79 aroA*.

It has been shown that the *G2 aroA* gene can confer a high tolerance to glyphosate in transgenic tobacco plants [22]. To investigate which *aroA* gene is the most suitable for the development of glyphosate resistant transgenic crops, we assessed the glyphosate tolerance of five genes in transgenic tobacco plants. Because EPSPS are located in chloroplasts [35], a signal peptide was added in front of EPSPS to traffic it to the chloroplast. Our results showed that including a signal peptide did improve the glyphosate tolerance plants expressing the *G2 aroA* gene in transgenic tobacco compared to that without a signal peptide (Figure S4). Integration of the *aroA* gene into the chloroplast genome is another way to locate the EPSPS in the chloroplast and to provide transgenic plants with glyphosate tolerance, and it has been shown that the *CP4 aroA* gene that was integrated in the chloroplast genome and expressed in chloroplast provided a high level of glyphosate tolerance in tobacco [36].

Glyphosate can inhibit plant growth by shutting down the shikimate pathway [15,27]. The WT tobacco plants only tolerated

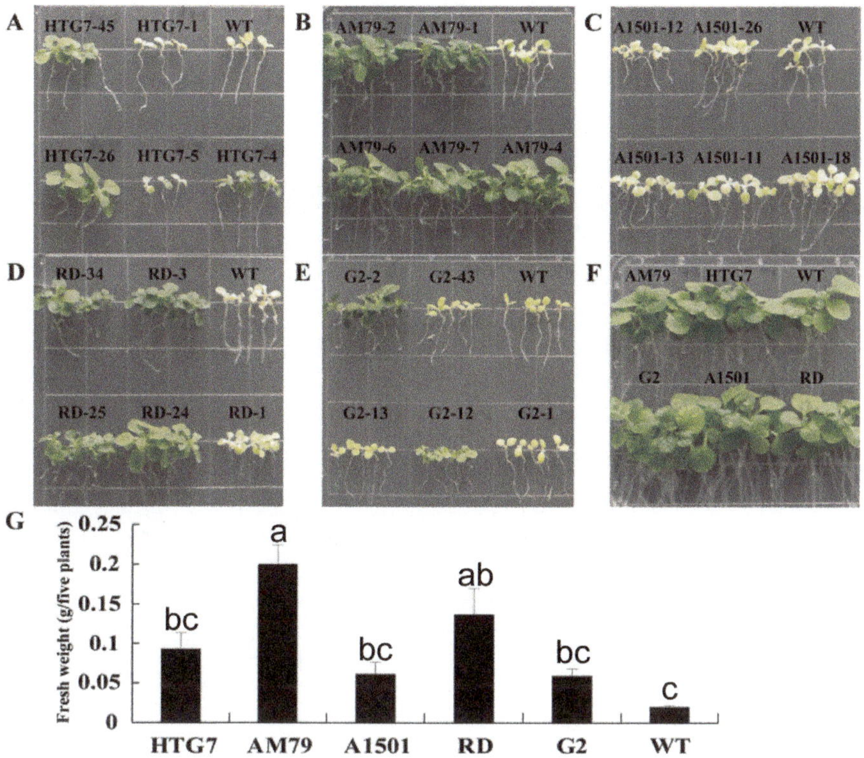

Figure 5. Glyphosate tolerance of transgenic tobacco seedlings on plates containing 1 mM glyphosate. T1 tobacco seeds were germinated on the MS medium containing 10 mg L^{-1} PPT and grown for 7 days. The live seedlings were transferred to MS medium on plates containing 1 mM glyphosate. Photographs were taken two weeks later, and the fresh weight was measured at the same time. (A) Photograph of tobacco harboring *HTG7 aroA* and WT grown on the medium containing 1 mM glyphosate. (B) Photograph of tobacco harboring *AM79 aroA* and WT grown on the medium containing 1 mM glyphosate. (C) Photograph of tobacco harboring *A1501 aroA* and WT grown on the medium containing 1 mM glyphosate. (D) Photograph of tobacco harboring *RD aroA* and WT grown on the medium containing 1 mM glyphosate. (E) Photograph of tobacco harboring *G2 aroA* and WT grown on the medium containing 1 mM glyphosate. (F) Photograph of tobacco plants grown on the medium without glyphosate. (G) Fresh weight of the tobacco plants. Data are shown as the average ± S.E. of seven to ten independent transgenic lines. Experimental data was tested by ANOVA analysis and different letter in each column means significant difference at $P<0.05$ level.

0.05 mM glyphosate. 0.1 mM glyphosate caused the WT plants to turn yellow, and concentrations higher than 0.1 mM could kill the WT plants (Figure 6). These findings are consistent with the results of other reports [13,22]. It has also been shown that transgenic tobacco plants expressing *G2 aroA* can tolerate 1 mM glyphosate [22], and our results with new transgenic lines confirmed this fact and indicated that the transgenic plants can tolerate up to 2 mM glyphosate. The glyphosate tolerance of plants expressing the other four *aroA* genes has not been evaluated in transgenic tobacco plants before. Here, we show that tobacco plants expressing *AM79 aroA* can grow well in medium containing 10 mM glyphosate, whereas for transgenic plants expressing the other four genes, 10 mM glyphosate leads to yellow leaves and shorter roots. These results indicate that *AM79 aroA* can provide transgenic plants with a higher glyphosate tolerance than the other four *aroA* genes can. The above conclusion was confirmed by the result that transgenic plants expressing *AM79 aroA* had the highest fresh weight after 15 days of growth on the medium containing 1 mM glyphosate, and also by the result that the transgenic plants expressing *AM79 aroA* had the highest survival rate after being sprayed with 6 L ha^{-1} Roundup® in a greenhouse.

Copy numbers of transgene in transgenic plants were investigated and the results showed that most transgenic plants had one or two copies (Figure 4). Even though some transgenic lines containing the same gene all have a single copy of transgene, different transcription level was observed (Figure S5), might due to the different insertion site in the genome, leading to the different glyphosate tolerance level of these lines. Different genes have similar transcription level, except that relative lower transcription level occurred for *A1501 aroA* and *G2 aroA* compared with other three *aroA* genes, which might be due to higher gene GC content of *A1501 aroA* (64.47% GC content) and *G2 aroA* (64.49% GC content). It have been reported that GC content is positively associated with transcript abundance [37], and the bacterial *cbnA* gene was not expressed in tobacco BY-2 cells under the control of 35S promoter, due to the high 65% GC content [38].

Treatment of plants with glyphosate inhibits EPSPS activity and leads to the accumulation of shikimate [1,39]. It has been observed that glyphosate treatment causes the accumulation of shikimate in glyphosate-insensitive tobacco cells, while there is a lack of shikimate accumulation in tolerant cells [27]. Glyphosate causes a marked increase of shikimate in both the control and tolerant alfalfa cell lines, but the accumulation of shikimate is lower in tolerant calli [40]. Consistent with above reports, treatment with a different concentration of glyphosate induced the accumulation of shikimate in WT and transgenic tobacco plants in this study. However, lower shikimate accumulation in transgenic plants over-expressing *AM79 aroA* indicated their higher glyphosate tolerance than the plants xpressing other *aroA* genes (Figure 8).

To better understand the possible mechanisms for the good glyphosate tolerance of AM79 EPSPS, we conducted sequence analysis (Figure S7). Amino acids 90–104 are strictly conserved in

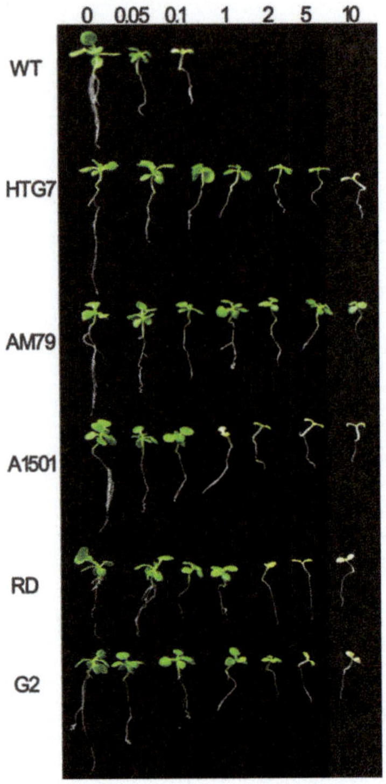

Figure 6. Glyphosate tolerance of transgenic tobacco under different glyphosate concentrations. Tobacco seeds of generation T1 were germinated on the MS medium containing 10 mg L^{-1} PPT and grown for 7 days at 100 µmol s^{-1} m^{-2} with a 16 h light/8 h dark period. The live seedlings were transferred to MS medium in plates containing different amounts of glyphosate and grown vertically for another two weeks.

class I EPSPS [12]. In this region, the normal active site exists in the site 106 (number according to *E. coli* EPSPS) of class I EPSPS is Pro while is Leu is type II EPSPS. However, the residue equivalent of this site in AM79 EPSPS is Phe114 (Figure S7). Glu354 in the CP4 EPSPS amino acid sequence plays an important role in the EPSPS catalytic action for its binding capacity in the complex with the substrate [4]. This active site preserved uniformity in all the EPSPS in the equivalent regions, but AM79 EPSPS has a His356 close to Glu357 before the active site, different from with the other EPSPS enzymes (Figure S7). Glyphosate binds EPSPS in the active site by occupying the PEP-binding site, causing the inhibition of EPSPS activity. Some reports revealed that certain adjacent amino acids beside the active site can drive a shift of the EPSPS structure, producing a higher glyphosate tolerance [12]. The special amino acids Ala107 and Phe114, and the unique amino acids Ala355, His356 in AM79 EPSPS, may take part in a similar function by inhibiting the bind of glyphosate to EPSPS. This structure basis might contribute to the superior glyphosate tolerance of AM79 EPSPS.

In summary, we evaluated the glyphosate tolerance of five bacteria *aroA* genes in *E. coli* and transgenic tobacco plants and confirmed that *AM79 aroA* showed a higher glyphosate tolerance than that conferred by the other four *aroA* genes, indicating that *AM79 aroA* is a good candidate for the development of herbicide-tolerant crops. The transformation of *AM79 aroA* to crops such as maize and soybean is underway.

Materials and Methods

Materials

The *E. coli* strain ER2799 (with the *aroA* deleted in its genome) [41] was from the Biotechnology Research Institute, Chinese Academy of Agricultural Sciences. *Agrobacterium tumefaciens* LBA4404, plasmid pCAMBIA3301, and wild-type tobacco plants (*Nicotiana tabacum* var. *Samsum*) were preserved in our laboratory. The low-copy plasmid pACYC184, T4 DNA ligase, and restriction enzymes were obtained from New England Biolabs.

Plasmid construction

Five *aroA* genes were amplified by PCR with the *Eco*RI site added on both sides of the gene and cloned into the *Eco*RI site of the pACYC184 plasmid to construct the vectors pACYC-HTG7, pACYC-AM79, pACYC-A1501, pACYC-RD and pACYC-G2. *CP4 aroA* gene (GenBank AB209952) was artificially synthesized and cloned into the *Bam*HI/*Sal*I site of the low-copy plasmid pACYC184 and transformed into the *E. coli* strain ER2799. All primers used in the research are listed in Table S1.

To express and purify the EPSPS proteins, five *aroA* genes were amplified by PCR adding *Bam*HI and *Hin*dIII at the 5′ end and 3′ end, respectively. The stop codon of each gene was knockout by PCR process for enhancing the binding capability by using two His tags. The five *aroA* genes were digested with *Bam*HI and *Hin*dIII and cloned into pET-28a (Novagen) digested with the same enzymes. The constructed plasmids were transformed into strain Rosetta (DE3).

To construct the plant transformation vectors, a 35S-gus-nos fragment digested from pBI121 with *Hin*dIII and *Eco*RI was cloned into the same site of pCAMBIA3301, and then the *gus* was replaced with a *G2 aroA* fragment digested with *Bam*HI and *Sac*I to construct the vector p3301-121G2. The signal peptide of the pea rib-1,5-bisphospate carboxylase (*rbcS*) small subunit was amplified using the PCR method, and *Xba*I and *Bam*HI were added to the upstream and downstream ends of the signal peptide, respectively. The signal peptide was then digested with *Xba*I and *Bam*HI and fused in front of *G2 aroA* to construct the plant transformation vector p3301-121spG2. The *G2 aroA* in vector p3301-121spG2 was replaced with *HTG7 aroA*, *AM79 aroA*, *A1501 aroA* and *RD aroA* to obtain the vectors p3301-121spHTG7, p3301-121spAM79, p3301-121spA1501 and p3301-121spRD, respectively. All plasmids used in the research are listed in Table S2.

Evaluation of glyphosate tolerance in *E. coli*

pACYC184, pACYC-HTG7, pACYC-AM79, pACYC-A1501, pACYC-RD and pACYC-G2 were transformed into *E. coli* ER2799 competent cells and plated on LB solid medium with 50 µg mL^{-1} tetracycline. The positive clones were identified using PCR amplification of the *aroA* genes. The single clone was inoculated in 5 mL LB liquid medium and grown overnight at 37°C until the OD_{600} reached approximately 0.6. All of the cultures were centrifuged, and the pellets were resuspended to $OD_{600} = 0.5$ with M9 liquid basic medium. Then the 500 µL cultures were subcultured to 200 mL M9 liquid medium containing 0, 20, 100, 150 and 200 mM glyphosate. The absorbance at OD_{600} was measured at 6, 8, 10, 12, 14, 16, 18, 20 and 22 h. The experiment was repeated three times. The empty strain ER2799 was detected as a control.

Enzyme assay

Single colony of the transformant was grown overnight in LB medium with 100 µg mL^{-1} kanamycin and 25 µg mL^{-1} chloromycetin at 37°C. 1 mL of overnight culture was added into

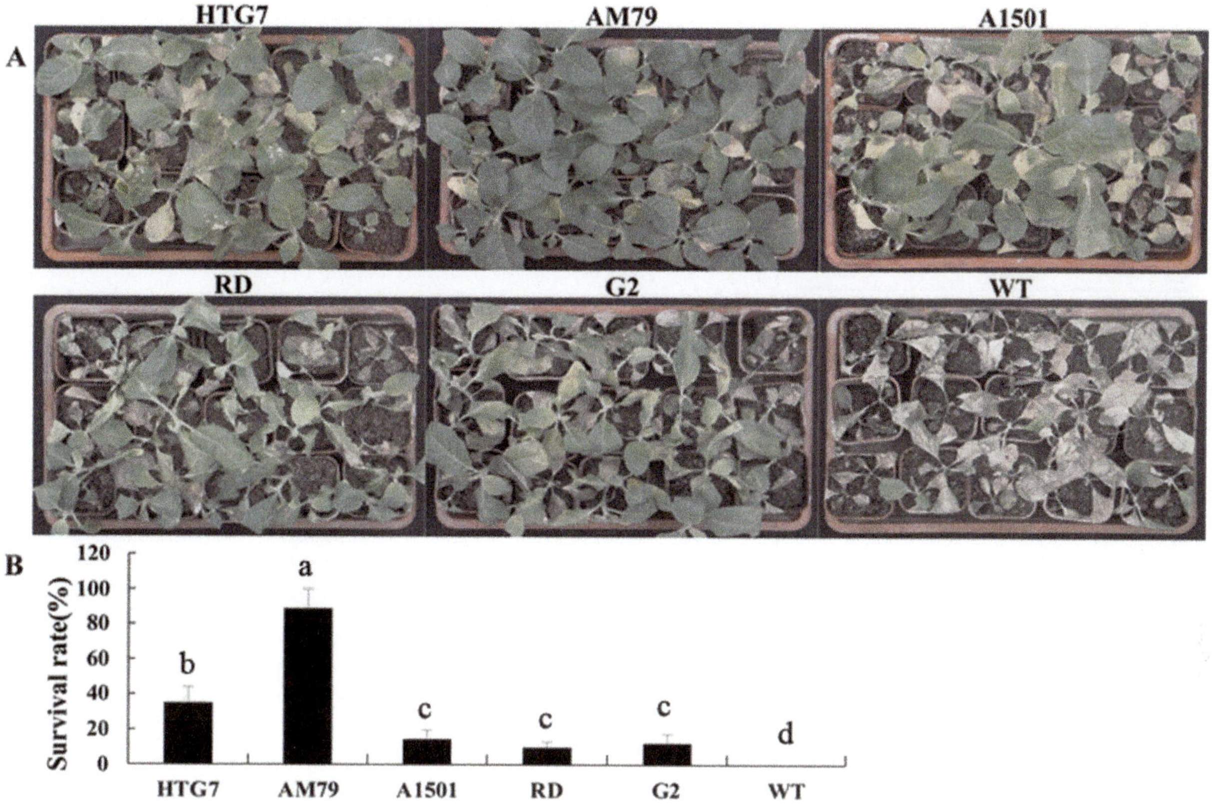

Figure 7. Glyphosate tolerance of transgenic tobacco plants in a greenhouse when sprayed with 6 L ha^{-1} Roundup®. T1 tobacco seeds were germinated on the MS medium containing 10 mg L^{-1} PPT and grown for 7 days at 100 μmol s^{-1} m^{-2} with a 16 h light/8 h dark period. The live seedlings were transferred to soil in pots and grown for another month. Six-to eight-leaf stage transgenic plants were sprayed with 6 L ha^{-1} Roundup®. Two weeks after treatment, injury was observed, and survival rate was measured. (A) Photograph of tobacco plants two weeks after glyphosate treatment. (B) Survival rate of tobacco plants. Data are shown as the average ± S.E. of seven to ten independent transgenic lines. Experimental data was tested by ANOVA analysis and different letter in each column means significant difference at $P<0.05$ level.

100 mL LB medium and grew to $OD_{600}=0.4$, then 0.5 mM isopropyl β-D-thiogalactoside (IPTG) was added and the cells were incubated for another 4 h at 28°C or for 12 h at 16°C. The proteins were purified using Ni^{2+}-NTA spin column (Qiagen) according to the described protocol. Protein samples were analyzed by SDS-PAGE electrophoresis.

The activity of EPSPS was assayed at 25°C in a 20 μl mixtures containing 50 mM HEPES-NaOH (pH 7.0), 2 mM dithiothreitol, 100 mM KCl, 1 mM S3P and varied concentrations of glyphosate and PEP [4,42].The reaction was initiated by the addition of enzyme (19 μg mL^{-1} HTG7 EPSPS, 19.2 μg mL^{-1} AM79 EPSPS, 21.2 μg mL^{-1} A1501 EPSPS, 8.08 μg mL^{-1} RD EPSPS, 26.4 μg mL^{-1} G2 EPSPS) and allowed to proceed for 3 min before the addition of 160 μl Lanzetta reagent. Color development was stopped after 3 min by adding 20 μl of 34% (w/v) sodium citrate, The change in absorbance at 655 nm was recorded 20 min later, and the amount of inorganic phosphate was calculated by comparison with phosphate standards. Enzymatic activity is expressed as increased amount of Pi (mmol) (min reaction time)$^{-1}$ (mg protein)$^{-1}$ (U mg^{-1}).The *Km* values for PEP were determined by fitting data to Michealis equation $V=V_{max}[S]/K_m+[S]$, where V is the velocity of the reaction, V_{max} is the maximum velocity, $[S]$ is the substrate(PEP) concentration, K_m is the Michaelis constant. The K_i is derived by determining the $K_{m(obs)}$ of PEP in the presence of increasing concentration of glyphosate. The date was fitting to the equation $K_{m(obs)}=(K_m/K_i)[I]+K_m$, where $K_{m(obs)}$ is the Michaelis constant for PEP in the presence of glyphosate, $[I]$ is the concentration of glyphosate, and K_m is the Michaeliss constant for PEP in the abence of glyphosate.

Tobacco transformation

Plant expression plasmids p3301-121spG2, p3301-121spHTG7, p3301-121spAM79, p3301-121spA1501 and p3301-121spRD were transferred into competent cells of the *A. tumefaciens* strain LBA4404 through freeze-thaw treatment. The transformed *A. tumefaciens* colonies were selected on YEB-agar plates containing 100 μg mL^{-1} of kanamycin and 125 μg mL^{-1} of streptomycin. The positive colonies were identified by PCR amplification of the inserted genes and used for the tobacco transformation as previously described [43] using phosphinothricin (PPT) as the selecting gene. The transgenic plants were confirmed by PCR amplification of the *aroA* gene.

Southern blot analysis

Genomic DNA was isolated from young leaves of tobacco plants using the CTAB method. 100 μg genomic DNA was digested by *Hin*dIII, which has only one recognition site within the plasmid, electrophoresed on 1% (w/v) agarose gel and transferred onto a $Hybond^{+}$ nylon membranes (Roche, Mannheim, Germany). The PCR fragment of the *aroA* gene was amplified using primers (Table S3), and was labeled using a DIG High Prime DNA Labeling and Detection Starter Kit II (Roche, Mannheim, Germany), and

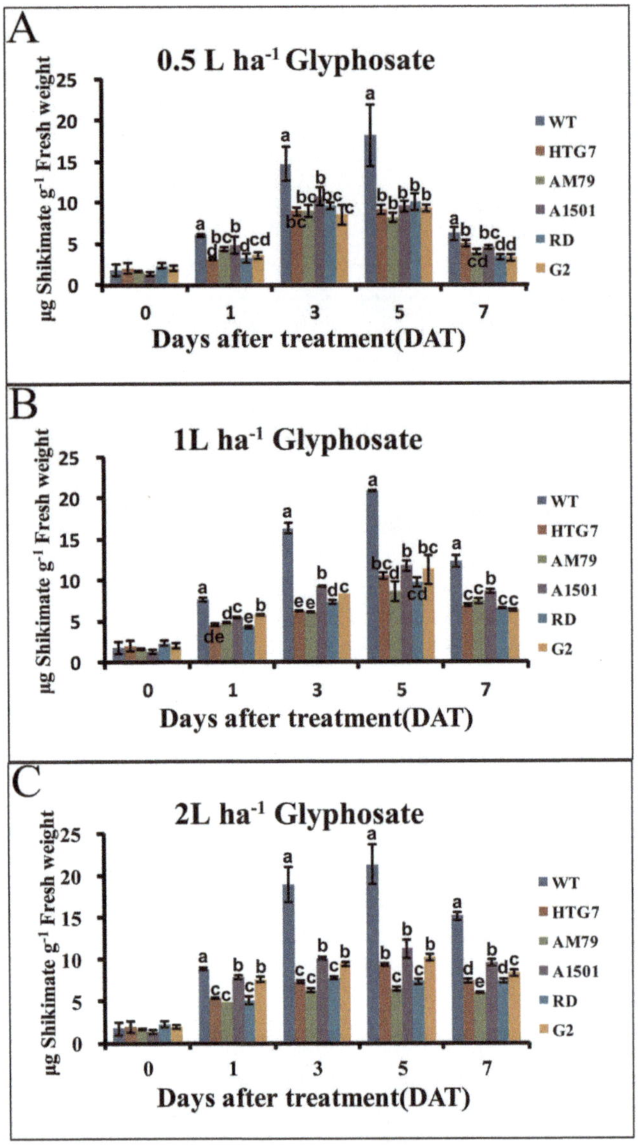

Figure 8. Shikimate content in the leaves of transgenic tobacco plants sprayed with different amounts of glyphosate. One-month-old plants grown in a green house were sprayed with 0.5, 1 or 2 L ha-1 Roundup®. Leaf samples were taken at 0, 1, 3, 5 and 7 d after glyphosate treatment for shikimate measurement. Data are shown as the average ±S.E. of three independent experiments. Experimental data was tested by ANOVA analysis and different letter in each column means significant difference at $P<0.05$ level.

hybridization and detection steps were performed according to the manufacturer's instructions.

RNA isolation and quantitative real-time RT-PCR

Total RNA from *E. coli* and tobacco leaves was isolated using Trizol reagent (Invitrogen, Carlsbad, CA, USA) and first-strand cDNA synthesis was performed with M-MLV reverse transcriptase (Promega) using oligo-dT primer. Specific primer for each gene was shown in Table S3. For quantitative real-time RT-PCR, 1 μl cDNA was mixed with 2×SYBR premix ExTaq (Takara), 0.2 μM forward primer, 0.2 μM reverse primer and 0.4 μl 50×ROX in 20 μl reaction mixture. Quantitative real-time RT-PCR was done using ABI 7300 system with the following protocol: 95°C 2 min, 40 cycles for 95°C 5 s, 58°C 30 s, 72°C 31 s. The relative transcript level was calculated using the $2^{-\Delta\Delta}$Ct method with tobacco actin gene or bacterial 16S rRNA gene as housekeeping genes.

Glyphosate tolerance analysis of transgenic plants

Heterozygous tobacco seeds of the T1 generation were germinated on MS medium containing 10 mg L^{-1} PPT and grown for 7 days at 100 μmol s^{-1} m^{-2} with a 16 h light/8 h dark period. The living seedlings with a similar size were transferred to MS medium in plates containing different amounts of glyphosate and grown vertically for another two weeks. Injury was observed, and the fresh weight was measured. One-week-old seedlings were transferred to soil and grown in a greenhouse for one month. Six-to-eight-leaf stage transgenic plants were sprayed with 6 L ha^{-1} Roundup®, injury was observed and survival rate was measured 2 weeks after treatment.

Shikimate measurement

One-month-old plants were sprayed with 0.5, 1 or 2 L ha^{-1} Roundup®. Leaf samples were taken at 0, 1, 3, 5 and 7 d after the glyphosate treatment for shikimate measurement. Approximately 0.5 g leaf was ground in liquid nitrogen, extracted in 1 mL of 0.25 N HCl, and then centrifuged at 12,000 rpm for 15 min at 4°C. The supernatant was collected and filtered through 0.45 μm nylon membrane. Shikimate concentration was determined by liquid chromatograph mass spectrometer (LC-MS) (LCMS-2010A, Shimadzu), equipped with an PTH-C18 column (250×4 mm, 5 μm)(Shimadzu, Japan) maintained at 40°C. The mobile phase was 0.1% trifluoroacetic acid (v/v) in 5% methanol, with a flow rate of 0.5 mL min^{-1} at 40°C. Shikimate was detected at 210 nm and quantified based on standard shikimate (Sigma-Aldrich).

Data analysis

Data were analyzed using ANOVA method in computer SAS software and the results were compared by using Duncan's multiple range tests.

Supporting Information

Figure S1 The transcription level of *aroA* genes in *E. coli* ER2799. *E. coli* ER2799 containing different *aroA* genes grew in M9 medium without or with 200 mM glyphosate for 16 h. Data are shown as mean Ct_{gene}/Ct_{16S} value ± S.E. for three independent biology replicates.

Figure S2 Enzyme kinetics analysis of five bacterial EPSPS proteins. SDS-PAGE electrophoresis of purified proteins (A), K_m and K_i measurement of HTG7 EPSPS (B), AM79 EPSPS (C), A1501 EPSPS (D), RD EPSPS (E) and G2 EPSPS (F). The observed K_m was measured at PEP concentrations ranging from 0 to 1 mM, and was plotted against the glyphosate concentration to obtain the *Km* and *Ki* for the enzyme.

Figure S3 PCR analysis of transgenic tobacco plants. PCR analysis of transgenic tobacco harboring *HTG7* (A), *AM79* (B), *A1501* (C), *RD* (D) and *G2* (E). M, DL 2000 plus DNA ladder; CK-, water as PCR control; WT, non-transgenic tobacco line; CK1, plasmid pACYC-HTG7 as a positive control; CK2, plasmid pACYC-AM79 as a positive control; CK3, plasmid pACYC-A1501 as a positive control; CK4, plasmid pACYC-RD as a positive control; CK5, plasmid pACYC-G2 as a positive control; 1–6, different transgenic tobacco lines.

Figure S4 Glyphosate tolerance analysis of the transgenic tobacco expressing *G2 aroA* with or without rbcS signal peptide. Photograph of four-to six-leave stage tobacco plants two weeks after 1 L ha^{-1} Roundup® treatment. Left, non-transgenic tobacco; middle, tobacco plants harboring plasmid p3301-121G2 without signal peptide; right, tobacco plants harboring plasmid p3301-121spG2 with signal peptide.

Figure S5 The transcription level of *aroA* genes in transgenic tobacco plants. Data are shown as mean *Ct* value ± SE for three independent biology replicates. The relative transcription level was analyzed using $2^{-\Delta\Delta Ct}$ method, and one line with the highest transcription level among different lines transformed with the same construct was normalized as 1.00. *Different letter means significant difference at $P<0.05$ level (Duncan's multiple range tests).

Figure S6 Glyphosate tolerance of *E. coli* expressing *AM79 aroA* or *CP4 aroA*. *AM79 aroA* or *CP4 aroA* was cloned into *Bam*HI/*Sal*I site of plasmid pACYC184. plasmids were transformed into *E. coli* ER2799 competent cells. M9 liquid medium was supplemented with different concentrations of glyphosate. OD_{600} was recorded 16 h after treatment. Data are shown as the average ± S.E. of three independent experiments.

Figure S7 Amino acid alignment of the EPSPS proteins and key active site residues analysis. Amino acid alignment of the five EPSPS proteins used in this study and the other three EPSPS proteins from *E. coli*, *Agrobacterium* sp. CP4 and *Streptococcus pneumoniae*. The classical key active sites residues of known EPSPS structures are marked using black dots. One dot indicates the universal active site, two dots show that the active sites are special in some amino acids and three dots indicate that the active site is unique in the amino acid sequence. The red frame marked I, II, III, IV, V indicates the five conserved domains in class II EPSPS enzymes. The underlined region indicates the general conserved domain in class I EPSPS enzymes.

Table S1 Oligonucleotides used for vector construction in this study.

Table S2 Strains and plasmids used in this study.

Table S3 Oligonucleotides used for Real-time RT-PCR and Southern blot analysis.

Acknowledgments

We thank Dr. Lu Gan in National Key Facility for Crop Gene Resources and Genetic Improvement (NAFCRI) of Chinese Academy of Agricultural Sciences (CAAS) for the help on shikimate determination using LC-MS method. We thank Dr. Georgina Hernández Chávez from Instituto de Biotecnología(UNAM) for the suggestion in the shikimate determination, and Dr. Yan Yang from University of Kansas for helpful discussion on enzyme assay.

Author Contributions

Conceived and designed the experiments: GYC YJL JHW GYW. Performed the experiments: GYC YJL. Analyzed the data: GYC YJL SXZ XWY RRC YWZ YL. Contributed reagents/materials/analysis tools: WL ML. Wrote the paper: GYC YJL GYW.

References

1. Barry G, Padgette SR (1992) Glyphosate tolerant 5-enolpyruvylshikimate-3-phosphate synthases. World Patent, WO 92/04449.
2. Della-Cioppa G, Bauer SC, Taylor ML, Rochester DE, Klein BK, et al. (1987) targeting a herbicide resistant enzyme from *Escherichia coli* to chloroplasts of higher plants. Nat Biotechnol 5: 579–584.
3. Stalker DM, Hiatt WR, Comai L (1985) A single amino acid substitution in the enzyme 5-enolpyruvylshikimate-3-phosphate synthase confers resistance to the herbicide glyphosate. J Biol Chem 260: 4724–4728.
4. Funke T, Han H, Healy-Fried ML, Fischer M, Schonbrunn E (2006) Molecular basis for the herbicide resistance of Roundup Ready crops. Proc Natl Acad Sci U S A 103: 13010–13015.
5. Steinrucken HC, Amrhein N (1980) The herbicide glyphosate is a potent inhibitor of 5-enolpyruvyl-shikimic acid-3-phosphate synthase. Biochem Biophys Res Commun 94: 1207–1212.
6. Bradshaw LD, Padgette SR, Kimball SL, Wells BH (1997) Perspectives on glyphosate resistance. Weed Technol 11: 189–198.
7. Comai L, Sen LC, Stalker DM (1983) An altered aroA gene product confers resistance to the herbicide glyphosate. Science 221: 370–371.
8. Comai L, Facciotti D, Hiatt WR, Thompson G, Rose RE, et al. (1985) Expression in plants of a mutant aroA gene from *Salmonella typhimurium* confers tolerance to glyphosate. Nature 317: 741–744.
9. Schonbrunn E, Eschenburg S, Shuttleworth WA, Schloss JV, Amrhein N, et al. (2001) Interaction of the herbicide glyphosate with its target enzyme 5-enolpyruvylshikimate 3-phosphate synthase in atomic detail. Proc Natl Acad Sci U S A 98: 1376–1380.
10. Priestman MA, Healy ML, Funke T, Becker A, Schonbrunn E (2005) Molecular basis for the glyphosate-insensitivity of the reaction of 5-enolpyruvylshikimate 3-phosphate synthase with shikimate. FEBS Lett 579: 5773–5780.
11. Healy-Fried ML, Funke T, Priestman MA, Han H, Schonbrunn E (2007) Structural basis of glyphosate tolerance resulting from mutations of Pro101 in *Escherichia coli* 5-enolpyruvylshikimate-3-phosphate synthase. J Biol Chem 282: 32949–32955.
12. Funke T, Yang Y, Han H, Healy-Fried M, Olesen S, et al. (2009) Structural basis of glyphosate resistance resulting from the double mutation Thr97->Ile and Pro101->Ser in 5-enolpyruvylshikimate-3-phosphate synthase from *Escherichia coli*. J Biol Chem 284: 9854–9860.
13. Zhou M, Xu H, Wei X, Ye Z, Wei L, et al. (2006) Identification of a glyphosate-resistant mutant of rice 5-enolpyruvylshikimate 3-phosphate synthase using a directed evolution strategy. Plant Physiol 140: 184–195.
14. Howe AR, Gasser CS, Brown SM, Padgette SR, Hart J, et al. (2002) Glyphosate as a selective agent for the production of fertile transgenic maize (*Zea mays* L.) plants. Mol Breeding 10: 153–164.
15. Baerson SR, Rodriguez DJ, Tran M, Feng Y, Biest NA, et al. (2002) Glyphosate-resistant goosegrass. Identification of a mutation in the target enzyme 5-enolpyruvylshikimate-3-phosphate synthase. Plant Physiol 129: 1265–1275.
16. Foster PL (2000) Adaptive mutation: implications for evolution. BioEssays 22: 1067–1074.
17. Belfiore NM, Anderson SL (2001) Effects of contaminants on genetic patterns in aquatic organisms: a review. Mutant Res 489: 97–122.
18. Lopez-Rodas V, Flores-Moya A, Maneiro E, Perdigones N, Marva F, et al. (2007) Resistance to glyphosate in the cyanobacterium *Microcystis aeruginosa* as result of pre-selective mutations. Evol Ecol 21: 535–547.
19. Powell HA, Kerby NW, Rowell P (1991) Natural tolerance of cyanobacteria to the herbicide glyphosate. New Phytol 119: 421–426.
20. Priestman MA, Funke T, Singh IM, Crupper SS, Schonbrunn E (2005) 5-Enolpyruvylshikimate-3-phosphate synthase from *Staphylococcus aureus* is insensitive to glyphosate. FEBS Lett 579: 728–732.
21. Tian YS, Xiong AS, Xu J, Zhao W, Gao F, et al. (2010) Isolation from *Ochrobactrum anthropi* of a novel class II 5-enopyruvylshikimate-3-phosphate synthase with high tolerance to glyphosate. Appl Environ Microbiol 76: 6001–6005.
22. Dun BQ, Wang XJ, Lu W, Zhao ZL, Hou SN, et al. (2007) Reconstitution of glyphosate resistance from a split 5-enolpyruvyl shikimate-3-phosphate synthase gene in *Escherichia coli* and transgenic tobacco. Appl Environ Microbiol 73: 7997–8000.
23. Liu Z, Pan Z, Xu Y, Dong Z, Yang Z, et al. (2006) Cloning and expression of a 5-enolpyruvylshikimate-3-phosphate synthase gene from *Halomonas variabilis*. DNA Seq 17: 208–214.
24. Liang A, Sha J, Lu W, Chen M, Li L, et al. (2008) A single residue mutation of 5-enoylpyruvylshikimate-3-phosphate synthase in *Pseudomonas stutzeri* enhances resistance to the herbicide glyphosate. Biotechnol Lett 30: 1397–1401.
25. Jin D, Lu W, Ping S, Zhang W, Chen J, et al. (2007) Identification of a new gene encoding EPSPs with high glyphosate resistance from the metagenomic library. Curr Microbiol 55: 350–355.

26. Lin M, Liang A, Lu W, Li L, Chen M, et al. (2007) An EPSPS synthase with high glyphosate resistance and encoded sequence. World Patent, WO/2009/059485.
27. Dyer WE, Weller SC, Bressan RA, Herrmann KM (1988) Glyphosate tolerance in tobacco (*Nicotiana tabacum* L.). Plant Physiol 88: 661–666.
28. He M, Yang ZY, Nie YF, Wang J, Xu P (2001) A new type of class I bacterial 5-enopyruvylshikimate-3-phosphate synthase mutants with enhanced tolerance to glyphosate. Biochim Biophys Acta 1568: 1–6.
29. Wang HY, Li YF, Xie LX, Xu P (2003) Expression of a bacterial aroA mutant, aroA-M1, encoding 5-enolpyruvylshikimate-3-phosphate synthase for the production of glyphosate-resistant tobacco plants. J Plant Res 116: 455–460.
30. Padgette S, Kolacz K, Delannay X, Re D, Lavallee B, et al. (1995) Development, identification, and characterization of a glyphosate-tolerant soybeans line. Crop Sci 35: 1451–1461.
31. Du W, Wallis NG, Mazzulla MJ, Chalker AF, Zhang L, et al. (2000) Characterization of *Streptococcus pneumoniae* 5-enolpyruvylshikimate 3-phosphate synthase and its activation by univalent cations. Eur J Biochem 267: 222–227.
32. Fitzgibbon JE, Braymer HD (1990) Cloning of a gene from *Pseudomonas* sp. strain PG2982 conferring increased glyphosate resistance. Appl Environ Microbiol 56: 3382–3388.
33. Castle LA, Siehl DL, Gorton R, Patten PA, Chen YH, et al. (2004) Discovery and directed evolution of a glyphosate tolerance gene. Science 304: 1151–1154.
34. Pedotti M, Rosini E, Molla G, Moschetti T, Savino C, et al. (2009) Glyphosate resistance by engineering the flavoenzyme glycine oxidase. J Biol Chem 284: 36415–36423.
35. Della-Cioppa G, Bauer SC, Klein BK, Shah DM, Fraley RT, et al. (1986) Translocation of the precursor of 5-enolpyruvylshikimate-3-phosphate synthase into chloroplasts of higher plants in vitro. Proc Natl Acad Sci U S A 83: 6873–6877.
36. Ye GN, Hajdukiewicz PT, Broyles D, Rodriguez D, Xu CW, et al. (2001) Plastid-expressed 5-enolpyruvylshikimate-3-phosphate synthase genes provide high level glyphosate tolerance in tobacco. Plant J 25: 261–270.
37. Stenoien HK, Stephan W (2005) Global mRNA stability is not associated with levels of gene expression in *Drosophila melanogaster* but shows a negative correlation with codon bias. J Mol Evol 61: 306–314.
38. Shimizu M, Kimura T, Koyama T, Suzuki K, Ogawa N, et al. (2002) Molecular breeding of transgenic rice plants expressing a bacterial chlorocatechol dioxygenase gene. Appl Environ Microbiol 68: 4061–4066.
39. Chang HS, Kim NH, Park MJ, Lim SK, Kim SC, et al. (2003) The 5-enolpyruvylshikimate-3-phosphate synthase of glyphosate-tolerant soybean expressed in *Escherichia coli* shows no severe allergenicity. Mol Cells 15: 20–26.
40. Binarova P, Cvikrova M, Havlick T, Eder J, Plevkova J (1994) Changes of shikimate pathway in glyphosate tolerant alfalfa cell lines with reduced embryogenic ability. Biol Plant 36: 65–73.
41. Chen LX, Pradhan S, Evans TC Jr.,(2001) Herbicide resistance from a divided EPSPS protein: the split *Synechocystis* DnaE intein as an in vivo affinity domain. Gene 263: 39–48.
42. Lanzetta PA, Alvarez LJ, Reinach PS, Candia OA (1979) An improved assay for nanomole amounts of inorganic phosphate. Anal Biochem 100: 95–97.
43. Horsch RB, Fry JE, Hoffmann NL, Eichholtz D, Rogers SG, et al. (1985) A simple and general method for transferring genes into plants. Science 227: 1229–1231.

17

Insight into the Wild Origin, Migration and Domestication History of the Fine Flavour Nacional *Theobroma cacao* L. Variety from Ecuador

Rey Gaston Loor Solorzano[1,2], Olivier Fouet[1], Arnaud Lemainque[3], Sylvana Pavek[3], Michel Boccara[1,4], Xavier Argout[1], Freddy Amores[2], Brigitte Courtois[1], Ange Marie Risterucci[1], Claire Lanaud[1]*

1 UMR AGAP, CIRAD, Montpellier, France, **2** EET-Pichilingue, INIAP, Quevedo El Empalme, Ecuador, **3** Centre National de Génotypage, CEA Institut de Génomique, Evry, France, **4** Cocoa Research Unit (CRU), University of the West Indies, St Augustine, Trinidad and Tobago

Abstract

Ecuador's economic history has been closely linked to *Theobroma cacao* L cultivation, and specifically to the native fine flavour Nacional cocoa variety. The original Nacional cocoa trees are presently in danger of extinction due to foreign germplasm introductions. In a previous work, a few non-introgressed Nacional types were identified as potential founders of the modern Ecuadorian cocoa population, but so far their origin could not be formally identified. In order to determine the putative centre of origin of Nacional and trace its domestication history, we used 80 simple sequence repeat (SSR) markers to analyse the relationships between these potential Nacional founders and 169 wild and cultivated cocoa accessions from South and Central America. The highest genetic similarity was observed between the Nacional pool and some wild genotypes from the southern Amazonian region of Ecuador, sampled along the Yacuambi, Nangaritza and Zamora rivers in Zamora Chinchipe province. This result was confirmed by a parentage analysis. Based on our results and on data about pre-Columbian civilization and Spanish colonization history of Ecuador, we determined, for the first time, the possible centre of origin and migration events of the Nacional variety from the Amazonian area until its arrival in the coastal provinces. As large unexplored forest areas still exist in the southern part of the Ecuadorian Amazonian region, our findings could provide clues as to where precious new genetic resources could be collected, and subsequently used to improve the flavour and disease resistance of modern Ecuadorian cocoa varieties.

Editor: Randall P. Niedz, United States Department of Agriculture, United States of America

Funding: The study was funded by the United States Department of State (U.S. Foreign Ministry); the U.S. Embassy, Quito; the U.S. Department of Agriculture (USDA-ARS); the Instituto Nacional Autonomo de Investigaciones Agropecuarias (INIAP-Ecuador); and the Centre de Coopération Internationale en Recherche Agronomique pour le Développement (CIRAD-France). The funders had no role in study design, data collection and analysis, decision to publish, or preparation of the manuscript.

Competing Interests: The authors have declared that no competing interests exist.

* E-mail: claire.lanaud@cirad.fr

Introduction

Theobroma cacao L. is the most economically important species of the *Theobroma* genus which is a member of the *Malvaceae* family. It is a diploid perennial species ($2n = 2\times = 20$) with a small genome ranging in size from 411 to 494 Mb [1]. Most of the *T. cacao* accessions are self incompatible due to a gameto-sporophytic incompatibility system [2–4]. However, some varieties, domesticated long ago, as the varieties Criollo from central America, Comun from Brazil, or Nacional from Ecuador, and some of their hybrid forms are self compatible.

Traditionnaly, two main genetic groups have been defined to classify cocoa accessions, based on morphological traits and geographical origins [5]: Criollo and Forastero. This classification also reflects the first commonly cultivated cocoa varieties: Criollo, first domesticated in Central America more than 2000 years ago, and a Lower Amazon Forastero variety (Amelonado type) domesticated in Brazil. Trinitario, a third group, corresponding to hybrids between Criollo and Forastero, was also recognised. However, the Forastero group also includes many other populations from all the Amazonian and Orinoco regions and presents a high diversity, as revealed by many genetic studies [6], [7], [8–16]. A new classification, identifying 10 genetic clusters, was more recently proposed [15].

The domestication of Criollo, originally cultivated by the Mayas in Central America was previously studied [17]. The authors concluded that the genetic base of the 'ancient' Criollo variety, cultivated in Central America before foreign cocoa introductions occurred during the 18th century, was very narrow.

Cocoa cultivation in Ecuador is very ancient, dating back to the pre-Columbian age before the Spanish colonization of this territory. Pizarro, during his first voyage in 1526 along the South American coasts (nowadays Ecuador), found evidence of small plantations of an apparently native cocoa tree [18], [19]. It is very likely that the 'Nacional' Ecuadorian cocoa population existed for several centuries prior to the arrival of the Europeans [20], but its origin in the Ecuadorian coastal region has never been clarified. Two hypotheses have been put forward: according to Allen and Lass [21], the Nacional cocoa variety could have originated from a local wild population, that has nowadays completely disappeared along with the original forest cover of the region, or Nacional could have been introduced in the coastal region from the

Amazonian area of Ecuador where wild cocoa is common [22–24]. There is no evidence that *T. cacao* and its products played any part in the lives of Ecuadorian coastal inhabitants during the Chorrera phase of the pre-Columbian Valdivia culture (2000 BC), contrary to the extensive amount of information available on this phenomenon with respect to Mesoamerican people, most notably the Mayas and Aztecs. In Ecuador, available information related to the history of Nacional cocoa cultivation is linked with the Spanish colonization history in this country.

Since the time when the first Spanish colonists began deforestation of the Ecuadorian coastal region, a large number of native cocoa trees have been reported [21–25], principally along the Guayas basin. Apparently, these colonists began to sow seeds of these native trees approximately 100 years after the discovery of America, and when the native Mesoamerican populations started to decline. The cocoa cultivation areas expanded, and the native Nacional variety rapidly became known worldwide due to its strong floral so-called "Arriba" aroma, exclusively generated by Nacional cocoa beans. The fine flavour quality of chocolate products obtained from Nacional cocoa beans has always been highly appreciated by chocolate manufacturers.

The native Nacional variety was the only one planted in Ecuador until the early 1890s, when foreign germplasm was first introduced in this country. In 1890, due to the quality traits of Nacional cocoa beans, Ecuador had a privileged position in the markets of Hamburg and London [26]. From 1910, foreign germplasm introductions progressively increased due to the appearance of two fungal diseases known as witches' broom (*Moniliophthora perniciosa*) and frosty pod (*Moniliophthora roreri*), which together devastated the native plantations.

A large genetic admixture between the native variety and foreign germplasm is currently found in modern Ecuadorian cocoa plantations [27]. The fine flavour cocoa aroma has decreased in this hybrid complex, and 25% of the Ecuadorian cocoa production was recently classified as 'bulk' cocoa by the International Cocoa Organization (ICCO). There is nowadays increasing demand for fine flavour cocoa, which presently represents 6.8% of the world cocoa production and for which Ecuador remains the main supplier [28].

In a previous work, a few non-introgressed Nacional types were identified within this hybrid population as potential representatives of the native Nacional variety [27]. Here we analyze their relationships with a wide range of wild cocoa genotypes covering a broad geographical range from upper and lower Amazonian regions [21], [22], [29], [30], [31], in order: a) to identify the putative centre of origin of Nacional, and b) to trace the domestication history of the Nacional cocoa variety.

Materials and Methods

Plant Material

A total of 176 individuals from different geographical origins were used for simple sequence repeat (SSR) analyses in this study (Table 1 and Table S1):

- Seven putative ancient Nacional variety individuals cultivated along the Ecuadorian coastal region [27].
- Sixty-five wild genotypes collected in the northern region A (LCT-EENa), central region B (LCT-EENb) and southern region C (LCT-EENc) of the upper Amazonian region of Ecuador (Fig. 1b) [21].
- Sixty-six wild genotypes collected in the upper Amazonian region of Peru (Fig. 1b) [22], [29]:IMC (13 ind.), Nanay (Na, 12 ind.), Parinari (Pa, 13 ind.), Morona (Mo, 8 ind.), Scavina (Sca, 8 ind.), Pound (P, 12 ind.).
- Eleven wild cocoa samples collected in French Guiana (Guy) [30].
- Three cultivated genotypes from the lower Amazonian region of Brazil (BA).
- Twenty-four Criollo (Cr) variety samples collected in Central America, from Venezuela to Mexico [32], were also used in this study.

The germplasm collection and country of origin is reported for each accession in Table S1 and Fig. 1. The wild cocoa accessions from Ecuador and Peru were selected from the CRU living collection on the basis of the geographical area sampled along the Amazonian provinces [21–22] (Fig. 1b).

Molecular Markers

Eighty SSR markers were chosen for these analyses (Table 2). Seventy four SSRs were isolated from expressed sequence tags (EST) and six from genomic sequences, and mapped [33–35]. They are distributed on all of the 10 cocoa chromosomes.

DNA Isolation and PCR Amplification

DNA samples were isolated from each individual as described previously [35]. For SSR analyses, PCR amplifications were performed as previously reported [35] and fragments were subsequently detected using a 'Megabase 1000' DNA analysis system (Molecular Dynamics/Amersham Life Science).

Statistical Analysis

Amplified SSR fragments were scored as alleles. The following genetic parameters were calculated for each population using GENETIX software [36]: unbiaised expected heterozygosity [37] (HE), observed heterozygosity (HO), proportion of polymorphic loci at which the frequency of the most common allele does not exceed 95% (P0.95) and 99% (P0.99), and mean number of alleles per locus.

We also estimated allelic richness (Rs) for each population, using a rarefaction approach to correct for unequal sample sizes [38] with the program FSTAT V. 2.9.3 [39].

Wright's *F*-statistics were estimated according to Weir and Cockerham [40] using the GENETIX software: F_{IS}, an estimate of heterozygote deficiency or excess, calculated for the whole population sample and for each population, and F_{ST}, the proportion of variance due to population differenciation, calculated for the whole population sample and between pairs of populations; their significance were assessed using a 1000-fold permutation test.

As most mutations in SSR involve the addition or subtraction of a small number of repeat units, according to the stepwise mutation model population, genetic differentiation was also estimated by R_{ST} [41]. Rho_{ST}, an unbiased version of R_{ST} in which allele sizes are transformed to standardize variances, was estimated for pairs of populations using GENEPOP software, WEB version 4.0.10 [42–44].

Analyses of genetic distance among pairs of populations [45] were carried out to determine the genetic similarity between the Nacional pool and the wild and cultivated genotypes using GENETIX.

In addition, a dendrogram was constructed using a dissimilarity index (simple matching) and the neighbour-joining (NJ) method

Table 1. Origin of the 14 groups of accessions used in this study.

Group name	Code (Fig. 1 and 2)	Origin	Geographical localities or areas	Sample size
Criollo	CR	Central America		22
VEN	VEN	Venezuela	Orinoco (Cuchivero- Padamo)	2
LCT-EENa	LCT-EENa	Ecuador	Amazonia (North region-A)	30
LCT-EENb	LCT-EENb	Ecuador	Amazonia (Centre region-B)	23
LCT-EENc	LCT-EENc	Ecuador	Amazonia (South region-C)	12
Nacional	SNA-B	Ecuador	Coast provinces	7
IMC	IMC	Peru	Iquitos	13
Morona	MO	Peru	Morona river	8
Nanay	NA	Peru	Nanay river	12
Parinari	PA	Peru	Tributaries of Marañon river	13
Pound	P	Peru	Nanay river	12
Scavina	SCA	Peru	Loreto, Ucayali river	8
BA	BA	Brazil		3
GUY	GU	French Guiana	Camopi - Tanpok	11
Total samples				176

Among them Criollo, Nacional and BA correspond to old cultivated varieties. The other groups include cocoa trees native of the mentioned regions.

with 500 bootstrap replicates. Neighbour-joining cluster analyses were carried out using DARWIN software 5.0 [46].

Finally, a paternity analysis was performed using the software CERVUS, version 2.0 [47], to identify the most likely ancestral population of the Nacional variety. An 80% confidence threshold was used for paternity assignment.

Results

Patterns of Genetic Diversity Among Cocoa Accessions

A total of 463 alleles were detected at the 80 analysed SSR loci across all 176 individuals sampled (Table 2). A wide range of allele numbers was generally obtained for each locus (from 2 to 15), with an average of 5.78 alleles per locus. Genetic diversity parameters were evaluated in each group. As shown in Table 3, the mean number of alleles obtained per locus and per accession group corresponding to wild accessions ranged from 1.1 (BA) to 4.45 (LCT-EENa). The allele richness, which takes the sample size into account, varies from 1.05 (CR) to 1.53 (LCT-EENa). For both parameters, the most polymorphic wild cocoa accessions were found in the centre and northern Amazonian regions (A and B) of Ecuador (LCT-EENa and LCT-EENb).

Significant F_{IS} values were observed in most of the populations with positive values showing an excess of homozygotes and negative values showing an excess of heterozygotes (Table 3). The highest positive values are observed in the BA and Criollo groups, which correspond to self compatible varieties. However, highly homozygous genotypes can also be identified in the other groups such as in the French Guyana, LCT-EEN and Nanay groups (Table S1).

The F_{IS} value for the whole sample set and over all loci was 0.2402, with a confidence interval comprised between 0.2155 and 0.2654 after a 1000-fold permutation test.

Genetic differentiation (F_{ST}) was estimated between pairs of populations (Table 4). Nearly all F_{ST} values were significant after permutation tests. Only 3 F_{ST} values were not significant: those between the population LCT-EENa from the northern part of the Ecuadorian upper Amazonian region and the two other centre and south regions of the Ecuadorian Amazonia (LCT-EENb and LCT-EENc), and the F_{ST} between the Pound and the Nanay populations.

The F_{ST} and Rho_{ST} values between cocoa populations were very similar, with slight variations depending on the pair-wise populations (Table 5). Only in the case of the Morona population, its differenciation from the Scavina, VEN and Nacional groups appeared higher when evaluated with the Rho_{ST} values compared to the F_{ST} values.

The F_{ST} value for the whole multilocus sample set was 0.4979, with a confidence interval comprised between 0.4782 and 0.5192 after a 1000-fold permutation test. The corresponding Rho_{ST} value for the whole sample set was 0.530.

The neighbour-joining tree (Fig. 2), clustered the majority of cocoa samples according to their geographical origins. Four main clusters were obtained in this analysis (Fig. 2):

- Cluster 1 included most of the accessions collected in the northern and central Amazonian regions of Ecuador.
- Cluster 2 included most of Peruvian accessions. It was also divided into two subgroups: One subgroup pooled the Scavina accessions; the other subgroup included most of the remaining Peruvian accessions along with wild accessions from French Guiana and Venezuela, and cultivated Lower Amazon samples from Brazil (BA).
- Cluster 3 included the Nacional cocoa accessions, the wild cocoa genotypes from the southern part of Ecuadorian Amazonia (region C) and the wild cocoa accessions from Peru collected along the Morona river.
- Cluster 4 included all Criollo accessions.

The NJ tree (Fig. 2) highlighted the close relationship between the Nacional pool and the wild cocoa genotypes (LCT-EEN) sampled in the Amazonian region C from Ecuador (Fig. 1b). The accessions sampled in Peru along the Morona river were also genetically close to the Nacional pool; the Morona river region is geographically close to region C of Ecuadorian Amazonia. The largest genetic diversity, expressed by the proportion of polymor-

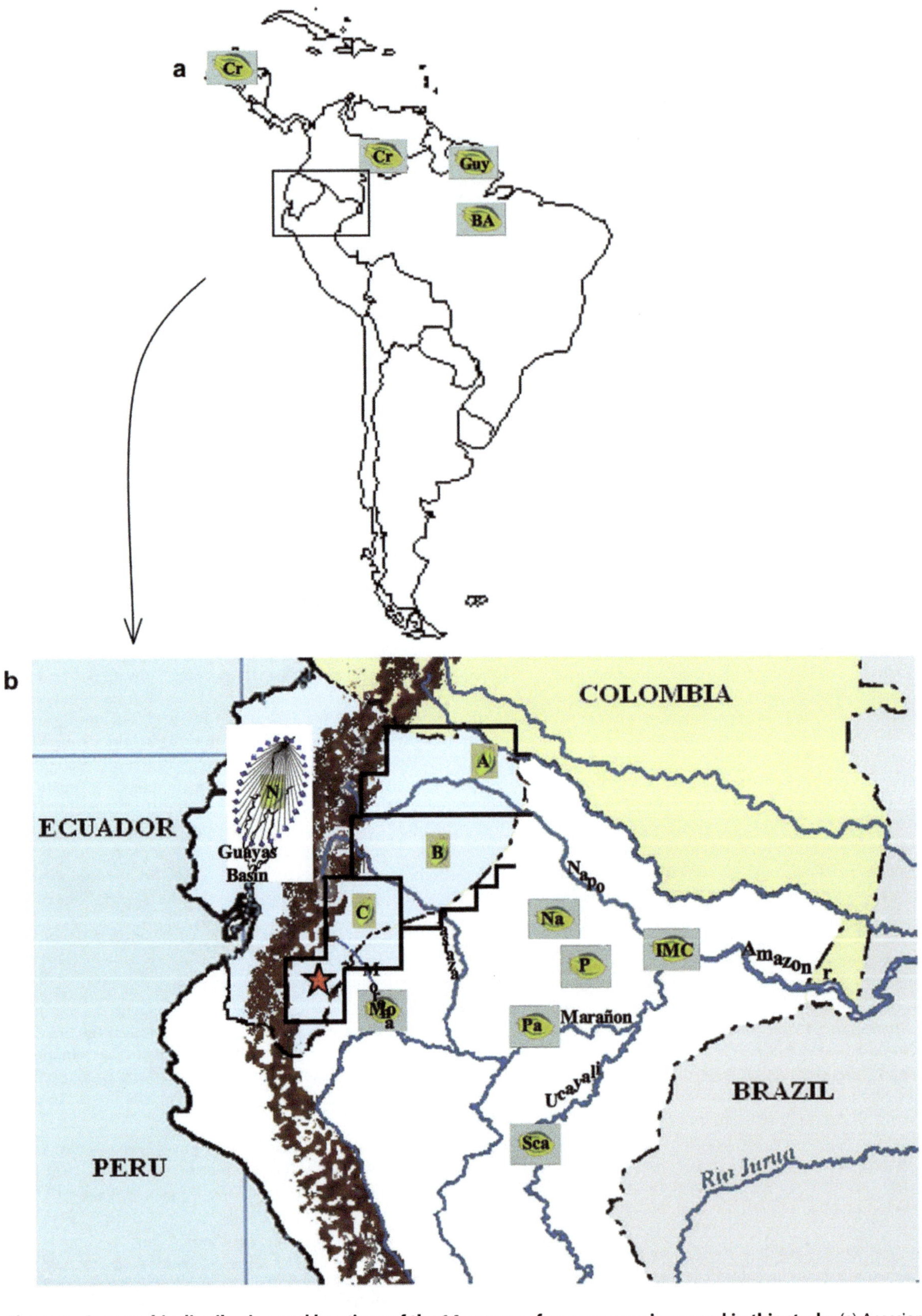

Figure 1. Geographic distribution and locations of the 14 groups of cocoa accessions used in this study. (a) Acessions from Central and South America excepted Ecuador and Peru; and (b) accessions from the Ecuadorian coast and the upper Amazonian region from Ecuador and Peru. The Zone of "Arriba" cocoa is delimited by blue dashes. The locations of Nangaritza and Yacuambi rivers are indicated by a red star. See "material and methods" for explanations of name codes.

Table 2. List of SSR loci (mTcCIRx) and number of alleles per locus identified in this study.

mTcCIR	Nb alleles/locus
1	7
6	11
15	10
26	10
37	11
60	8
292	4
293	5
294	8
297	2
298	6
299	2
302	13
304	6
305	3
308	5
309	4
312	3
316	7
322	15
323	9
324	2
325	7
326	7
327	3
329	10
331	10
333	3
336	7
337	3
339	3
341	2
342	4
344	10
348	3
350	4
352	13
356	4
359	5
361	6
364	2
366	8
369	8
373	5
374	6
375	4
376	6
378	6
379	5
380	4
382	11
383	6
387	4
391	7
392	8
393	8
394	12
396	3
397	2
398	2
400	6
402	3
404	2
405	3
411	7
413	4
417	4
418	2
419	2
420	4
422	13
426	4
427	4
429	8
432	5
435	5
436	3
438	2
446	6
447	4
80	463
	Average 5.78

phic loci, the mean number of alleles per locus, and the allele richness, was found in the LCT-EEN accessions from Ecuador, and particularly in those originated from regions A and B (Fig. 1b).

Genetic Distance between the Native Nacional Genotypes and Wild Populations Present in the Amazonian Regions

To refine the previous analysis, coefficients of genetic distance were calculated for pair-wise comparisons of the 14 cocoa groups [45]. A matrix of genetic distance values for all groups is presented in Table 4. The highest genetic distance (2.445) was obtained between the cultivated Criollo and the Lower Amazon group from Brazil (BA), while the most similar populations were LCT-EENa and LCT-EENb, with a genetic distance of 0.039. In the case of the Nacional pool, the lowest genetic distance was noted with the

Table 3. Genetic diversity parameters revealed by SSR markers within each group analyzed.

Group name	HE	HO	P(0.95)	P(0.99)	NMA/ locus	Allelic richness	F_{IS}
Criollo	0.050	0.006	0.150	0.175	1.250	1.050	0.883*
VEN	0.152	0.056	0.225	0.225	1.237	1.152	0.719*
LCT-EENa	0.533	0.317	0.950	0.975	4.450	1.532	0.410*
LCT-EENb	0.485	0.358	0.950	0.962	4.150	1.485	0.266*
LCT-EENc	0.432	0.271	0.787	0.812	2.637	1.432	0.385*
Nacional	0.114	0.094	0.312	0.312	1.375	1.114	0.186*
IMC	0.391	0.479	0.800	0.862	2.700	1.391	−0.236*
Morona	0.321	0.314	0.700	0.700	2.000	1.321	0.034
Nanay	0.234	0.242	0.650	0.750	2.150	1.233	−0.039
Parinari	0.312	0.304	0.637	0.750	2.337	1.312	0.028
Pound	0.399	0.292	0.850	0.900	3.200	1.399	0.277*
Scavina	0.343	0.359	0.750	0.750	2.125	1.343	−0.040
BA	0.068	0.000	0.125	0.125	1.125	1.068	1.000*
Guy	0.114	0.064	0.287	0.325	1.337	1.114	0.457*

HE: Unbiased gene diversity (Nei, 1978) [37]; HO: Observed heterozygosity; Proportion of polymorphic loci at which the frequency of the most common allele does not exceed 95% (0.95) and 99% (0.99); and mean number of alleles per locus (NMA/locus).

groups from the LCT-EENc region (0.250) and the Morona region (0.307).

Subsequent analyses of genetic distance were carried out to identify the genetic distances between the Nacional pool and wild individuals collected along the neighbouring rivers within each Amazonian region of Ecuador (A, B, C), including all accessions collected in the upper Amazonian region of Peru (data not shown). This new approach allowed us to identify a few LCT-EENc samples (LCT-EEN85, LCT-EEN86 and LCT-EEN91), collected along the adjacent rivers Yacuambi and Nangaritza, as being wild cocoa trees genetically closest to the Nacional pool originally cultivated in the coast region. A key feature of some of the wild cocoa trees from these adjacent rivers is their low heterozygosity level (e.g. 16% for LCT-EEN 91). Note also that the Amazonian region C is geographically close to the Guayas basin (Fig. 1b) where the first Nacional cocoa plantations were established.

Potential Wild Ancestors of the Nacional Variety

A paternity analysis was performed to identify the most likely representatives of the ancestral population from which the Nacional variety originated. All putative tested parents were from the south Amazonian region C of Ecuador, closest to the Nacional genotypes.

Our results (Table 6) indicated that the LCT-EEN85, LCT-EEN86 and LCT-EEN91 genotypes were the most likely parents of Nacional individuals. These findings were clearly in line with our previous results, suggesting that the native Nacional pool likely descended from wild cocoa trees growing in the vicinity of the Nangaritza and Yacuambi rivers in the southern part of Amazonian region C.

Discussion

Throughout the history of cocoa cultivation in Ecuador, substantial genetic changes, including reproductive behaviour, have occurred as a consequence of foreign germplasm introductions and genetic admixture associated with natural and human selection. The first study of genetic diversity of modern Ecuadorian *T. cacao* accessions was undertaken by Lerceteau et al. [6], [7]. However, the origin of Nacional cocoa variety, its genetic relationships with wild cocoa trees from Amazonia, and its domestication history had yet to be clarified.

In this study, we used 80 SSR markers to analyse relationships between representatives of the putative founders of the Nacional variety and wild populations from Amazonia in order to identify its wild genetic origin and elucidate the domestication events that occurred before the first Nacional plantations were set up in Ecuador in the Guayas Basin.

Table 4. Coefficients of genetic distance (Nei, 1972) [45] calculated for pair-wise comparisons of the 14 groups of *T. cacao* accessions studied.

	VEN	LCT-EENa	LCT-EENb	LCT-EENc	Nacional	IMC	Morona	Nanay	Parinari	Pound	Scavina	BA	Guy
Criollo	2.141	0.910	0.929	1.079	1.395	1.435	1.293	1.574	1.468	1.461	1.352	2.445	1.611
VEN		0.735	0.892	1.098	1.276	0.385	0.977	0.380	0.402	0.322	1.172	0.233	0.535
LCT-EENa			0.039	0.143	0.492	0.468	0.446	0.619	0.517	0.493	0.631	0.724	0.689
LCT-EENb				0.144	0.582	0.534	0.493	0.715	0.618	0.588	0.652	0.861	0.789
LCT-EENc					0.250	0.632	0.431	0.792	0.718	0.673	0.673	1.110	0.930
Nacional						0.748	0.307	1.017	0.900	0.855	0.779	1.300	1.226
IMC							0.571	0.269	0.312	0.148	0.765	0.416	0.473
Morona								0.906	0.669	0.718	0.615	0.928	0.991
Nanay									0.422	0.051	0.967	0.412	0.586
Parinari										0.303	0.775	0.329	0.304
Pound											0.742	0.347	0.475
Scavina												1.218	0.955
BA													0.557

Table 5. Genetic differenciation between pairs of populations.

Population	Criollo	VEN	LCT-EENa	LCT-EENb	LCT-EENc	Nacional	IMC	Morona	Nanay	Parinari	Pound	Scavina	BA	Guy
Criollo	–	**0.9386**	**0.5804**	**0.6283**	**0.7392**	**0.9107**	**0.7711**	**0.8354**	**0.8552**	**0.8137**	**0.7743**	**0.8279**	**0.9403**	**0.9110**
VEN	0.9144	–	**0.3359**	**0.4112**	**0.4876**	**0.8299**	**0.3545**	**0.5888**	**0.4967**	**0.4134**	**0.2515**	**0.6078**	**0.5133**	**0.7298**
LCT-EENa	0.6004	0.3522	–	*0.0081*	*0.0880*	**0.3582**	**0.2802**	**0.2750**	**0.3824**	**0.3240**	**0.2801**	**0.3310**	**0.3864**	**0.4436**
LCT-EENb	0.6480	0.4269	−0.0018	–	**0.0999**	**0.4303**	**0.3328**	**0.3261**	**0.4462**	**0.3891**	**0.3392**	**0.3714**	**0.4608**	**0.5135**
LCT-EENc	0.7719	0.4362	0.0710	0.0638	–	**0.3191**	**0.3929**	**0.3336**	**0.5165**	**0.4568**	**0.3901**	**0.4149**	**0.5523**	**0.6105**
Nacional	0.9294	0.7884	0.3935	0.4274	0.2812	–	**0.5710**	**0.4727**	**0.7304**	**0.6551**	**0.5820**	**0.6386**	**0.8615**	**0.8442**
IMC	0.7607	0.3062	0.2615	0.2757	0.3451	0.5740	–	**0.4309**	**0.3334**	**0.3204**	**0.1483**	**0.4734**	**0.4279**	**0.5174**
Morona	0.8821	0.6739	0.3154	0.3223	0.3490	0.5671	0.4580	–	**0.6123**	**0.5075**	**0.4557**	**0.4740**	**0.6377**	**0.7147**
Nanay	0.8734	0.5561	0.3501	0.3856	0.4592	0.6960	0.2938	0.6278	–	**0.4703**	*0.0659*	**0.6140**	**0.5670**	**0.6736**
Parinari	0.8483	0.4635	0.2828	0.3341	0.3726	0.5897	0.2909	0.5107	0.4055	–	**0.3031**	**0.5233**	**0.4328**	**0.4887**
Pound	0.7847	0.2825	0.2506	0.2779	0.3197	0.5512	0.1341	0.4693	0.0536	0.2538	–	**0.4549**	**0.3557**	**0.5120**
Scavina	0.8536	0.6008	0.4253	0.4403	0.4528	0.6416	0.4629	0.5525	0.6045	0.5131	0.4194	–	**0.6632**	**0.6929**
BA	0.9389	0.6263	0.4661	0.5238	0.5532	0.8310	0.4772	0.7368	0.6762	0.5198	0.4736	0.6716	–	**0.7752**
GUY	0.9340	0.8182	0.4525	0.5148	0.5821	0.8285	0.4920	0.7632	0.6088	0.3947	0.4547	0.7219	0.8216	–

It was estimated by F_{ST} (Weir and Cockerham, 1984) [40], based on variance in allele frequencies (in bold in the table), and Rho_{ST}, an estimator of R_{ST} (Slatkin, 1995) [41] taking the variance in allele sizes into account (values below the diagonal). Non-significant F_{ST} values are printed in italic (after a 1000-fold permutation test).

The accessions studied here cover most of the geographic origins of known *T. cacao* accessions, except for the Upper Amazonian Brazilian accessions, which were not very represented in this study, but for which no specific cluster was identified in the last proposed classification [15]. In our study, the diversity was first structured in four main groups, corresponding to different geographic origins, and with subdivisions within some of these four main groups:

- The first group (cluster 1) included mostly accessions from the northern and central regions of Ecuadorian Amazonia (regions A and B), but without any structuring between them. This is in agreement with the very low genetic distance (0.039) and low genetic differentiation (F_{ST} of 0.0081 and Rho_{ST} of −0.0018) between these 2 groups.
- The second group (cluster 2) included most of the Peruvian accessions, i.e. mainly those collected in the northern region of Peru, associated with wild accessions collected in French Guiana and in the Orinoco region of Venezuela, and with some accessions from Ecuador. The cultivated genotypes from Lower Amazonia in Brazil were also included in this group.

The Upper Amazonian region has always been considered as the primary centre of origin and diversity of *T. cacao* L. species [5], [48]. The high genetic similarity observed between some wild cocoa accessions from the Upper Amazonian region (Parinari or Pound) and the group from French Guiana, suggests that during *T. cacao* evolution the extension of cocoa populations towards the eastern part of South America and until the French Guyana region, could have started from this Parinari Upper Amazonian region.

- The third group (cluster 3) included most of the accessions from the southern Amazonian region of Ecuador, the accessions collected in Peru along the Morona river, close to this southern Ecuadorian region, and the Nacional accessions. In this group, a close genetic relationship was found among some wild accessions from the southern part of Ecuadorian Amazonia (region C), and the Nacional pool from the coastal region. These results were supported by the genetic distance assessment, as well as by a parentage analysis. Our results suggest that the Amazonian region located in the Province of Zamora Chinchipe could be the centre of domestication of the Nacional variety at the origin of the first Nacional cocoa plantation that was set up on the Guayas riverbanks. Later, from this latter plantation, cocoa cultivation could have expanded along the Guayas tributary rivers (upper waters), which is now the 'Arriba' cocoa production region in the provinces of Guayas, Los Ríos and Manabí (Fig. 1b).
- The fourth group (cluster 4) included all Criollo accessions collected from Mexico to Venezuela, reflecting its narrow genetic base. In a previous study, some accessions from Colombia (EBC) were identified as being the closest to Criollo [17]. It has been suggested that Colombia is the centre of origin of Criollo. In our study, we did not have access to EBC Colombian accessions, but the higher genetic similarity and lower genetic differenciation (FST and RhoST values) observed between Criollo and accessions from the northern Amazonian region of Ecuador showed the same geographic trend as previously noted [17].

This migration and domestication hypothesis is supported by the geographical proximity between the first cocoa plantation set up along the Guayas riverbanks [26] and the wild cocoa collected along the Nangaritza, Yacuambi and Zamora riverbanks in the province of Zamora Chinchipe (Fig. 1b, region C). The morphological similarity between the Nacional variety and the native cocoa populations located in the Ecuadorian jungle (fruit and seed) close to the Amazonian cities of Archidona and Macas in region C, and already reported [49], further supports this hypothesis.

Inbreeding coefficients (F_{IS}) demonstrate a clear and significant excess of homozygotes in some groups. It is the case for the old domesticated and self compatible varieties Criollo and BA varieties, but also for some wild genotypes such as those of the VEN, GUY and LCT-EEN groups. Highly homozygous wild genotypes were identified in each region sampled. This latter result

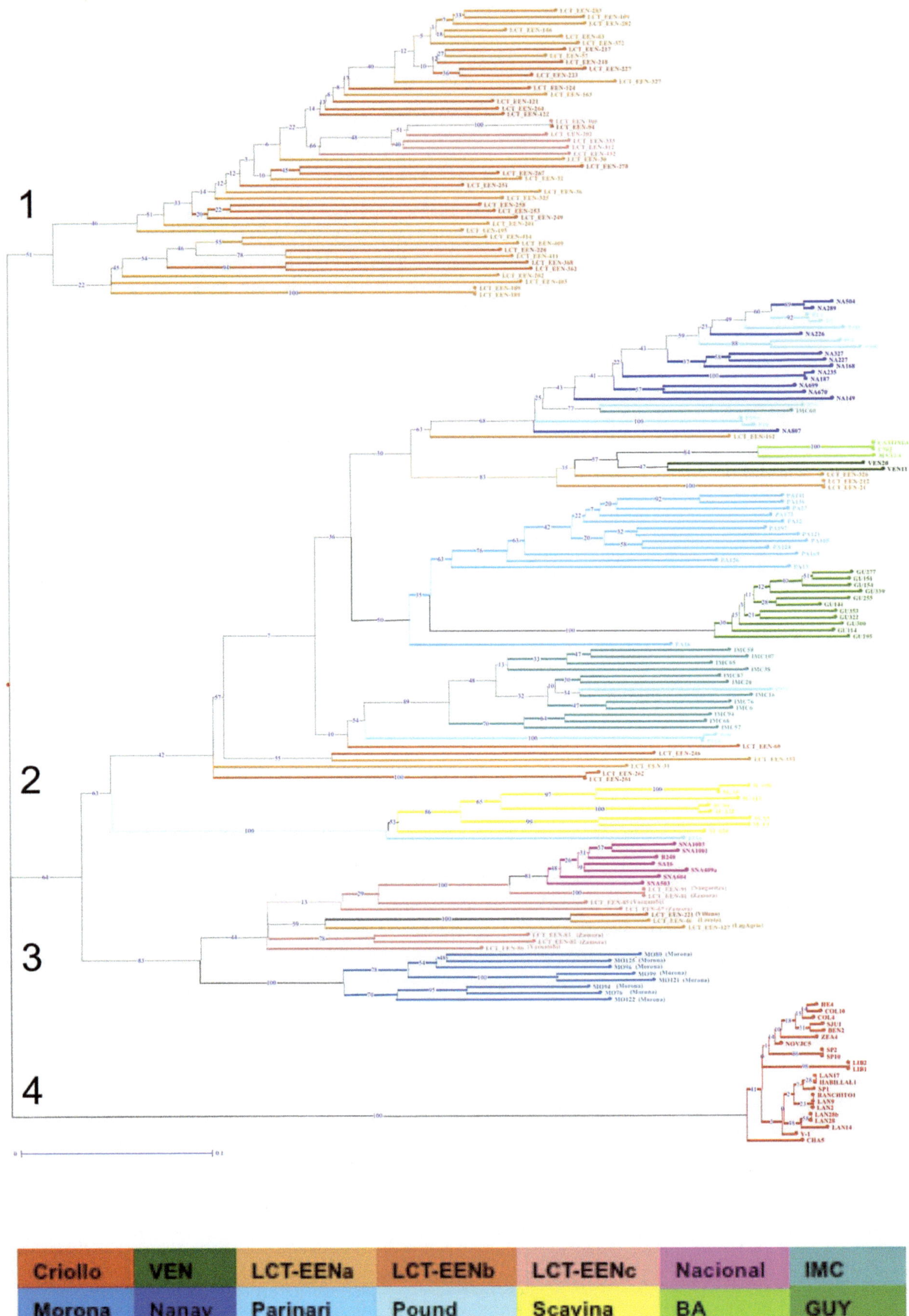
1
2
3
4
Criollo
VEN
LCT-EENa
LCT-EENb
LCT-EENc
Nacional
IMC
Morona
Nanay
Parinari
Pound
Scavina
BA
GUY

Figure 2. Genetic relationships among the wild and cultivated cocoa accessions. Neighbour-joining tree from a dissimilarity matrix, constructed with 500 bootstraps (the bootstrap values are indicated in blue in each branch), and representing genetic relationships among the whole set of wild and cultivated cocoa accessions analysed in this study. GU (French Guiana), MO (Morona), NA (Nanay), PA (Parinari), P (Pound), SCA (Scavina), SNA (Nacional). The numbers 1 to 4 indicate the cluster numbers reported in the text. Information about the rivers present in the areas where the wild genotypes closest to Nacional were collected (cluster 3), is reported. See Table 1 for code identification.

suggests that self compatible alleles already existed in these wild populations and that natural selection could have already eliminated part of the genetic burden due to inbreeding, thus facilitating the process of domestication of some wild populations.

Most representatives of the native Nacional variety have a high level of homozygosity, probably associated with their self compatibility [50–52]. The question is whether the domestication process has fixed and selected, by selfing, highly homozygous genotypes, or whether natural selection, leading to increased homozygosity, was already under way in wild populations before this domestication event. A common domestication feature is the reduction of genetic diversity in crops relative to wild progenitors. The severity of genetic loss ascribed to bottleneck effects varies greatly among crop species. Some authors indicated that this reduction results from two major forces [53]: first, most domestication events are thought to have involved initial populations of small size (relative to wild ancestors) that contained a narrow level of genetic diversity; the second factor to have an impact on crop genomes is the selection in favour of agronomic traits that distinguish crops from their ancestors. Humans usually apply selection pressure on the ancestral gene pool to select for favourable traits, thus increasing or fixing favourable alleles at genes controlling these traits.

The particular characteristics of some cocoa varieties could have facilitated the selection and migration of specific cocoa materials. From this standpoint, the specific aromatic flavour of chocolate produced from Nacional cocoa beans, which can already be detected on the bean pulp, even without a fermentation step, could have been one of the criteria used by the primitive human communities to choose the cocoa mother tree for further seed sowing. Indeed, it is possible that travelling merchants transporting cocoa pods along the roads used the fresh pulp only for their own refreshment and nutrition, but without consuming the cocoa beans, thereby introducing the cocoa tree into a new environment [54]. An extensive flavour aroma study in wild genotypes combined with an evolutionary analysis of this trait variation could help to identify candidate genes responsible for flavour in the aromatic Nacional variety.

Table 6. Most probable ancestors of Nacional individuals.

Individual	Prob. Non-exclusion	First parent	Second parent
Sa16	6.937×10^{-7}	LCT-EEN91	LCT-EEN86
SNA409a	4.072×10^{-6}	LCT-EEN86	LCT-EEN91
SNA503	5.139×10^{-6}	LCT-EEN91	LCT-EEN85
SNA604	4.001×10^{-6}	LCT-EEN91	LCT-EEN85
SNA1001	1.065×10^{-6}	LCT-EEN86	LCT-EEN85
SNA1003	6.188×10^{-7}	LCT-EEN86	LCT-EEN85
B240	3.893×10^{-6}	LCT-EEN91	LCT-EEN86

The two most likely parents of Nacional cocoa individuals were identified with the Cervus 2.0 software parentage analysis. All putative parents tested were from the southern Amazonian region C of Ecuador.

Many crop species have been domesticated for thousands of years. Archeological evidence, based mainly on the presence of theobromine in pottery residue, revealed that cocoa was used in the early formative period in Mesoamerica, dating back to 600 BC until 1900 BC [55–58].

Cocoa domestication seems to have occurred much more recently in most of the other cultivating countries, despite the fact that wild cocoa populations exist in their areas, but no archeological evidence has been found so far in these countries. In Ecuador, information on the domestication of its native cocoa plantations dates back only four centuries.

There is no evidence of human dispersal of this species before the first cocoa plantations were planted by the Spanish in the Ecuadorian coastal region. Knowledge of the dispersal mechanism involved in the long-distance migration process is thus essential to explain our results. Several hypotheses have been put forward to explain the cocoa migration process, such as: the transport of fruits or seeds by birds, animals or humans, but so far none of these has been formally confirmed. It was suggested that the Nacional cocoa variety arrived via the old Inca roads and was planted by the native people of that time along the coastal regions [49]. However, archaeological evidence suggests that these old roads were not built by the Incas but rather by pre-Columbian native people, who inhabited the coastal, Andean and Amazonian regions of Ecuador [59] thousands of years before the arrival of the Incas in the Ecuadorian regions.

Recently, a ceremonial formative site was discovered on the eastern slopes of the Andes in the southern part of Ecuadorian Amazonia, and dating back to the 3rd millennium BC. The cultural remains showed a high degree of development of these societies, with ceramics and marine shells (*Spondyle* and *Strombus*), which is evidence that commercial exchanges with coastal people occured in this region [60], [61]. This site is located in the province of Zamora Chinchipe (region C), where the putative ancestors of the Nacional variety were identified. Archaeological evidence also showed that these exchanges went as far as a place presently known as La Cueva de los Tayos, which is located in the province of Morona Santiago [59].

Unfortunately, contrary to the Mayas who used hieroglyphs and represented cocoa, the more primitive people that inhabited Ecuador did not use writing symbols, so there is little evidence of cocoa domestication and use before Spanish arrival. However, archaeological evidence about contacts and exchanges of products between pre-Columbian peoples from coastal, sierran and Amazonian regions of Ecuador, dating back 3000 years BC [60], [62], could provide further explanations on the origin of cocoa trees in the Ecuadorian coastal region.

From our results, future cocoa expeditions should be carried out to confirm that the southern Amazonian area of Ecuador, could be the putative centre of origin of the Nacional variety, and to collect new germplasm with the same Nacional flavour specificities adapted for fine flavour cocoa improvement in Ecuador.

Supporting Information

Table S1 Origin, collection, % heterozygosity and status of the 176 *T. cacao* accessions analysed in the present study. CATIE: Centro Agronomico Tropical de

Investigación y Ensenanza (Costa Rica); CIC: Centro de Investigación Caribia (Colombia); CIRAD/Mpl: Centre de Coopération internationale en recherche agronomique pour le développement (France)/Montpellier Centre; CNRA: Centre National de Recherches Agronomiques (Côte d'Ivoire); CRU: Cocoa research Unit – University of the West Indies (Trinidad and Tobago); EET-P: Estación Experimental Tropical-Pichilingue (Ecuador); FONAIAP: Fondo Nacional de Investigaciones Agropecuarias (Venezuela); INTA: Instituto Nicaraguense de Tecnologia Agropecuario (Nicaragua); INIFAP: Instituto Nacional de Investigaciones Forestales. Agrícolas y Pecuarias (Mexico). The % of heterozygosity was evaluated by 80 SSR (as reported in Table 2).

Acknowledgments

We thank Xavier Perrier for his assistance with the Darwin software, Christian Leclerc for his valuable comments to improve this paper, and David Manley for revising the English.

Author Contributions

Conceived and designed the experiments: RGLS CL FA. Performed the experiments: RGLS OF AL SP MB. Analyzed the data: RGLS CL XA BC AMR AL. Contributed reagents/materials/analysis tools: RGLS FA MB. Wrote the paper: RGLS CL.

References

1. Argout X, Salse J, Aury JM, Guiltinan M, Droc G, et al. (2011) The genome of *Theobroma cacao*. Nature Genetics 43, 101–108. doi:10.1038/ng.736.
2. Knight R, Rogers HH (1955) Incompatibility in *Theobroma cacao*. Heredity 9: 69–77.
3. Cope FW (1958) Incompatibility in *Theobroma cacao*. Nature 181: 279.
4. Cope FW (1962) The mechanism of pollen incompatibility in *Theobroma cacao* L. Heredity 17: 157–182.
5. Cheesman E (1944) Notes on the nomenclature, classification and possible relationships of cocoa populations. Tropical Agriculture 21: 144–159.
6. Lerceteau E, Flipo S, Pétiard V, Crouzillat D (1997) Genetic differentiation among Ecuadorian *Theobroma cacao* L. accessions using molecular and morphological analyses. Euphytica 95: 77–87.
7. Lerceteau E, Robert T, Petiard V, Crouzillat D (1997) Evaluation of the extent of genetic variability among *Theobroma cacao* accessions using RAPD and RFLP markers. Theoretical and Applied Genetics 95: 10–19.
8. Laurent V, Risterucci AM, Lanaud C (1993) Chloroplast and mitochondiral DNA diversity in *Theobroma cacao*. Theoretical and Applied Genetics, 87 (1–2): 81–88.
9. Laurent V, Risterucci AM, Lanaud C (1994) Genetic diversity in cocoa revealed by cDNA probes. Theoretical and Applied Genetics, 88 (2): 193–198.
10. N'Goran JKA, Laurent V, Risterucci AM, Lanaud C (1994) Comparative genetic diversity studies of *Theobroma cacao* L. using RFLP and RAPD markers. Heredity, 73: 589–597.
11. Sounigo O, Umaharan R, Christopher Y, Sankar A, Ramdahin S (2005) Assessing the Genetic diversity in the International Cocoa Genebank, Trinidad (ICG, T) using isozyme electrophoresis and RAPD. Genetic Resources and Crop Evolution 52: 1111–1120.
12. Bekele FL, Bekele I, Butler DR, Bidaisee GG (2006) Patterns of morphological variation in a sample of cacao (*Theobroma cacao* L.) germplasm from the International Cocoa Genebank, Trinidad. Genetic Resources and Crop Evolution 53: 933–948.
13. Sereno ML, Albuquerque PSB, Vencovsky R, Figueira A (2006) Genetic Diversity and Natural Population Structure of Cacao (*Theobroma cacao* L.) from the Brazilian Amazon Evaluated by Microsatellite Markers. Conservation Genetics 7: 13–24.
14. Zhang D, Arevalo-Gardini E, Mischke S, Zúñiga-Cernades L, Barreto-Chavez A, Adriazola del Aguila J (2006) Genetic diversity and structure of managed and semi-natural populations of cocoa (*Theobroma cacao*) in the Huallaga and Ucayali Valleys of Perú, Ann. Bot. (Lond.) 98: 647–655.
15. Motamayor JC, Lachenaud P, Da Silva e Mota JW, Loor R, Kuhn DN, et al. (2008) Geographic and genetic population differentiation of the Amazonian chocolate tree (*Theobroma cacao* L.) PLoS One, 3 (10): e3311 (8p.).
16. Lachenaud P, Zhang D (2008) Genetic diversity and population structure in wild stands of cacao trees (*Theobroma cacao* L.) in French Guiana. Ann. For. Sci. 65–310 (7p). DOI: 10.1051/forest:2008011.
17. Motamayor JC, Risterucci AM, Lopez PA, Ortiz CF, Moreno A, Lanaud C (2002) Cacao domestication I: the origin of the cacao cultivated by the Mayas. Heredity 89: 380–38.
18. Stevenson WB (1829) Historical and descriptive narrative of twenty year's residence in South America. Vol II Londres.
19. Prescott WH (1847) History of the conquest of Peru with a preliminary view of the civilization of Incas. London.
20. Bartley BGD (2005) The genetic diversity of cacao and its utilization. CABI Publishing, London, UK.
21. Allen BJ, Lass AR (1983) London cocoa trade Amazon project. Final report phase 1. Cocoa Growers' Bulletin 34: 1–72.
22. Pound FJ (1938) Cacao and witchbroom disease (*Marasmius perniciosus*) of South America. In H Toxopeus (ed.) Reprint Archives of cocoa research Vol 1 Amer Cacao Res. Institute.
23. Soria JV (1970) Principal varieties of cocoa cultivated in tropical America. Cocoa Growers' Bulletin 19: 12–21.
24. Vera BJ (1987) Antecedentes históricos. Manual del cultivo de cacao. INIAP, EET-Pichilingue. Quevedo, Ecuador, pp: 6–9.
25. Crawford LR (1980) El Ecuador en la época cacaotera. Respuestas locales al auge y colapso en el ciclo monoexportador. Editorial universitaria. Universidad Central del Ecuador.
26. Preuss P (1901) Expedition nach Central- und Sudamerika 1899/1900. Kolonial-Wirtschaftlichen Komitees, Berlin.
27. Loor Solorzano RG, Risterucci AM, Courtois B, Fouet O, Jeanneau M, et al. (2009) Tracing the native ancestors of modern *Theobroma cacao* L. population in Ecuador. Tree genetics and genomes, 5 (3): 421–433. [20100330]. doi:10.1007/s11295-008-0196-3.
28. International Cocoa Organization (2012) ICCO Quarterly Bulletin of Cocoa Statistics, Vol. XXXVIII - No.2, Cocoa year 2010/11 (may 2012).
29. Pound FJ (1945) A note about the cacao populations of South America. Report and Proceedings Cocoa Research Conference, London. Colonial 192: 95–7. (Reprinted 1982 in Arch. Cocoa Res 1: 93–97).
30. Lachenaud P, Sallée B (1993). Les cacaoyers spontanés de Guyane. Localisation, écologie et morphologie. Café cacao thé (Paris), 37, n°2: 101–114.
31. Lachenaud P, Mooleedhar V, Couturier C (1997) Les cacaoyers spontanés de Guyane: Nouvelles prospections. Plantations, Recherche, Développement 4: 25–32.
32. Motamayor JC (2001) Etude de la diversité génétique et de la domestication des cacaoyers du groupe Criollo (*Theobroma cacao* L.) à l'aide de marqueurs moléculaires. Thèse de doctorat, Université Paris XI, Orsay, France, 176p.
33. Fouet O, Allegre M, Argout X, Jeanneau M, Lemainque A, et al. (2011) Structural characterization and mapping of functional EST-SSR markers in *Theobroma cacao*, Tree Genet. Genomes, 99, 1–19. Doi:10.1007/s11295-011-0375-5.
34. Lanaud C, Larmande P, Risterucci AM, Fargeas D (2001) A high density cocoa genetic map enriched with microsatellite markers. 13th international cocoa research conference, Kota Kinabalu, Sabah, Malaysia, 9–14 oct 2000, p255–265.
35. Risterucci AM, Grivet L, N'Goran JAK, Pieretti I, Flament MH, et al. (2000) A high-density linkage map of *Theobroma cacao* L. Theoretical and Applied Genetics 101: 948–955.
36. Belkhir K, Borsa P, Chikhi L, Raufaste N, Bonhomme F (2004) GENETIX 4.05, logiciel sous Windows TM pour la génétique des populations, Laboratoire Génome, Populations, Interactions; CNRS UMR 5000; Université Montpellier II, Montpellier (France).
37. Nei M (1978) Estimation of average heterozygosity and genetic distance from a small number of individuals. Genetics 89: 583–590.
38. El Mousadik A, Petit RJ (1996) High level of genetic differentiation for allelic richness among populations of the argan tree [Argania spinosa (L.) Skeels] endemic to Morocco. Theoretical and Applied Genetics 92: 832–839.
39. Goudet J (1995) FSTAT (Version 1.2): A computer program to calculate F-statistics. J. Hered. 86: 485–486.
40. Weir BS, Cockerham CC (1984) Estimating F-statistics for the analysis of population structure. Evolution 38: 1358–1370.
41. SLATKIN M (1995) A measure of population subdivision based on microsatellite allele frequencies. Genetics 139: 457–462.
42. Raymond M, Rousset F (1995) GENEPOP (version 1.2): population genetics software for exact tests and ecumenicism. J. Heredity 86: 248–249.
43. Rousset F (2008) Genepop'007: a complete reimplementation of the Genepop software for Windows and Linux. Mol. Ecol. Resources 8: 103–106.
44. Michalakis Y, Excoffier L (1996) A generic estimation of population subdivision using distances between alleles with special interest to microsatellite loci. Genetics 142: 1061–1064.
45. Nei M (1972) Genetic distance between populations. Am Nat 106: 283–292.
46. Perrier X, Flori A, Bonnot F (2003) Data analysis methods. In: Hamon P, Seguin M, Perrier X, Glaszmann JC, Ed., Genetic diversity of cultivated tropical plants. Enfield, Science Publishers. Montpellier, France 43–76.
47. Marshall TC, Slate J, Kruuk LEB, Pemberton JM (1998) Statistical confidence for likelihood-based paternity inference in natural populations. Molecular Ecology, 7, 639–655.

48. Catrecasas J (1964) Cacao and its allies: a taxonomic revision of the genus *Theobroma*. Bulletin of the United States, National Museum, Smithsonian Institution (Washintong) 35: 379–614.
49. Vera BJ (1993) Botánica del cacao. Manual del cultivo de cacao 2da edición. INIAP, EET-Pichilingue. Quevedo, Ecuador, pp: 10–15.
50. Quiroz J (1990) Estudio de la compatibilidad en algunos cultivares de cacao (*Theobroma cacao* L.). Tesis de Ing. Agr. Universidad Técnica de Babahoyo. Babahoyo, Ecuador. 30p.
51. Pastorelly RD (1992) Evaluación de algunas características del cacao tipo Nacional, en la zona de Tenguel. Tesis Ing. Agr. Universidad Agraria del Ecuador. Guayaquil, Ecuador, p7.
52. Loor RG (1998) Obtención de híbridos de cacao tipo Nacional provenientes de materiales de alta productividad y resistentes a enfermedades. Tesis de Ingeniero Agrónomo. Universidad Técnica de Manabí, Portoviejo, Ecuador. 67p.
53. Tanksley SD, McCouch SR (1997) Seed banks and molecular maps: unlocking genetic potential from the wild. Science 277: 1063–1066.
54. Stone D (1984) Pre-Columbian migration of *Theobroma cacao* Linnaeus and Manihot esculenta Crantz from Northern South America into Mesoamerica: a partial hypothetical view. In: Pre-Columbian Plant Migration. Cambridge, MA, Harvard University Press. 67–79, 81–83.
55. Hurst WJ, Tarka SM, Powis TG, Valdez F, Hester RT (2002) Archaeology: cacao usage by the earliest Maya civilization. Nature 418: 289–290.
56. Henderson JS, Joyce RA, Hall GR, Hurst WJ, McGovern PE (2007) Chemical and archaeological evidence for the earliest cacao beverages. Proceedings of the National Academy of Sciences of the United States of America 104: 18937–18940.
57. Powis TG, Hurst WJ, Rodriguez MC, Ponciano Ortiz C, Blake M et al. (2007) Oldest chocolate in the New World. Antiquity 81 (314): project galley online http://www.antiquity.ac.uk/projgall/powis/index.html.
58. Powis TG, Cyphers A, Gaikwad NW, Grivetti L, Cheong K (2011) Cacao use and the San Lorenzo Olmec. Proceedings of the National Academy of Sciences of the United States of America108: 8595–8600.
59. Salazar E (1995) Entre mitos y fabulos. El Ecuador aborigen, Quito, Corporación Editora Nacional.
60. Valdez F, Guffroy J, de Saulieu G, Hurtado J, Yépez A (2005) Découverte d'un site cérémoniel formatif sur le versant oriental des Andes. Comptes Rendus de l'Academie des Sciences de l'Institut de France. Palevol 4: 369–374.
61. Valdez F (2008) Inter-Zonal Relationships in Ecuador (Chapter 43). In Handbook of South American Archaeology. Silverman, H. et W. Isbell eds., Kluwer Academic Publishers, Pp. 865–887, Chicago.
62. Ayala ME (1983) Nueva historia del Ecuador. Editorial Grijalbo Ecuatoriana.

High Level Resistance against Rhizomania Disease by Simultaneously Integrating Two Distinct Defense Mechanisms

Ourania I. Pavli[1]*, Anastasia P. Tampakaki[2], George N. Skaracis[1]

1 Department of Crop Sciences, Agricultural University of Athens, Athens, Greece, **2** Department of Agricultural Biotechnology, Agricultural University of Athens, Athens, Greece

Abstract

With the aim of achieving durable resistance against rhizomania disease of sugar beet, the employment of different sources of resistance to *Beet necrotic yellow vein virus* was pursued. To this purpose, *Nicotiana benthamiana* transgenic plants that simultaneously produce dsRNA originating from a conserved region of the BNYVV replicase gene and the $HrpZ_{Psph}$ protein in a secreted form (SP/$HrpZ_{Psph}$) were produced. The integration and expression of both transgenes as well as proper production of the harpin protein were verified in all primary transformants and selfed progeny (T1, T2). Transgenic resistance was assessed by BNYVV-challenge inoculation on T2 progeny by scoring disease symptoms and DAS-ELISA at 20 and 30 dpi. Transgenic lines possessing single transformation events for both transgenes as well as wild type plants were included in inoculation experiments. Transgenic plants were highly resistant to virus infection, whereas in some cases immunity was achieved. In all cases, the resistant phenotype of transgenic plants carrying both transgenes was superior in comparison with the ones carrying a single transgene. Collectively, our findings demonstrate, for a first time, that the combination of two entirely different resistance mechanisms provide high level resistance or even immunity against the virus. Such a novel approach is anticipated to prevent a rapid virus adaptation that could potentially lead to the emergence of isolates with resistance breaking properties.

Editor: Alfredo Herrera-Estrella, Cinvestav, Mexico

Funding: The study was funded by the Laboratory of Plant Breeding and Biometry and the Laboratory of General and Agricultural Microbiology. No other funders supported this work. The funders had no role in study design, data collection and analysis, decision to publish, or preparation of the manuscript.

Competing Interests: The authors have declared that no competing interests exist.

* E-mail: raniapavli@yahoo.gr mailto:gskaracis@aua.gr

Introduction

Sugar beet (*Beta vulgaris* L. ssp. *vulgaris*) is one of the most important industrial crop species, occupying globally a cultivated area of approximately 8.1 million hectares spread over 41 countries [1]. Worldwide, the economic viability of sugar beet cultivation is largely depended on the successful protection against pathogens, of which Rhizomania, a viral root disease caused by *Beet necrotic yellow vein virus* (BNYVV), exerts a very high impact. The disease causes severe economic losses as a consequence of a dramatic reduction in root yield, sugar content and purity, especially when infections occur early in the growing season [2]. In most of the rhizomania infested areas, partially resistant sugar beet varieties have replaced susceptible cultivars since they could lose up to 80–90% of their potential sugar yield [3].

BNYVV is a rod-shaped virus with a multipartite plus ssRNA genome, consisting of four genomic messenger-sense RNAs, with some isolates harbouring a fifth RNA segment [4–6]. RNAs 1 and 2 carry all necessary information for housekeeping functions, whereas the natural infection process requires the host-specific function of additional proteins, directly involved in pathogenesis and vector transmission, encoded by the small RNA species [7]. On the basis of its molecular characteristics, BNYVV has been classified in three major pathotypes, referred to as A, B, and P [8,9]. BNYVV isolates containing a fifth RNA species are generally considered as more aggressive than those containing RNAs 1–4 [4,10,11]. A is the most widespread type, found in most EU countries, USA, China and Japan, whereas the other two types present a more limited spread [12,13].

Rhizomania incidence and severity cannot be considerably reduced by preventive cultural practices such as rotation, avoidance of excessive soil moisture and early plantings. Consequently, the only substantial means to ensure a viable crop production in rhizomania incidence areas is the use of varieties specifically bred as resistant to the disease [14]. In this respect, coping with rhizomania to date has mainly been based on cultivars endowed with the *Rz1* gene ("Holly" source), a dominant gene conferring sufficiently high levels of protection against BNYVV [15–17]. In terms of generating high level disease resistance, the potential of various genetic engineering approaches has been further explored. Although successful at varying levels, all such methodologies have to date exploited the introgression of a single transgene with the perspective of engineering resistance based on either the expression of gene products of pathogen [18] or non-pathogen origin [19,20] or the RNA-silencing signaling route [21–23].

However, recent changes in field and molecular BNYVV epidemiology, as manifested by the recent emergence of resistance-breaking mutants for the *Rz1*-mediated resistance, resulting from high selective pressure to overcome this gene [1,13,24–31], probably reflect an ensuing new endemic disease development. Such an evolution would require major adjustments in mainstream breeding programs if they were to keep providing a durable crop protection through the use of appropriate cultivars. Due to the limited availability of useful natural genetic sources of resistance against the prevailing virus strains [32] as well as the recent emergence of novel resistance breaking strains, all relevant breeding activities become of paramount importance.

In this study we explored the possibility to integrate two entirely different functional defense mechanisms in order to generate transgenic resistance against rhizomania disease of sugar beet. Towards this direction, we produced transgenic *N. benthamiana* plants simultaneously expressing two distinct transgenes: the $hrpZ_{Psph}$ from *P. syringae*. pv. *phaseolicola*, which has been previously proved as a valuable tool for achieving rhizomania resistance, and a conserved region of the BNYVV replicase gene arranged as inverted repeat, an arrangement known to act as a strong silencing inducer.

Materials and Methods

Plasmids and bacterial strains

Agrobacterium tumefaciens strain C58C1 harboring the binary plant expression vector construct pBin.Hyg.Tx-SP/$hrpZ_{Psph}$ [33] and *A. tumefaciens* strain GV3101 carrying the intron-hairpin construct GW-IR3 [23] were used to transform *N. benthamiana* plants. The former plasmid contains the *hrpZ* gene from *P. syringae*. pv. *phaseolicola* NPS3121 (approx. 1 kb) fused in-frame with the signal peptide from the tobacco pathogenesis-related protein, whereas the latter carries a fragment (459 bp) of the BNYVV RNA 1-encoded replicase gene, which upon transcription is processed as dsRNA.

Bacterial cells were grown at 28°C in liquid LB selection medium containing: rifampicine (50 μg ml^{-1}), carbenicillin (100 μg ml^{-1}) and kanamycin (50 μg ml^{-1}) for strain C58C1; rifampicine (20 μg ml^{-1}), gentamycin (25 μg ml^{-1}) and spectinomycin (50 μg ml^{-1}) for GV3101, for 2 days or until OD_{600} = 0.6–1 was reached. Cultures were centrifuged and then resuspended to a final concentration of 10^7 cfu/ml. Cell suspensions were mixed in equal volumes and the resulting mixture was used as inoculum for plant transformation.

For the generation of transgenics carrying only one transgene, used as controls, cell suspensions were separately used for plant transformation.

Plant transformation and evaluation of transgenics

For plant transformation, leaf discs from healthy *N. benthamiana* plants (5–6 weeks-old) were used as explants, according to the protocol described by Horsch et al. [34]. Following selection of transformants with respect to antibiotic resistance to hygromycin (30 μg ml^{-1}) and kanamycin (150 μg ml^{-1}) for the transforming plasmids pBin.Hyg.Tx-SP/$hrpZ_{Psph}$ and GW-IR3 respectively, regenerated shoots were rooted and transferred to soil.

Transgene integration and absence of Ti plasmid-originated sequences was examined by a multiplex PCR assay, using specific primers to amplify the 995 bp, 459 bp and 590 bp fragments of $hrpZ_{Psph}$ [20], the BNYVV-derived transgene [23] and *virG* of *A. tumefaciens* [20]. Transgene expression was verified by RT-PCR using abovementioned primers. Plants that were PCR-positive for transgene integration and expression were subsequently examined for the accumulation of SP/$HrpZ_{Psph}$ by immunoblot blot analysis as previously described [20]. Plants that met all above requirements were selfed and thereof progeny seeds (T1, T2) were selected against hygromycin and kanamycin. Resistant plants were grown *in vitro* for shoot and root formation and were subsequently transferred to soil. Representative T2 transgenic plants were further assessed on the basis of BNYVV-derived transgene expression and SP/$HrpZ_{Psph}$ protein accumulation.

Virus inoculations

For foliar rub-inoculations of *N. benthamiana* plants, heavily BNYVV-infected sugar beet plants were exploited as virus source. In order to evaluate resistance under high infectivity pressure, a previously characterized as highly infectious virus isolate, due to the presence of specific aa at key positions of the BNYVV p25 protein, was used [31].

Transgenic lines simultaneously expressing SP/$HrpZ_{Psph}$ and BNYVV-derived dsRNA were assessed for resistance to the BNYVV. In addition, transgenic lines carrying only one transgene, either SP/$hrpZ_{Psph}$ [20] or BNYVV-dsRNA (this study), were also included in the experiments as comparative controls. In total, 30 transgenic *N. benthamiana* T0 lines were evaluated for BNYVV resistance: i) ten doubly transformed plants expressing SP/$HrpZ_{Psph}$ and BNYVV-derived dsRNA, ii) ten lines expressing SP/$HrpZ_{Psph}$ and iii) ten lines expressing the virus-derived transgene. Twelve T2 plants from each transgenic line were challenge-inoculated with BNYVV and were evaluated for resistance in two consecutive resistance assays.

Plants at the stage of 4–6 leaves were mechanically inoculated by first dusting at least 3 leaves with carborundum, in order to facilitate virus entrance, and then applying a virus-infected leaf extract. The inoculum source was obtained as described in Pavli et al. [23]. Inocula were tested on non-transgenic control *N. benthamiana* plants to confer infectivity. Tissue from non-transgenic uninoculated plants was accordingly processed and used as experimental negative control.

Evaluation of virus resistance

Assessment of virus resistance in transgenic *N. benthamiana* was performed on the basis of macroscopically scoring disease symptoms and measuring virus titers. Plants were regularly monitored and symptoms were scored during the 30 day period post inoculation (dpi). The relative amount of virus in inoculated plants was estimated by DAS-ELISA (Adgen Phytodiagnostics) at 20 and 30 dpi, according to the supplier's instructions. Replicate leaf tissue samples were homogenized in 1:3 extraction buffer. Samples scoring absorbance values (405 nm) at least three times the respective measurements of the negative controls were considered as positive.

Results and Discussion

Generation of transgenic *N. benthamiana* plants simultaneously expressing SP/$hrpZ_{Psph}$ and BNYVV replicase-derived transgene

The simultaneous expression of two distinct transgenes has been pursued for a first time as a means to develop high level and stable resistance against this most devastating disease of the sugar beet crop. In this context, a conserved region originating from the BNYVV replicase gene, arranged as inverted repeat, as well as the harpin SP/$HrpZ_{Psph}$ from *P. syringae* pv. *phaseolicola* have been simultaneously produced in transgenic *N. benthamiana* plants.

A. tumefaciens-mediated transformation of *N. benthamiana* plants with two distinct plasmids, pBin.Hyg.Tx-SP/$hrpZ_{Psph}$ and GW-IR3, was carried out by a standard leaf disc method. Sixteen (16) independent transgenic lines (T0) expressing SP/$hrpZ_{Psph}$ along with BNYVV-derived dsRNA were developed and ten of them were randomly selected and selfed to produce T1 and T2 lines.

Primary double transgenics (T0) and selfed progeny (T1, T2) were phenotypically normal at a macroscopic level, presenting however an increased vigor in comparison to the GW-IR3-transformed lines or to the non-transgenic plants of the same age. This phenotype has been previously ascribed to the expression of SP/*hrpZ*$_{Psph}$ [20]. In addition, such enhanced growth properties have been reported to occur both upon external application of purified harpins [35–38] and transgene expression [39–41].

Proper transgene integration and absence of *A. tumefaciens* was verified by a multiplex PCR assay, targeting regions of *hrpZ*$_{Psph}$, BNYVV-replicase gene and *virG* of the disarmed Ti plasmid, in all primary transgenics (T0) and selfed progeny (T2) tested. Figure 1 shows the presence of *hrpZ* and BNYVV-replicase region amplicons in representative T2 double transgenic plants as well the presence of the corresponding amplicons in transgenic plants carrying single transgenes. At the same time, the absence of *virG* amplicons confirms the transgenic nature of plants grown under antibiotic selection (data not shown). Transgene expression was verified, though at varying levels, by means of a RT-PCR assay, in more than 90% of the samples examined (Figure 2). More specifically, transgene expression was detected in 57 out of 60 T2 double transgenics and of these, 44 plants showed high expression levels whereas, 11 and 2 plants showed moderate and low transgene expression levels, respectively. In case of single transgenic plants, 54 were found positive for the BNYVV-derived transgene expression, and of these 27, 16 and 11 plants showed high, moderate and low expression levels, respectively.

Immunoblot analysis of all primary transgenics (T0) and T2 transgenic lines tested indicated the production of three immuno-reactive bands corresponding to SP/HrpZ, HrpZ and a truncated form of the protein (Figure 3). This result demonstrated that the protein was indeed produced and the signal peptide was properly processed and it is also consistent with a previous study [20].

N. benthamiana plants expressing SP/HrpZ$_{Psph}$ and BNYVV-derived dsRNA show high-level resistance to BNYVV inoculation

The ability of the two distinct transgenes, SP/*hrpZ*$_{Psph}$ and BNYVV-derived dsRNA, in conferring resistance to rhizomania disease was evaluated by BNYVV-challenge inoculation of double transgenic plants (T2 progeny) of *N. benthamiana*, which represents a model crop that can be readily exploited for the purpose of assessing rhizomania resistance as previously described [19–21]. Resistance was assessed by visual symptom observation as well as by measurement of virus titers using ELISA at 20 and 30 dpi.

In order to examine resistance of double transformed lines in comparison to lines carrying single transgenes, thirty transgenic *N. benthamiana* lines T0 were evaluated for BNYVV resistance: i) ten double transformed plants expressing both SP/HrpZ$_{Psph}$ and BNYVV-derived dsRNA, ii) ten lines expressing SP/HrpZ$_{Psph}$ and iii) ten lines expressing the BNYVV-derived transgene. Six progeny plants (T2), for each of the ten transgenic lines (T0) per group, were evaluated for virus infection. Challenge inoculation experiments were repeated twice by choosing different T2 plants in each experiment. In total, twelve T2 plants for each of the ten T0 lines (in total 120 T2 plants per group) were evaluated for resistance.

In non-transgenic plants, symptoms of leaf curling and sporadic chlorosis appeared at 10–14 dpi, whereas all of them exhibited systemic disease symptoms i.e. severe mosaic, occasional leaf distortion and general stunting, at 20–22 dpi. In addition, ELISA readings were in agreement with the abovementioned visual assessments (Table 1). Instead, transgenic plants expressing the *SP/hrpZ*$_{Psph}$ were highly resistant to BNYVV in terms of both symptom development and virus accumulation. Such results are in agreement with previous observations [20] and further substantiate the conclusion that the endogenous expression of SP/*hrpZ*$_{Psph}$ confers enhanced resistance against rhizomania. It is worth noting however, that the SP/HrpZ$_{Psph}$-based resistance reported in this study, although quantitatively similar to the one earlier demonstrated [20], was assessed in T2 progeny, as opposed to our previous experiments conducted on T1 transgenic plants. Transgenic plants expressing a fragment of the virus replicase gene, which is transcribed in dsRNA, were challenge-inoculated and assessed for resistance to the BNYVV. Four out of the ten transgenic lines (T0) tested showed high level resistance with respect to symptom development and virus content, pointing to a mechanism analogous to that resulting in resistant sugar beet hairy roots transgenically expressing the BNYVV replicase-derived dsRNA [23]. In contrast, the T2 progeny of the remaining six lines (T0) segregated in varying levels of virus infection ranging from complete resistance to susceptibility (Table 1). It is worth mentioning, that highly resistant lines as well as resistant plants originating from segregating lines were characterized by high transgene expression levels whereas, susceptible plants showed low transgene expression levels. Such findings indicate that line performance is correlated to the corresponding level of the BNYVV replicase transgene expression.

Upon challenge with BNYVV, all T2 transgenic plants expressing both the SP/*hrpZ*$_{Psph}$ and the virus-derived transgene were characterized as highly resistant or even immune to infection. On the basis of symptom expression, the majority of T2 plants (85.0%) remained completely symptomless throughout the period

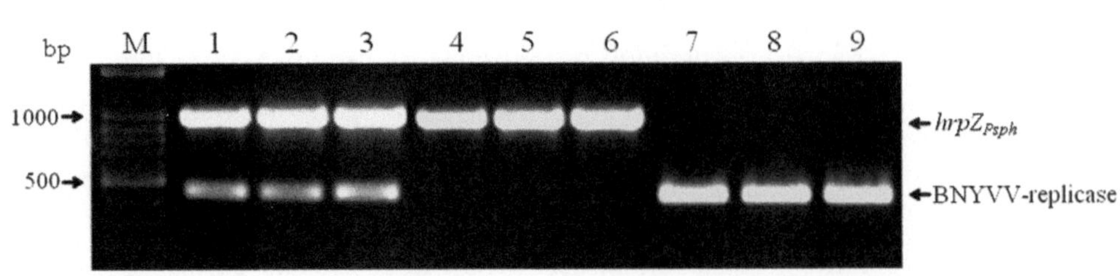

Figure 1. Amplification products obtained by multiplex PCR on total genomic DNA from *A. tumefaciens*-transformed *N. benthamiana* plants. Lane M: Marker in bp (Gene Ruler Ladder mix, Fermentas). Lanes 1, 2, 3: Representative T2 double transgenic plants, carrying the 995 bp fragment of *hrpZ*$_{Psph}$ and the 459 bp fragment of the BNYVV-replicase gene. Lanes 4, 5, 6: Representative T2 transgenic plants, carrying the 995 bp fragment of *hrpZ*$_{Psph}$. Lanes 7, 8, 9: Representative T2 transgenic plants, carrying the 459 bp fragment of the BNYVV-replicase gene. The 590 bp amplicon corresponding to *virG* of *A. tumefaciens* could not be obtained, thus verifying the absence of Ti plasmid sequences in antibiotic resistant *N. benthamiana* transformants.

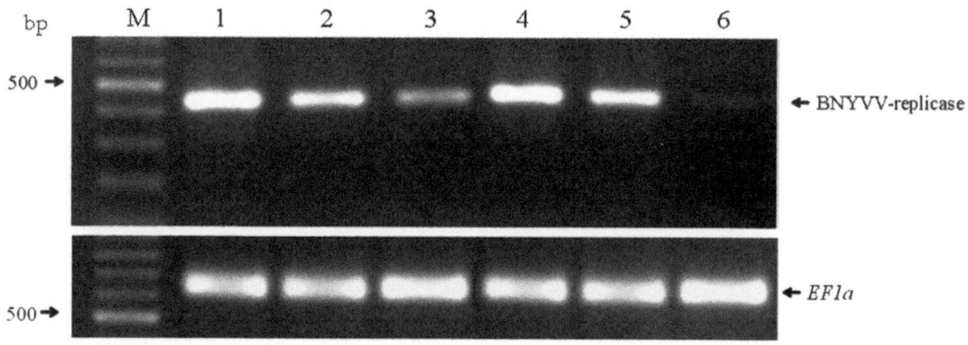

Figure 2. RT-PCR amplification products obtained from transgenic *N. benthamiana* plants expressing the BNYVV replicase-derived transgene. Lane M: Marker in bp (Gene Ruler Ladder mix, Fermentas). Lanes 1, 2, 3: Representative T2 double transgenic plants, expressing the *hrpZ*$_{Psph}$ and the BNYVV replicase-derived transgenes, presenting high, moderate and low resistance levels, respectively. Lanes 4, 5, 6: Representative T2 GW-IR3-transformed plants presenting high, moderate and low resistance levels, respectively. The *EF1a* gene transcript is used as loading control for equal amounts of RNA.

of observation, whereas the remaining plants (15.0%) showed either leaf curling (13.33%) or mild stunting (1.67%), which were though delayed by approximately 15 days compared to the non-transgenic ones (Table 1) (Figure 4). More specifically, all the T2 plants deriving from eight of the ten T0 lines (96 plants) were entirely symptomless and virus-free as evidenced by visual scoring and ELISA assay, respectively (80.0%). The T2 progeny of the remaining two T0 lines were also partially resistant with a portion of them (6 plants) remaining symptomless (25.0%), while the rest 18 plants (75.0%) manifested mild to moderate disease symptoms and relatively low virus titers. At 20 dpi, the great majority of transgenic plants (91.4%) were negative to infection according to ELISA values (data not shown). At 30 dpi, 83.33% of these plants remained completely virus-free while 16.67% presented very low or marginally positive values (Table 1). These findings highlight the effectiveness of the simultaneous expression of the SP/*hrpZ*$_{Psph}$ and the replicase transgenes in conferring high level resistance or even immunity against BNYVV. At the same time, it has been demonstrated that such resistance is superior to the SP/*hrpZ*$_{Psph}$-based resistance [20] or to the RNA silencing-mediated resistance described in this study (data presented in Table 1). Collectively, these data support the conclusion that although the single expression of abovementioned transgenes resulted in enhanced resistance against the disease, their co-expression in transgenic plants led to superior performance, presumably due to additive transgene effects.

Previously, it has been shown that the BNYVV-resistant phenotype is positively correlated with harpin accumulation to the plant cell exterior and it was further speculated that the observed resistance may be attributed to a primed state of plants expressing the SP/*hrpZ*$_{Psph}$ [20]. Furthermore, an intriguing observation was the development of a localized necrosis only in the SP/*hrpZ*$_{Psph}$-expressing leaves in the virus inoculation area. Authors suggested that this necrosis might be attributed to an augmentation or synergistic effects of defense responses elicited by the extracellularly targeted harpin and virus infection. It is interesting that double transformants did not develop an analogous necrosis in the virus inoculation area. It is likely that the primed state of SP/HrpZ$_{Psph}$-expressing plants mounts a first basal level of defense which may be subsequently enhanced by the RNA-silencing signaling route. In this context, it is tempting to speculate that such network interference or combined action of the two defense mechanisms may result in high disease resistance and absence of visible cell death.

A thorough investigation of the transcriptomic profile of defense-related genes in plants expressing SP/HrpZ$_{Psph}$ and BNYVV-derived dsRNA is anticipated to shed light in the resistance observed in this study. Furthermore, it is important to investigate the occurrence of siRNAs prior and at different time intervals post inoculation in order to determine the timing of the RNA silencing activation.

Conclusively, our findings highlight the potential of integrating two genetically distinct defense mechanisms as a novel tool to achieve high level and stable resistance against rhizomania disease of sugar beet. Such "co-targeting" approaches may be extended to a large number of viruses and/or other biotic agents and to a wide variety of cultivated crop species.

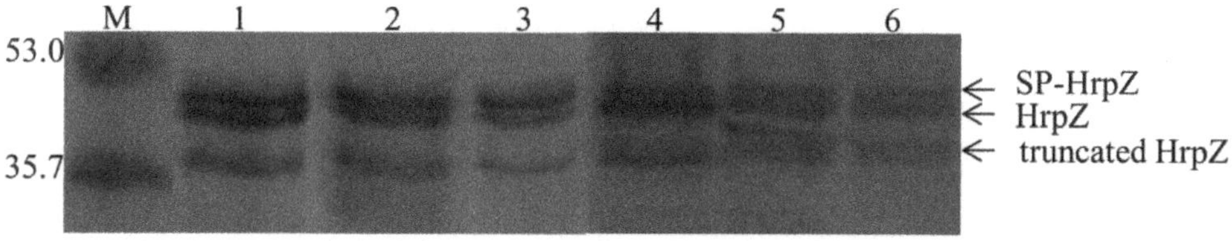

Figure 3. Western blot analysis of SP/HrpZ$_{Psph}$ in protein extracts from leaves of double and single transgenic *N. benthamiana* plants. Lane M: Marker in kDa (Broad range pre-stained SDS marker, Biorad). Lanes 1, 2, 3: Representative T2 double transgenic plants transformed with the plasmids pBin.Hyg.Tx-SP/*hrpZ*$_{Psph}$ and GW-IR3 showing high, moderate and low resistance levels, respectively. Lanes 4, 5, 6: Representative T2 plants transformed with the plasmid pBin.Hyg.Tx-SP/*hrpZ*$_{Psph}$ showing high, moderate and low resistance levels, respectively. The lower band represents the truncated form of the harpin (approx. 2 kDa smaller than the full-length HrpZ$_{Psph}$). Equal amounts of total protein (30 ug) were loaded.

Table 1. Symptom severity and virus titer of transgenic *N. benthamiana* T2 plants expressing SP/HrpZ$_{Psph}$ and BNYVV-derived dsRNA in comparison to T2 plants expressing either SP/HrpZ$_{Psph}$ or BNYVV-derived dsRNA, 30 days after challenge with BNYVV.

Plant type	T2 plants tested	T0 origin	[1]Symptom development			[2]Virus content		
			Score	Number of plants		Score	Number of plants	
		8	–	96[3]	80.0[4]	–	96	80.0
SP/HrpZ$_{Psph}$			–	6	5.0	–	4	3.33
&	120					+	2	1.67
BNYVV-dsRNA		2	+	16	13.33	+	12	10.0
						++	4	3.33
			++	2	1.67	++	2	1.67
		7	–	84	70.0	–	78	65.0
						+	6	5.0
			–	10	8.33	+	10	8.33
SP/HrpZ$_{Psph}$	120		+	18	15.0	+	16	13.33
		3				++	2	1.67
			++	8	6.67	++	5	4.17
						+++	3	2.5
		4	–	48	40.0	–	40	33.33
						+	8	6.67
			–	9	7.5	–	6	5.0
BNYVV-dsRNA	120					+	3	2.5
		6	+	35	29.17	+	27	22.5
						++	8	6.67
			++	28	23.33	++	23	19.16
						+++	5	4.17
Non-transgenics	30		+++	30	100[3]	+++	30	100

[1]– absence of symptoms, + leaf curling, ++ faint mosaic, mild stunting, +++ severe mosaic, leaf distortion, general stunting.
[2]– – value indicative of virus absence, + close to the positive threshold (three times the negative control), ++ half reading of positive control (non-transgenic), +++ equal to the positive control.
[3]numbers denote number of plants in corresponding categories.
[4]numbers denote percent of plants in corresponding categories.

Figure 4. Symptoms of *Beet necrotic yellow vein virus* in *N. benthamiana* at 30 days post inoculation. A) Non-transgenic plant. B) T2 transgenic plant expressing fragment of the BNYVV-replicase gene. C) T2 transgenic plant expressing the *hrpZ$_{Psph}$* transgene. D) T2 double transgenic plant expressing the *hrpZ$_{Psph}$* and a fragment of the BNYVV-replicase gene. The transgenic plant expressing the *hrpZ$_{Psph}$* as well as the double transgenic plant, expressing both the *hrpZ$_{Psph}$* and the BNYVV-replicase transgene, are symptomless. In contrast, the BNYVV dsRNA-expressing plant and the non-transgenic control plant exhibit symptoms of general stunting and/or leaf curling.

Author Contributions

Conceived and designed the experiments: OIP APT GNS. Performed the experiments: OIP. Analyzed the data: OIP APT GNS. Contributed reagents/materials/analysis tools: APT GNS. Wrote the paper: OIP APT GNS.

References

1. Rush CM, Liu H-Y, Lewellen RT, Acosta-Leal R (2006) The continuing saga of rhizomania of sugar beets in the United States. Phytopathology 90: 4–15.
2. Rush CM (2003) Ecology and epidemiology of Benyviruses and plasmodiophorid vectors. Ann Rev Phytopathol 41: 567–592.
3. Casarini B (1999) Le avversità: loro natura, prevenzione e lotta. *In* La barbabietola negli ambienti mediterranei. Casarini B, Biancardi E, Ranalli P (eds.) Bologna, Italy: Edagricole. pp 273–421.
4. Tamada T, Shirako Y, Abe H, Saito M, Kigushi T, et al. (1989) Production and pathogenicity of isolates of *beet necrotic yellow vein virus* with different numbers of RNA components. J Gen Virol 70: 3399–3409.
5. Kiguchi T, Saito M, Tamada T (1996) Nucleotide sequence analysis of RNA-5 of five isolates of *beet necrotic yellow vein virus* and the identity of a deletion mutant. J Gen Virol 77(4): 575–580.
6. Koenig R, Haeberle AM, Commandeur U (1997) Detection and characterization of a distinct type of *beet necrotic yellow vein virus* RNA 5 in a sugar beet growing area in Europe. Arch Virol 142: 1499–1504.
7. Tamada T (1999) Benyviruses. In: Webster RG, Granoff A (eds) Encyclopedia of Virology (2nd edn). London, UK: Academic Press. pp 154–160.
8. Koenig R, Luddecke P, Haeberle AM (1995) Detection of *beet necrotic yellow vein virus* strains, variants and mixed infections by examining single-strand conformation polymorphisms of immunocapture RT-PCR products. J Gen Virol 76: 2051–2055.
9. Koenig R, Lennefors BL (2000) Molecular analysis of European A, B and P type sources of *beet necrotic yellow vein virus* and detection of the rare P type in Kazakhstan. Arch Virol 145: 1561–1570.
10. Tamada T, Kusume T, Uchino H, Kigushi T, Saito M (1996) Evidence that *beet necrotic yellow vein virus* RNA-5 is involved in symptom development of sugar beet roots. In: Sherwood JL, Rush CM (eds) Proceedings of the 3rd Symposium of the International Working Group on Plant Viruses with Fungal Vectors. Denver, American Society of Sugar Beet Technologists, Dundee, UK, pp 49–52
11. Heijbroek W, Musters PMS, Schoone AHL (1999) Variation in pathogenicity and multiplication of *Beet necrotic yellow vein virus* in relation to the resistance of sugar beet cultivars. Eur J Plant Pathol 105:397–405.
12. Kruse M, Koenig R, Hoffmann A, Kaufmann A, Commandeur U, et al. (1994) Restriction fragment length polymorphism analysis of reverse transcription-PCR products reveals the existence of two major strain groups of *beet necrotic yellow vein virus*. J Gen Virol 75: 1835–1842.
13. Schirmer A, Link D, Cognat V, Moury B, Beuve M, et al. (2005) Phylogenetic analysis of isolates of *Beet necrotic yellow vein virus* collected worldwide. J Gen Virol 86: 2897–2911.
14. Biancardi E, Lewellen RT, DeBiaggi M, Erichsen AW, Stevanato P (2002) The origin of rhizomania resistance in sugar beet. Euphytica 127: 383–397.
15. Lewellen RT, Skoyen IO, Erichsen AW (1987) Breeding sugarbeet for resistance to rhizomania: Evaluation of host-plant reactions and selection for and inheritance of resistance, in: Proceedings of the 50th Congress of the IIRB. Brussels, Belgium: International Institute for Beet Research. pp 139–156.
16. Lewellen RT, Biancardi E (1990) Breeding and performance of rhizomania resistant sugar beet, in: Proceedings of the 53rd Congress of the IIRB. Brussels, Belgium: International Institute for Beet Research. pp 79–87.
17. Scholten OE, Jansen RC, Keizer LCP, De Bock TSM, Lange W (1996) Major genes for resistance to beet necrotic yellow vein virus (BNYVV) in *Beta vulgaris*. Euphytica 91: 331–339.
18. Mannerlöf M, Lennefors B-L, Tenning P (1996) Reduced titer of BNYVV in transgenic sugar beets expressing the BNYVV coat protein. Euphytica 90: 293–296.
19. Fecker LF, Koenig R, Obermeier C (1997) *Nicotiana benthamiana* plants expressing *beet necrotic yellow vein virus* (BNYVV) coat protein-specific scFv are partially protected against the establishment of the virus in the early stages of infection and its pathogenic effects in the late stages of infection. Arch Virol 142: 1857–1863.
20. Pavli OI, Kelaidi GI, Tampakaki AP, Skaracis GN (2011) The *hrpZ* gene of *Pseudomonas syringae* pv. *phaseolicola* enhances resistance to rhizomania disease in transgenic *Nicotiana benthamiana* and sugar beet. PLoS ONE 6(3): e17306.
21. Andika IB, Kondo H, Tamada T (2005) Evidence that RNA silencing-mediated resistance to *Beet necrotic yellow vein virus* is less effective in roots than in leaves. MPMI 18: 194–204.
22. Lennefors B-L, Savenkov EI, Bensefelt J, Wremerth-Weich E, van Roggen P, et al. (2006) dsRNA-mediated resistance to *Beet Necrotic Yellow Vein Virus* infections in sugar beet (*Beta vulgaris* L. ssp. *vulgaris*). Mol Breed 18: 313–325.
23. Pavli OI, Panopoulos NJ, Goldbach R, Skaracis GN (2010). BNYVV-derived dsRNA confers resistance to rhizomania disease of sugar beet as evidenced by a novel transgenic hairy root approach. Transgenic Res 19: 915–922.
24. Tamada T, Miyanishi M, Kondo H, Chiba S, Han CG (2002) Pathogenicity and molecular variability of *Beet necrotic yellow vein virus* isolates from Europe, Japan, China, and the United States. In: Proceedings of the 5th Symposium of the International Working Group on Plant Viruses with Fungal Vectors. Denver, American Society of Sugar Beet Technologists, Zurich, Switzerland. pp. 13–16.
25. Chiba S, Kondo H, Miyanishi M, Andika IB, Han C, et al. (2011) The Evolutionary History of *Beet necrotic yellow vein virus* Deduced from Genetic Variation, Geographical Origin and Spread, and the Breaking of Host Resistance. Mol Plant Microbe Interact 24(2): 207–218.
26. Chiba S, Miyanishi M, Andica IB, Kondo H, Tamada T (2008) Identification of amino acids of the beet necrotic yellow vein virus p25 protein required for induction of the resistance response in leaves of *Beta vulgaris* plants. J Gen Virol 89: 1314–1323.
27. Acosta-Leal R, Rush CM (2007) Mutations associated with resistance-breaking isolates of Beet necrotic yellow vein virus and their allelic discrimination using TaqMan technology. Phytopathology 97: 325–330.
28. Acosta-Leal R, Bryan BK, Smith JT, Rush CM (2010) Breakdown of host resistance by independent evolutionary lineages of *Beet necrotic yellow vein virus* involves a parallel C/U mutation in its *p25* gene. Phytopathology 100: 127–133.
29. Acosta-Leal R, Fawley MW, Rush CM (2008) Changes in the intraisolate genetic structure of *Beet necrotic yellow vein virus* populations associated with plant resistance breakdown. Virology 376: 60–68.
30. Koenig R, Loss S, Specht J, Varrelmann M, Luddecke P, et al. (2009) A single U/C nucleotide substitution changing alanine to valine in the beet necrotic yellow vein virus p25 protein promotes increased virus accumulation in roots of mechanically inoculated, partially resistant sugar beet seedlings. J Gen Virol 90: 759–763.
31. Pavli OI, Prins M, Goldbach R, Skaracis GN (2011) Efficiency of *Rz1*-based rhizomania resistance and molecular studies on BNYVV isolates from sugar beet cultivation in Greece. Eur J Plant Pathol 130: 133–142.
32. Grimmer MK, Trybush S, Hanley S, Francis SA, Karp A, et al. (2007) An anchored linkage map for sugar beet based on AFLP, SNP and RAPD markers and QTL mapping of a new source of resistance to *Beet necrotic yellow vein virus*. Theor Appl Genet 114: 1151–1160.
33. Tampakaki AP, Panopoulos NJ (2000) Elicitation of hypersensitive cell death by extracellularly targeted $HrpZ_{Psph}$ produced in planta. Mol Plant Microbe Interact 13(12): 1366–1374.
34. Horsch R, Fry J, Hoffman N, Eichholtz D, Rogers S, et al. (1985) A simple and general method for transferring genes into plants. Science 227: 1229–1231.
35. Dong H-P, Yu H, Bao Z, Guo X, Peng J, et al. (2005) The ABI2-dependent abscisic acid signaling controls HrpN-induced drought tolerance in *Arabidopsis*. Planta 221: 313–327.
36. Ren H, Song T, Wu T, Sun L, Liu Y, et al. (2006) Effects of a biocontrol bacterium on growth and defense of transgenic rice plants expressing a bacterial type-III effector. Ann Microbiol 56: 281–287.
37. Chen L, Zhang SJ, Zhang SS, Qu S, Ren X, et al. (2008) A fragment of the *Xanthomonas oryzae* pv. *oryzicola* harpin $HpaG_{Xooc}$ reduces disease and increases yield of rice in extensive grower plantings. Phytopathology 98: 792–802.
38. Huo R, Wang Y, Ma L-L, Qiao J-Q, Shao M, et al. (2010) Assessment of inheritance pattern and agronomic performance of transgenic rapeseed having $harpin_{Xooc}$-encoding *hrf2* gene. Transg Res DOI 10.1007/s11248-010-9365-x
39. Oh CS, Beer SV (2007) AtHIPM, an Ortholog of the apple HrpN-Interacting Protein, is a negative regulator of plant growth and mediates the growth-enhancing effect of HrpN in *Arabidopsis*. Plant Physiol 145: 426–436.
40. Shao M, Wang J, Dean RA, Lin Y, Gao X, et al. (2008) Expression of a harpin-encoding gene in rice confers durable nonspecific resistance to *Magnaporthe grisea*. Plant Biotechnol J 6: 73–81.
41. Conrath U, Beckers GJM, Flors V, Garcia-Agustin P, Jakab G, et al. (2006) Priming: Getting ready for battle. Mol Plant Microbe Interact 19: 1062–1071.

19

Construction of Chromosome Segment Substitution Lines in Peanut (*Arachis hypogaea* L.) Using a Wild Synthetic and QTL Mapping for Plant Morphology

Daniel Fonceka[1], Hodo-Abalo Tossim[2], Ronan Rivallan[1], Hélène Vignes[1], Elodie Lacut[1], Fabien de Bellis[1], Issa Faye[3], Ousmane Ndoye[2], Soraya C. M. Leal-Bertioli[4], José F. M. Valls[4], David J. Bertioli[5], Jean-Christophe Glaszmann[1], Brigitte Courtois[1], Jean-François Rami[1]*

1 CIRAD, UMR AGAP, Montpellier, France, **2** ISRA/Ceraas, Thiès, Sénégal, **3** ISRA, Centre National de Recherche Agronomique, Bambey, Sénégal, **4** Embrapa Recursos Genéticos e Biotecnologia, Brasilia, Distrito Federal, Brazil, **5** Universidade de Brasília, Brasília, Distrito Federal, Brazil

Abstract

Chromosome segment substitution lines (CSSLs) are powerful QTL mapping populations that have been used to elucidate the molecular basis of interesting traits of wild species. Cultivated peanut is an allotetraploid with limited genetic diversity. Capturing the genetic diversity from peanut wild relatives is an important objective in many peanut breeding programs. In this study, we used a marker-assisted backcrossing strategy to produce a population of 122 CSSLs from the cross between the wild synthetic allotetraploid (*A. ipaënsis*×*A. duranensis*)^{4x} and the cultivated Fleur11 variety. The 122 CSSLs offered a broad coverage of the peanut genome, with target wild chromosome segments averaging 39.2 cM in length. As a demonstration of the utility of these lines, four traits were evaluated in a subset of 80 CSSLs. A total of 28 lines showed significant differences from Fleur11. The line×trait significant associations were assigned to 42 QTLs: 14 for plant growth habit, 15 for height of the main stem, 12 for plant spread and one for flower color. Among the 42 QTLs, 37 were assigned to genomic regions and three QTL positions were considered putative. One important finding arising from this QTL analysis is that peanut growth habit is a complex trait that is governed by several QTLs with different effects. The CSSL population developed in this study has proved efficient for deciphering the molecular basis of trait variations and will be useful to the peanut scientific community for future QTL mapping studies.

Editor: James C. Nelson, Kansas State University, United States of America

Funding: This study was funded by the Cirad and the Generation Challenge Programme grants number G3005.05, G4008.49, and G4007.13.03. The funders had no role in study design, data collection and analysis, decision to publish, or preparation of the manuscript.

Competing Interests: The authors have declared that no competing interests exist.

* E-mail: jean-francois.rami@cirad.fr

Introduction

Wild relatives of crop plants are sources of useful agriculturally interesting traits that can be tapped to improve the cultivated species [1,2]. Historically, the transfer of traits from wild to cultivated species was reported to be a labor intensive and time-consuming process that required several backcrosses and was often unsuccessful [3,4]. The advent of map-based molecular markers and the related quantitative trait locus (QTL) mapping approach allowed the development of new molecular breeding strategies to efficiently exploit the positive alleles remaining in the wild species. The development of interspecific introgression line (IL) libraries monitored by molecular markers allows the representation of the entire wild species genome in a set of lines each of which carries one or few wild donor segments in the genetic background of a cultivated species [3,5]. These lines are especially useful for mapping complex traits, as they are permanent populations that can be phenotyped in several environments. Moreover, the effects of the alleles of the wild donor parent are evaluated in a homogeneous cultivated genetic background that reduces the interactions between donor alleles [6,7]. Finally, the introgression lines are useful starting points for breeding [2] and for mapping genes underlying QTLs [6,8].

Since the pioneering work in tomato [9], several interspecific introgression line libraries with different names but the same genetic construction have been produced for many crops: Chromosome Segment Substitution Lines (CSSLs) for rice [5,6], Backcross Inbred Lines (BILs) for lettuce [10], Backcross Recombinant Inbred Lines (BCRILs), Introgression Lines (ILs), Near Isogenic Lines (NILs) or QTL-NILs for tomato and barley [9,11–14].

Peanut (*Arachis hypogaea L.*) is an allotetraploid ($2n = 4x = 40$) tropical legume that is native to South America and has an AB genome. The speciation of cultivated peanut was associated with a change in ploidy level (diploid to allotetraploid), creating reproductive isolation from its wild relatives. This speciation process superimposed with domestication has contributed to the narrow the genetic basis of cultivated peanut.

Peanut wild relatives have long been identified as important sources of interesting traits such as disease resistance [15,16]. However, the genetic barrier between wild and cultivated peanut species caused by ploidy differences has hampered the use of wild diploid relatives in peanut breeding programs. Nevertheless,

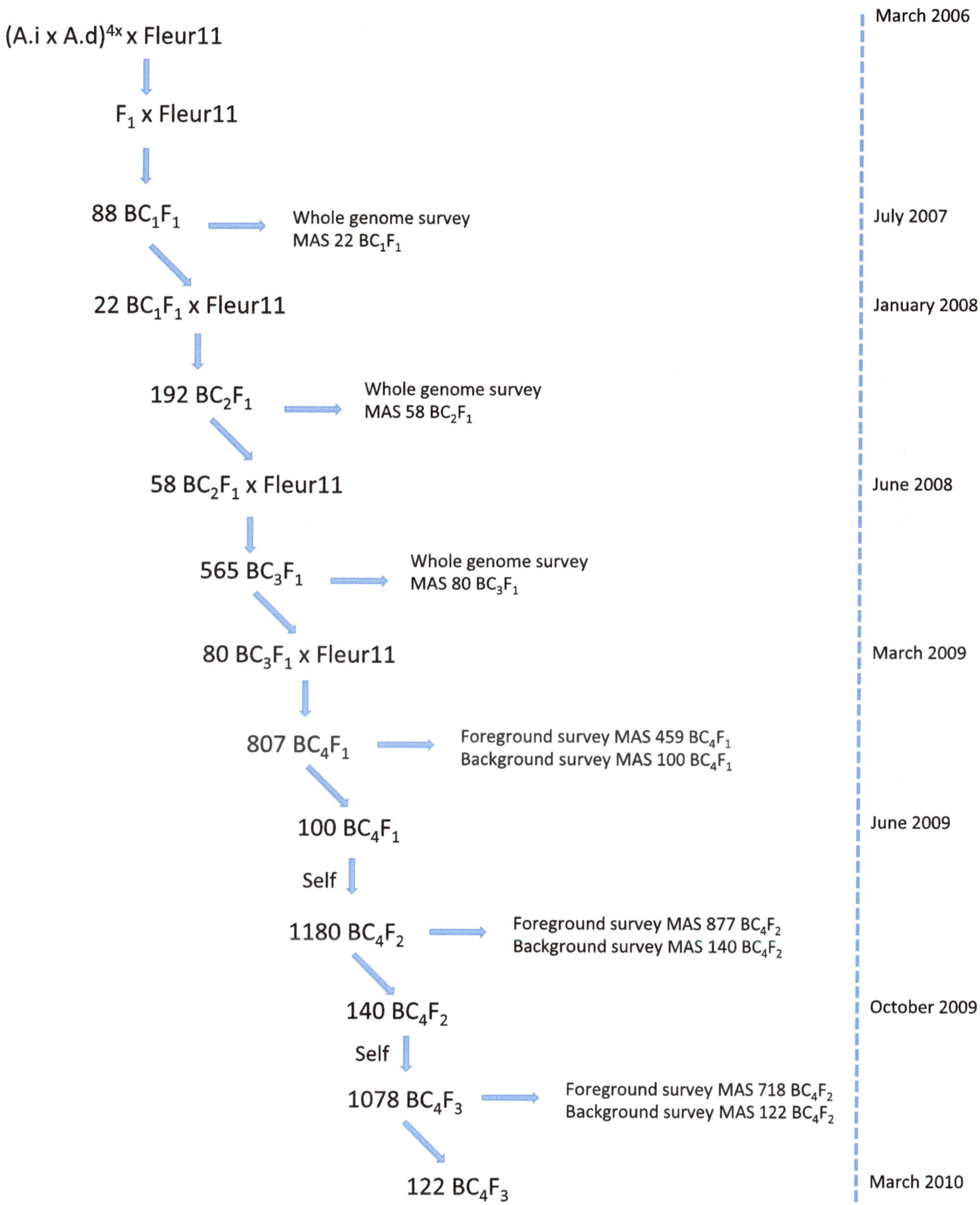

Figure 1. Breeding scheme for the development of the CSSL population.

disease resistance genes have been successfully transferred from wild diploid peanut species to cultivated tetraploid species [17–19].

In the last decade, tremendous efforts have been invested in the development of molecular tools for genetic studies in peanut. Thousands of Simple Sequence Repeat (SSR) markers have been

Table 1. Selected morphological characteristics of the parents.

Species	Accession	Genome	Ploidy level	Cycle	Growth Habit	Plant Height (cm)	Length of the secondary axis (cm)
A. ipaënsis	KG30076	B	2n	Annual	Prostrate	45	50
A. duranensis	V14167	A	2n	Annual	Prostrate	20–30	140
A. hypogaea	Fleur11	AB	4n	Annual	Erect	16–22	20–25

developed and used for genetic map construction and QTL mapping, mainly in intraspecific crosses [20,21]. In peanut interspecific cultivated×wild crosses, molecular markers have been used as to evaluate the extent of wild chromosome introgressions into the genetic background of cultivated species *a posteriori* [22], or to identify introgressed wild genomic regions that carry disease resistance genes [4,23]. However, there has been little progress with respect to the integration of molecular markers during the process of alien gene transfer in peanut breeding programs. Thus, there is still a gap to fill by systematizing the use of molecular markers during alien gene transfer and, more generally, by developing permanent interspecific QTL mapping populations such as CSSLs in peanut.

We have recently developed an SSR-based genetic linkage map using a BC_1F_1 population derived from the cross between the cultivated Fleur11 variety and the synthetic allotetraploid (*A. ipaënsis*×*A. duranensis*)^{4x} [24]. We have also produced an advanced-backcross QTL (AB-QTL) population from the same cross and identified several wild alleles that contribute positive variation to complex traits [25]. Here, as a follow-up to these studies, we report the development of a CSSL population. As a first example of the utility of this population, we have mapped several QTLs involved in peanut morphology. The potential of this CSSL population for deciphering the molecular basis of complex traits is also discussed.

Materials and Methods

Plant Material

The parents of this study were the synthetic wild allotetraploid [*A. ipaënsis* KG30076 (diploid BB genome)×*A. duranensis* V14167 (diploid AA genome)]^{4x} [26] used as the donor parent and the cultivated Fleur11 variety used as the recurrent parent. This synthetic allotetraploid, which combines the A and B genomes of the most probable wild ancestor of cultivated peanut was kindly provided by EMBRAPA in the framework of the Generation Challenge Program (GCP) consortium. Selected agro-morphological characteristics of the wild diploid parents *A. ipaënsis* and *A. duranensis* and the cultivated Fleur11 variety are shown in Table 1. A BC_4F_3 population was produced by four successive backcross generations of selected BC_s individuals (from BC_1F_1 to BC_4F_1) to the Fleur11 variety followed by two generations of self-fertilization (from BC_4F_1 to BC_4F_3). All crosses were performed under greenhouse conditions at the Centre d'Etudes Régional pour l'Amelioration de l'Adaptation à la Sécheresse (CERAAS, Senegal) between 2006 and 2010. The Fleur11 variety was used as female parent to produce the F_1, BC_1F_1 and BC_2F_1 generations and as male to produce the BC_3F_1 and BC_4F_1 generations. The day before each crossing experiment, the plants used as female parent were manually emasculated. The crosses were performed early in the morning from 6 am to 10 am by capping the pollen sacs of the flowers of the male parent on the pistils of the female parent. Flowers that had not been pollinated were removed manually. The crossing scheme and the timeline used to generate the CSSL population are shown in Figure 1.

DNA Isolation

At each generation, 3 to 30 seeds from selected BC_n (depending on the number of seeds available and on the number of wild chromosome segments to be derived from each individual) were sown in large deep pots in the greenhouse. One seed was sown per pot. Young leaves were harvested from 10–15 day old plants and immediately stored at 4°C on ice before drying. DNA was extracted from 20 mg of dried leaves following a slightly modified mixed alkyltrimethylammonium bromide (MATAB) protocol [27]. Briefly, leaves were ground in quartz sand using a mortar and pestle or using a Mixer Mill (MM301, Retsch) and dissolved in 750 µL of MATAB buffer at 74°C. The samples were incubated for 20 minutes at 74°C and cooled for 5 minutes at room temperature. Then, 750 µL of chloroform-isoamyl alcohol (CIA) (24:1) was added to each sample. All samples were homogenized by gentle shaking before centrifugation at 12,000 rpm for 20 minutes. The supernatant was collected and the DNA was precipitated with 600 µL of isopropanol. After centrifugation, pellets were washed with 300 µL of 70% ethanol, air dried and dissolved in 500 µL of Tris-EDTA.

Table 2. Comparison between the observed and expected numbers of heterozygous loci in the population. χ^2: Chi-square goodness of fit test at 1 ddl.

	All lines			Selected lines		
Generation	Observed	Expected	χ^2	Observed	Expected	χ^2
BC_1F_1	6630	6496	5.5**	1631	1625	0.04
BC_2F_1	5283	5052	14.08***	1438	1649.5	36.15***
BC_3F_1	10701	10056.62	47.18***	1353	1508.87	18.40***
BC_4F_1	5149	4131.12	267.51***	884	899.75	0.58

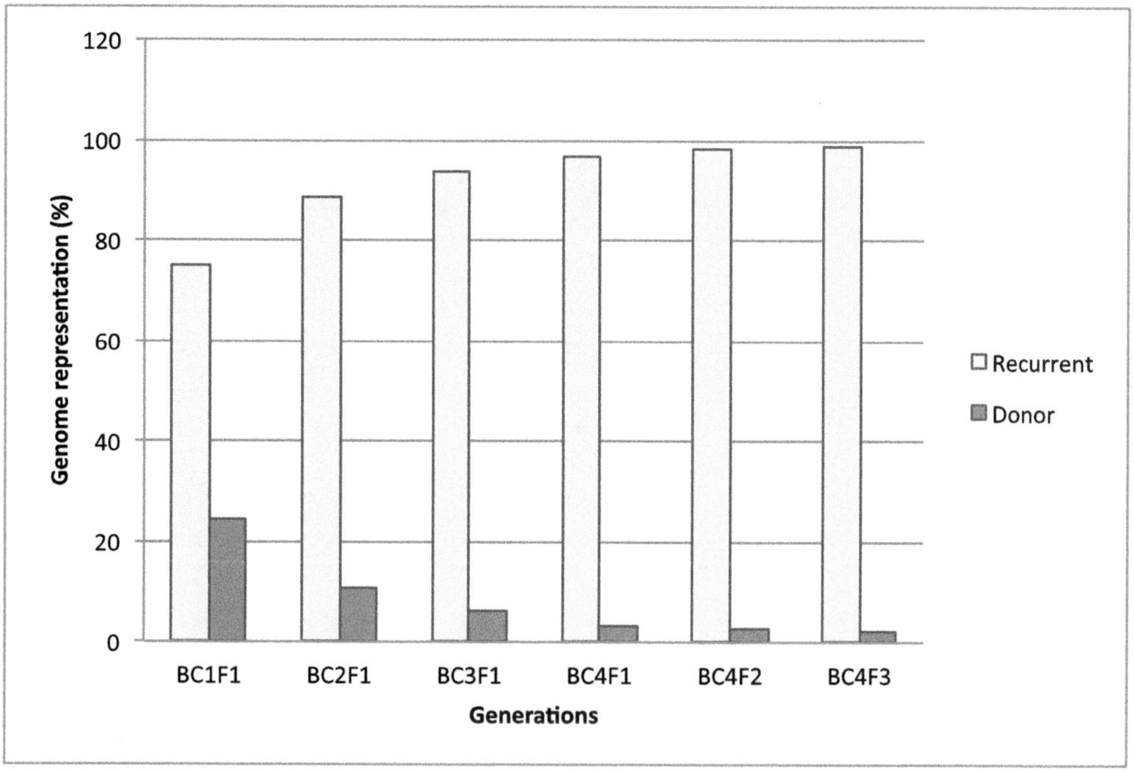

Figure 2. Evolution of donor and recurrent genome representation during the development of the CSSL population.

Microsatellite Analysis

An SSR-based genetic linkage map based on 88 BC_1F_1 individuals was produced previously [24]. A framework map of 115 SSR markers that amplified 147 loci was derived from this map. Compared to the initial map, the framework map offered regular coverage of all of the linkage groups, with an average of one marker per 12 cM. These 115 SSR markers were used to genotype individuals in each backcross generation.

PCR amplifications were performed in a MJ Research PTC-100 TM thermocycler (Waltham, MA, USA) or in an Eppendorf Mastercycler using 25 ng of DNA in a 10 µl final volume comprising of buffer (10 mM Tris-HCl (pH 8), 100 mM KCl, 0.05% w/v gelatin, and 2.0 mM $MgCl_2$), 0.1 µM of the M13-tailed primer, 0.1 µM of the other primer, 160 µM of dNTPs, 1 U of Taq DNA polymerase (Life Technologies, USA), and 0.1 µM of IR700- or IR800-labeled fluorescent M13 primer (MWG, Germany). Touchdown PCR was performed as follows: initial denaturation at 95°C for 1 min; 10 cycles of 94°C for 30 s, Tm (+5°C, −0.5°C/cycle) for 1 min, and 72°C for 1 min; 25 cycles of 94°C for 30 s, Tm for 1 min, and 72°C for 1 min; and a final elongation step at 72°C for 8 min. IR700 or IR800-labeled PCR products were diluted 7-fold and 5-fold, respectively, subjected to electrophoresis in a 6.5% polyacrylamide gel and then sized using the Infra-Red (IR) fluorescence scanning system of the Li-COR sequencer (LI-COR, USA). Images were analyzed using the Jelly 0.1 software (Rami, unpublished) and exported as a data table.

Marker-assisted Selection of Individuals at Each Backcross Generation

All 115 SSR markers of the framework map were used to genotype a total of 88, 192, and 565 individuals in the BC_1F_1, BC_2F_1 and BC_3F_1 generations respectively. At each generation, the genotyping data were used to select the individuals used to produce further backcross generations. Selection was performed using the CSSL-finder software [28]. To identify a minimum subset of individuals providing an optimal coverage of the cultivated peanut genome with chromosome segments from the wild donor, we selected for individuals with multiple target wild chromosome segments of 30 cM length each, forming contiguous overlapping segments between successive lines, and with the best possible recovery of the genetic background of Fleur11. Using these criteria, 22 individuals were selected and backcrossed in the BC_1F_1 generation, 58 in the BC_2F_1 generation, and 80 in the BC_3F_1 generation.

In the BC_4F_1 generation, 807 individuals were genotyped. The genotyping of these individuals was performed in two steps. Because genotypic data were available for the 80 BC_3F_1 individuals chosen for developing the BC_4F_1 generation, the chromosomal regions that segregated in BC_4F_1 individuals were known. Based on this information, foreground selection was first applied to confirm the presence of the target wild chromosome segments in the BC_4F_1 individuals; 459 individuals were selected. Background selection was then performed on this subset of individuals at specific regions that still contained undesirable wild segments. A new genotypic dataset that summarized the genotypic information produced in the BC_3F_1 generation and those from the foreground and the background selection produced in the BC_4F_1 generation was computed. New criteria were added to CSSL-finder to allow redundancy between individuals (i.e., multiple individuals carrying a wild chromosome segment that covers the same genomic region) and to derive only one target wild chromosome segment per line. Finally, using these new criteria, we selected 100 of the 459 BC_4F_1 individuals for self-fertilization.

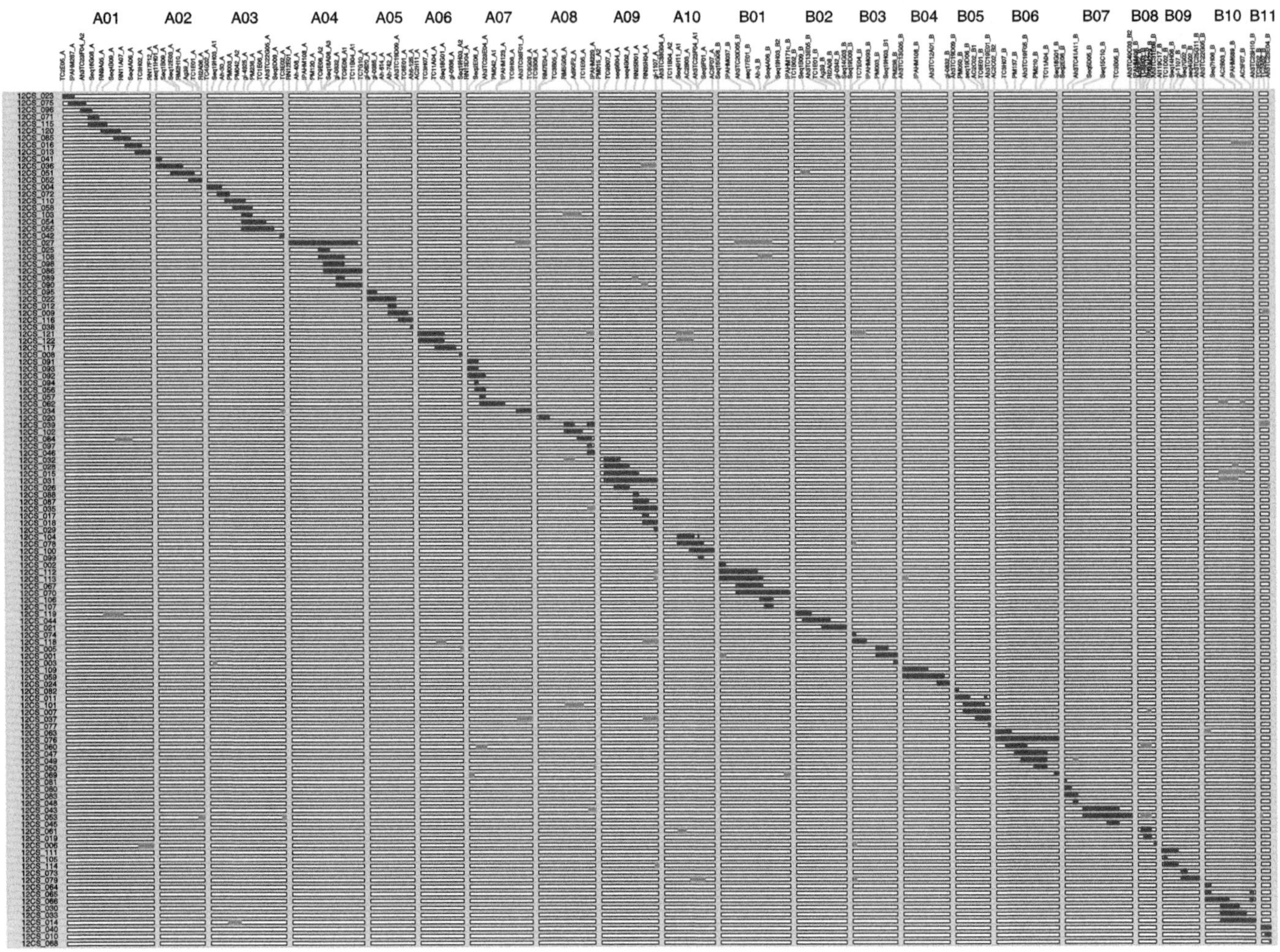

Figure 3. Graphical genotypes of the 122 CSSLs. The 21 linkage groups of the map are named from a01 to a10 (A-genome) and from b01 to b11 (B-genome) and are represented vertically. The marker names are shown on each linkage group. The 122 CSSL are depicted horizontally. The black areas represent the wild target chromosome segments, while the white areas represent the Fleur11 genetic background, and the grey areas represent the wild supernumerary chromosome segments.

In the BC_4F_2 generation, 1180 individuals were genotyped using the foreground selection approach that allowed the identification of 877 individuals carrying a target wild chromosome segment. Background selection was then performed on this subset of individuals. The selection of a final subset of BC_4F_2 individuals to be self-fertilized was performed manually. We first selected individuals that carried a unique homozygous or heterozygous target wild chromosome segment in a homogeneous Fleur11 genetic background. This first set of lines did not offer sufficient coverage of the genetic map. In order to cover all regions, we also selected individuals that carried one homozygous or heterozygous target wild chromosome segment and only one other heterozygous wild undesirable chromosome segment elsewhere in the genome. We finally selected among the other plants that had multiple wild chromosome segments, the best individuals with three, four or five segments that covered the genomic regions that had not yet been included. A total of 140 individuals were self-fertilized to produce the BC_4F_3 generation.

In the BC_4F_3 generation, 1078 individuals were genotyped using the foreground selection approach, which allowed the identification of 718 individuals that carried homozygous target wild chromosome segment. Background selection on this subset of individuals allowed the final selection of the 122 BC_4F_3 individuals that were used as our final CSSL population.

Phenotyping

A subset of 86 CSSLs was selected to provide a minimum set of lines with optimal genome coverage. Six lines could not be included in the phenotyping experiment due to limited seed quantities (12CS_117, 059, 080, 083, 061 and 066) leading to small uncovered regions in linkage groups A06, B04, B07, B08 and B10. The introgressions represented by the final subset of 80 lines covered 88.7% of the genetic map. A field trial involving the 80 selected lines and Fleur11 used as control was conducted from September to December 2011 at the Centre National de Recherche Agronomique (CNRA) in Bambey (14.42°N and 16.28°W), Senegal. The experimental design was an alpha-lattice design with 6 replicates and 6 blocks per replicate. The blocks contained 15 rows of 3 meters each. The 80 CSSL and Fleur11 plants (represented 10 times in each replicate) were arranged in rows of 10 plants. The spacing was 30 cm between plants and 50 cm between rows. Flower color (YF) (orange versus yellow) was rated visually. Peanut growth habit (GH) is usually recorded on a 1–6 scale based on the IBPGR and ICRISAT (1992) descriptors.

Table 3. Genotypic features of the progeny of each generation.

Generation	Number of genotyped lines	Number of selected lines	% homozygous Fleur 11 in selected lines	% heterozygous in selected lines	% homozygous AiAd	Mean size of introgressed segments (cM)	Min size of introgressed segments (cM)	Max size of introgressed segments (cM)	Mean number of introgressed segments per line
BC_1F_1	88	22	48.68	50.99	0	40.36	2.33	99.65	17.07
BC_2F_1	192	58	77.16	21.59	0	35.63	2.35	73.92	11
BC_3F_1	565	80	87.31	10.1	1.23	39.23	11.3	115.6	6.43
BC_4F_1	807	100	93.66	6.13	0.24	39.15	6.5	115.6	4.06
BC_4F_2	1180	140	96.17	2.4	1.4	39.5	5.7	115.6	2.12
BC_4F_3	1078	122	97.28	0.14	2.24	39.24	5.7	115.6	1.5

This recording is well adapted for germplasm description but more difficult to use for QTL mapping because of the qualitative nature of the resulting data. Therefore, in our study, the GH trait was represented by the ratio of the length of the creeping part of a given lateral branch to the total length of the lateral branch. This phenotyping method yielded quantitative values that were more suitable for QTL mapping analyses. Main stem height (PH) was measured from the cotyledonary axil to the terminal bud. Finally, plant spread (PS) was measured as the distance between the widest branch tips. All morphological traits were recorded at 55–60 days after sowing.

Statistical Analysis

Introgression analysis. The frequency of homozygous cultivated Fleur11 and heterozygous wild/cultivated genotypes at each locus was determined for all lines and for the selected lines at each backcross generation (BC_1F_1, BC_2F_1, BC_3F_1 and BC_4F_1). The observed homozygous/heterozygous segregation ratio was compared with the expected 0.5, 0.25, 0.125 and 0.0625 ratios in BC_1F_1, BC_2F_1, BC_3F_1 and BC_4F_1, respectively, using a chi-square test. The percentage of wild genome at each generation was estimated using the ratio of the total length of the introgressed wild segments to the total length of the genetic map. The total length of the introgressed segments was calculated as the sum of consecutive intervals having heterozygous or homozygous wild genotype plus half the size of each flanking interval with a recurrent parent homozygous genotype. Graphical genotypes were drawn using a script written in R (R Development Core Team, 2010).

QTL mapping. Basic statistical analyses were performed for each trait. For QTL detection, an analysis of variance (ANOVA) was first performed with the SAS software to estimate the genotype and replicate effects on each trait following a standard linear model with genotype, replication, block and interaction effects, as follows:

$$Y_{ijk} = \mu + G_i + r_j + b_{jk} + e_{ijk}$$

with Y_{ijk} = the observed value for a given trait, μ = mean value of the population, G_i = genotype effect, r_j = replicate effect, b_{jk} = block within replicate effect and e_{ijk} = residual error. When the analysis revealed a significant difference between genotypes, a Dunnett's multiple comparison [29] of differences between the CSSLs and Fleur11 was conducted using the GLM procedure in SAS software. We accepted a QTL in a wild chromosome segment when the trait value of a given line was significantly different from that of Fleur11 ($P<0.05$). QTLs with effects in the same direction on overlapping or flanking segments between contiguous lines were considered to be the same. The relative effect of a wild introgression in a given line was calculated using the least square means (LSMEANS) output of the GLM procedure as follows:

$$RE = \frac{LsMean(IL) - LsMean(Fleur11)}{LsMean(Fleur11)} \times 100$$

Results

Population Development

The F_1 hybrids between the cultivated Fleur11 variety and the synthetic wild allotetraploid (*A. ipaënsis* KG30076×*A. duranensis* V14167)^{4x} exhibited a good level of fertility. The rate of success of

Table 4. Variation and descriptive statistics of the traits measured in the CSSL population and the recurrent parent.

	Fleur11						80 CSSLs					
Trait	Nb Plots	Nb plants	Mean	Min	Max	CV(%)	Nb Plots/IL	Nb plants	Mean	Min	Max	CV(%)
GH	60	458	0.00	0.00	0.00	-	6	4169	0.11	0.00	0.87	-
PS	60	458	43.16	34.17	54.40	11.63	6	4169	47.52	31.00	80.40	18.07
PH	60	458	15.23	8.60	21.15	18.39	6	4169	14.83	7.20	24.07	20.90

the hybridization in the backcross experiments was also high, ranging from 75% to 80%. However, individuals resulting from inadvertent self-fertilization, identified by molecular analysis, were observed in some cases. Fleur11 is a short cycle variety, and the interval from sowing to the appearance of the first flowers was 21 days on average. The short duration between sowing and flowering required us to limit the number of individuals to be genotyped per backcross family at each generation. On average, 9.6 individuals (min = 3, max = 30) were genotyped per family and per generation. A wide range of phenotypic variation for traits related to growth habit, pod and seed shape and size was observed between the families in the first backcross generations. We were careful to avoid selection based on plant phenotype during all the process of development of the libraries. However, as we advanced in the backcross generations, most of the families exhibited a phenotype similar to that of the recurrent parent, indicating a progressive recovery of the genetic background of the cultivated variety.

Heterozygosity and reduction of the wild genome representation in successive backcross generations. When considering all lines, the observed number of heterozygous loci at the backcross generations BC_1F_1, BC_2F_1, BC_3F_1 and BC_4F_1 was significantly ($P<0.01$) higher than expected (Table 2). This could be due to the limited number of lines that were genotyped at each generation. The strategy of population development using marker assisted selection allowed us to significantly ($P<0.01$) reduce the number of heterozygous loci in BC_2F_1 and BC_3F_1 generations. This strategy had no significant effect in BC_1F_1, due to the low number of lines (22) that were selected, or in BC_4F_1, because the heterozygous loci were mainly the target wild chromosome segments. The residual heterozygosity in the CSSLs population was 0.14%.

The average donor genome contribution decreased drastically from BC_1F_1 to BC_4F_1 (24.6% to 3.0%) and remained stable from BC_4F_1 to BC_4F_3 (3.0% to 2.3%), during the fixation of target wild chromosome segments (Figure 2).

Number and size of the introgressed segments; genome coverage. Marker assisted selection was applied over six generations, from BC_1F_1 to BC_4F_3, allowing the selection of a set of 122 CSSLs with fixed target wild chromosome segments. The average number of wild donor chromosome segments varied from 17 in BC_1F_1 to 1.5 in the CSSL population (Table 3). Among the 122 CSSLs, 72 lines carried one unique donor segment, 37 lines carried two segments, ten lines carried three segments, one line carried four segments and two lines carried five segments. The estimated length of the wild donor segments varied slightly in the different backcross generations and was, on average, 39.2 cM in the CSSL population (Table 3). Among the 122 CSSLs, 42 lines were selected to provide redundancy in some regions of the genome in which several QTLs of agronomic interest had been mapped in a previous study [25]. This was for example the case for LGs a04, a07, a09, b05, b06 and b10 (Figure 3). These lines could be considered potential NILs for these genomic regions.

In the CSSL population, most of the target wild chromosome segments overlap, except those between successive lines in LGs a03, a06, a07 and a08 and in LGs b03, b07 and b09 (Figure 3). The introgressed segments in the whole CSSL population covered 92.1% of the genetic map. Detailed introgression information for each of the 122 CSSL lines is given in supporting information Table S1.

Phenotypic Analysis

Phenotypic measurements were performed on a subset of 80 CSSLs. Three quantitative traits (GH, PH and PS) and one qualitative trait (YF) were evaluated. The plant growth habit (GH) was measured as the ratio of the length of the creeping part of a given lateral branch to the total length of the lateral branch. The values of the ratio were 0 for an erect growth habit, between 0.16 and 0.5 for decumbent phenotypes, and greater than 0.5 for procumbent phenotypes. The values of the PH and PS traits varied from 7.0 cm to 24.07 cm and from 31.0 cm to 80.4 cm, respectively (Table 4).

For each trait, the mean of the CSSL population was similar to the mean of the recurrent Fleur11 parent (Table 4). The coefficient of variation for the PS trait was higher in the CSSL population than in Fleur11 (18.0% versus 11.0%).

QTL Detection

For all traits combined, Dunnett's test revealed a total of 28 lines with significant differences from Fleur11 (Table 5). Among these lines, fifteen had one unique wild chromosome segment, eleven had two wild segments and three had three wild segments. In cases for which a given line had more than one wild segment, it was not obvious how to assign the QTL to a single chromosome segment. We were, however, able to resolve most of these cases by comparison with lines having overlapping segments and similar phenotypes. When the QTL assignment on a single chromosome segment was still ambiguous, QTLs were considered putative in all segments and consequently not assigned. Finally, with these adjustments, 42 QTLs were detected for all traits. In the following paragraphs, we detail the QTL detection per trait.

Growth habit (GH). A total of 20 lines had a significantly different GH compared with Fleur11. The analysis of the line×trait associations allowed the identification of 14 genomic regions containing QTLs. Eleven QTLs were unambiguously mapped to nine linkage groups: a01 (two QTLs), a03 (two QTLs) and LGs a05, a06, a07, b04, b05, b06 and b11 (one QTL each) (Figure 4). For two QTLs, the most likely locations were proposed. The lines carrying these QTLs (12CS_104 and 12CS_033) have a procumbent phenotype and carried each a single wild introgression on LGs a10 and b10 (Figure 4). On LG a10, the wild chromosome segment of line 12CS_104 fully overlaps with that of

Table 5. Significant line×trait associations observed in the CSSL population.

CSSL Name	Plant nb	Chromosome	GH (%)	PH (%)	PS (%)	YF
12CS_023	55	A01	41.8***	ns	23.5***	0
12CS_016	52	A01	36.9***	−25.7***	18.2**	0
12CS_052	56	A02	ns	17.7*	ns	0
12CS_004	50	A03	20.4***	ns	26.9***	0
12CS_042	48	A03	17.5**	−17.7*	ns	0
12CS_022	54	A05	76.5***	−28***	64.5***	0
12CS_012	51	A05	47.6***	−32.6***	31.9***	0
12CS_009	52	A05	49.9***	−20.9***	41.9***	1
12CS_116	54	A05	39.4***	ns	39.1***	1
12CS_008	57	A06	22.7***	ns	17.2**	0
12CS_091	46	A07	16.4**	−25***	ns	0
12CS_039	52	A08	ns	22.1***	ns	0
12CS_084	48	A08	22.7***	ns	18.6**	0
12CS_104	60	A10	60.1***	ns	53.4***	0
12CS_119	53	B02	ns	19.3**	ns	0
12CS_021	46	B02	ns	19.6**	ns	0
12CS_001	53	B03	ns	21.3***	ns	0
12CS_024	50	B04	36.7***	ns	42.3***	0
12CS_011	55	B05	45.8***	−20.7***	34.8***	0
12CS_007	56	B05	35.2***	ns	25.9***	0
12CS_076	56	B06	29.7***	ns	17.3**	0
12CS_047	57	B06	55.9***	−35.6***	32.6***	0
12CS_050	54	B06	48.4***	−20.4**	40.3***	0
12CS_069	57	B06	ns	31.4***	21.7***	0
12CS_079	53	B09	ns	−18.5**	ns	0
12CS_033	58	B10	60.5***	−16.1*	70.7***	0
12CS_014	53	B10	ns	18.2**	17.3**	0
12CS_010	54	B11	29.0***	ns	33.3***	0

The values for GH (growth habit), PH (plant height) and PS (plant spread) represent the differences between the CSSL line and Fleur11, relative to the value of Fleur11. For the categorical trait YF (flower color), the value 1 corresponds to the yellow flower phenotype. Plant nb: number of plants measured. Chromosome: chromosome carrying the target segment.

12CS_078 and on LG b10, the wild chromosome segment of line 12CS_033 fully overlaps with that of 12CS_014. However, 12CS_078 and 12CS_014 have an erect GH. If we assume that the estimated size of the segments could differ from their real size, because the precise site of recombination is unknown, the most likely locations of the QTLs on LG a10 and b10 are within the regions upstream of the wild segments. The assignment of the remaining QTL was more complicated. For instance, line 12CS_084 carried two wild segments, one of which is located on LG a01 and overlaps with the QTL-carrying segment in line 12CS_016. Thus, these two lines could carry the same QTL. Another wild segment is located on LG a08. This segment partly overlaps with the segments in lines 12CS_101 and 12CS_039. However, these lines have erect GHs. The QTL conferring spreading GH could also be restricted to the non-overlapping region on LG a08. Therefore, this QTL has been considered putative in the a08 linkage group.

In most of the lines, wild alleles at the GH QTLs conferred a decumbent phenotype (ratios between 0.16 and 0.5). However, four lines 12CS_022, 12CS_104, 12CS_047 and 12CS_033, contain wild segments on LG a05, a10, b06 and b10 that conferred a procumbent phenotype (Table 5). Interestingly, 12CS_022 partly overlaps with 12CS_012, 12CS_009 and 12_CS_116, which each have decumbent phenotypes. These results suggest that two QTLs with different effects could be responsible for GH in these regions, one conferring a procumbent phenotype and the other conferring a decumbent phenotype. Alternatively, these lines might share one common QTL but differ at one additional QTL that could be responsible for growth habit differences.

Plant height (PH). A total of 18 lines had main stem height significantly different to that of Fleur11. The analysis of the line×trait associations allowed the identification of 15 genomic regions containing QTLs. Thirteen QTLs were located on ten linkage groups: two QTLs on LG b02, two QTLs on LG b06, 2 QTLs on LG b10 and one QTL each on LGs a01, a03, a05, a07, b03, b05 and b09 (Figure 4). Two QTLs on LG a02 and a08 were considered as putative.

The wild alleles at PH QTLs on LG a01, a03, a05, a07, b05 and b09 were associated with a shorter main stem height, while the two QTLs on LG b02 and one QTL on LG b03 and were associated with an increase in the main stem height (Table 5). Two

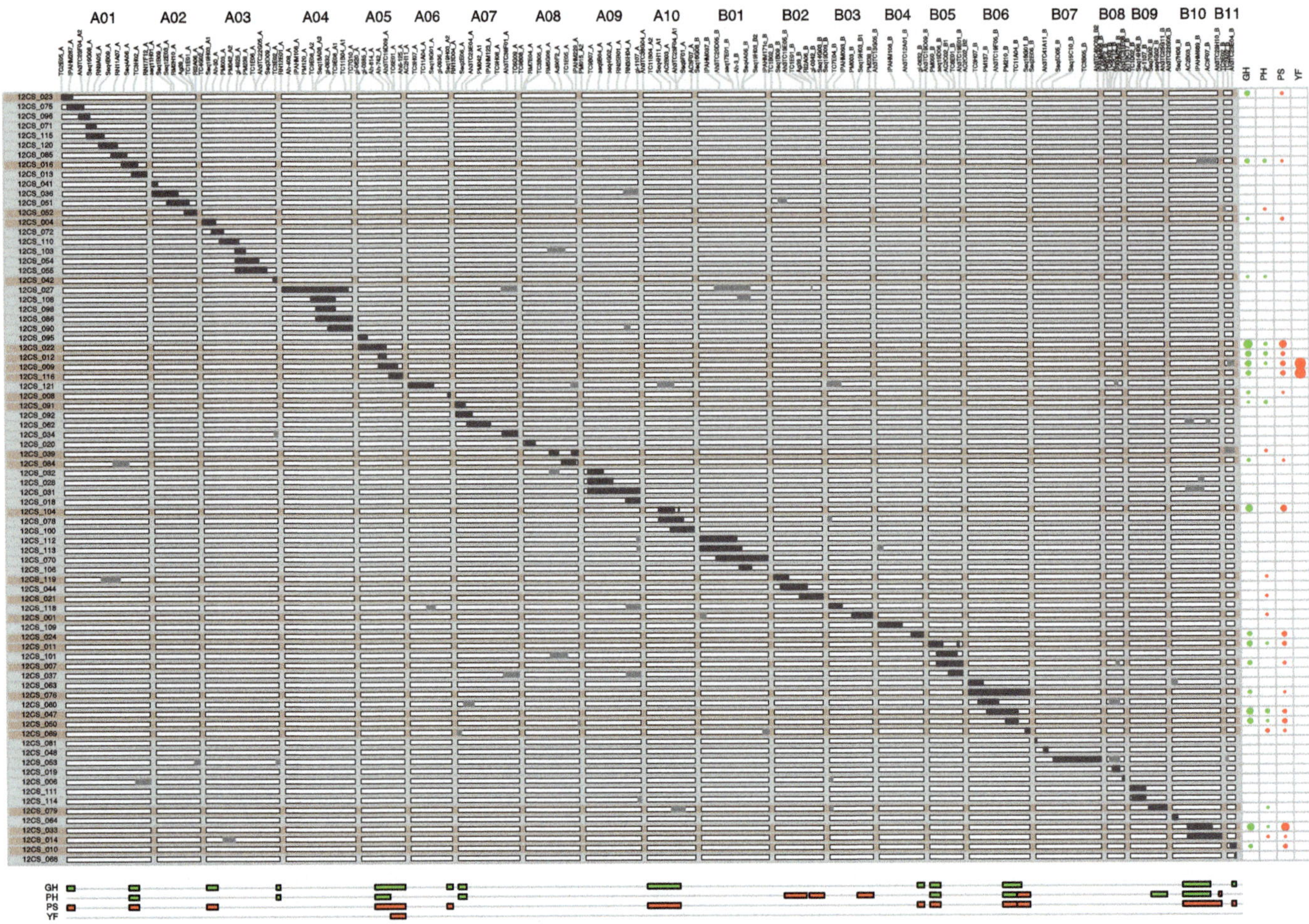

Figure 4. Graphical genotypes of the 80 CSSLs, highlighting significant lines×traits significant association and QTL positions. Graphical genotypes are represented in the same way as in Figure 3. Lines showing significant differences from Fleur11 are highlighted in orange. Relative line effects for each trait are depicted as circles on the right of the graphic. The size of each circle is proportional to its relative effect. Horizontal bars at the bottom of the graphic indicate deduced QTL positions. The colors of the circles and bars represent the direction of the effect. Red = increasing effect, green = decreasing effect.

QTLs with opposing effects were also mapped on LG b06 and b10, respectively.

Plant spread (PS). A total of 20 lines had plant spread values significantly different from that of Fleur11 (Table 5). The line×trait associations allowed the identification of 12 QTLs in 10 linkage groups. Two QTLs were mapped on LG a01, two on LG b06 and one QTL each on LGs a03, a05, a06, a10, b04, b05, b10 and b11 (Figure 4). The plant-spread QTLs co-segregated either with plant growth habit QTLs or with plant height QTLs.

At all QTLs the wild alleles were involved in increasing plant spread. As observed for GH, several lines that have overlapping chromosome segments had also different plant spread. This was the case for successive lines on LG a05 (12CS_022, 12CS_012, 12CS_009 and 12CS_016) and successive lines on LG b10 (12CS_033 and 12CS_014). These results suggest that QTLs with different effects or more than one QTL were responsible for differences in plant spread in these LGs.

Flower color (YF). Two lines, 12CS_009 and 12CS_016, had yellow flowers while Fleur11 had orange flowers. These two lines are successive with contiguous overlapping segments on LG a05 (Figure 4). Thus the genomic region involved in flower color is restricted to the overlapping region between the fragments on LG a05.

Discussion

Population Development

The CSSL population we have developed offers broad coverage of the genome of cultivated peanut especially in the context of the large genome size (c. 2800 Mb/1C and 20 linkage groups). The level of coverage of the donor genome of our CSSL population compares favorably to that of sets of introgression lines (IL) in other crop species. In tomato, Eshed and Zamir [9] reported the development of 50 ILs providing a complete genome coverage using the wild species *Lycopersicon pennellii* as donor. In two other studies of tomato IL development, the level of genome coverage was 85% and 97.6%, using *L. hirsutum* [11] and *L. lycopersicoides* [30], respectively, as donors. In barley, 86.6% genome coverage was reported in a set of 59 ILs developed from the cross between the wild species *Hordeum vulgare ssp. spontaneum* and a spring barley cultivar [31]. Lower levels of genome coverage have also been reported. In peanut, a set of 46 ILs covering 30% of the peanut diploid genome has been developed from a cross between the allotetraploid cultivated species *A. hypogaea* and the wild diploid species *A. cardenasii* [22]. In wheat, a set of 97 ILs covering 37.7% of the donor genome has been produced from the cross between a synthetic hexaploid genotype and a Chinese cultivar [32]. In the present study, most of the CSSLs (62.3%) contained a single wild fragment in a homogeneous cultivated genetic background. For the lines that contained more than one fragment, additional backcrossing efforts will be necessary to derive lines that harbor unique wild chromosome segments.

QTL Mapping

In the present study, as a first demonstration of the utility of the CSSL population, four traits were evaluated in a subset of 80 lines. The line×trait associations allowed the identification of a total of 42 QTLs. Among these QTLs, 37 have been unambiguously assigned to a given genomic region notwithstanding the presence of few additional wild chromosome segments in some lines. Probable locations have been proposed for 2 additional QTLs. This was the case for the GH QTLs on LGs a10 and b10, for which the uncovered regions upstream the wild chromosome segments in lines 12CS_104 and 12CS_033 were proposed as the most probable locations of the QTLs. Among the whole CSSL population (122 lines), one line having wild chromosome segments that cover the QTL region on LG b10 has been identified. This line will be used to confirm the location of the QTL. Finally, three QTLs were putatively assigned to different genomic regions. Additional backcrossing efforts will be necessary for the lines carrying these QTLs to eliminate supernumerary wild chromosome segments and clarify the QTL positions.

Growth habit is a complex trait governed by several QTLs with different effects. Peanut growth habit has long been studied, mainly in cultivated×cultivated crosses, and one to four genes have been proposed as responsible for this trait [33]. Genic-cytoplasmic interactions involving at least two cytoplasm types and two genes have also been described [34]. In our recently published study, six QTLs for growth habit have been mapped in an interspecific AB-QTL design involving the same parents as those used in the present study [25]. In the present study, we mapped 14 QTLs, five of which were common with those mapped in the previous study. The difference in the number of QTLs mapped in the two studies may be related to the higher efficiency of QTL detection using the CSSL population. Indeed, in the AB-QTL population, some QTLs, especially those with smaller effects, were likely masked by QTLs with greater effects. Different numbers of mapped QTLs between population types (CSSL versus AB-QTLs) have also been reported in tomato [9] and in rice [35]. In summary, a greater number of QTLs are involved in peanut growth habit in interspecific crosses than in intraspecific crosses. Moreover, in the CSSL population, we observed that lines might have different wild segments carrying QTLs but similar phenotypes. Analyzing these results in an evolutionary context and assuming that the wild species (*A. duranensis* and *A. ipaënsis*) used in this study are the most probable ancestors of cultivated peanut [36,37], raises new questions about the evolution of cultivated peanut during its domestication. Which QTLs/genes involved in GH have been maintained in the cultivated species? Which ones have been lost? Are similar phenotypes explained by the same or different QTLs? Mapping GH QTLs in intraspecific crosses involving various cultivated varieties and comparing them with QTLs mapped in the CSSL population will shed new light on this issue.

QTL effects vary between lines with overlapping wild chromosome segments. In the present study, QTLs with effects in the same direction but of different magnitudes as well as QTLs with opposite effects have been detected in successive lines with overlapping wild chromosome segments. For example, this was the case for the successive lines 12CS_022, 12CS_012, 12CS_009 and 12CS_016, which have different GH effects on LG a05 and for the successive lines 12CS_047, 12CS_050 and 12CS_069, which carry PH QTLs with opposite effects on LG b06 (Table 5). The lines 12CS_047 and 12CS_050 have spreading GH with reduced PH, while 12CS_069 has erect GH with increased PH. Interestingly, on LG b06, the line 12CS_076 carries a wild chromosome segment that overlaps with those of 12CS_047, 12CS_050 and 12CS_069 (Figure 4). This line has spreading GH and PH similar to that of Fleur11. The PH phenotype of 12CS_076 could be caused by QTLs with opposite effects in this LG corresponding to segments of lines 12CS_050 and 12CS_069. Surprisingly, the relative effect of the GH QTL is of lower magnitude in line 12CS_076 than in 12CS_047 or 12CS_050 (Table 5). This result suggests that the QTL involved in increased PH could also have a pleiotropic effect on GH.

Assuming that the QTLs involved in main stem height increase have antagonistic effects on those involved in spreading GH, one could expect that the introgression of these QTLs in cultivated runner-type peanut will modify their growth habit. This could be further confirmed by the introgression of wild segments carrying QTLs involved in increasing PH into the genetic background of a runner-type peanut variety.

Conclusion

In this study, we produced a population of 122 CSSLs with the aim of broadening the gene pool of cultivated peanut. This population has been efficiently used to dissect the genetic architecture of several morphological traits. Plant growth habit for example, which has long been described as a simple trait governed by one to four genes, was found to be affected by several QTLs with different effects. Our findings confirm the efficiency of the CSSL population type for QTL mapping.

This CSSL population is a public resource, available upon request, and it will be helpful to the peanut scientific community to better understand the molecular bases of the variation of agriculturally interesting traits.

Supporting Information

Table S1 Introgression information for the 122 CSSL lines. Rows highlighted in grey correspond to the subset of 80 lines characterized in this study.

Author Contributions

Conceived and designed the experiments: DF JCG BC JFR. Performed the experiments: DF HAT RR HV EL FdB IF ON. Analyzed the data: DF BC JFR. Contributed reagents/materials/analysis tools: IF ON SCMLB JFMV DJB. Wrote the paper: DF SCMLB DJB JCG BC JFR.

References

1. Tanksley SD, McCouch SR (1997) Seed banks and molecular maps: unlocking genetic potential from the wild. Science 277: 1063–1066.
2. Gur A, Zamir D (2004) Unused natural variation can lift yield barriers in plant breeding. PLoS Biol 2: e245.
3. Zamir D (2001) Improving plant breeding with exotic genetic libraries. Nat Rev Genet 2: 983–989.
4. Garcia GM, Tallury SP, Stalker HT, Kochert G (2006) Molecular analysis of *Arachis* interspecific hybrids. Theor Appl Genet 112: 1342–1348.
5. Wan XY, Wan JM, Su CC, Wang CM, Shen WB, et al. (2004) QTL detection for eating quality of cooked rice in a population of chromosome segment substitution lines. Theor Appl Genet 110: 71–79.

6. Yano M (2001) Genetic and molecular dissection of naturally occurring variation. Curr Opin Plant Biol 4: 130–135.
7. Grandillo S, Tanksley SD, Zamir D (2007) Exploitation of natural biodiversity through genomics. In: Varshney RK, Tuberosa R, editors. Genomics-Assisted Crop Improvement. Springer Netherlands. 121–150.
8. Alpert KB, Tanksley SD (1996) High-resolution mapping and isolation of a yeast artificial chromosome contig containing fw2.2: A major fruit weight quantitative trait locus in tomato. Proc Natl Acad Sci 93: 15503–15507.
9. Eshed Y, Zamir D (1995) An introgression line population of *Lycopersicon pennellii* in the cultivated tomato enables the identification and fine mapping of yield-associated Qtl. Genetics 141: 1147–1162.
10. Jeuken MJ, Lindhout P (2004) The development of lettuce backcross inbred lines (BILs) for exploitation of the *Lactuca saligna* (wild lettuce) germplasm. Theor Appl Genet 109: 394–401.
11. Monforte AJ, Tanksley SD (2000) Development of a set of near isogenic and backcross recombinant inbred lines containing most of the *Lycopersicon hirsutum* genome in a *L. esculentum* genetic background: A tool for gene mapping and gene discovery. Genome 43: 803–813.
12. Tanksley SD, Nelson JC (1996) Advanced backcross QTL analysis: a method for the simultaneous discovery and transfer of valuable QTLs from unadapted germplasm into elite breeding lines. Theor Appl Genet 92: 191–203.
13. Frary A, Doganlar S, Frampton A, Fulton T, Uhlig J, et al. (2003) Fine mapping of quantitative trait loci for improved fruit characteristics from *Lycopersicon chmielewskii* chromosome 1. Genome 46: 235–243.
14. Von Korff M, Wang H, Léon J, Pillen K (2004) Development of candidate introgression lines using an exotic barley accession (*Hordeum vulgare ssp. spontaneum*) as donor. Theor Appl Genet 109: 1736–1745.
15. Holbrook CC, Stalker HT (2010) Peanut Breeding and Genetic Resources. In: Janick J, editor. Plant Breeding Reviews. John Wiley & Sons, Inc. 297–356.
16. Simpson CE (2001) Use of Wild *Arachis* Species: Introgression of Genes into *A. hypogaea* L. Peanut Science 28: 114–116.
17. Stalker (1984) Utilizing *Arachis cardenasii* as a source of Cercospora leafspot resistance for peanut improvement. Euphytica 33: 529–538.
18. Simpson CE, Starr JL (2001) Registration of "COAN" Peanut. Crop Sci 41: 918.
19. Simpson CE, Starr JL, Church GT, Burow MD, Paterson AH (2003) Registration of "NemaTAM" Peanut. Crop Sci 43: 1561.
20. Pandey MK, Gautami B, Jayakumar T, Sriswathi M, Upadhyaya HD, et al. (2011) Highly informative genic and genomic SSR markers to facilitate molecular breeding in cultivated groundnut (*Arachis hypogaea*). Plant Breeding. Available: http://onlinelibrary.wiley.com/doi/10.1111/j.1439-0523.2011.01911.x/Accessed 22 November 2011.
21. Pandey MK, Monyo E, Ozias-Akins P, Liang X, Guimarães P, et al. (2011) Advances in Arachis genomics for peanut improvement. Biotechnol Adv. Available: http://onlinelibrary.wiley.com/doi/10.1111/j.1439-0523.2011.01911.x/Accessed 15 February 2012.
22. Garcia GM, Stalker HT, Kochert G (1995) Introgression analysis of an interspecific hybrid population in peanuts (*Arachis hypogaea* L.) using RFLP and RAPD markers. Genome 38: 166–176.
23. Nagy ED, Chu Y, Guo Y, Khanal S, Tang S, et al. (2010) Recombination is suppressed in an alien introgression in peanut harboring Rma, a dominant root-knot nematode resistance gene. Mol Breeding 26: 357–370.
24. Fonçéka D, Hodo-Abalo T, Rivallan R, Faye I, Sall MN, et al. (2009) Genetic mapping of wild introgressions into cultivated peanut: a way toward enlarging the genetic basis of a recent allotetraploid. BMC Plant Biol 9: 103.
25. Fonceka D, Tossim H-A, Rivallan R, Vignes H, Faye I, et al. (2012) Fostered and left behind alleles in peanut: interspecific QTL mapping reveals footprints of domestication and useful natural variation for breeding. BMC Plant Biology 12: 26.
26. Favero AP, Simpson CE, Valls JFM, Vello NA (2006) Study of the Evolution of Cultivated Peanut through Crossability Studies among *Arachis ipaënsis*, *A. duranensis*, and *A. hypogaea*. Crop Sci 46: 1546–1552.
27. Risterucci AM, Grivet L, N'Goran JAK, Pieretti I, Flament MH, et al. (2000) A high-density linkage map of *Theobroma cacao* L. Theor Appl Genet 101: 948–955.
28. Lorieux M (2007) CSSL Finder. Available: http://mapdisto.free.fr/CSSLFinder/Accessed 10 October 2010.
29. Dunnett CW (1985) Multiple comparisons between several treatments and a specified treatment. Lect Notes Stat 35: 39–47.
30. Chetelat RT, Meglic V (2000) Molecular mapping of chromosome segments introgressed from *Solanum lycopersicoides* into cultivated tomato (*Lycopersicon esculentum*). Theor Appl Genet 100: 232–241.
31. Schmalenbach I, Körber N, Pillen K (2008) Selecting a set of wild barley introgression lines and verification of QTL effects for resistance to powdery mildew and leaf rust. Theor Appl Genet 117: 1093–1106.
32. Liu S, Zhou R, Dong Y, Li P, Jia J (2006) Development, utilization of introgression lines using a synthetic wheat as donor. Theor Appl Genet 112: 1360–1373.
33. Singh F, Oswalt DL (1991) Genetics and Breeding of Groundnut. Available: http://ec2-50-19-248-237.compute-1.amazonaws.com/2754/Accessed 4 June 2012.
34. Ashri A (1968) Genic-cytoplasmic interactions affecting growth habit in peanuts, *A. hypogaea* II. a Revised Model. Genetics 60: 807–810.
35. Tian F, Li DJ, Fu Q, Zhu ZF, Fu FC, et al. (2006) Construction of introgression lines carrying wild rice (*Oryza rufipogon* Griff.) segments in cultivated rice (*Oryza sativa* L.) background and characterization of introgressed segments associated with yield-related traits. Theor Appl Genet 112: 570–580.
36. Kochert G, Stalker HT, Gimenes MA, Galgaro ML, Lopes CR, et al. (1996) RFLP and cytogenetic evidence on the origin and evolution of allotetraploid domesticated peanut, *Arachis hypogaea* (Leguminosae). Am j bot 83: 1282–1291.
37. Seijo G, Lavia GI, Fernandez A, Krapovickas A, Ducasse DA, et al. (2007) Genomic relationships between the cultivated peanut (*Arachis hypogaea*, Leguminosae) and its close relatives revealed by double GISH. Am J Bot 94: 1963–1971.

20

Genetic Variation within Native Populations of Endemic Silkmoth *Antheraea assamensis* (Helfer) from Northeast India Indicates Need for *In Situ* Conservation

Y. Tunginba Singh[1,3], Sudeshna Mazumdar-Leighton[1]*, Mahaswetta Saikia[1], Prashant Pant[1], Sochanngam Kashung[1], Kartik Neog[2], Rajen Chakravorty[2], Suresh Nair[3], Javaregowda Nagaraju[4], Cheerukeri Raghavendra Babu[5]

1 Department of Botany, University of Delhi, Delhi, India, **2** Central Muga and Eri Research and Training Institute, Ladoigarh, Assam, India, **3** International Centre for Genetic Engineering & Biotechnology, New Delhi, India, **4** Center for DNA Fingerprinting & Diagnostics, Hyderabad, Andhra Pradesh, India, **5** Center for Management of Degraded Ecosystems, University of Delhi, Delhi, India

Abstract

A. assamensis is a phytophagous Lepidoptera from Northeast India reared on host trees of Lauraceae family for its characteristic cocoon silk. Source of these cocoons are domesticated farm stocks that crash frequently and/or wild insect populations that provide new cultures. The need to reduce dependence on wild populations for cocoons necessitates assessment of genetic diversity in cultivated and wild populations. Molecular markers based on PCR of Inter-simple sequence repeats (ISSR) and simple sequence repeats (SSR) were used with four populations of wild insects and eleven populations of cultivated insects. Wild populations had high genetic diversity estimates ($H_i = 0.25$; $H_S = 0.28$; $H_E = 0.42$) and at least one population contained private alleles. Both marker systems indicated that genetic variability within populations examined was significantly high. Among cultivated populations, insects of the Upper Assam region ($H_i = 0.19$; $H_S = 0.18$; $H_E = 0$) were genetically distinct ($F_{ST} = 0.38$ with both marker systems) from insects of Lower Assam ($H_i = 0.24$; $H_S = 0.25$; $H_E = 0.3$). Sequencing of polymorphic amplicons suggested transposition as a mechanism for maintaining genomic diversity. Implications for conservation of native populations in the wild and preserving in-farm diversity are discussed.

Editor: Thomas Flatt, Vetmeduni Vienna Institute of Population Genetics, Austria

Funding: This research was supported by grant (BT/PR5233/PBD/19/116/2004-7) from Department of Biotechnology, Government of India to SML. The funders had no role in study design, data collection and analysis, decision to publish, or preparation of the manuscript.

Competing Interests: The authors have declared that no competing interests exist.

* E-mail: smazumdar@botany.du.ac.in

Introduction

Antheraea assamensis is a silk-producing Lepidoptera indigenous to Northeast India and Indo-Burma [1]. Its cocoon silk has been used for centuries to weave ceremonial fabric that is golden yellow in color, lustrous and resistant to UV radiation [2,3]. After *Antheraea yamamai*, the cocoon silk of *A. assamensis* is touted as the second most expensive silk in the world and its cultivation contributes significantly to livelihoods of indigent tribal peoples. *A. assamensis* larvae are reared outdoors, on trees of *Persea bombycina* and *Litsea monopetala* (family Lauraceae) for commercial silk production [4]. Popular practice relies on collection of wild moths from nearby forests and rejuvenations of cultivated populations. However, it is difficult to maintain cultivated stocks which crash repeatedly leading to more collections from the wild. Despite extensive efforts since 1800s, attempts to cultivate the insect outside its natural host range have failed. Varied anthropogenic developments in the area, especially felling of host trees, have led to shrinking and fragmented habitat for the insect [5]. It is thus imperative to develop tools to study natural versus cultivated populations of this economically important, endemic Lepidoptera.

This paper investigates genetic variability in fifteen populations of *A. assamensis* moths originating from cultivated populations or wild moths from forests. Cultivated populations were collected from two regions of Northeast India, Upper Assam and Lower Assam [6,7]. Wild populations were collected from a third topographic region, the Shillong plateau (Figure 1). Two molecular marker systems based on PCR of simple sequence repeats (SSR) and inter simple sequence repeats (ISSR) were employed. ISSR markers were developed from repeat sequences within introns of gene families available for *Antheraea* species in NCBI database [8]. SSR markers used in this study have been described previously [9]. Genetic diversity and relatedness among insect populations from this region has been examined earlier with single marker systems and smaller datasets [9,10,11]. SSR markers provide single-locus population data based upon sizes of amplicons obtained with unique PCR primers that flank repeat sequences. SSR markers are "co-dominant" enabling assessment of heterozygote alleles in a population [12]. Multi-locus population data based on presence/absence of amplicons is obtained using ISSR primers designed from sequence repeats [13,14]. ISSR markers are "dominant" as they cannot distinguish heterozygote alleles, resulting in an under-representation of heterozygosity estimates [15]. Data obtained with both markers systems were evaluated for standard population genetic estimates and patterns of genetic variation among and within populations of wild and cultivated

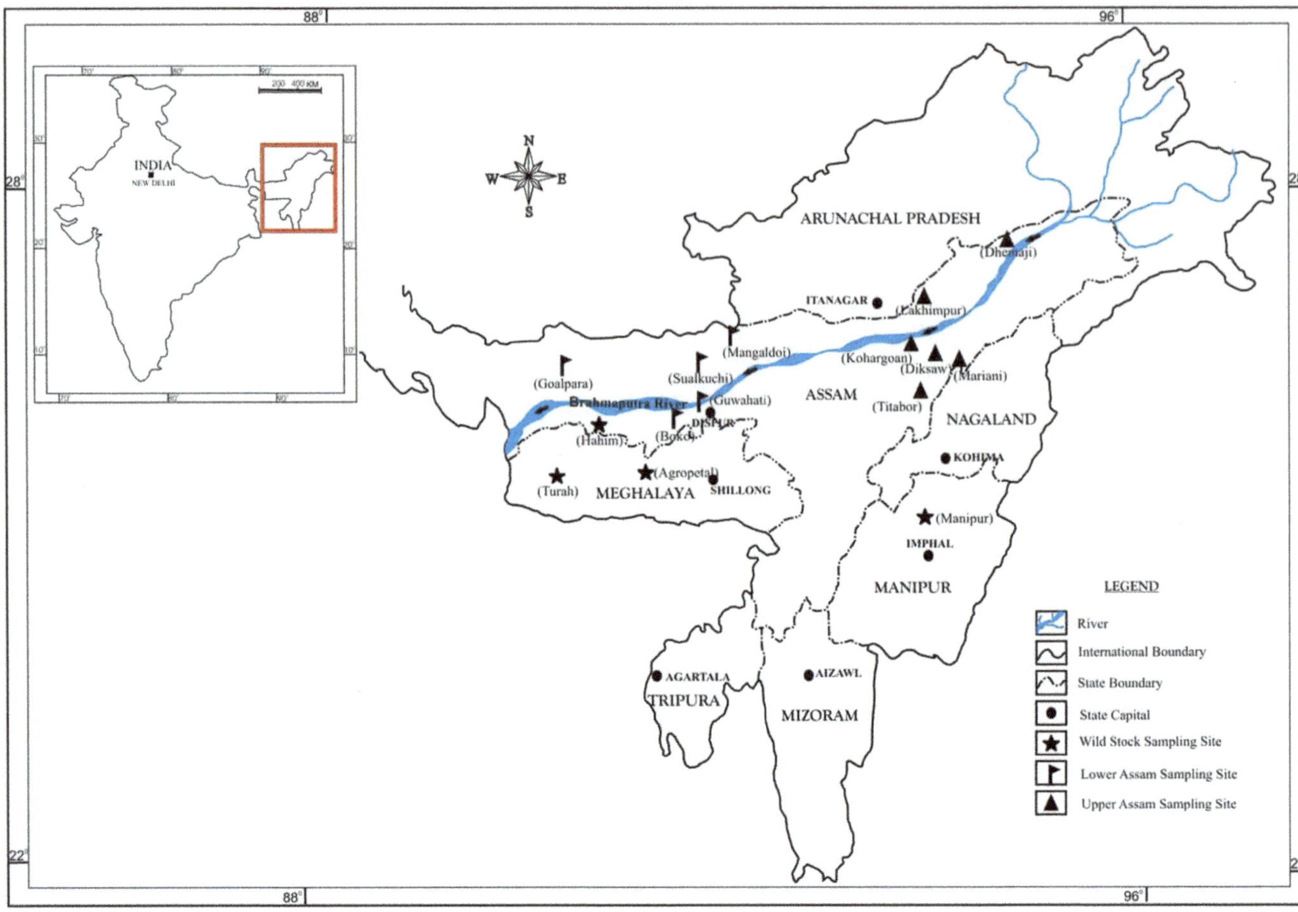

Figure 1. A map image showing the distribution of collection sites of *A. assamensis*. Names on the map correspond to collection sites in Table 1.

moths from different regions. Sequences and genomic organization of several ISSR amplicons were also obtained to determine the nature of genomic loci that contribute to population variations in *A. assamensis*. Results obtained are discussed for their relevance to conservation strategies.

Results

Genetic Diversity Estimates for A. assamensis Populations Using SSR and ISSR Markers

High genetic diversity estimates were obtained for the full data set comprising of fifteen populations (Table 1) using both marker systems. Table 2 and 3 showed that Nei's gene diversity ($H_i = 0.37 \pm 0.02$) estimated for all populations using all ISSR primers was similar to Nei's expected heterozygosity value ($H_E = 0.37 \pm 0.25$) obtained from SSR data. Both ISSR and SSR primers yielded variable numbers and sizes of amplicons. Bayesian heterozygosity estimates (H_S) obtained with all ISSR primers was more circumspect than H_i. However, trends observed with the frequentist (H_i) and Bayesian estimates (H_S) were consistent. For instance, AhISSR5 showed lowest percentage polymorphism, lowest H_i and H_S values among all the ISSR primers tested (Table 2). An excess of homozygotes was observed with each SSR primer (Table 3). Table 4 shows that several cultivated populations from Upper Assam region had low estimates of genetic diversity (H_i and H_S). Nei's heterozygosity estimates could not be estimated for insect populations from this region, as a single allele (monomorphic locus) was amplified with each SSR primer tested. A wild population from the Shillong plateau (Agropetal) showed the highest percent polymorphism, highest number of haplotypes, H_i and H_T estimates using both marker systems (Table 4). Two private alleles were detected in this population with Ahsat020 primer (not shown). Populations with highest H_S values were Hahim (a wild population from the Shillong plateau) and Mangaldoi (a cultivated population from Lower Assam).

Partitioning Genetic Variation among and within Populations Using ISSR and SSR Markers

AMOVA showed that majority of genetic variation in the full data set detected by both markers was found within the fifteen populations (64% with ISSR markers, $F_{ST} = 0.36$ and 82% with SSR markers, $F_{ST} = 0.18$ at P-value = 0, Table S1). Comparison of the group of eleven cultivated populations versus the group of four wild populations using ISSR marker data (Table 5A) indicated that while most of genetic variation (57.54%) resided within the populations examined, 17.94% genetic variation was found between the two groups, $F_{CT} = 0.18$. Results with SSR marker data (Table 5C) showed that most of the genetic variation (79.58%) was found within cultivated versus wild populations, but the genetic variation among these groups was only 5.1%, $F_{CT} = 0.05$. When all populations were grouped into three according to region (cultivated populations from Upper Assam

Table 1. Details of collection sites (populations) of *A. assamensis* from NE India.

No.	Collection Site (Acronym)	Number of moths(M = male, F = female)	Latitude/Longitude	Altitude (m)	Geographical region
CULTIVATED POPULATIONS					
1	Boko (BK)	M = 5, F = 9	26°03′51.7″N 91°16′52.82″E	48	Lower Assam
2	Mangaldoi (MD)	M = 7, F = 9	26° 38′29.98″N 92°03′16.57″E	89	Lower Assam
3	Guwahati (NB)	M = 7, F = 6	26°09′34.62″N 91°41′27.23″E	176	Lower Assam
4	Sualkuchi (SK)	M = 10, F = 8	26°31′10.61″N 91°30′42.66″E	71	Lower Assam
5	Goalpara (GP)	M = 12, F = 4	26°17′54.49″N 90°32′28.82″E	44	Lower Assam
6	Mariani (MR)	M = 2, F = 16	26°45′06.07″N 94°24′43.02″E	120	Upper Assam
7	Dhemaji (DM)	M = 3, F = 10	27° 39′49.08″N 94°55′41.02″E	215	Upper Assam
8	Diksaw (D)	M = 10, F = 10	26°46′35.73″N 94°07′23.42″E	96	Upper Assam
9	Kohargaon (KG)	M = 12, F = 6	26°45′29.59″N 93°58′00.43″E	86	Upper Assam
10	Titabor (TB)	M = 7, F = 13	26°16′13.44″N 93°51′18.96″E	142	Upper Assam
11	Lakhimpur (LK)	M = 8, F = 7	27°13′54.10″N 94°05′24.33″E	92	Upper Assam
WILD POPULATIONS					
12	Hahim (HA)	M = 15, F = 6	25°54′20.65″N 91°03′29.02″E	331	Shillong Plateau
13	Manipur (MN)	M = 6, F = 12	25°16′07.48″N 93°56′26.99″E	1941	Shillong Plateau
14	Turah (TR)	M = 14, F = 6	25°41′35.54″N 90°26′23.99″E	464	Shillong Plateau
15	Agropetal (AG)	M = 10 F = 12	25°41′35.54″N 90°26′23.99″E	464	Shillong Plateau

region, cultivated populations from Lower Assam region; wild populations from Shillong plateau region), approximately 20% of the total genetic variation ($F_{CT} = 0.2$) was observed between the three groups with both markers (Table 5B1, 5D1). Interestingly, when the two groups of cultivated populations from Upper Assam and Lower Assam regions were compared amongst themselves, approximately 62% within-population genetic variation and F_{ST} of 0.38 was observed with both markers (Table 5B2, 5D2). However, the two markers yielded very different results for partitioning of genetic variation among populations within the two groups of cultivated populations. The F_{SC} value was 0.26 with ISSR data (Table 5B2) but only 0.004 with SSR data (Table 5D2).

Genetic Structure of Wild and Cultivated *A. assamensis* Populations from Different Regions

PCO analyses using Euclidean distances among 207 ISSR fingerprints for cultivated and wild populations from different regions (Figure 2) accounted for 17% and 10% of the total genetic variance on principal axis 1 and principal axis 2, respectively. Figure 2 also shows that while cultivated insects from Upper Assam were mostly distinct from wild insects of the Shillong

Table 2. A primer-wise comparison of all *A. assamensis* populations for percentage polymorphism (%P), Nei's gene diversity (H_i), and Bayes heterozygosity estimate (H_S).

ISSR Primer	Sequence	No. of amplicons (% P)	Nei's H_i	Bayes H_S	Source & nature of amplicons cloned	GenBank Accession Numbers
AhISSR2	$(GAA)_6C$	6 (100)	0.3±0.03	0.29±0.02	NB-9_{1316P}, SK-7_{684M}, SK-7_{643M}	GQ273748, EU123529, EU123530
AhISSR4	$(TCG)_6C$	7 (100)	0.3±0.04	0.25±0.02	NB-11_{893M}	GQ273750
AhISSR5	$(CGA)_6T$	4 (75)	0.28±0.06	0.18±0.02	NB-5_{437P}, TR-4_{547M}, MD-8_{547M}, DM-1_{547M}	GQ273749, GQ273755, GQ273756, GQ273757
AhISSR11	$C(GATA)_5$	8 (100)	0.3±0.02	0.2±0.01	GP-16_{609M}, GP-16_{825M}	GQ273753, GQ273754
AhISSR12	$(GATA)_5T$	11 (100)	0.39±0.02	0.28±0.01	MN-1_{1064M}	GQ273752
AhISSR13	$(GATA)_5A$	6 (100)	0.37±0.01	0.28±0.01	GP-13_{1249P}, GP-14_{607M}	EU123527, EU123528
AhISSR110	$(GATA)_5C$	8 (100)	0.39±0.02	0.28±0.01	TR-4_{1456P}, TR-4_{1452P}, HA-3_{1400P}	DQ872512, DQ872513, GQ273751
All primers			0.37±0.02	0.24±0.01		

GenBank Accessions for cloned and sequenced amplicons are provided. The population and moth from which the amplicon was obtained are indicated in uppercase. Amplicon size and nature of the locus is shown in subscript. The subscript "P" refers to a polymorphic locus (amplicon present in some individuals of a particular population), while the subscript "M" refers to a monomorphic locus (amplicon present in all individuals of that population).

Table 3. Estimates of heterozygosity calculated from SSR data. N_a is observed number of alleles (amplicons produced) and N_e is the effective number of alleles; H_E and H_O refer to expected and observed heterozygosity values; Asterisk (*) denotes significant deviation from HWE at $p<0.05$.

SSR Primers	Repeat motif	N_a	N_e	H_E	H_O
AaSat020F	$(TCGTG)_5$	5	1.1	0.13	0.06*
AaSat044	$(AT)_{20}$	3	1.38	0.38	0.24*
AaSat065F	$(AT)_{11}N_n(AT)_{11}$	4	2.16	0.62	0.47*
All primers			1.5±0.5	0.37±0.25	0.26*±0.2

Plateau, some insects from Lower Assam were similar to wild insects suggesting limited mixing/gene flow. Figure 2 inset shows results of PCO analyses comparing cultivated populations from two regions, Upper Assam and Lower Assam where the first principal component axis explained 14% of the genetic variance between the two groups. This result was supported by previous pair-wise estimates of Nei's genetic distance obtained from ISSR marker data comprising of fifty loci (Table S2). Among hierarchical clustering methods, UPGMA provided the best fit for ISSR data with fifty loci and indicated clustering of populations based upon source (wild and cultivated) as well as region (Figure S1). A standardized Mantel statistic of $r = 0.4675$ was obtained for the association between geographic and Nei's genetic distances with $P = 0.001$ (for 999 replicated randomizations). This correlation suggested that the some of the populations differentiated *in situ* rather than having differentiated elsewhere and migrating to their current locations.

In order to further explore the results described above, a Bayesian approach for clustering related individuals into discrete units (or K), was applied. For ISSR data with 50 loci, and $K = 3$, discrete groups of populations comprising wild insects from Shillong Plateau region, cultivated insects from Upper Assam and Lower Assam regions were retrieved with both admixture and no-admixture models (Figure 2B). Results from the admixture model showed "mixing" among three cultivated populations of Lower Assam, i.e. Boko, Guwahati and Mangaldoi with insects from wild populations (Figure 2B). It is unclear if the observed admixing was due to migration or anthropogenic introduction of wild insects into cultivated populations. Two cultivated populations from Lower Assam (Goalpara and Sualkuchi) were discrete. Application of SSR data gave the most cohesive results at $K = 2$ and $K = 3$ (Figure 2C), where cultivated insects of Upper Assam were distinct from all other insects.

Genomic Variability in Populations of *A. assamensis* and Motifs of Transposable Elements

Several ISSR amplicons were cloned into pDrive vector and their sequences submitted to GenBank (Table 2). These sequenced products originated from amplicons from loci that were (i) monomorphic ($NB11_{893M}$) for Upper Assam (UA), Lower Assam (LA) and Shillong plateau regions (SP); (ii) polymorphic for LA and SP but absent from UA ($NB5_{437P}$); (iii) polymorphic ($TR4_{1452P}$, $TR4_{1456P}$, $HA3_{1450P}$) for UA, LA and SP; (iv) polymorphic ($GP13_{1249P}$) and monomorphic ($GP14_{607M}$) for moths of the same population; and (v) polymorphic ($NB9_{1316P}$) and monomorphic ($SK7_{684M}$, $SK7_{643M}$) for moths from different populations. Amplicons representing loci that were (i) monomorphic from the same moth of a population ($GP16_{825M}$, $GP16_{609M}$) and (ii)

Table 4. Population-wise comparison of percentage polymorphism (%P), Nei's gene diversity (H_I) and Bayesian heterozygosity (H_S) values determined from ISSR-PCR data and Nei's average gene heterozygosity (H_T) values determined from SSR-PCR data. n.d. = not detectable. Hap: Number of haplotypes.

Populations	ISSR			SSR	
	% (P)	Nei's H_I	Bayesian H_S	Hap	H_T
CULTIVATED POPULATIONS (*Lower Assam Region*)					
Boko	80	0.29±0.19	0.29±0.01	6	0.25
Mangaldoi*	82	0.32±0.18	**0.31±0.02**	9	0.28
Guwahati*	78	0.31±0.25	0.30±0.02	6	0.32
Sualkuchi*	52	0.20±0.21	0.22±0.01	5	0.20
Goalpara*	54	0.19±0.20	0.21±0.02	5	0.26
CULTIVATED POPULATIONS (*Upper Assam Region*)					
Mariani	52	0.18±0.20	0.19±0.02	1	n. d.
Dhemaji*	58	0.25±0.23	0.28±0.02	1	n. d.
Diksaw	64	0.22±0.19	0.21±0.01	1	n. d.
Kohargaon	44	0.15±0.20	0.21±0.02	1	n. d.
Titabor	**30**	**0.09±0.17**	**0.14±0.01**	**1**	**n. d.**
Lakhimpur	40	0.14±0.20	0.17±0.02	1	n. d.
WILD POPULATIONS (*Shillong Plateau Region*)					
Hahim*	64	0.35±0.18	**0.31±0.01**	7	0.36
Manipur*	58	0.23±0.22	0.27±0.02	4	0.29
Turah*	78	0.27±0.20	0.26±0.02	8	0.36
Agropetal	**90**	**0.38±0.16**	**0.26±0.02**	**13**	**0.54**

Asterisks (*) indicate populations from which selected loci were cloned and sequenced. Populations with highest and lowest genetic diversity estimates are shown in bold.

monomorphic ($MD8_{547M}$ of LA, $TR4_{547M}$ of SP and $DM1_{547M}$) were also analyzed to determine if such monomorphic loci contain the same sequence. About 10 kbp of the *A. assamensis* genome was sequenced. Sequence analyses of these cloned loci revealed that without exception, primers based on the same repeat sequence but containing different mononucleotide anchor amplified unique, non-redundant loci across the *A. assamensis* genome.

Results obtained after analyzing the data with Spectral Repeat Finder (SRF) indicated the presence of interspersed, imperfect repeats ranging from 2 to 10-mers with a low spectral frequency of.0005 to.001. The $TR4_{1456P}$ locus contained the largest copy number (103) of such a repeat. Amplicons obtained with tri-nucleotide repeat primers had 40%GC compared with primers based on tetra-nucleotide repeat sequence (34%GC). Most sequences were AT-rich and contained stop codons in all three frames especially at the 5′ end. Sequence of the $TR4_{1456P}$ amplicon contained twenty copies of a TATAAAAT sequence while $MN1_{1064M}$ locus contained a 40 bp AT-rich region. No tandem, perfect simple sequence repeats were observed in any amplicon using either SRF or mREP softwares. Analyses with RepeatMasker and Consensus softwares (Figure 3) revealed the occurrence of a 109 bp sequence resembling the SAKE (L2) LINE element from *B. mori* in $SK7_{643M}$ loci; a 55 bp sequence resembling EnSpm DNA transposable element from *Triticum aestivum* in $GP16_{609M}$; a 57 bp sequence resembling an *A. mylitta* transposase psuedogene Anm4 in $GP13_{1249P}$; a 50 bp sequence resembling a Helitron from *Anopheles gambiae* and a 166 bp

Table 5. Partitioning by AMOVA of genetic variation in *A. assamensis* moths from (A) cultivated and wild populations, (B1) three regions of Northeast India and (B2) cultivated populations from 2 regions of Northeast India using ISSR marker data; (C) cultivated and wild populations, (D1) three regions of Northeast India and (D2) cultivated populations from 2 regions of Northeast India using SSR marker data.

	Source of variation	d.f.	Variance components	Percentage of variation	Fixation indices	P-value*
A.	Among groups	1	1.6	17.94	$F_{CT}=0.18$	0.00196±0.00136
	Among populations within groups	13	2.18	24.52	$F_{SC}=0.3$	0
	Within populations	192	5.12	*57.54*	$F_{ST}=0.43$	0
	Total	206	8.9			
B1.	Among groups	2	1.72	20.28	$F_{CT}=0.20$	0
	Among populations within groups	12	1.64	19.33	$F_{SC}=0.24$	0
	Within populations	192	5.12	*60.39*	$F_{ST}=0.40$	0
	Total	206	8.48			
B2.	Among groups	1	1.27	16.65	$F_{CT}=0.17$	0.0001±0.00009
	Among populations within groups	9	1.67	21.77	$F_{SC}=0.26$	0
	Within populations	140	4.71	*61.58*	$F_{ST}=0.38$	0
	Total	150	7.65			
C.	Among groups	1	0.02	5.09	$F_{CT}=0.05$	0.11632±0.00964
	Among populations within groups	13	0.07	15.33	$F_{SC}=0.16$	0
	Within populations	363	0.37	*79.58*	$F_{ST}=0.20$	0
	Total	377	0.47			
D1.	Among groups	2	0.10	19.97	$F_{CT}=0.2$	0.001±0.001
	Among populations within groups	12	0.01	2.91	$F_{SC}=0.04$	0.008±0.003
	Within populations	363	0.37	*77.12*	$F_{ST}=0.23$	0
	Total	377	0.48			
D2.	Among groups	1	0.15	37.82	$F_{CT}=0.38$	0
	Among populations within groups	9	.001	0.27	$F_{SC}=0.004$	0.336±0.015
	Within populations	233	0.25	*61.9*	$F_{ST}=0.38$	0.003±0.002
	Total	243	0.4			

*Significance tests were performed for all P-value estimates at 1023 permutations.

sequence resembling a non-LTR retrotransposon, Jockey from the Lepidoptera, *Lymantria dispar* in $TR4_{1456P}$, $TR4_{1452P}$ and $HA3_{1450P}$ loci. A simple sequence repeat $(CTAAG)_n$ was identified in $TR4_{1456P}$ and $TR4_{1452P}$ close to the Jockey-like stretch. About 50% of the ISSR amplicons sequenced in this study were associated with remnants/footprints of transposable elements. Furthermore, the $(GATA)_n$ primer binding regions along with an additional 12 to 28 bp were found to resemble hAT and the Harbinger elements from zebra fish in the sequences of $GP16_{825M}$ and $GP13_{1249P}$ loci. During searches of the GenBank databases, the $(GATA)_n$ primer binding sequences along with 6–12 bases were found to resemble micro-satellite regions from a wide variety of organisms. BLAST hits of the EST databases often indicated that the 3′ ends of the amplicons contained regions of coding sequences suggesting that intron-exon junctions may be spanned during amplification with these ISSR primers. This may explain an observation made earlier that the stop codons were generally clustered in the 5′ends of the sequences. In the case of $GP13_{1249P}$ and $SK7_{643P}$ sequences obtained with primer AhISSR13 and AhISSR5, a 130 bp sequence was common in both amplicons immediately proximal to the forward primer sequence. In fact, several sequenced amplicons obtained with the same primers contained perfectly conserved sequences immediately proximal to the forward primers, suggesting the occurrence of multiple copies of repetitive regions in the *A. assamensis* genomes. Finally, we observed identical sequences within a monomorphic locus of 547 bp amplified from several individuals belonging to the three population clusters. This result supported a basic premise of ISSR-PCR that shared bands (scored as 1) may represent homologous loci.

Southern Blot with a Repetitive Element from a Polymorphic Locus as Probe Hybridizes to Lepidopteran Genomes and is Organized Differently in Populations of *A. assamensis*

An amplicon of 800 bp with sequence similarity to lepidopteran transposable elements was obtained by nested amplification from

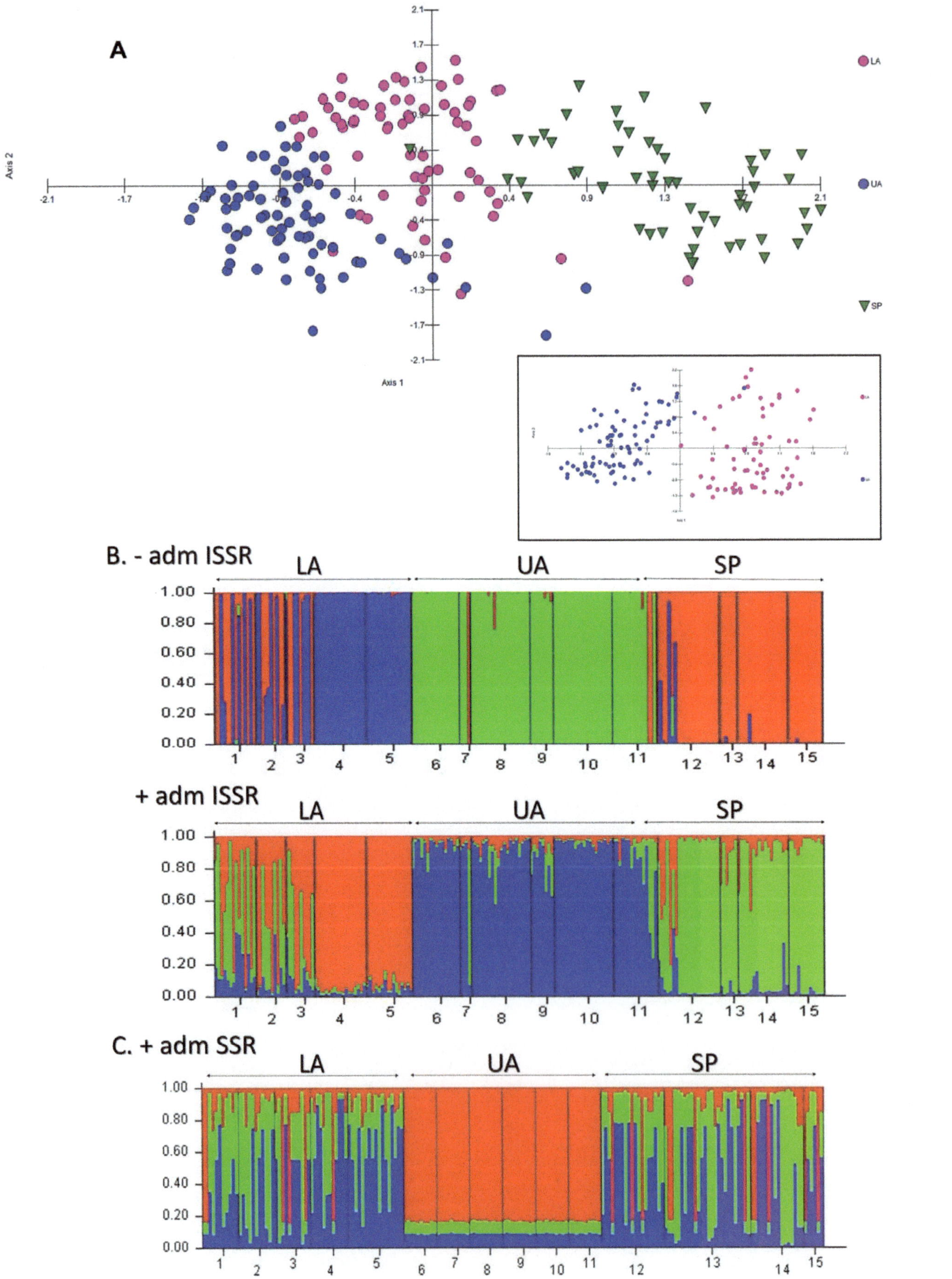
A
Axis 1
Axis 2
LA
UA
SP
B. - adm ISSR
LA
UA
SP
1.00
0.80
0.60
0.40
0.20
0.00
1 2 3 4 5 6 7 8 9 10 11 12 13 14 15
+ adm ISSR
LA
UA
SP
1.00
0.80
0.60
0.40
0.20
0.00
1 2 3 4 5 6 7 8 9 10 11 12 13 14 15
C. + adm SSR
LA
UA
SP
1.00
0.80
0.60
0.40
0.20
0.00
1 2 3 4 5 6 7 8 9 10 11 12 13 14 15

Figure 2. Analyses of sampled populations using PCO and STRUCTURE. Genetic relationships were determined using (A) PCO analyses of 207 ISSR fingerprints obtained for fifteen populations collected from 3 regions. Designations for regions are: Upper Assam or UA (cultivated populations), Lower Assam or LA (cultivated populations) and Shillong plateau or SP (wild populations). Inset shows PCO analyses of ISSR fingerprints from eleven cultivated populations from Upper Assam and Lower Assam regions. Putative population genetic structure (B) was obtained with Admixture (+adm ISSR) and without Admixture (-adm ISSR) settings at K = 3 using ISSR data with STRUCTURE [59]. Each vertical bar represents a moth distributed into 1 of K colored clusters. Populations 1 (BK), 2 (MD), 3(NB), 4 (GP) and 5 (SK) are cultivated populations from Lower Assam (LA); populations 6 (MR), 7 (DM), 8 (D), 9 (KG), 10 (TB) and 11 (LK) are cultivated populations from Upper Assam (UA); populations 12 (HA), 13 (MN), 14 (TR) and 15 (AG) are wild populations from Shillong plateau (SP). Population acronyms are expanded in Table 1. Putative population genetic structure (C) was obtained with Admixture (+adm SSR) setting at K = 3 for 15 populations (189 insects) using SSR data as above.

the $TR4_{1456P}$ locus (Figure 3) and used as a probe in Southern hybridization with *Hind* III and *Hae* III digested DNA from *A. assamensis*, *A. proylei*, *A. mylitta*, *B. mori* and *S. cynthia* (Figure 4A). All four individuals of *A. assamensis* examined (TR3 and TR4 from Turah of SP, BK12 from Boko of LA and KG8 from Kohargaon of UA) contained the 1456 bp amplicon. Low stringency salt washes showed hybridization of the probe with heterologous sequences from *A. proylei*, *A. mylitta* but not *B. mori* and *S. cynthia* indicating that the probe was specific for *Antheraea* species (Figure 4B). PCR amplification with Fp2/Ah16 primers (used to generate the probe) also supported this result as no amplification products were obtained with *B. mori* and *S. cynthia* (not shown). High stringency washes revealed that the probe (with an internal restriction site for *Hae* III) hybridized to multiple but distinct loci in the genomes of individual *A. assamensis* that came from different populations (Figure 4C). These results suggested that these repetitive elements occurred in multiple copies, were interspersed in the *Antheraea* genomes and distributed differently in moths of different populations. Nei's genetic distances between Boko (a cultivated population from Lower Assam region) and Kohargaon (a cultivated population from Upper Assam region) was 0.1644, Boko and Turah (a wild population from the Shillong plateau region) was 0.1839; Turah and Kohargaon was 0.3011 (Table S2).

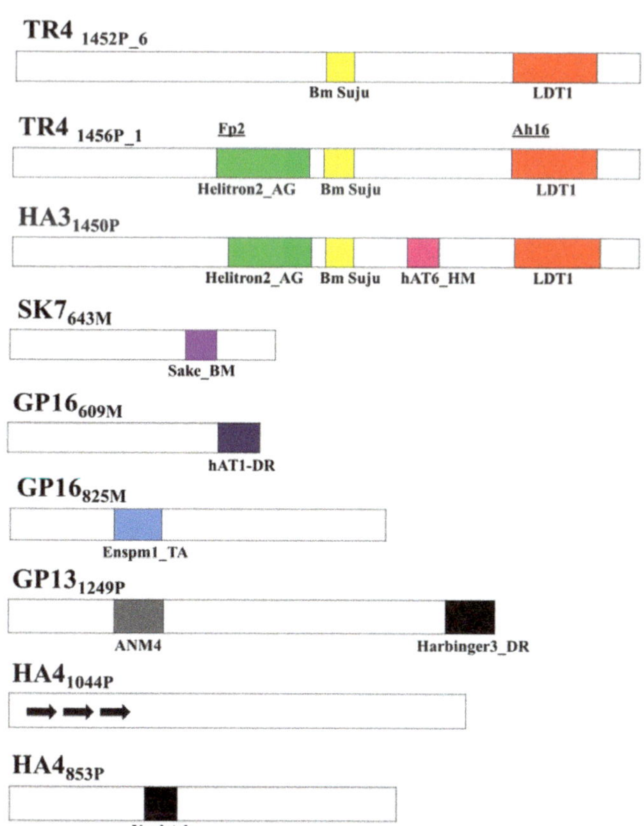

Figure 3. A map showing positions of motifs resembling putative transposable elements (boxes shown in different colors) within sequenced ISSR amplicons. Abbreviations are as described in Table 2. Arrows denote tandem repeats. The bold lines ( ) indicate position of nested primers (Fp2 and Ah16) within amplicon EU872512.

Discussion

High Genetic Diversity Detected in *A. assamensis* Using SSR and ISSR Markers

Cocoon silk of *A. assamensis* is a major cottage industry of economic and cultural significance in North-eastern India [16]. Rearing of cultivated populations depends on re-invigoration by insects collected from wild populations. Combined use of two marker systems with same set of insect populations can improve the reliability of information obtained for devising robust conservation strategies. Estimated gene diversity indices in *A. assamensis* ISSR markers were high at 0.37±0.02 (H_i) using frequentist methods and 0.24±0.01 (H_S) while using Bayesian principles (Table 2). Even though actual values differed, trends between estimates obtained with frequentist approach (H_i) and Bayesian approach (H_S) were similar. Similar discrepancies in numerical values have been observed earlier and attributed to the different approaches for these measures [17]. Expected heterozygosity (H_E) estimated from SSR data was also high at 0.37±0.23 (Table 3). Similar high estimates of gene diversity have been reported for smaller data sets of *A. assamensis* using molecular markers [9,10,11], suggesting that intra-specific diversity is high and the insect germplasm has sufficient genetic potential to envisage well-designed conservation efforts. Average heterozygosity (H_O) observed in Lepidoptera using allozyme analysis is 0.127 [18], ranging from a low of 0 for *E. mana* to a high of 0.46 for *O. brumata* [30/19]. Results from this study (Table 3) resembled previous report where the observed heterozygosity with SSR markers was usually lower than expected due to null alleles and/or Wahlund's effect [9]. High genetic diversity, percent polymorphism and low gene flow have been reported for silk-producing insects cultivated in India, such as Indian Eri silk moth, *Samia cynthia;* Indian Tussah silk moth, *Antheraea mylitta* and races of *Bombyx mori* [20–26].

High Genetic Diversity in Wild Populations, Cultivated Populations from Different Regions Differ in Diversity Estimates

High gene diversity estimates for wild populations from the Shillong plateau with both marker systems ($Hi = 0.25±0.15$; $Hs = 0.28±0.01$ with ISSR markers and $H_E = 0.42±0.25$ with SSR markers) indicated the need to preserve forests in North-east India as repositories of diverse germplasm of *A. assamensis*. The highest number of haplotypes (12), and two private SSR alleles was detected in the Agropetal population which contained native

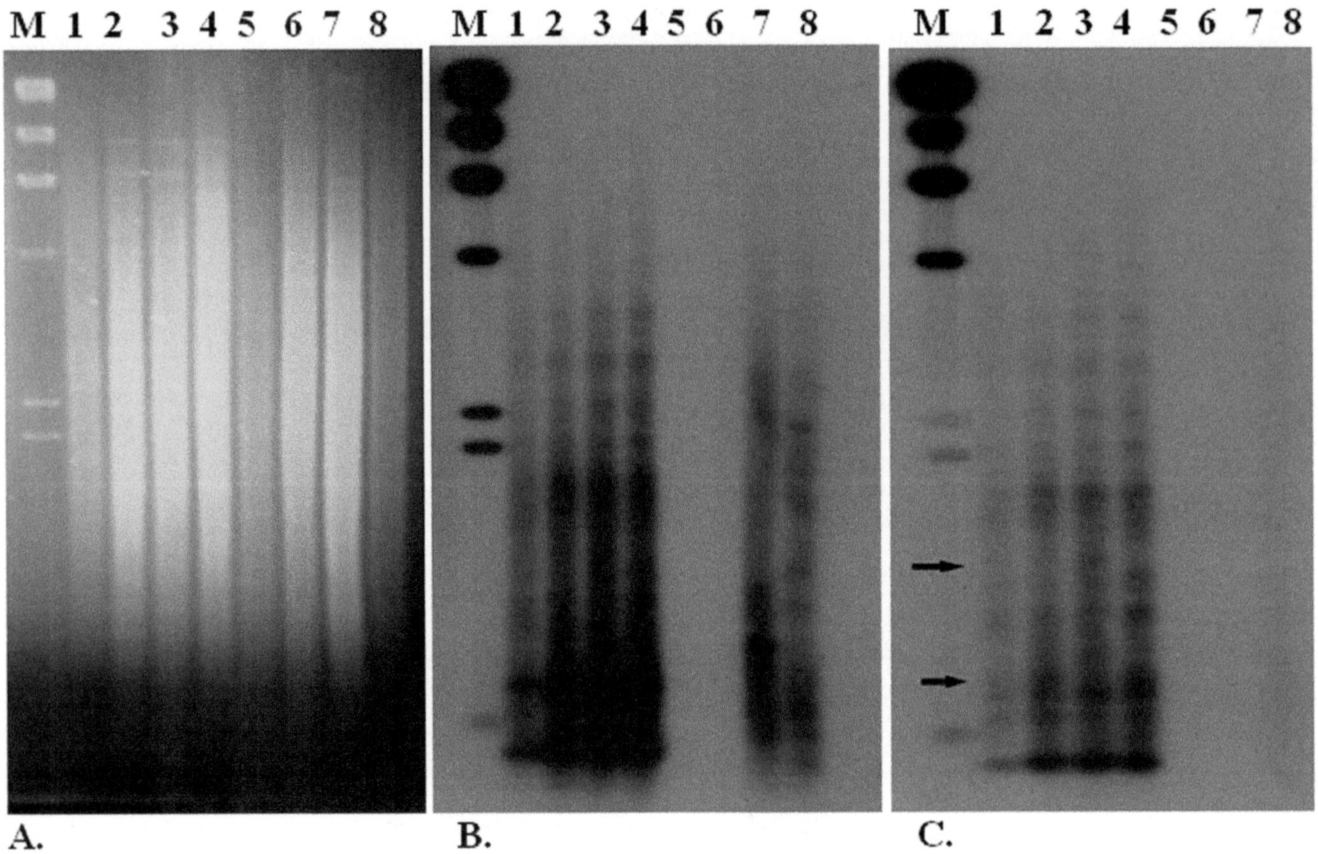

Figure 4. Southern hybridization of various silkmoths using a putative transposable element amplicon from *A. assamensis* as probe. Genomic DNAs (A) of *A. assamensis moths* BK12 (lane 1), TR3 (lane 2), TR4 (lane 3), KG8 (lane 4); *B. mori* (lane5), *S. cynthia* (lane 6), *A. mylitta* (lane 7) and *A. proylei* (lane 8) were digested with *Hae* III. Southern hybridization of the digested DNAs with an 800 bp amplicon containing a putative mariner-like element are shown after (B) a low stringency wash and (C) a high stringency wash. Lane M shows λ DNA restricted with *Hind* III as a size marker. Arrows denote hybridization signals that distinguish moths from different populations.

insects. Similar reports have been reported earlier for wild populations from this region using single markers systems [9,10,11]. Genetic diversity estimates observed in all cultivated populations were $H_i = 0.24 \pm 0.17$; $H_S = 0.21 \pm 0.01$ for ISSR markers and $H_E = 0.21 \pm 0.11$ with SSR markers. While several observations for cultivated populations held between the two marker systems, some did not. For example, in contrast to the ISSR data, cultivated populations of Upper Assam were monomorphic with all eleven SSR markers (Table 4). Nevertheless, both genetic marker systems clearly indicated that cultivated populations from Upper Assam were the most homogenous and had lowest estimates of genetic diversity ($H_i = 0.19 \pm 0.17$; $H_S = 0.18 \pm 0.01$; $H_E = 0$). Both Nei's H_i and Bayesian H_S values were the lowest for Titabor (a cultivated population from Upper Assam) which also had the lowest percentage polymorphism and least number of polymorphic bands (Table 4). Interestingly, in a report with 6 populations of *S. cynthia* collected from sites similar to our study, moths from Titabor also showed the lowest genetic heterozygosity values [25]. The significance of this observation is unclear but could refer to local rearing practices promoting inbreeding or reduced gene flow from other populations. In contrast to Upper Assam, cultivated populations from Lower Assam had higher genetic diversity estimates in this study ($H_i = 0.24 \pm 0.16$, $H_S = 0.25 \pm 0.01$ with ISSR markers and $H_E = 0.3 \pm 0.21$ with SSR markers). Reasons for these observations are unclear, but could be recent introduction of wild parents to invigorate cultivated populations of Lower Assam. The values from Upper Assam populations could be a snapshot of genetic diversity values in populations containing more inbred insects, prior to a stock invigoration. More work is needed to understand and monitor the genetic consequences of introducing wild insects into cultivated populations.

Genetic Variation is High within Wild and Cultivated Insect Populations from Different Regions

Both molecular marker systems indicated that genetic variation was partitioned within individuals of the fifteen populations rather than among groups of populations based upon topography or origin (Table 5). The import of high within-population genetic variation is unclear. It could represent a genetic strategy of an endemic insect species to survive in its habitat. Occurrence of high genetic variation within populations is of particular significance in developing conservation strategies for animals to survive under changing environmental conditions or biotic stress [27]. It can also be of significance for development of pure breeding lines at different locations [28]. Comparison of patterns of genetic variation among the group of wild populations and the group of all cultivated populations is interesting, as this distinction likely underlines the use of insects from wild populations to invigorate cultivated populations. Interestingly, a $F_{CT} = 0.18$ value was observed among these two groups with ISSR markers (Table 5A) and a low $F_{CT} = 0.05$ was observed with SSR markers (Table 5C).

The SSR markers gave better results for among groups of populations based on region/topography. F_{CT} values were similar for both markers (0.2) when three groups of populations from different regions (wild populations from Shillong plateau, cultivated populations from Lower Assam and cultivated populations from Upper Assam) were examined (Table 5B1, 5D1). Topography and associated factors may thus be a factor influencing patterns of genetic variation observed in this insect. Genetic variation observed among and within cultivated populations from Upper Assam region versus Lower Assam region (Table 5B2, 5D2) could be influenced by low genetic diversity estimates and preponderance of monomorphic loci in Upper Assam populations (Table 4). Reasons for distinctions between cultivated populations of Upper and Lower Assam regions remain, but do not include choice of host plant species. All larvae of cultivated populations examined in this study were reared on natural stands of *P. bombycina*, a highly out-bred species. Further work is needed to investigate the role of host plant genotype (if any) on population genetics of cultivated insect populations.

Population Structure of Wild and Cultivated Populations of *A. assamensis* from Different Regions

Principal coordinate analyses (Figure 2) and AMOVA (Table 5) indicated that insects of wild populations from the Shillong Plateau were distinct from insects of almost all cultivated populations from Upper Assam. Insects of the cultivated populations from Upper Assam were distinct from those of Lower Assam. While these results suggested that both source of the insects (i.e. wild versus cultivated) and regional topography were possible factors influencing genetic variation observed between populations of *A. assamensis*, some mixing of genotypes especially between wild populations and the geographically proximal cultivated populations of Lower Assam was also indicated. Moderate genetic distance estimates (Table S2) and its concurrence with geographic distance among populations observed in this study may be attributed to low migration and mixing of the genotypes on a limited geographical scale. The observation that genetic differentiation increased with increasing geographic distance further supports the contention that the wild moths do not migrate far [11,29,30]. Analysis with STRUCTURE using ISSR data indicated a putative population structure where wild populations and cultivated populations from Upper Assam and Lower Assam formed discrete clusters (Figure 2B). Populations of Lower Assam were further organized into at least two sub-clusters, one of which showed mixing with wild populations due to migration and/or anthropogenic intervention. Limited migration has been reported among wild populations of the insect [11]. In this study, SSR data could not be used to distinguish wild populations from cultivated populations from Lower Assam (Figure 2C), reflecting AMOVA results (low $F_{CT} = 0.05$, between cultivated and wild populations) observed earlier (Table 5C).

In contrast with the genetic data, little geographic structure was observed with phenotypic data analyzed for the fifteen populations in this study. Sexual dimorphism for selected traits (moth weight, cocoon weight and wing-length) was evident (Figure S2). Positive correlations were observed between traits such as moth weight and shell weight, an economic trait (Figure S3). In a 2-way (sex and locality) ANOVA of the data, the interaction was not statistically significant, indicating that sex-specific differences were not locality-related. However, attempts to hierarchically cluster the populations failed, suggesting that the morphological traits considered here were not sufficiently distinct to group the insects based upon source or region. Sexual dimorphism and intra-specific variation in venation patterns and vein-thickness of *A. assamensis* moths has been reported recently [31]. Our results suggest that measured phenotypic traits may be under stabilizing selection with similar optima in all populations, since the low rate of migration would be sufficient to allow the populations to diverge by selection or genetic drift if stabilizing selection were not the same at all locations. The nature of this stabilizing selection is a matter of conjecture and could well be constraints imposed by a common life history [32]. Alternatively, factors such as reduced gene flow between some populations could be responsible.

Sequence Analyses of Polymorphic Loci in *A. assamensis* and Southern Hybridization Reveal Motifs from Several Transposable Elements

Amplicons representing several polymorphic and monomorphic loci from moths within the same and different populations were sequenced (Table 2). About 50% of the sequenced amplicons representing polymorphic loci contained regions ranging from 28 to 166 bp with high sequence similarity to known class II transposable elements (Figure 3). In the case of $TR4_{1456P}$ amplicon, sequence motifs of more than one mobile genetic element were detected. The association of transposable elements with population genetic variation is well known for animals and plants [33,34,35]. Most of these studies cite the occurrence of transposable elements and remnants of transposition events in introns consistent with our design of intron repeat-based ISSR primers yielding amplicons containing putative transposable elements (Table S3). Associations of simple sequence microsatellite repeats (as used for design of ISSR primers) with transposable elements have been reported in other organisms such as rice [36,37]. Transposable elements like *Bilbo* have been shown to maintain genetic variability within populations and influence population structure of *Drosophila* species [38,39]. In *D. melanogaster*, transposable elements (some of which occur in introns) which show population heterogeneity have been attributed a major role in adaptation to local temperate climates [40]. Transposable elements are known to constitute 35% of the *Bombyx mori* genome [41]. Thus, they may be strongly implicated in population genetic variations and evolution of Lepidoptera. Southern hybridization studies were carried out with DNA from five sericigenous Lepidoptera with a 800 bp probe containing a 166 bp region resembling a non-LTR retrotransposon Jockey element from *Lymantria dispar*. The probe hybridized with DNA from *A. assamensis*, *A. mylitta* and *A. proylei* but not *B. mori* or *S. cynthia*, suggesting that it was specific for *Antheraea* species. In *A. assamensis*, the probe bound to multiple but identical loci within *Hin*d III digested DNAs of four moths from different populations (not shown). In the case of *Hae* III digested DNAs of the same four *A. assamensis* moths, the probe hybridized to multiple loci, some of which were distinct for moths from different populations (Figure 4) suggesting that sequences contained in the probe might be involved in origin of polymorphic loci in *A. assamensis*. The probe was a fragment of an amplicon obtained with AhISSR 110 primer whose sequence was based on a $(GATA)_n$ repeat. Our Southern hybridization result is reminiscent of the autosomal, non-allelic hypervariable Bkm DNA loci in the moth *Ephestia kuehniella* known to be associated with mobile genetic elements [42]. Insertion of class II transposable element belonging to Helitron-like superfamily has recently been shown to interrupt micro-satellite loci in *Ostrinia nubilalis*, *Pectinophora gossypiella* and *B. mori* [43]. Further research is required to investigate the nature and roles of transposable elements in Antheriids as they may have be relevant for breeding experiments and the ability of progeny of diverse populations to adapt and survive in a particular region. Local micro-climate and disease/predator loads in localities from where the collections for this study may be variable and pose unique challenges to the survival of the species, especially for native populations.

In conclusion, the use of two molecular markers as described in this study strongly indicates that wild populations need to be preserved in their natural habitats as invaluable germplasm for invigorating cultivated populations. Our results coincide with other reports based on single marker systems [9,10,11] which provide snapshots of allele frequencies in different populations of *A. assamensis*. Further research is necessary to understand the role of transposable elements and other genomic mechanism(s) that can influence genetic variability within populations of *A. assamensis*. Taken together, the importance of monitoring the history and genetic consequences of introduction of wild moths (mixing) during invigoration of cultivated stocks is paramount. Molecular markers can be useful for such efforts. Both wild and cultivated populations of *A. assamensis* are endemic and restricted to Northeast India [16]. If considered as a meta-population, then dynamic strategies and different priorities may be necessary for conservation of native insects and management of cultivated insects.

Materials and Methods

Collection Sites and Morphometric Measurements of Insects

Cocoons of *A. assamensis* were randomly collected from forests, silk farmers, state, and central government insect rearing depots in Northeast India. To the best of our knowledge, no collection was made from a restricted or protected area. A total of 256 cocoons were collected representing fifteen collection sites (henceforth referred to as populations). The location, altitude and latitude of these fifteen sites were ascertained from governmental maps and Google earth (Table 1). Cocoons used in this study were collected from *P. bombycina* trees, except those collected from Hahim which were obtained from larvae fed on *P. bombycina* and *L. monopetala*. Cocoons were weighed and placed singly in rearing chambers. Each emergent, fully eclosed moth was given an id tag; its sex, fresh weight, shell weight (after moth eclosion), wing length [44] dimensions recorded. Non- parametric Mann-Whitney U tests were performed using SPSS version 14.0 (IBM Inc., USA) to compare all males with all females for moth weight, forewing length, hind wing length, cocoon weight and shell weight. Individuals of each sex from every population were tested for significant interactions by 2-way ANOVA before proceeding with multiple comparisons using the LSD and Tukey's HSD test for significance at alpha 0.5. Attempts were also made to derive dendrograms using hierarchical clustering with squared Euclidean distance measures for the morphological parameters studied.

Isolation of DNA from Moths, Primer Design, and PCR Amplifications

High molecular weight, nuclear genomic DNA was isolated from an abdominal half of each moth using a proteinase K, SDS-based nuclei extraction method [45]. The remaining abdominal half was frozen in liquid nitrogen and stored at –20°C. The quality and quantity of the isolated DNA was ascertained by standard spectro-photometry and 0.8% agarose gel electrophoresis followed by staining with ethidium bromide [46]. For designing anchored ISSR primers, intron sequences of *Antheraea* species available in the public domain NCBI sequence database were collated and searched for sequence repeats using the software mREPS [47]. A total of thirty ISSR primers were designed based on predominant di-, tri-and tetra-nucleotide repeats identified by the above search. Repeats that would yield amplicons in the size range of 250 bp to 2 kbp were selected. Each primer was provided with a unique 3′ or 5′ mononucleotide anchor (Table 2). The primers were synthesized at IDT Inc., USA. PCR amplifications were performed with a profile of 40 cycles of DNA denaturation at 94°C for 30 seconds, primer annealing at 42 to 55°C for 30 seconds (depending upon the T_m of the primer used, Table S3) and primer extension at 72°C for 1 minute using an ABI 2720 Thermal cycler. The total reaction volume was 25 µl which contained 0.5 U of Taq DNA polymerase in buffer containing 3 mM magnesium chloride provided by the manufacturer, (Promega Inc., USA), 200 µM of each dNTPs, 30 ng/µl of each primer, and 10 ng of genomic DNA from each moth as template. PCR products were electrophoresed on a 1.5% agarose gel and visualized by ethidium bromide staining. PCR amplifications were performed at least in duplicate to ensure repeatability. Eleven SSR primers were synthesized commercially and applied as recommended in Arunkumar et al., 2010 [9]. Amplification profiles comprised of an initial denaturation at 94°C for 3 minutes followed by 35 cycles of DNA denaturation at 94°C for 20 seconds, primer annealing at recommended temperatures for 10 seconds, primer extension at 72°C for 45 seconds and a final extension at 72°C for 10 minutes using an ABI 2720 Thermo-cycler. Total reaction volume was 10 µl with 0.5 U of *Taq* DNA polymerase in buffer containing 1.5 mM magnesium chloride provided by the manufacturer (Promega Inc., USA), 100 µM of each dNTPs, 5 pmol of each primer, and approximately 10 ng of genomic DNA from each moth as template. All reagents were of molecular biology grade and purchased from Sigma Aldrich Inc., USA, except for the primers, which were synthesized by IDT Inc., USA.

Population Genetic Analyses of A. assamensis DNAs

Amplification products obtained consistently with each ISSR primer for each insect representing the fifteen populations were scored for the presence (1) or absence (0) of bands. Only repeatable bands in the range of 0.25 to 2.5 kbp were scored. Rare alleles were pruned from the dataset [8]. The number of polymorphic loci and percentage polymorphic loci (P) was determined for each primer at single- and multi-population level. Appropriate estimates of Nei's gene diversity (H_i) were obtained using two free-wares: POPGENE version 1.32 and TFPGA version 1.3 [48,49]. Estimations were done assuming Hardy Weinberg equilibrium conditions (Inbreeding coefficient, $F_{IS} = 0$), moderate ($F_{IS} = 0.5$) and high disequilibrium ($F_{IS} > 0.5$). A Bayesian approach for estimating total average panmictic heterozygosity (H_S) was determined reiteratively using Hickory version 1.03 [50] with the f full, f = 0, theta = 0, f-free models and default sampling parameters (Burn-in = 50,000, sample = 250,000 and thin = 50) with three runs. As the deviation information criteria (DIC) were low and statistically similar for f full, f = 0 and f free models, only those values obtained with the f free model are described here. The use of the f free model is touted to be particularly suited for data obtained using dominant markers [51]. SSR-PCR products were manually scored for the presence and size of amplicons. Data from only those individuals for which all three loci could be consistently scored were used. Calculations for observed number of alleles (Na); effective number of alleles (Ne); numbers of haplotypes; estimates of observed, expected and average heterozygosity estimates (H_O, H_E and H_T) and pair-wise tests for linkage disequilibrium [52] were performed with the SSR-PCR data using Arlequin 3.5 [53] and default parameters.

Data obtained was also converted into haplotype frequencies by AMOVA PREP [54] and CONVERT [55]. Total genetic variation (Φ-statistics) was estimated within populations and among populations using by AMOVA [53,56] using Arlequin 3.1 [53]. Patterns of genetic variation among and within populations comprising groups based on source (wild or cultivated) and region (Upper Assam, Lower Assam and Shillong plateau) were compared. Two groups of cultivated populations from Upper Assam and Lower Assam were also compared. Matrices of inter-population geographic distances (in km) and Nei's genetic distances estimated from the ISSR data were used to perform a Mantel test using TFPGA and 999 randomized permutations to determine statistical significance [57]. Principal coordinate (PCO) analysis was conducted with Euclidean distances between all ISSR fingerprints of insects from wild and cultivated populations from different regions, using MVSP ver. 3.1 [58]. To detect population structure and genetic admixtures by a Bayesian approach, STRUCTURE 2.3.4. [59,60] was used with data from ISSR markers and SSR markers. Independent runs at multiple K values (K is the number of inferred clusters) were used for the ISSR data; Multiple independent runs at $K = 1$ up to $K = 15$ were attempted for the SSR data. Settings used were burn-in period = 10^5; no. of MCMC reps after burn-in = 10^5; Ancestry model with and without admixture; and correlated allele frequency between populations was assumed. Most likely value for assigning K clusters was determined from an *ad-hoc* statistic (ΔK) derived from second order rate of change in likelihood function was used derived from second order rate of change with respect to K of the likelihood function [61]. Where applicable in the above estimations, standard assumptions made for dominant marker data were adopted. These included assuming that each shared ISSR band represented homologous loci and absence of bands could be attributed to lack of primer binding sites in the DNA of some individuals of the populations tested due to chromosomal rearrangements, transposition events and mutations or deletions of pertinent repeat sequences.

Cloning, Sequencing and Analyses of Selected ISSR-PCR Amplicons

Selected amplicons obtained from individual moths of various populations (Table 2) were gel purified and cloned into the pDrive vector according to the manufacturer's instructions (Qiagen Inc., USA). Reasons for selecting these amplicons for sequencing are described in the Results section. Plasmid DNA was isolated from positive colonies using Qiagen plasmid purification kit, and sequenced on both strands at Macrogen Inc., South Korea. At least three clones were sequenced to confirm each sequence. The sequences were examined with an online software Spectral Repeat Finder or SRF [62] which uses Fourier transformation to search sequences for repeat patterns, regions of repeats in a sequence, copy number and percentage nucleotide content. The sequences were also analyzed for the presence of tandem repeats using mREPS [47]. The sequences were used to search the annotated nucleic acid databases at GenBank using BLAST-n [63], and the repeat libraries using RepeatMasker [64] and Censor [65] using default parameters. Sequences have been submitted to the NCBI database as accessions DQ872512, DQ872513, DQ872514, EU123527, EU123528, EU123529, EU123530, GQ273748, GQ273749, GQ273750, GQ273751, GQ273752, GQ273753, GQ273754, GQ273755, GQ273756 and GQ273757.

Southern Hybridizations Using an Amplicon Containing a Repetitive Element as Probe

Southern hybridizations were performed after Mazumdar-Leighton and Broadway [66]. High molecular weight nuclear genomic DNA isolated from single moths of *A. assamensis* (BK-12 from LA region, TR-3 from SP, TR-4 from SP, KG-8 from UA), *B. mori*, *S. cynthia A. proylei* and *A. mylitta* was digested completely with the restriction enzyme *Hin*d III and *Hae* III (New England Biolabs, Inc., MA, USA). The probe used was an 800 bp amplicon obtained by nested PCR (using primers Fp2: 5′CTATGTCATC-CACTGTACTGTAATT3' and Ah16: 5′TTTCACGAACCCA-TACA GGC3') from a plasmid clone DQ872512 containing a polymorphic locus $TR4_{1456P}$ obtained with primer AhISSR110. This locus occurred in the dataset with a frequency of 37% and showed high sequence similarity to lepidopteran transposable elements. The 800 bp amplicon contained an internal restriction site for *Hae* III but lacked sites for *Hin*d III. Low stringency washes (4X SSC +0.1%SDS) were done twice for 30 minutes at 65°C before overnight autoradiography at −80°C. Subsequent high stringency washes (2XSSC +0.1%SDS, 0.5X SSC +0.1%SDS) for 30 minutes each at 65°C was followed by 2 days autoradiography at −80°C. Primers Fp2 and Ah16 used for nested PCR and amplification of the probe were also used to amplify DNA from the heterologous insects, *B. mori*, *S. cynthia*, *A. proylei* and *A. mylitta*. These PCRs were performed as described earlier except that a primer annealing temperature of 60°C was used.

Supporting Information

Figure S1 A tree depicting clusters among 15 populations of *A. assamensis* based on Nei's genetic distances. The number of loci supporting each node and proportion of similar replicates (bootstrap) calculated using TFPGA is indicated in parentheses. Population color codes are the same as Figure 2A.

Figure S2 Mean (±SD) weight of moth weight (MW), cocoon weight (CW) and shell weight (SW) of male (M) □ and female (F) ■ *A. assamensis*.

Figure S3 Relation between moth weight and cocoon weight in male (Δ) and female ( ) *A. assamensis*. The r^2 values are also indicated.

Table S1 Partitioning of genetic variation for the full set of *A. assamensis* moths using (A) ISSR marker data and (B) SSR marker data. *Significance tests were performed at 1023 permutations. **Bayesian analogue (θii value) was 0.32+0.02.

Table S2 Matrix showing pair-wise estimates of Nei's genetic distance estimates based on 50 loci from ISSR data. Wild populations from Shillong plateau are shown in bold, cultivated populations from Lower Assam region are italicized. Populations showing the highest and lowest genetic distance estimates are italicized in bold.

Table S3 Details of thirty ISSR primers designed from intronic repeats within Antheraea genes and their ability to amplify lepidopteran genomic DNAs.

Acknowledgments

A resource lab instrumentation loan from Bio-Rad Inc., India, is gratefully acknowledged. Ph.D. fellowships from Department of Biotechnology (to MS and PP) and Council of Scientific and Industrial Research, Govt. of India, (to YTS and SK) are acknowledged.

Author Contributions

Conceived and designed the experiments: SML CRB RC. Performed the experiments: YTS MS PP SK KN. Analyzed the data: YTS PP SK SML. Contributed reagents/materials/analysis tools: SN JN. Wrote the paper: SML PP YTS MS SK. Collection of samples: MS KN YTS.

References

1. Hugon T (1837) Remarks on the silkworms and silks of Assam. J Asiatic Soc Bengal 6: 21–38.
2. Chowdhury SN (2001) Sericulture and weaving (an overview). Designer Graphics Press Dibrugarh, Assam, India.
3. Good IL, Kenoyer JM, Meadow RH (2009) New evidence for early silk in the Indus civilization. Archaeometry 50: 457–466.
4. Bindroo BB, Singh NT, Sahu AK, Chakravorty R (2006) Muga silkworm host plants. Indian Silk 44: 13–17.
5. Singh KC and Chakravorty R (2006) Seri-biodiversity of North-Eastern India-an update. In JP Handique and MC Kalita (eds.) Biodiversity, Conservation and Future concern. Gauhati University, Gauhati, India. 8–19.
6. Purkait B (2004) Hydrogeology. In: Singh VP, Sharma N, Ojha SCP, editors. The Brahmaputra basin water resources. Springer. 113.
7. Sarma JN (2005) Fluvial process and morphology of the Brahmaputra river in Assam, India. Geomorphology 70: 226–256.
8. Singh YT (2010) A study of *Antheraea assamensis* (Helfer): Population biology, genetics and genome. Ph D Thesis. University of Delhi, India.
9. Arunkumar KP, Kafiyathullah L, Nagaraju J (2009) Microsatellite markers for the Indian golden silkmoth, *Antheraea assama* (Saturniidae: Lepidoptera). Mol Eco Resour 9: 268–270.
10. Neog K, Singh HR, Unni B, Sahu AK (2010) Analysis of genetic diversity of muga silkworm (*Antheraea assamensis*, Helfer; Lepidoptera : Saturniidae) using RAPD-based molecular markers. Afr J Biotechnol 9: 1746–1752.
11. Arunkumar KP, Sahu AK, Mohanty AR, Awasthi AK, Pradeep AR, et al. (2012) Genetic Diversity and Population Structure of Indian Golden Silkmoth (*Antheraea assama*). PLoS ONE 7(8): e43716.
12. Zietkiewicz E, Rafalski A, Labuda D (1994) Genome fingerprinting by simple sequence repeat (SSR) - anchored polymerase chain reaction amplification. Genomics 20: 176–183.
13. Abbot P (2001) Individual and population variation in invertebrates revealed by inter-simple sequence repeats (ISSRs). J Insect Sci 1: 8–10.
14. Behura SK (2006) Molecular marker systems in insects: current trends and future avenues. Mol Ecol 15: 3087–3113.
15. Bornet B, Muller C, Paulus F, Branchard M (2002). Highly informative nature of inter simple sequence repeat (ISSR) sequences amplified using tri- and tetra-nucleotide primers from DNA of cauliflower (*Brassica oleraceae* var. *botrytis* L.). Genome 45: 890–896.
16. Beeson CFC (1941) The ecology and control of the forest insects of India and the neighboring countries. Vasant Press, Dehra Dun, India.
17. Bahulikar RA, Stanculescu D, Preston CA, Baldwin IT (2004) ISSR and RFLP analysis of the temporal and special population structure of the post- fire annual, *Nicotiana attenuata*, in SW Utah. BMC Ecol 4: 12.
18. Nevo E, Beiles A, Ben-Shlomo R (1984) The evolutionary significance of genetic diversity: ecological, demographic and life history correlates. In: Mani GS, editor. Lecture Notes in Biomathematics (Evolutionary Dynamics of Genetic Diversity). Springer-Verlag, Berlin. 13–213.
19. Dongen SV, Backeljau T, Matthysen E, Dhont AA (1998) Genetic population structure of the winter moth (*Operophtera brumuta* L) (Lepidoptera, Geometridae) in a fragmented landscape. Heredity 80: 92–100.
20. Nagaraja GM, Nagaraju J (1995) Genome fingerprinting of the silkworm, *Bombyx mori*, using random arbitrary primers. Electrophoresis 16: 1633–1638.
21. Reddy KD, Nagaraju J, Abraham EG (1999) Genetic characterization of the silk worm *Bombyx mori* by simple sequence repeat (SSR)-anchored PCR. Heredity 83: 681–687.
22. Chaterjee SN, Tanushree T (2004) Molecular profiling of silkworm biodiversity in India. Genetika 40: 1618–1627.
23. Pradeep AR, Chaterjee SN, Nair CV (2005) Genetic differentiation induced by selection in an inbred population of the silkworm *Bombyx mori*, revealed by RAPD and ISSR marker systems. J Appl Genet 46: 291–298.
24. Kar PK, Vijayan K, Mohandas TP, Nair CV, Saratchandra B, et al. (2005) Genetic variability and genetic structure of wild and semi-domestic populations of tasar silkworm (*Antheraea myllita*) ecorace Daba as revealed by ISSR markers. Genetica 125: 173–183.
25. Vijayan K, Anuradha HJ, Nair CV, Pradeep AR, Awasthi AK, et al. (2006) Genetic diversity and differentiation among populations of the eri silkworm, *Samia cynthia ricini*, revealed by ISSR markers. J Insect Sci 30: 1536–2442.
26. Saha M, Mahendran B, Kundu SC (2008) Development of random amplified polymorphic DNA markers for tropical tasar silkworm *Antheraea mylitta*. J Econ Entomol 101: 1176–1182.
27. Toro MA, Caballero A (2005) Characterization and conservation of genetic diversity in subdivided populations. Philos Trans R Soc Lond B Biol Sci 360: 1367–1378.
28. Notter DR (1999) The importance of genetic diversity in livestock populations of the future. J Anim Sci 77: 61–69.
29. Oliver CG (1972) Genetic and phenotypic differentiation and geographic distance in four species of Lepidoptera. Evolution 26: 221–241.
30. White TCR (1972) The distribution, dispersal and host range of *Antheraea eucalypti* (Lepidoptera: Saturniidae) in New Zealand. Pac Insects 14: 669–673.
31. Nath R, Devi D (2009) Venation pattern and shape variation in wing of *Antheraea assamensis* (Lepidoptera: Saturnidae) of Assam, India. Int J Trop Insect Sci 29: 70–78.
32. Allen CE, Zwaan BJ, Brakefield PM (2011) Evolution of sexual dimorphism in the Lepidoptera. Annu Rev Entomol 56: 445–464.
33. Kidwell MG, Lisch D (1997) Transposable elements as source of variation in animals and plants. Proc Natl Acad Sci USA 94: 7704–7711.
34. Bennett EA, Coleman LE, Tsui C, Pittard WS, Devine SE (2004) Natural genetic variation caused by transposable elements in humans. Genetics 168: 933–951.
35. Huang X, Lu G, Zhao Q, Liu X, Han B (2008) Genome-wide analysis of transposon insertion polymorphisms reveals intraspecific variation in cultivated rice. Plant Physiol 148: 25–40.
36. Temnykh S, Declerck G, Lukashova A, Lipovich L, Cartinhour S, et al. (2001) Computational and experimental analysis of microsatellites in rice (*Oryza sativa* L): Frequency, length variation, transposon associations, and genetic marker potential. Genome Res 11: 1441–1452.
37. Méglécz E, Petenian F, Danchin E, D'Acier AC, Rasplus JY, et al. (2004) High similarity between flanking regions of different microsatellites detected within each of two species of Lepidoptera: *Parnassius apollo* and *Euphydryas aurinia*. Mol Ecol 13: 1693–1700.
38. Biémont C, Cizeron G (1999) Distribution of transposable elements in *Drosophila* species. Genetica 105: 43–62.
39. Lepetit D, Brehim A, Fouillet P, Biémont C (2002) Insertion polymorphism of retrotransposable elements in populations of insular, endemic species *Drosophila madeirensis*. Mol Ecol 11: 347–354.
40. Gonzalez J, Lenkov K, Lipatov M, Macpherson JM, Petrov DA (2008) High rate of recent transposable element-induced adaptation in *Drosophila melanogaster*. PLoS Biol 6: e251.
41. Osanai-Futahashi M, Suetsugu Y, Mita K, Fujiwara H (2008) Genome-wide screening and characterization of transposable elements and their distribution analysis in the silkworm, *Bombyx mori*. Insect Biochem Mol Biol 38: 1046–1057.
42. Traut W (1987) Hypervariable Bkm DNA loci in a moth, *Ephestia kuehniella*: Does transposition cause restriction fragment length polymorphism? Genetics 115: 493–498.
43. Coates BS, Sumerford DV, Hellmich RL, Lewis LC (2010) A Helitron-like transposon superfamily from Lepidoptera disrupts (GAAA)n microsatellites and is responsible for flanking sequence similarity within a microsatellite family. J Mol Evol 70: 275–288.
44. Betts CR, Wootton RJ (1988) Wing shape and flight behavior in butterflies (Lepidoptera: Papilionoidea and Hesperioidea): A preliminary analyses. J Exp Biol 138: 271–288.
45. Henry JM, Raina AK, Ridgway RL (1990) Isolation of high molecular-weight DNA from insects. Anal Biochem 185: 147–150.
46. Sambrook J, Fritsch EF, Maniatis T (1989) Molecular cloning: A laboratory manual. Cold Spring Harbor, USA: Cold Spring Harbor Laboratory Press.
47. Kolpakov R, Bana G, Kucherov G (2003) mREPS: Efficient and flexible detection of tandem repeats in DNA. Nucleic Acids Res 31: 3672–3678.
48. Yeh FC and Boyle TJB (1997) Population genetic analysis of co-dominant and dominant markers and quantitative traits. Belgian Journal of Botany 129: 157.
49. Miller MP (1997) Tool for population genetic analyses (TFPGA) version 1.3. North Arizona University, USA: Department of Biological Sciences.
50. Holsinger KE, Lewis PO (2003) Hickory: A package for analysis of population genetic data. Department of Ecology and Evolutionary Biology, U-3043, University of Connecticut, CT, USA.
51. Holsinger KE, Lewis PO, Dey DK (2002) A Bayesian approach to inferring population structure from dominant markers. Mol Ecol 11: 1157–1164.
52. Nei M (1987) *Molecular Evolutionary Genetics*. Columbia University Press, New York.
53. Excoffier L, Lischer H (2010) Arlequin suite ver 3.5: A new series of programs to perform population genetics analyses under Linux and Windows. Mol Ecol Resour 10: 564–567.
54. Miller MP (1998) AMOVA-PREP 1.01: A program for the preparation of the AOVA input files from dominant marker raw data. North Arizona University, USA: Department of Biological Sciences.
55. Glaubitz JC (2004) CONVERT: A user-friendly program to reformat diploid genotypic data for commonly used population genetic software packages. Mol Ecol Notes 4: 309–310.
56. Weir BS (1996) Genetic data analysis II. Methods for discrete population genetic data. Sinauer Associates, USA.
57. Sokal RR, Rohlf FJ (1995) Biometry. New York, USA: WH Freeman and Company.
58. Kovach WL (2007) MVSP - A Multi-Variate Statistical Package for Windows, ver. 3.1. Kovach Computing Services, Pentraeth, Wales, U.K.

59. Pritchard JK, Stephens P, Donnelly P (2000) Inference of population structure using multilocus genotype data. *Genetics* 155: 945–959.
60. Falush D, Stephens M, Pritchard JK (2003). Inference of population structure: Extensions to linked loci and correlated allele frequencies. Genetics, 164: 1567–1587.
61. Evanno G, Regnaut S, Goudet J (2005) Detecting the number of clusters of individuals using the software STRUCTURE: a simulation study. Mol Ecol 14: 2611–2620.
62. Sharma D, Isaac B, Raghava GPS, Ramaswamy R (2004) Spectral Repeat Finder (SRF): Idenitifcation of repetitive sequences using Fourier transformation. Bioinformatics 20: 1405–1412.
63. Altschul SF, Gish W, Miller W, Myers EW, Lipman DJ (1990) Basic local alignment search tool. J MolBiol 215: 403–410.
64. Smit AFA, Hubley R, Green P (1996–2004) RepeatMasker Open-3.0. http://www.repeatmasker.org/cgi-bin/WEBRepeatMasker (accessed October 2011).
65. Kohany O, Gentles AJ, Hankus L, Jurka J (2006) Annotation, submission and screening of repetitive elements in Repbase: RepbaseSubmitter and Censor. BMC Bioinf 7: 474.
66. Mazumdar-Leighton S, Broadway RM (2001) Identification of six chymotrypsin cDNAs from larval midgets of *Helicoverpa zea* and *Agrotis ipsilon* feeding on the soybean (Kunitz) trypsin inhibitor. Insect Biochem Mol Biol 31: 633–644.

21

Abundant Microsatellite Diversity and Oil Content in Wild *Arachis* Species

Li Huang[1]◑, Huifang Jiang[1]◑, Xiaoping Ren[1], Yuning Chen[1], Yingjie Xiao[2], Xinyan Zhao[1], Mei Tang[1], Jiaquan Huang[1], Hari D. Upadhyaya[3], Boshou Liao[1]*

1 Key Laboratory of Biology and Genetic Improvement of Oil Crops, Ministry of Agriculture, Oil Crops Research Institute of the Chinese Academy of Agricultural Sciences, Wuhan, China, **2** National Key Laboratory of Crop Genetic Improvement and National Center of Plant Gene Research (Wuhan), Huazhong Agricultural University, Wuhan, China, **3** International Crops Research Institute for the Semi-Arid Tropics, Patancheru, Andhra Pradesh, India

Abstract

The peanut (*Arachis hypogaea*) is an important oil crop. Breeding for high oil content is becoming increasingly important. Wild *Arachis* species have been reported to harbor genes for many valuable traits that may enable the improvement of cultivated *Arachis hypogaea*, such as resistance to pests and disease. However, only limited information is available on variation in oil content. In the present study, a collection of 72 wild *Arachis* accessions representing 19 species and 3 cultivated peanut accessions were genotyped using 136 genome-wide SSR markers and phenotyped for oil content over three growing seasons. The wild *Arachis* accessions showed abundant diversity across the 19 species. *A. duranensis* exhibited the highest diversity, with a Shannon-Weaver diversity index of 0.35. A total of 129 unique alleles were detected in the species studied. *A. rigonii* exhibited the largest number of unique alleles (75), indicating that this species is highly differentiated. AMOVA and genetic distance analyses confirmed the genetic differentiation between the wild *Arachis* species. The majority of SSR alleles were detected exclusively in the wild species and not in *A. hypogaea*, indicating that directional selection or the hitchhiking effect has played an important role in the domestication of the cultivated peanut. The 75 accessions were grouped into three clusters based on population structure and phylogenic analysis, consistent with their taxonomic sections, species and genome types. *A. villosa* and *A. batizocoi* were grouped with *A. hypogaea*, suggesting the close relationship between these two diploid wild species and the cultivated peanut. Considerable phenotypic variation in oil content was observed among different sections and species. Nine alleles were identified as associated with oil content based on association analysis, of these, three alleles were associated with higher oil content but were absent in the cultivated peanut. The results demonstrated that there is great potential to increase the oil content in *A. hypogaea* by using the wild *Arachis* germplasm.

Editor: Khalil Kashkush, Ben-Gurion University, Israel

Funding: The authors are grateful to the research grants from the National Basic Research Program of China (2011CB109300) and Crop Germplasm Program (NB09-2130135-4) and the National Science Fund of China (No. 31271764). The funders had no role in study design, data collection and analysis, decision to publish, or preparation of the manuscript.

Competing Interests: The authors have declared that no competing interests exist.

* E-mail: lboshou@hotmail.com

◑ These authors contributed equally to this work.

Introduction

The genus *Arachis* originated in South America and contains at least 80 species that have been classified into nine taxonomic sections based on morphological variation, geographical distribution and cross-compatibility [1,2]. Some *Arachis* species are used as forage for animal production in Asia and Africa. The most economically important species in the genus is the cultivated peanut *Arachis hypogaea*. Peanuts are widely grown in more than 100 countries and are used to produce food and edible oil for human consumption. Because the amount of peanut oil used in developing countries is increasing, more than 60% of peanuts produced worldwide are crushed for edible oil (http://faostat.fao.org/faostat/collections?subset=agriculture 2010). Thus, enhancing the oil content of peanut cultivars is becoming an increasingly important breeding objective in most developing countries.

While most wild *Arachis* species are diploid (such as AA or BB, $2n = 2x = 20$), the cultivated peanut is an allotetraploid (AABB, $2n = 4x = 40$). A wild allotetraploid species in the *Arachis* section, *A. monticola* (AABB, $2n = 4x = 40$), is thought to be the direct wild tetraploid ancestor of *A. hypogaea* [3,4]. Cytological and molecular studies have shown that *A. duranensis* (AA) and *A. ipaënsis* (BB) have a high similarity to *A. hypogaea*. It is probable that *A. duranensis* and *A. ipaënsis* are the wild diploid progenitors of *A. hypogaea*, which may have arisen from a single hybridization event between *A. duranensis* and *A. ipaënsis* followed by chromosome duplication [3,5]. Tetraploid species are also found in the sections *Extranervosae* and *Rhizomatosae*. These tetraploid species are believed to have evolved independently [6].

Wild *Arachis* species are widely distributed throughout a large region of South America and show extensive morphological variability depending upon their distinct environments. Perennial peanuts are characterized by tuberiform hypocotyls and tuberous roots for adaptation to upland areas (as in the sections *Erectoides* and *Extranervosae*). In contrast, annual species typically have fibrous root systems and reproductive systems adapted to lowland areas (as in the section *Arachis*) [7]. *Arachis* species are autogamous, disperse seeds underground and are geographically isolated from one another; because of these characteristics, hybrid sterility is often encountered when producing interspecific hybrids between species in the same section or in different sections [8]. This reproductive barrier among wild *Arachis* has inhibited the gene flow among these plants and may have facilitated the process of *Arachis* speciation. It is critical to pinpoint the genetic regions underlying the speciation process and determine the mechanism of divergent selection that shaped the different adaptive traits in wild *Arachis* species. Natural gene exchange between wild diploid species and cultivated peanut may have been further limited due to genomic re-arrangement during polyploidization [9]. In addition, domestication events have greatly reduced the genetic diversity in the cultivated peanut. Successive self-pollination and the use of a few elite breeding lines with little exotic germplasm in breeding programs resulted in a narrow genetic base of cultivated peanut germplasm [10,11]. Thus, there is a potential to mine novel variants in wild species and transfer them into cultivated peanut. Wild *Arachis* species possess genetic variability in pest and disease resistance traits, which could be used to improve the cultivated peanut. These traits include resistance to peanut stunt virus (PSV) [12], peanut stripe virus (PStV) [13], nematodes [14], early leafspot [15], late leafspot [16], rust [4], bacterial wilt [17], and spotted spider mites [18]. The traits in some wild *Arachis* species that confer resistance to pests and disease have been successfully transferred into cultivated peanuts [19,20].

High oil content in a high-yielding genetic background is a key objective of peanut breeding worldwide. Limited information is available regarding oil content variation in wild *Arachis* species and the relationship between oil content and genetic variation across the entire genome. Upadhyaya et al. reported that the range of oil content in wild *Arachis* accessions was 45%–55% at ICRISAT in Hyderabad, India [21]. Oil content is a quantitative trait controlled by many genes that have small effects and show high genotype × environment interactions. It is therefore of interest to investigate the variation in oil content among wild *Arachis* accessions in China and assess the value of these accessions in breeding. The characterization of the population structure and phylogenic relationships of wild *Arachis* accessions is essential to evaluate the level of differentiation among species and sections as well as to investigate the relationships between allelic variation and oil content in wild *Arachis* species. This information will be useful for identifying wild *Arachis* accessions that are ideal donor parents to enhance the cultivated peanut and to broaden the diversity of germplasm in peanut breeding.

In the present study, a collection of 3 cultivated peanut accessions and 72 wild *Arachis* accessions (representing 19 species from 5 different sections) was assembled. The *Arachis* collection was genotyped with 136 SSR primers and phenotyped for oil content. The objectives of the study were (a) to evaluate genetic diversity among accessions of different species and sections within the genus *Arachis*, (b) to infer the population structure and phylogenic relationship of the *Arachis* accessions and (c) to assess the variation of oil content in *Arachis* accessions and detect alleles associated with oil content.

Results

The Genetic Diversity of 75 *Arachis* Accessions

The 75 *Arachis* accessions used in this study belong to 20 species from 5 sections. These accessions were evaluated for allelic diversity using 136 SSR markers (Table 1). The wild *Arachis* accessions were highly diverse. In total, 944 alleles were identified (6.94 alleles per marker), and the Shannon-Weaver diversity index was 0.4130. Among the 19 wild *Arachis* species, *A. duranensis* exhibited the highest diversity, with a Shannon-Weaver diversity index of 0.3522. The cultivated *Arachis* (*A. hypogaea*) accessions exhibited relatively low diversity, with only 309 alleles identified (2.27 alleles per marker) and a Shannon-Weaver diversity index of 0.0662 (Table 1).

Unique Alleles Among *Arachis* Species and AMOVA

The number of unique alleles is an effective indicator of the genetic differentiation within a population [22]. Species-specific alleles were detected based on amplification events in different species to assess genetic differentiation. In total,129 unique alleles in different species were detected using 75 SSR markers (Table 1). This suggests that the wild *Arachis* accessions have high genetic diversity. Among these 75 SSR markers, 28 markers amplified multiple unique species-specific alleles, 25 markers were able to amplify alleles across all the 19 wild species. Different amplification events using the same SSR marker in different species reflected the distinct mutational histories of multiple alleles from the same microsatellite region. Different numbers of unique alleles were amplified in different species. *A. rigonii* exhibited the largest number of unique alleles (75), which suggested that *A. rigonii* has a high level of differentiation compared to other species. Several species, including *A. chacoense*, *A. helodes*, *A. kuhlmannii*, *A. cryptopotamica*, *A. appressipila* and *A. chiquitana*, did not exhibit any species-specific alleles, which may be because only a single accession was studied for each of these species.

AMOVA was performed to further investigate genetic differentiation among *Arachis* sections and species (Table 2). Approximately 7.34% ($P<0.001$) of the total molecular variation was attributed to genetic differentiation between the sections; 27.51% ($P<0.001$) was attributed to genetic differentiation among species within sections. This indicates that genus *Arachis* possesses wide diversity within both sections and species.

The Alleles of Cultivated Peanut and its Wild Relatives

Determining the differences in genetic structure between *A. hypogaea* and its diploid and tetraploid wild relatives is necessary to understand the evolution of the cultivated peanut. Because the probable B-genome diploid wild progenitor *A. ipaënsis* was not included in this study, only *A. duranensis* (AA), *A. monticola* (AABB) and *A. hypogaea* (AABB) were considered in the analysis. In the comparison of *A. hypogaea* with its wild relatives, six evolutionary modes were described for microsatellite alleles during the domestication of cultivated peanut (Table 3): (i) Five alleles (0.66%) were amplified in *A. hypogaea* but not in *A. duranensis* and *A. monticola*, suggesting that these microsatellite regions emerged after the domestication of *A. hypogaea*; (ii) Eight alleles (1.05%) were amplified in *A. monticola* but not in *A. duranensis* and *A. hypogaea*. These alleles most likely emerged after the polyploidization event due to genomic rearrangements but were lost after the domestication event; (iii) Two hundred and ten alleles (27.63%) were amplified in *A. duranensis* but not in *A. monticola* and *A. hypogaea*. These alleles most likely emerged during the speciation of *A. duranensis* but were lost after the polyploidization event; (iv) Two hundred and thirty-three alleles (30.66%) were amplified in *A. duranensis* and *A. monticola* but not in *A. hypogaea*. These alleles most likely emerged during the speciation of *A. duranensis* but were lost in the domestication event; (v) Sixteen alleles (2.11%) were amplified in *A. duranensis* and *A. hypogaea* but not in *A. monticola*, indicating that the alleles most likely emerged independently in the speciation of *A. duranensis* and during the domestication of cultivated peanut;

Table 1. Summary of genetic diversity and unique alleles among different *Arachis* species.

	Section	Species	N	Allele No.	Allele No./marker	Shannon-Weaver diversity index	Unique alleles
Wild *Arachis*	*Arachis*	*A. batizocoi*	4	558	4.10	0.29	3
		A. cardenasii	3	450	3.31	0.22	2
		A. chacoense	2	413	3.04	0.12	0
		A. correntina	3	451	3.32	0.17	1
		A. duranensis	23	750	5.51	0.35	12
		A. helodes	1	329	2.42	0.00	0
		A. hoehnei	2	421	3.10	0.17	2
		A. kuhlmannii	1	282	2.07	0.00	0
		A. monticola	6	532	3.91	0.20	3
		A. stenosperma	3	449	3.30	0.18	1
		A. villosa	6	517	3.80	0.19	5
	Erectoides	*A. cryptopotamica*	1	331	2.43	0.00	0
		A. oteroi	2	413	3.04	0.11	1
		A. paraguariensis	5	579	4.26	0.27	4
	Heteranthae	*A. pusilla*	2	406	2.99	0.14	4
	Procumbentes	*A. appressipila*	2	418	3.07	0.15	0
		A. chiquitana	1	310	2.28	0.00	0
		A. rigonii	4	535	3.93	0.28	75
	Extranervosae	*A. macedoi*	1	220	1.62	0.00	16
Cultivated groundnut	*Arachis*	*A. hypogaea*	3	310	2.28	0.07	3

(vi) Two hundred and eighty-eight alleles (37.89%) were amplified in *A. duranensis*, *A. monticola* and *A. hypogaea*, indicating that these alleles were highly conserved during the evolution of cultivated peanut (Table 3).

Population Structure and Phylogenic Analyses

The population structure of the 75 *Arachis* accessions, representing 20 species belonging to 5 sections, was analyzed using 136 SSR markers and the model-based software STRUCTURE. The structure analysis was performed by setting the possible number of groups (k), ranging from 1 to 10, with 5 replications for each k. The LnP(D) value increased continuously with the increase of k and peaked at $k = 3$ (Figure S1). Accordingly, the 75 *Arachis* accessions were classified into three clusters. Cluster I comprised 45 accessions from the *Arachis*, *Procumbentes*, *Heteranthae* and *Extranervosae* sections. There were 30 accessions from *Arachis* (11 *A. duranensis*, 3 *A. cardenasii*, 3 *A. correntina*, 3 *A. stenosperma* s, 2 *A. batizocoi*, 2 *A. chacoense*, 2 *A. hoehnei*, 1 *A. helodes*, 1 *A. kuhlmannii*, 1 *A. villosa* and 1 *A. monticola*), 8 accessions from *Erectoides* (5 *A. paraguariensis*, 2 *A. oteroi*, and 1 *A. cryptotamica*), 4 accessions from *Procumbentes* (2 *A. appressipila*, 1 *A. chiquitana* and 1 *A. rigonii*), 2 accessions from *Heteranthae* (*A. pusilla*), and 1 accession from *Extranervosae* (*A. macedoi*). Cluster II consisted of 28 accessions from *Arachis*, including 13 *A. duranensis*, 5 *A. monticola*, 5 *A. villosa*, 2 *A. batizocoi* and 3 *A. hypogaea* accessions. Cluster III contained 3 accessions, WH10026, WH10058, and WH4367, belonging to the species *A. rigonii* in the section *Procumbentes* (Figure 1). The phylogenic analysis also grouped the 75 *Arachis* accessions into 3 clusters (I, II and III) corresponding to the structure analysis with

Table 2. Analysis of molecular variance among sections and species.

Source of variation	DF	Var component	Variation (%)
Among sections	4	10.92	7.34**
Among species within sections	14	40.97	27.51**
Among accessions within species	125	97.07	65.16**
Total	143	148.98	

**P<0.001, for 1000 permutations.

Table 3. Summary of six evolutionary modes of microsatellite alleles amplified in the cultivated groundnut and its wild relatives.

Mode	*A. duranensis*	*A. monticola*	*A. hypogaea*	n	%
i	–	–	+	5	0.66
ii	–	+	–	8	1.05
iii	+	–	–	210	27.63
iv	+	+	–	233	30.66
v	+	–	+	16	2.11
vi	+	+	+	288	37.89

Note: "+" indicates that the SSR alleles are amplified in the specific species, "–" indicates that the SSR alleles aren't amplified in the specific species.

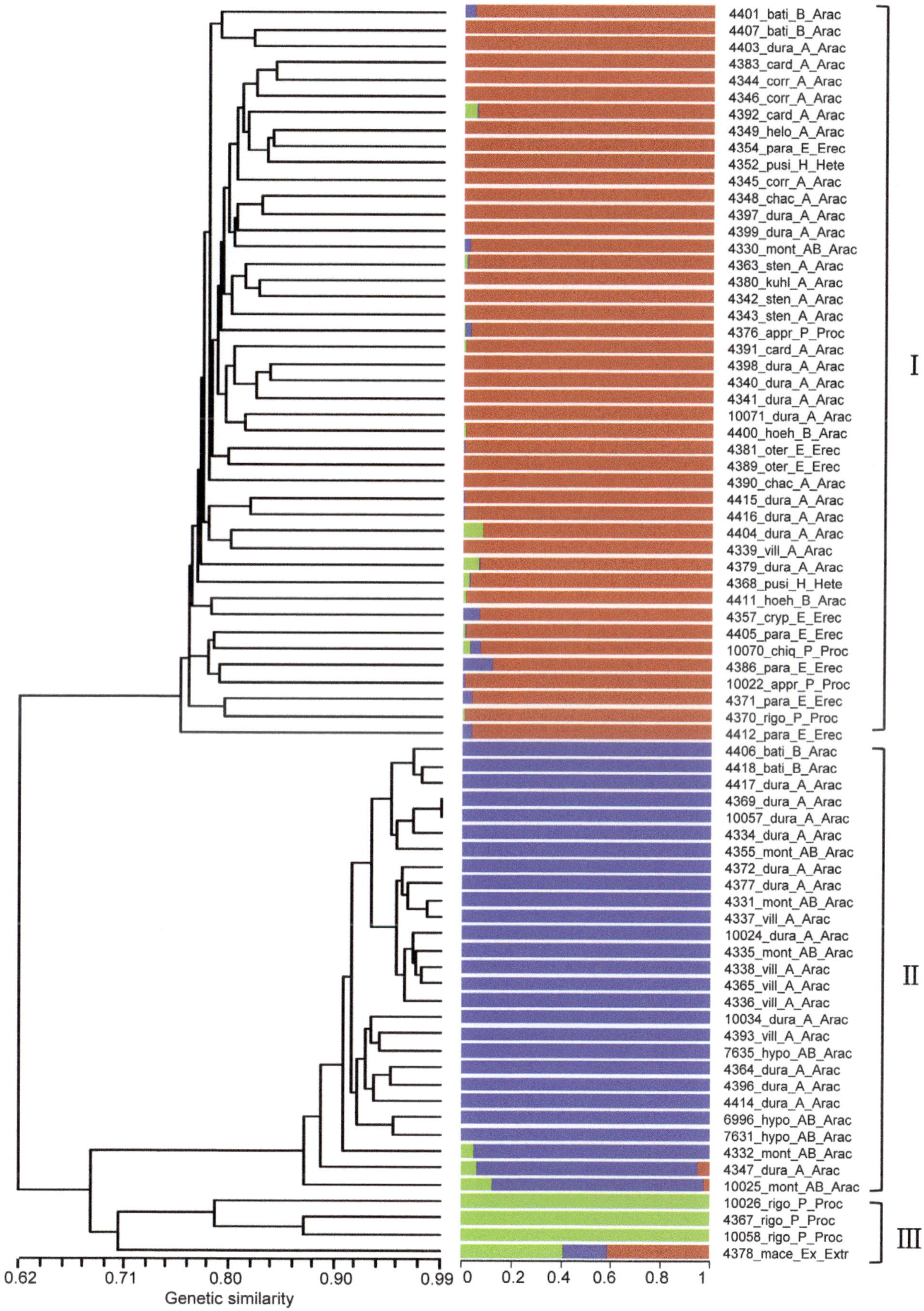
4401_bati_B_Arac
4407_bati_B_Arac
4403_dura_A_Arac
4383_card_A_Arac
4344_corr_A_Arac
4346_corr_A_Arac
4392_card_A_Arac
4349_helo_A_Arac
4354_para_E_Erec
4352_pusi_H_Hete
4345_corr_A_Arac
4348_chac_A_Arac
4397_dura_A_Arac
4399_dura_A_Arac
4330_mont_AB_Arac
4363_sten_A_Arac
4380_kuhl_A_Arac
4342_sten_A_Arac
4343_sten_A_Arac
4376_appr_P_Proc
4391_card_A_Arac
4398_dura_A_Arac
4340_dura_A_Arac
4341_dura_A_Arac
10071_dura_A_Arac
4400_hoeh_B_Arac
4381_oter_E_Erec
4389_oter_E_Erec
4390_chac_A_Arac
4415_dura_A_Arac
4416_dura_A_Arac
4404_dura_A_Arac
4339_vill_A_Arac
4379_dura_A_Arac
4368_pusi_H_Hete
4411_hoeh_B_Arac
4357_cryp_E_Erec
4405_para_E_Erec
10070_chiq_P_Proc
4386_para_E_Erec
10022_appr_P_Proc
4371_para_E_Erec
4370_rigo_P_Proc
4412_para_E_Erec
4406_bati_B_Arac
4418_bati_B_Arac
4417_dura_A_Arac
4369_dura_A_Arac
10057_dura_A_Arac
4334_dura_A_Arac
4355_mont_AB_Arac
4372_dura_A_Arac
4377_dura_A_Arac
4331_mont_AB_Arac
4337_vill_A_Arac
10024_dura_A_Arac
4335_mont_AB_Arac
4338_vill_A_Arac
4365_vill_A_Arac
4336_vill_A_Arac
10034_dura_A_Arac
4393_vill_A_Arac
7635_hypo_AB_Arac
4364_dura_A_Arac
4396_dura_A_Arac
4414_dura_A_Arac
6996_hypo_AB_Arac
7631_hypo_AB_Arac
4332_mont_AB_Arac
4347_dura_A_Arac
10025_mont_AB_Arac
10026_rigo_P_Proc
4367_rigo_P_Proc
10058_rigo_P_Proc
4378_mace_Ex_Extr
I
II
III
0.62
0.71
0.80
0.90
0.99
Genetic similarity
0
0.2
0.4
0.6
0.8
1

Figure 1. Dendrogram and population structure of 75 wild and cultivated *Arachis* accessions. The 75 *Arachis* accessions were classified into three clusters by structure analysis, I, II, and III, basically corresponding to the phylogenic dendrogram. Red, blue and green corresponds to cluster I, II, and III, respectively. The proportion of each color of the horizontal bar represents the assignment possibilities to the specific cluster. The names of accessions and taxonomical information are given next to the horizontal bars, starting with the accession number followed by an abbreviated form of species name followed by respective genomes and sections. (Abbreviated species names: appr: *A. appressipila;* bati: *A. batizocoi;* card: *A. cardenasii;* chac: *A. chacoense;* chiq: *A. chiquitana;* corr: *A. correntina;* cryp: *A. cryptopotamica;* dura: *A. duranensis;* helo: *A. helodes;* hoeh: *A. hoehnei;* hypo: *A. hypogaea;* kuhl: *A. kuhlmannii;* mace: *A. macedoi;* mont: *A. monticola;* oter: *A. oteroi;* para: *A. paraguariensis;* pusi: *A. pusilla;* rigo: *A. rigonii;* sten: *A. stenosperma;* vill: *A. villosa;* Abbreviated section names: Arac: *Arachis;* Hete: *Heteranthae;* Proc: *Procumbentes;* Erec: *Erectoides;* Extr: *Extranervosae*).

only one exception; WH4378 was assigned to cluster I in the structure analysis but classified in cluster III in the phylogenic analysis (Figure 1).

The dendrogram for sections and species of *Arachis* was constructed based on Nei's distance to analyze pair-wise relationships. *Arachis* and *Procumbentes* were clustered together with a genetic distance of 0.054. *Extranervosae* stood alone, showing a relatively large distance from other sections (*Procumbentes*: 0.162, *Arachis*: 0.164, *Erectoides*: 0.206, *Heteranthae*: 0.245) (Figure 2A). *A. macedoi* (in section *Extranervosae*) was highly differentiated from other species and had the largest distance from *A. cryptopotamica* (0.315; Figure 2B). *A. villosa* and *A. monticola* were the most closely related, with the shortest distance between two species (0.017). Cultivated peanut accessions were clustered with the probable diploid progenitor (*A. duranensis*, AA, distance as 0.079) and tetraploid progenitor (*A. monticola*, AABB, distance as 0.043), as well as with two other wild species from the sections *Arachis*, *A. batizocoi* (BB, distance: 0.099) and *A. villosa* (AA, 0.044) (Figure 2B).

Oil Content Variation in Wild *Arachis*

The oil content of the 72 wild *Arachis* accessions was tested over three years, from 2008 to 2010. A significant phenotypic variation of oil content was observed in the 72 wild *Arachis* accessions, ranging from 51.44% to 62.90%, 51.40% to 62.79% and 54.22% to 63.34% in 2008, 2009 and 2010, respectively (Figure 3A). The oil content showed a normal distribution in the 72 wild *Arachis* accessions, and the phenotypic distributions were similar in the three years. The peak oil content distribution in 2010 was slightly higher than those in 2008 and 2009 (Figure 3A) due to the weather. Among the 72 wild *Arachis* accessions, most lines had oil contents ranging from 55% to 58%, and an *A. rigonii* accession (WH10026) had the highest oil contents (62.90%, 62.79% and 61.10%) in all three years. In addition, 7 accessions (WH4347, WH4377, WH10034, WH4330, WH10025, WH4376, and WH4367) not only had oil contents of more than 57% in each season but also had coefficients of variation lower than 0.01.

The average oil content of the wild *Arachis* accessions was compared to assess the influence of species differentiation on trait performance. The accessions from section *Procumbentes* had an average oil content of 57.65% (ranging from 54.31% to 62.26%), higher than that in other sections (Figure 3B). Among the 19 wild *Arachis* species, *A. rigonii* showed the highest average oil content, at 58.62%. *A. kuhlmannii* had the lowest average oil content at 54.56% (Figure 3C). Of the 19 wild species evaluated, 7 had higher oil content than the average of 72 wild *Arachis* accessions (56.69%): *A. chacoense* (56.70%), *A. monticola* (57.57%), *A. villosa* (57.75%), *A. cryptopotamica* (56.69%), *A. oteroi* (57.18%), *A. chiquitana* (56.70%), and *A. rigonii* (58.62%) (Figure 3C).

Association Analysis of Oil Content in *A. duranensis*

Based on the *Q* model, an association analysis of oil content was performed to determine which SSR alleles of the 23 *A. duranensis* accessions (AA) were most highly correlated with oil content (Table 4). A total of nine alleles were found to be significantly associated with oil content ($P<0.01$). Four alleles, with sizes of 320 bp, 330 bp, 340 bp and 350 bp, were amplified from the marker PM204. These were significantly associated with oil content, and the 320 bp allele accounted for the largest phenotypic variation in oil content (31.67%; $P=5.43\times10^{-4}$).

The accessions were grouped based on the presence or absence of associated alleles to assess the combined effect of the associated alleles on oil content (Table 4). Among the nine associated alleles, five were linked to increased oil content. The 330 bp allele amplified from the marker PM204 correlated with the highest increase in phenotypic oil content, likely enhancing the oil content by approximately 1.76%. The remaining four alleles were linked to decreases in oil content. The 270 bp allele amplified from marker 2A5 correlated with the largest decrease in oil content, likely decreasing the oil content by 0.88%. Out of the nine alleles associated with oil content, six alleles were shared between *A. hypogaea* and its wild relatives; four of six alleles were linked to decreased oil content and two alleles were linked with increased oil content. The remaining three alleles were exclusively amplified in wild relatives but not in *A. hypogaea,* and all were linked to increased oil content (Table 4).

Discussion

Allele Diversity and Genetic Differentiation among Species and Sections within the Genus *Arachis*

The evaluation of the genetic diversity in *Arachis* germplasm is crucial for the efficient exploitation of the valuable alleles present in wild species during cultivated peanut improvement, which has been demonstrated in several previous studies [23,24,25]. In the past, only a few dozen markers were used to assess genetic diversity. In the present study, we investigated the genetic diversity of 72 wild *Arachis* and 3 *A. hypogaea* accessions using 136 genome wide SSR markers. A total of 944 alleles (6.94 alleles per marker and a Shannon-Weaver diversity index of 0.4130) were detected in the wild *Arachis* species. The number of alleles was higher than that in *A. hypogaea* (Table 1), indicative of the low polymorphism in *A. hypogaea* [25,26]. The large difference of genetic diversity between the cultivated peanut accessions and the wild species can likely be attributed to (a) domestication events that greatly reduced genetic diversity of the cultivated peanut [10], (b) limited natural gene exchange between diploid wild species and cultivated peanut due to genomic rearrangement during the polyploidization event [9,27], and (c) the founder effect in improvement of cultivated peanut caused by the use of few elite breeding lines and little exotic germplasm in breeding programs [11]. This finding illustrates the potential in introgressing genetic diversity from wild *Arachis* to broaden the genetic base of cultivated peanut breeding. *A. duranensis* of the section *Arachis* exhibited the highest diversity among the 19 wild *Arachis* species studied (Table 1), indicating the great potential of its use in cultivated peanut improvement [8].

The presence of alleles that are unique to a specific population is an effective indicator of genetic differentiation in populations [22,28]. In the present study, 75 SSR markers amplified a total of

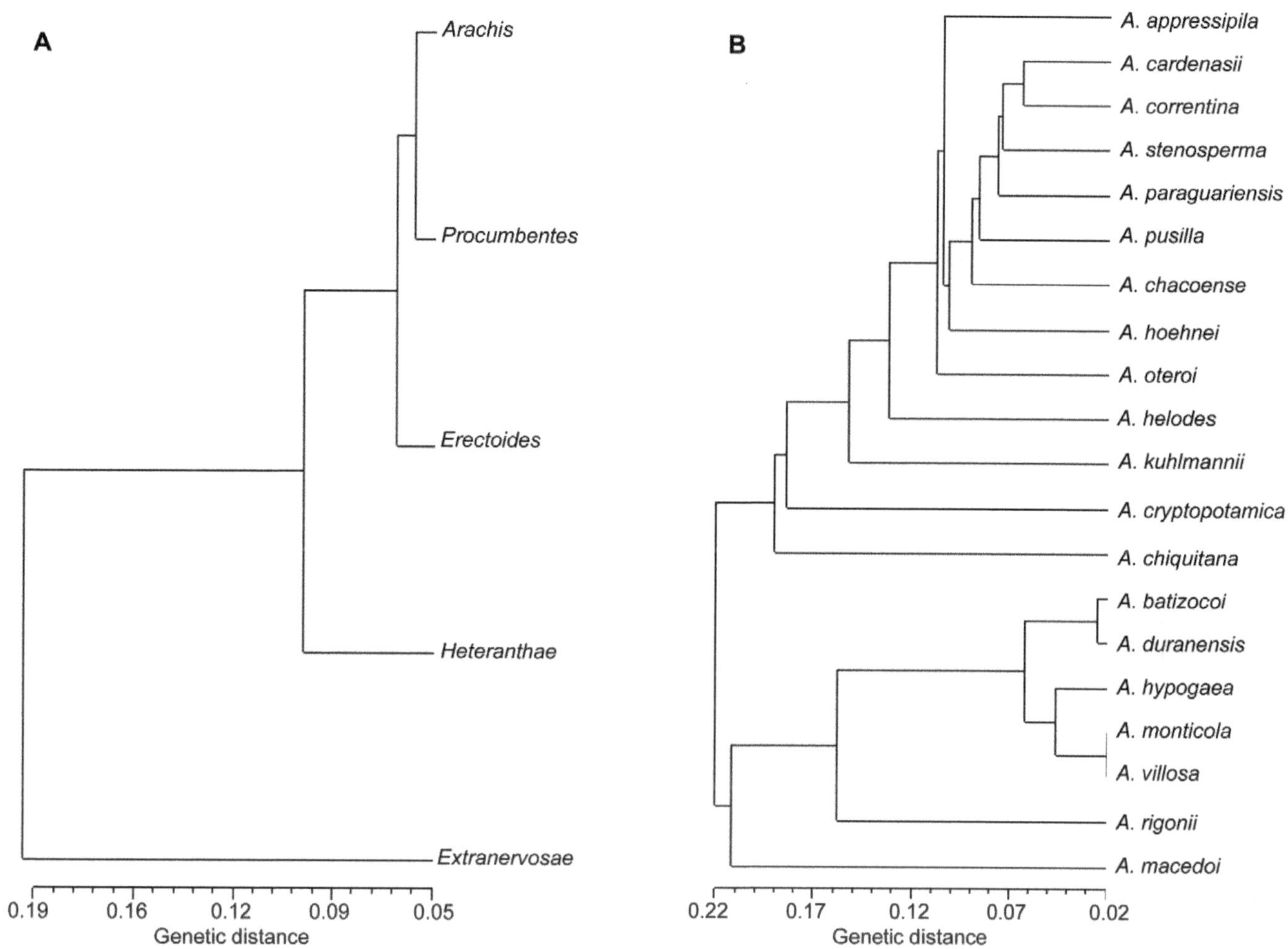

Figure 2. Nei's distance based dendrogram showing the genetic relations among a) sections and b) species of *Arachis*.

129 unique species-specific alleles in the genus *Arachis* (Table 1). Different numbers of unique alleles were observed in different species, reflecting the different levels of speciation and diversity of these species [22]. A significant level of genetic differentiation among sections and species of wild *Arachis* ($P<0.001$, Table 2) was demonstrated by AMOVA, which confirmed the distinct nature of different *Arachis* species. *A. rigonii,* of the section *Procumbentes,* originated in Eastern Bolivia and northern Argentina, the origin center of wild *Arachis* [29], and contained the largest number of unique alleles (75, Table 1). This suggests that the alleles specific to *A. rigonii* are likely native variants that were lost after natural selection or domestication events. It was previously observed that the majority of Arachis species harboring species-specific alleles had originated in Brazil [30]. These unique alleles would be useful markers for efforts to broaden the genetic base of breeding pools and enhance economically important traits, i.e., by introgressing the specific alleles into the cultivated peanut.

Relationships among the 75 *Arachis* Accessions

Arachis species are highly differentiated from each other, and the hybrids between most species are sterile [1]. It is easier to transfer valuable alleles between species that are closely related. Thus, the evaluation of the genetic relationships among various *Arachis* species is necessary for the successful and efficient exploitation of the genetic diversity that exists within this genus. The present study assessed the relationships among 75 accessions representing 20 species from the sections *Arachis*, *Erectoides*, *Extranervosae*, *Heteranthae* and *Procumbentes*. The 75 *Arachis* accessions were generally grouped into 3 clusters corresponding to their species, sections and genome types in a structural analysis, and this grouping correlated nearly perfectly with the classification based on phylogenic analysis (Figure 1). Only one accession (WH4378) was assigned to different clusters in the structural analysis (cluster I) and the phylogenic analysis (cluster III). This could be due to the statistical bias of the model-based structure analysis, as WH4378 was the only accession from section *Extranervosae* in this study. Most accessions of *A. monticola* (AABB) were clustered together with *A. hypogaea* (AABB) (Figure 1), indicating the high similarity between these two tetraploid species. This supports the hypothesis that *A. monticola* is the direct ancestor of *A. hypogaea* [3,4]. The accessions of *A. villosa* (AA) and *A. batizocoi* (BB) were also grouped with the accessions of *A. hypogaea* (Figure 1). This suggests that these two wild diploid species are also closely related to the cultivated peanut. The accessions of *A. diogoi* and *A. batizocoi* showed the least genetic distance from *A. hypogaea* after the presumed diploid progenitors *A. duranensis* and *A. ipaënsis* [30]. This finding indicates that it may be possible to transfer valuable alleles from *A. villosa* and *A. batizocoi* into the cultivated species through direct hybridization. The three *A. rigonii* accessions of section *Procumbentes* exist in a lone cluster (Figure 1), suggesting a high level of

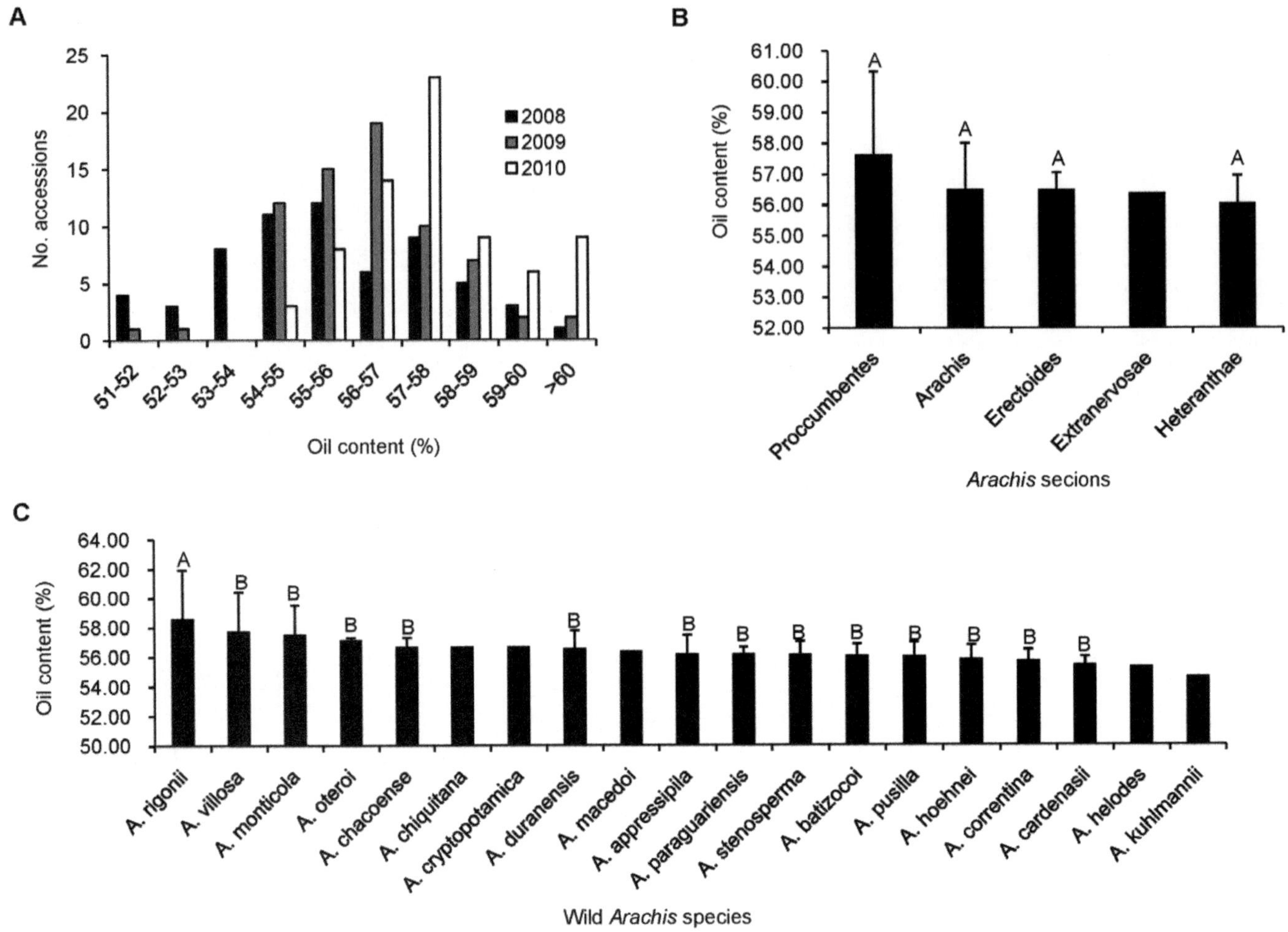

Figure 3. Phenotypic variability of oil content among accessions, species and sections of *Arachis*. a) The distribution of oil content of the 72 wild *Arachis* accessions across three consecutive years from 2008 to 2010; b) Oil content variation among five sections of *Arachis*; c) Oil content variation among twenty species of *Arachis*. The oil content comparisons among sections and species were based on the average value of three years. The capital letters above the bars for the sections and species indicate the significance of difference.

Table 4. Detailed information of SSR marker alleles showing significant associations with oil content in the accessions of *Arachis duranensis*.

SSR primer	Associated alleles (bp)[a]	P value	Variance (%)[b]	Effects[c]	Variety[d]
PM204	320	0.0005429	31.69	−0.18	WH4416
	330	0.0041	23.94	+1.76	WH4377
	340	0.002	26.84	+1.57	WH4377
	350	0.0095	20.29	+1.68	WH4377
2A5	270	0.0027	25.58	−0.85	WH4398
	280	0.0078	21.16	+0.03	WH4396
11H1	570	0.0018	28.87	−0.13	WH4416
3B8	***440***	0.0024	26.09	+0.15	WH10034
TC9E8	680	0.0071	21.58	−0.77	WH4416

[a]Bold and italics are associated SSR alleles exclusively amplified in wild *Arachis* species of section *Arachis*, but not *Arachis hypogaea*.
[b]Percentage of phenotypic variance explained by the associated SSR allele.
[c]Oil content changes of the accessions with the presence of associated alleles relative to the absence of associated alleles. Positive value indicates presence of the associated alleles increase oil content, whereas negative value indicates presence of the associated alleles decrease oil content.
[d]The accessions harboring the associated alleles, which exhibited the highest oil content for the positive effects, and the lowest oil content for the negative effects.

differentiation between *A. rigonii* and other species. This is consistent with the finding that *A. rigonii* had the greatest number of species-specific alleles (Table 2). Furthermore, the remaining *Procumbentes* accessions clustered together with the sections *Arachis*, *Erectoides*, and *Heteranthae*, indicating that the genetic variation within this section may be so high that accessions from the same section cannot be clustered together (Table 3) [30]. This pattern also suggests that the species of different sections dispersed in sympatric habitats are most likely clustered together [29].

Evolutionary Inferences for the Cultivated Peanut

The two *Arachis* species *A. duranensis* (AA) and *A. ipaënsis* (BB) are believed to be the wild diploid progenitors of *A. hypogaea* (AABB), which experienced an interspecific hybridization event followed by chromosome duplication [3,5]. The wild tetraploid *Arachis* species *A. monticola* (AABB) was proposed to be the direct ancestor of *A. hypogaea* because a high level of similarity between the two species was shown in molecular marker and cytological analyses [3,4]. Investigation of the genetic differences between the cultivated peanut and its diploid and tetraploid wild relatives is of great importance for inferring the evolutionary history of cultivated peanut. In the present study, we found six evolutionary modes for the alleles of the cultivated peanut. The majority of alleles were represented by evolutionary modes iii, iv and vi (Table 3), indicating that most alleles emerged during the speciation of the progenitor diploid wild *Arachis* species but were lost after the polyploidization and domestication events. More alleles were associated with mode iv than with mode i (Table 3), suggesting that directional selection and the hitchhiking effect played more important roles than did novel mutations in the domestication of the cultivated peanut [31]. This finding implies that many alleles that exist in wild *Arachis* relatives have been lost in the cultivated peanut because of domestication by ancient breeders. The effective utilization of these ancient alleles in wild relatives will be important the improvement of the cultivated peanut.

The Potential of Wild *Arachis* Germplasm in Enhancing Oil Content in Cultivated Peanut

Wild *Arachis* germplasm has been shown to harbor high-level resistance to foliar and viral diseases, which is not observed in the cultivated peanut [12,15,16]. Some wild *Arachis* materials have been used successfully to develop new peanut varieties that are resistant to pests and disease [19,20,32]. In the present study, the average oil content of 19 wild *Arachis* species was 56.69%; seven species had oil content higher than 56.69% (Figure 3C). This was significantly higher than the oil content of the cultivated peanut germplasm, which averaged 50.76% over 6390 accessions [33]. These results suggest that the wild *Arachis* accessions are great potential resources for enhancing the oil content of the cultivated peanut, in addition to conferring resistance to pest and disease.

Association analysis is an efficient approach for identifying alleles that are correlated with target traits [34]. In the present study, nine alleles were found to be significantly associated with oil content ($P<0.01$, Table 4). The SSR marker 2A5 amplified the 270 bp and 280 bp alleles in the wild *Arachis* accessions, which account for 25.58% and 21.16% of the phenotypic variance in oil content, respectively (Table 4). The 240 bp and 250 bp alleles of 2A5 were amplified in the high and low oil content materials of cultivated peanut with a match rate of 88.9% and 95.0%, respectively [35]. These alleles had previously been demonstrated to be associated with oil content, which confirmed the reliability of the association analysis results. These same markers could amplify multiple associated alleles with different effects (Table 4) [35], which may be due to the functional differentiation of orthologous genomic regions among evolutionarily related species [36]. Among the nine associated alleles, the 340 bp and 350 bp alleles from PM204 and the 440 bp allele from 3B8 were amplified exclusively in the accessions of *A. duranensis* and *A. monticola* and not in the accessions of *A. hypogaea* (Table 4). This suggests that these alleles had undergone directional selection or selective hitchhiking that greatly reduced their genetic variability [37]. Three wild species had specific alleles associated with increased oil content (Table 4), and introgression of these alleles into the cultivated peanut may be a useful strategy for improving oil content. The accessions of *A. duranensis* are highly cross-compatible with *A. hypogaea* [1]; two *A. duranensis* accessions, WH4377 and WH10034, are of particular value. WH4377 harbored the 340 bp and 350 bp alleles from PM204, and WH10034 harbored the 440 bp allele from 3B8. Thus, these three alleles correlated with increased oil content could be introgressed into cultivated varieties.

Materials and Methods

Plant Materials and Oil Content Phenotyping

A collection of 72 wild *Arachis* accessions was assembled. These accessions represented 19 species from the sections *Arachis*, *Erectoides*, *Extranervosae*, *Heteranthae* and *Procumbentes*. Three widely-grown peanut cultivars (*A. hypogaea*) of China, Zhonghua8, Zhonghua12 and Baisha1016, were also included for comparison with their wild *Arachis* relatives. Detailed information for the 75 *Arachis* accessions is listed in Table S1.

The 72 wild *Arachis* accessions were planted in the Wild *Arachis* Nursery of the Oil Crops Research Institute of the Chinese Academy of Agricultural Sciences in Wuhan, China. Each accession was planted in a single row, with 10 plants in each row, 10 cm between plants within each row and 30 cm between rows. Oil content was tested with the Soxhlet extraction method using fresh dried mature seeds with intact testa harvested between 2008 to 2010.

SSR Genotyping

Genomic DNA was extracted from young leaves collected from eight to ten plants of each accession using a modified cetyl-trimethyl ammonium bromide (CTAB) method [38]. The integrity and quality of the DNA was evaluated on a 1% agarose gel by comparison with known concentrations of uncut lambda DNA standard.

A total of 136 SSR markers from different resources were used to genotype the 75 *Arachis* accessions. SSR markers with the prefixes pPGPseq, pPGSseq, PM, Ah, IPAHM, PMc, Lec, EM, AC, gi, RN, TC, and AHBGS were obtained from the literature [4,39,40,41,42,43,44,45,46,47]. SSR markers with the prefixes XY and POCR were developed by our laboratory [48]. PCR reactions followed the protocol described by Chen et al. [38]. PCR products were visualized on a 6% polyacrylamide gel followed by silver staining. The fragment sizes of the PCR products were estimated by comparison with a 50 bp DNA ladder. Each polymorphic fragment was scored as '1' or '0' according to the presence or absence of amplification.

Statistical Analysis

Genetic and oil content variation. The total number of alleles, number of alleles per marker, and the Shannon-Weaver diversity index were analyzed using the PopGene 1.32 program [49] to evaluate the genetic variability within the wild *Arachis* sections and species and cultivated peanut. Unique alleles were used to evaluate the genetic variants exclusive to the specific section or species. The phenotypic variation of the oil content in the 72 wild *Arachis* accessions and the difference in oil content among sections and species were analyzed using SAS 8.02 [50].

AMOVA. To investigate the genetic differentiation among the 72 wild *Arachis* accessions, analysis of molecular variance (AMOVA) was performed using the software package Arlequin 3.1 [51] with 1,000 permutations and the sum of squared size differences as the molecular distance.

Population structure and phylogenic analyses. The population structure of the 75 *Arachis* accessions was analyzed using 136 SSR markers and the program STRUCTURE 2.2 [52]. Five independent simulations were performed for values of k (the number of groups) ranging from 1 to 10. For each simulation, 10,000 iterations before a burn-in length of 10,000 MCMC (Markov Chain Monte Carlo) replications were conducted with the admixture and related frequency models. The optimal k value was determined by the posterior probability [LnP(D)]. Accessions were assigned to a corresponding group based on their maximum membership probabilities, as described by Remington et al. [53]. A UPGMA dendrogram was constructed to better depict the relationship among the 75 *Arachis* accessions using NTSYS 2.0 [54]. Based on Nei's (1973) genetic distance [55], a dendrogram of the sections and species of *Arachis* was also constructed using PowerMarker 3.51 [56].

Association analysis of oil content. A complex genetic structure may lead to a high possibility of false positives in association mapping [57]. Because wild *Arachis* species are highly differentiated [1,2], 23 accessions of *A. duranensis* were selected to perform an association analysis to investigate SSR alleles highly correlated with oil content. The mean values of oil content over three consecutive years rather than values in a single year were used to increase the statistical power of the association mapping. The Q model, controlling the population structure of 23 *A. duranensis* accessions derived from the structure analysis, was used to perform the GLM procedure in the package TASSEL 2.1 [58]. The SSR allele is significantly associated with oil content when the P value is less than 0.01. The R^2 value indicates the percentage of phenotypic variance explained by the associated allele. Allele effects on oil content were estimated based on the difference in the oil content of accessions with the allele compared to accessions without the allele. A positive value indicates that the associated allele increases oil content, whereas a negative value indicates the associated allele decreases oil content.

Supporting Information

Figure S1 Estimation of LnP(D) in the 75 accessions of *Arachis*. The bar indicates standard deviation.

Table S1 List of 75 *Arachis* accessions representing 20 species and 5 sections.

Author Contributions

Conceived and designed the experiments: HJ LH BL. Performed the experiments: XR YC XZ MT JH. Analyzed the data: LH YX. Contributed reagents/materials/analysis tools: HJ YX. Wrote the paper: HJ LH BL HDU.

References

1. Krapovickas A, Gregory WC (1994) Taxonomía del género *Arachis* (Leguminosae). Bonplandia 8: 1–186.
2. Valls JFM, Simpson CE (2005) New species of *Arachis* (leguminosae) from Brazil, Paraguay and Bolivia. Bonplandia 14: 35–63.
3. Kochert G, Stalker HT, Gimenes M, Galgaro L, Lopes CR, et al. (1996) RFLP and cytogenetic evidence on the origin and evolution of allotetraploid domesticated peanut, *Arachis hypogaea* (Leguminosae). Am J Bot 83: 1282–1291.
4. Moretzsohn M, Hopkins M, Mitchell S, Kresovich S, Valls J, et al. (2004) Genetic diversity of peanut (*Arachis hypogaea* L.) and its wild relatives based on the analysis of hypervariable regions of the genome. BMC Plant Biol 4: 11.
5. Seijo JG, Lavia GI, Fernández A, Krapovickas A, Ducasse D, et al. (2004) Physical mapping of the 5S and 18S–25S rRNA genes by FISH as evidence that *Arachis duranensis* and *A. ipaensis* are the wild diploid progenitors of *A. hypogaea* (Leguminosae). Am J Bot 91: 1294–1303.
6. Smartt J, Stalker H (1982) Speciation and cytogenetics in *Arachis*. In: Pattee HE, Young CT (eds) Peanut Science and Technology, American Peanut Research and Education Society, Yoakum, TX. 21–49.
7. Gregory WC, Krapovickas A, Gregory MP, Summerfield RJ, Bunting AH (1980) Structure, variation, evolution, and classification in *Arachis*. In: Advances in legume science Edited by: Summerfield RJ, Bunting AH Kew, England: Royal Botanical Gardens; 1980: 469–481.
8. Gregory MP, Gregory WC (1979) Exotic germplasm of *Arachis* L. interspecific hybrids. J Heredity 70: 185–193.
9. Soltis DE, Soltis PS (1999) Polyploidy: recurrent formation and genome evolution. Trends Ecol Evol 14: 348–352.
10. Halward TM, Stalker HT, Larue EA, Kochert G (1991) Genetic variation detectable with molecular markers among unadapted germplasm resources of cultivated peanut and related wild species. Genome 34: 1013–1020.
11. Upadhyaya HD, Gowda C, Buhariwalla H, Crouch J (2006) Efficient use of crop germplasm resources: identifying useful germplasm for crop improvement through core and mini-core collections and molecular marker approaches. Plant Genet Resour 4: 25–35.
12. Subrahmanyam P, Ghanekar AM, Nolt BL (1985) Resistance to groundnut disease in wild *Arachis* species. In Proceedings of the International Workshop on Cytogenetics of Arachis Edited by: Moss JP, Patancheru: ICRISAT; 1985: 49–55.
13. Xu Z, Chen K, Zhang Z (1987) Evaluation for PStV resistance in wild *Arachis* accessions. Chin Oil Crop Sci: 91–93.
14. Starr J, Schuster G, Simpson C (1990) Characterization of the resistance to *Meloidogyne arenaria* in an interspecific *Arachis* spp. Hybrid. Peanut Sci 17: 106–108.
15. Ouedraogo M, Smith O, Simpson C, Smith D (1994) Early and late leaf spot resistance and agronomic performance of nineteen interspecific derived peanut lines. Peanut Sci 21: 99–104.
16. Pande S, Rao J (2001) Resistance of wild *Arachis* species to late leaf spot and rust in greenhouse trials. Plant Disease 85: 851–855.
17. Chen B, Jiang H, Ren X, Liao B, Huang J (2008) Identification and molecular traits of *Arachis* species with resistance to bacterial wilt. Acta Agri Boreali-Sinica 23: 170–175.
18. Johnson D, Wynne J, Campbell W (1977) Resistance of wild species of *Arachis* to the twospotted spider mite, *Tetranychus Urticae* 1, 2, 3. Peanut Sci 4: 9–11.
19. Isleib T, Holbrook C, Gorbet D (2001) Use of plant introductions in peanut cultivar development. Peanut Sci 28: 96–113.
20. Mallikarjuna N, Senthilvel S, Hoisington D (2011) Development of new sources of tetraploid *Arachis* to broaden the genetic base of cultivated groundnut (*Arachis hypogaea* L.). Genet Resour Crop Evol 58: 889–907.
21. Upadhyaya HD, Dwiveid SL, Nadaf HL, Singh S (2011) Phenotyic diversity and identification of wild *Arachis* accessions with useful agronomic and nutritional traits. Euphytica 182: 103–115.
22. Chen S, Nelson MN, Ghamkhar K, Fu T, Cowling WA (2008) Divergent patterns of allelic diversity from similar origins: the case of oilseed rape (*Brassica napus* L.) in China and Australia. Genome 51: 1–10.
23. Angelici C, Hoshino AA, Nóbile PM, Palmieri DA, Valls JFM, et al. (2008) Genetic diversity in section *Rhizomatosae* of the genus *Arachis* (Fabaceae) based on microsatellite markers. Genet Mol Biol 31: 79–88.
24. Barkley NA, Dean RE, Pittman RN, Wang ML, Holbrook CC, et al. (2007) Genetic diversity of cultivated and wild-type peanuts evaluated with M13-tailed SSR markers and sequencing. Genet Res 89: 93–106.
25. Gimenes MA, Lopes CR, Valls JFM (2002) Genetic relationships among *Arachis* species based on AFLP. Genet Mol Biol 25: 349–354.
26. Kochert G, Halward T, Branch W, Simpson C (1991) RFLP variability in peanut (*Arachis hypogaea* L.) cultivars and wild species. Theor Appl Genet 81: 565–570.
27. Young ND, Weeden NF, Kochert G (1996) Genome mapping in legumes (Family *Fabaceae*) genome mapping in plants Austin, TX: Landes Biomedical Press Paterson. 211–277.
28. Younas M, Xiao Y, Cai D, Yang W, Ye W, et al. (2012) Molecular characterization of oilseed rape accessions collected from multi continents for exploitation of potential heterotic group through SSR markers. Mol Biol Rep 39: 5105–5113.
29. Valls J, Rao VR, Simpson C, Krapovickas A (1985) Current status of collection and conservation of South American groundnut germplasm with emphasis on wild species of *Arachis*. In: Moss JP (ed) Proceedings of an international workshop

on cytogenetics of *Arachis*, 31 Oct-2 Nov 1983. ICRISAT Centre, Patancheru, Andhra Pradesh, India, 15–35.
30. Koppolu R, Upadhyaya HD, Dwivedi SL, Hoisington DA, Varshney RK (2010) Genetic relationships among seven sections of genus *Arachis* studied by using SSR markers. BMC Plant Biol 10: 15.
31. Yamasaki M, Tenaillon MI, Vroh Bi I, Schroeder SG, Sanchez-Villeda H, et al. (2005) A large-scale screen for artificial selection in maize identifies candidate agronomic loci for domestication and crop improvement. Plant Cell 17: 2859.
32. Mallikarjuna N, Sastri D (2002) Morphological, cytological and disease resistance studies of the intersectional hybrid between *Arachis hypogaea* L. and *A. glabrata* Benth. Euphytica 126: 161–167.
33. Jiang H, Ren × (2006) Genetic diversity of peanut resource on morphological characters and seed chemical components in China. Chin Oil Crop Sci 28: 421–426.
34. Hasan M, Friedt W, Pons-Kühnemann J, Freitag N, Link K, et al. (2008) Association of gene-linked SSR markers to seed glucosinolate content in oilseed rape (*Brassica napus* ssp. *napus*). Theor Appl Genet 116: 1035–1049.
35. Huang L, Zhao X, Zhang W, Fan Z, Ren X, et al. (2011) Identification of SSR markers linked to oil content in peanut (*Arachis hypogaea* L.) through RIL population and natural population. Acta Agron Sin 37: 1967–1974.
36. Wang J, Long Y, Wu B, Liu J, Jiang C, et al. (2009) The evolution of *Brassica napus* FLOWERING LOCUST paralogues in the context of inverted chromosomal duplication blocks. BMC Evol Biol 9: 271.
37. Olsen KM, Caicedo AL, Polato N, McClung A, McCouch S, et al. (2006) Selection under domestication: evidence for a sweep in the rice waxy genomic region. Genetics 173: 975.
38. Chen B, Jiang H, Liao B, Ren X, Huang J, et al. (2008) Genetic diversity analysis of *Arachis* germplasm by SSR. J Trop Subtrop Bot 16: 296–303.
39. Ferguson M, Burow M, Schulze S, Bramel P, Paterson A, et al. (2004) Microsatellite identification and characterization in peanut (*A. hypogaea* L.). Theor Appl Genet 108: 1064–1070.
40. He G, Meng R, Newman M, Gao G, Pittman R, et al. (2003) Microsatellites as DNA markers in cultivated peanut (*Arachis hypogaea* L.). BMC Plant Biol 3: 3.
41. He G, Meng R, Gao H, Guo B, Gao G, et al. (2005) Simple sequence repeat markers for botanical varieties of cultivated peanut (*Arachis hypogaea* L.). Euphytica 142: 131–136.
42. Moretzsohn MC, Barbosa AVG, Alves-Freitas DMT, Teixeira C, Leal-Bertioli SCM, et al. (2009) A linkage map for the B-genome of *Arachis* (*Fabaceae*) and its synteny to the A-genome. BMC Plant Biol 9: 40.
43. Proite K, Leal-Bertioli SCM, Bertioli DJ, Moretzsohn MC, Da Silva FR, et al. (2007) ESTs from a wild *Arachis* species for gene discovery and marker development. BMC Plant Biol 7: 7.
44. Gimenes M, Hoshino A, Barbosa A, Palmieri D, Lopes C (2007) Characterization and transferability of microsatellite markers of the cultivated peanut (*Arachis hypogaea*). BMC Plant Biol 7: 9.
45. Liang X, Chen X, Hong Y, Liu H, Zhou G, et al. (2009) Utility of EST-derived SSR in cultivated peanut (*Arachis hypogaea* L.) and *Arachis* wild species. BMC Plant Biol 9: 35.
46. Cuc L, Mace E, Crouch J, Quang V, Long T, et al. (2008) Isolation and characterization of novel microsatellite markers and their application for diversity assessment in cultivated groundnut (*Arachis hypogaea*). BMC Plant Biol 8: 55.
47. Hopkins MS, Casa AM, Wang T, Mitchell SE, Dean RE, et al. (1999) Discovery and characterization of polymorphic simple sequence repeats (SSRs) in peanut. Crop Sci 39: 1243–1247.
48. Tang M, Chen Y, Ren X, Huang L, Zhou X, et al. (2012) Genetic diversity of *Arachis* accessions by EST-SSR from cultivated peanut (*Arachis hypogaea* L.). Acta Agron Sin38: 1221–1231.
49. Yeh F, Boyle T (1997) Population genetic analysis of codominant and dominant markers and quantitative traits. Belg J Bot 129: 157.
50. SAS Institute (1999) SAS/STAT User's Guide, Version 8. SAS Institute, Cary.
51. Excoffier L, Laval G, Schneider S (2005) Arlequin version 3.0: an integrated software package for population genetics data analysis. Evol Bioinform Online 1: 47–50.
52. Pritchard JK, Stephens M, Donnelly P (2000) Inference of population structure using multilocus genotype data. Genetics 155: 945–959.
53. Remington DL, Thornsberry JM, Matsuoka Y, Wilson LM, Whitt SR, et al. (2001) Structure of linkage disequilibrium and phenotypic associations in the maize genome. Proc Natl Acad Sci USA 98: 11479–11484.
54. Rohlf F (2006) NTSYS-PC, numerical taxonomy and multivariate analysis system. Exeter Software, Setauket, N.Y.
55. Nei M (1973) Analysis of gene diversity in subdivided populations. Proc Natl Acad Sci USA 70: 3321–3323.
56. Liu K, Muse SV (2005) PowerMarker: an integrated analysis environment for genetic marker analysis. Bioinformatics 21: 2128–2129.
57. Flint-Garcia SA, Thornsberry JM, Buckler Iv ES (2003) Structure of linkage disequilibrium in plants. Ann Rev Plant Biol 54: 357–374.
58. Bradbury PJ, Zhang Z, Kroon DE, Casstevens TM, Ramdoss Y, et al. (2007) TASSEL: software for association mapping of complex traits in diverse samples. Bioinformatics 23: 2633–2635.

22

Metagenomic Analysis of the Microbiota from the Crop of an Invasive Snail Reveals a Rich Reservoir of Novel Genes

Alexander M. Cardoso[1]*, Janaína J. V. Cavalcante[1], Maurício E. Cantão[2,3], Claudia E. Thompson[2], Roberto B. Flatschart[1], Arnaldo Glogauer[2], Sandra M. N. Scapin[1], Youssef B. Sade[1], Paulo J. M. S. I. Beltrão[1], Alexandra L. Gerber[2], Orlando B. Martins[1,4], Eloi S. Garcia[1,6], Wanderley de Souza[1,5]*, Ana Tereza R. Vasconcelos[2]*

1 Instituto Nacional de Metrologia, Qualidade e Tecnologia, Duque de Caxias, Rio de Janeiro, Brazil, **2** Laboratório Nacional de Computação Científica, Petrópolis, Rio de Janeiro, Brazil, **3** Empresa Brasileira de Pesquisa Agropecuária - CNPSA, Concórdia, Santa Catarina, Brazil, **4** Instituto de Bioquímica Médica, Universidade Federal do Rio de Janeiro, Rio de Janeiro, Rio de Janeiro, Brazil, **5** Instituto de Biofísica Carlos Chagas Filho, Universidade Federal do Rio de Janeiro, Rio de Janeiro, Rio de Janeiro, Brazil, **6** Instituto Oswaldo Cruz, Fundação Oswaldo Cruz, Rio de Janeiro, Rio de Janeiro, Brazil

Abstract

The shortage of petroleum reserves and the increase in CO_2 emissions have raised global concerns and highlighted the importance of adopting sustainable energy sources. Second-generation ethanol made from lignocellulosic materials is considered to be one of the most promising fuels for vehicles. The giant snail *Achatina fulica* is an agricultural pest whose biotechnological potential has been largely untested. Here, the composition of the microbial population within the crop of this invasive land snail, as well as key genes involved in various biochemical pathways, have been explored for the first time. In a high-throughput approach, 318 Mbp of 454-Titanium shotgun metagenomic sequencing data were obtained. The predominant bacterial phylum found was *Proteobacteria*, followed by *Bacteroidetes* and *Firmicutes*. *Viruses*, *Fungi*, and *Archaea* were present to lesser extents. The functional analysis reveals a variety of microbial genes that could assist the host in the degradation of recalcitrant lignocellulose, detoxification of xenobiotics, and synthesis of essential amino acids and vitamins, contributing to the adaptability and wide-ranging diet of this snail. More than 2,700 genes encoding glycoside hydrolase (GH) domains and carbohydrate-binding modules were detected. When we compared GH profiles, we found an abundance of sequences coding for oligosaccharide-degrading enzymes (36%), very similar to those from wallabies and giant pandas, as well as many novel cellulase and hemicellulase coding sequences, which points to this model as a remarkable potential source of enzymes for the biofuel industry. Furthermore, this work is a major step toward the understanding of the unique genetic profile of the land snail holobiont.

Editor: Olle Terenius, Swedish University of Agricultural Sciences, Sweden

Funding: This work was supported by Fundação Carlos Chagas Filho de Amparo à Pesquisa do Estado do Rio de Janeiro (FAPERJ) and Conselho Nacional de Desenvolvimento Científico e Tecnológico (CNPq). The funders had no role in study design, data collection and analysis, decision to publish, or preparation of the manuscript.

Competing Interests: The authors have declared that no competing interests exist.

* E-mail: amcardoso@bioqmed.ufrj.br (AMC); wsouza@biof.ufrj.br (WdS); atrv@lncc.br (ATRV)

Introduction

The use of renewable energy sources in the global energy matrix is important for two main reasons. Firstly, our fossil fuel reserves are diminishing, and the lack of readily available substitutes may result in an energy shortage in the not-so-distant future. Secondly, burning of fossil fuels release large amounts of greenhouse gases (carbon dioxide, nitrogen oxides, sulfur hexafluoride, hydrofluorocarbons and perfluorocarbons) aggravating global warming [1–2].

Brazil is a world leader in renewable energy. Here, fuel ethanol has been produced from sugarcane since the 1970s. This was an important step toward the use of sustainable fuels for vehicles with Otto-cycle engines. By law, anhydrous ethanol has also been added to the gasoline sold in gas stations at a level of at least 20% (E20). Other countries have adopted a similar strategy [1,3]. Despite these mandates, the competitiveness of ethanol versus gasoline on the global market is still low, and more efficient production will be important to ensure its widespread acceptance.

In traditional ethanol production, sugarcane processing results in solid waste called bagasse, which is composed mainly of lignocellulose. Although it can be burned in industrial plants for heating purposes, a large fraction remains unused. Bagasse has a large energy potential, up to 19 MJ/kg [4], but most organisms cannot readily ferment cellulose and lignin. Improved methods to degrade bagasse and other agriculture residues into fermentable sugars would reduce waste and supply the refinery with more sources to produce ethanol without increasing the cultivation area and accompanying costs. Some organisms produce enzymes that can efficiently degrade cellulose. Their identification, improvement, and production for industrial application may contribute to making second-generation ethanol a competitive commodity [5–6]. The relevance of the search for new lignocellulolytic enzymes and microorganisms is demonstrated by the increasing number of projects focusing on the symbiotic and mutualistic microbiota from the digestive tracts of animals such as termite (*Nasutitermes* sp.), cow (*Bos taurus*), wallaby (*Macropus eugenii*), and giant panda (*Ailuropoda melanoleuca*) [7–11].

Throughout evolution, the digestive systems of animals adapted mechanical, physiological, and chemical methods to improve use of the available diet. In herbivores, enzymes play a vital role in the degradation of raw vegetable fiber, converting it to small units. The resulting oligomers and monomers become available for absorption by the gut epithelium or support the growth of resident microbes, which also produce enzymes that take part in the digestive process or contribute to the supply of nutritional and defense factors essential to the host organism's health [12].

The giant land snail *Achatina fulica* is a pulmonate mollusk native to sub-Saharan Africa. Due to human interference, prolificacy, physiological adaptability, and ability to eat a broad range of plant species (Table S1), *Achatina* has reached a near-global distribution and can be found close to cities, roads and farms. Because of its voracious appetite and the speed with which it spreads, it is now considered to be the most destructive terrestrial gastropod worldwide [13–14]. Despite that, there are surprisingly few studies focusing on this easy-to-handle snail.

The first studies on animal-produced cellulolytic enzymes were performed on snails (*Helix sp.*) at the end of the 19th century [15]. In addition to their own enzymatic repertoire, land snails retain a microbiota that specializes in the rapid hydrolysis of lignocellulosic plant biomass, which contributes to their extraordinary digestive efficiency (up to 80%) [16]. However, the composition and ecological role of the microbiota of land snails, including *A. fulica*, remain largely unexplored.

Unlike other strategies for novel enzyme and microorganism identification, metagenomic analysis offers some clear benefits as a powerful alternative to culture-dependent methods. With the increase in throughput and the development of fast searching algorithms, metagenomic analysis is becoming easier and more efficient. The resulting data may also be continuously submitted to further prospection and evaluation. Thus, metagenomics improves the access to the full genetic potential of a given environment [17].

Here, for the first time we investigated the microbial enzymes within the crop of the giant snail, focusing on those enzymes that are potentially relevant to the degradation of plant biomass into fermentable sugars, using high-throughput sequencing and bioinformatic approaches. Comparison of the carbohydrate active enzymes (CAZy) in the snail crop to a diversity of other biomes including arthropoda, herbivores, omnivores, soil, and aquatic metagenomes using hierarchical clustering revealed the snail's similarity to the insect pest beetle *Dendroctonus ponderosae*, which indicates similar metabolic capabilities. Furthermore, this work provides insights into the land snail microbiome.

Materials and Methods

Sample Collection

Seven wild adult *Achatina fulica* snails were collected by hand in the early morning during the rainy season in March 2008. The snails were collected from a habitat with low evidence of impact from human activities within the city of Rio de Janeiro (22°49′18.83′S; 43°31′30.01′W) in a place covered by humid tropical evergreen broadleaf forest vegetation, the so-called Mata Atlântica [18]. The snails were washed with sterile distilled water and then transported to the laboratory. The crop fluid was collected by cannulation of the mouth-esophagus with a #24 needle-less scalp vein set attached to a syringe by a Luer lock connector [19].

DNA Extraction

For community metagenome sequencing, total DNA was isolated according to a protocol specific for high molecular weight DNA extraction from environmental samples [20]. Briefly, the pooled crop fluid (15 ml) was filtered first through a 3.0-μm filter and then through a 0.2-μm Sterivex (Millipore) filter to separate free-living microbes from larger organisms, viruses and particles. Total nucleic acid was isolated from the Sterivex by cell lysis with Proteinase K and SDS followed by phenol-chloroform extraction. In addition, the non-filtered crop fluid microbiotic DNA was extracted using the Power Soil DNA Extraction Kit (MoBio), according to the manufacturer's recommended protocol. DNA concentration, purity, and the overall integrity were checked on a Nanodrop (Thermo Scientific) and by agarose gel electrophoresis.

Pyrosequencing and Sequence Processing Workflow

Two libraries were prepared by following the instructions from the Rapid Library Preparation Method Manual - GS FLX Titanium Series (454-Roche): one from 500 ng of DNA extracted from the snails' filtered digestive juice (crop fluid) and another from 500 ng of DNA extracted from non-filtered juice. Titration, emulsion PCR, and sequencing steps were performed according to the manufacturer's protocol. A two-region 454 sequencing run was performed on a 70×75 PicoTiterPlate (PTP) using the Genome Sequencer FLX System (Roche). Each region was loaded with one of the library preparations. Before sequence analysis and assembly, artificially replicated sequences that are generated as an inherent artifact of 454-based pyrosequencing were identified and removed using the Replicates software [21]. The remaining reads were further filtered to remove short sequences (fewer than 180 bp) or sequences with Phred quality ≤20 using the LUCY program [22].

Taxonomic Distribution and Functional Analysis

MEGAN4 [23] was used to analyze the taxonomic distribution of each read hit of the BLASTX algorithm [24] against the NCBI-NR protein database. Meta-RNA software [25] was employed to perform 16S rRNA identification and the sequences were classified with RDP Classifier software based on the RDP Database [26]. We analyzed the functional composition of the assembled metagenome data by comparison with the SEED [27] and COG databases [28] available on the MG-RAST server [29] using an e-value ≤1e−5 as a cutoff. Functional comparison against public metagenomes was also performed. To identify statistically significant and biologically meaningful differences between the snail crop and other microbiomes, we employed the two-way Fisher's exact test with a Benjamin-Hochberg FDR multiple test correction within the Statistical Analysis of Metagenomic Profiles, STAMP [30]. Principal coordinates analysis (PCoA) and Rarefaction curve of annotated species richness were calculated in MG-RAST [28].

Further functional analysis was executed using the KEGG database [31] using MEGAN4 software [23]. The functional annotation and analysis of all contig sequences were also performed using the System for Automated Bacterial Integrated Annotation, SABIA [32]. An automatic annotation was performed using the following criteria: ORFs with BlastP hits on databases KEGG, NCBI-nr, or UniProtKB/Swiss-Prot, subject and query coverage ≥60%, and positives ≥60% were assigned "valid". ORFs with no BlastP hits found on databases NCBI-nr, KEGG,

UniProtKB/Swiss-Prot, TCDB, and Interpro or not included in the criteria above were assigned as "hypothetical".

Carbohydrate-active enzyme domains assigned by the CAZy database [33] were searched in predicted protein sequences using HMMER 3.0 software [34] against the Pfam 26.0 database [35] using a cutoff of 1e−4. To compare the snail crop CAZyome to other sequenced metagenomes, a comparative analysis was performed using the carbohydrate enzyme content as a comparative metric [36–37].

Comparative Metric Analysis

Host-associated and environmental metagenomes were downloaded from the IMG/M (Integrated Microbial Genomes with Microbiome Samples) database [38], including metagenomes from plant, arthropod, mammal, water, and soil samples. The number of genes in Pfam (obtained through the Pfam count statistical data downloaded for each metagenome) was calculated using a self-written bash script using a BLASTP e-value of 1e−04. The correspondence between Pfam and CAZy was automatically reached using another self-written bash script, resulting in a matrix with the number of reads found in different CAZy families in each metagenome. A new matrix, with each row corresponding to a metagenome and each column corresponding to a CAZy ID, was obtained. The proportion of each CAZy with respect to the total number of annotated CAZy families in that metagenome was calculated, and the resultant value was directed to the appropriate cell of the matrix. Trends in the CAZy family frequencies were examined in SPSS using Principal Component Analysis (PCA) as well as hierarchical clustering [39].

The main idea of the PCA method is to form a new variable (or variables) from a data set, which should contain as much of the variability of that data set as possible, reducing it to a smaller set of uncorrelated variables that will be used in further analysis. Descriptive statistics, such as the mean and standard deviation, were calculated for all variables (CAZy families). The analysis was performed on standardized data based on a correlation matrix. Consequently, the importance of a variable was not determined by the relative size of its variance. The criteria for the number of components to extract included an eigenvalue greater than 1, which indicates that a principal component explains more of the variance than would be expected by chance. The communalities (amount of variance in a variable accounted by all the components) were computed as well as the share of variance of original data explained by each and all components. Hierarchical clustering using Euclidean distances with Ward linkage was performed on standardized data generated from both the proportion matrix and the principal component values obtained by the PCA. The hierarchical clustering analysis allowed us to identify relatively homogeneous clusters of cases based on measured features. The results are displayed in a dendrogram, a hierarchical tree diagram, which shows the linkage points at increasing levels of dissimilarity.

Results and Discussion

Snail Anatomy

In *A. fulica* (as in many pulmonate gastropods) the food, scraped by the radula and ingested by the buccal mass, is mixed with the secretions of the salivary gland and accumulates in the crop (ingluvius), a distensible muscular compartment (Figure 1). The crop and stomach is filled via two cannaliculi with the juice produced by the digestive gland (also called hepatopancreas or midgut gland). A muscular valve restricts the passage of the crop contents to the stomach [40]. The medial part of the gut is surrounded by the digestive gland, which secretes more enzymes into the midgut lumen and also absorbs nutrients. The epithelium of the digestive tube is ciliated along almost its entire length, allowing the food to mix with digestive juices and helping transport the alimentary mass. The ciliated epithelia also allow the microbial flora to anchor on the digestive tube [41]. The unabsorbed part of the alimentary mass (bolus) is compacted and passed directly into the rectum.

The snail's digestion is primarily extracellular [40]. Mapping reducing sugar-producing activity against carboxymethyl cellulose (CMC), avicel, and micronized bagasse [42] along each morpho-functional section of the digestive tube revealed that the crop and the digestive gland have the highest net and specific enzyme activity. To assess the taxonomic, genetic, and functional potential of the microorganisms within *A. fulica*'s crop, a massive random shotgun sequencing effort was performed.

Overview of the *A. fulica* Metagenome

A total of 1,297,598 reads were produced after raw data processing on 454-Titanium pyrosequencer. The Replicates package identified and removed 188,709 reads as replicas. From the remaining 1,108,889 reads, LUCY removed 198,922 low-quality reads, allowing the use of 909,967 sequences for further taxonomic and functional categorization (Table S2). From these sequences, 1,671 were identified as 16S rDNA using the MG-RAST server against RDP, which corresponds to 0.18% of the total sequence obtained. Despite the advantages over 16S libraries, which can show biases in amplification (PCR), reads of 16S rRNA obtained from random DNA shotgun sequencing are not indicated for binning [43]. Other research has reported low assignment values when working with analogous methodologies: 0.06% for the Amazon River metagenome [44] and an average of 0.15% for the bovine rumen metagenome [8].

In addition, another group distribution was performed by MEGAN4, which is software based on the NCBI taxonomy database [23]. Approximately 82% (743,954) of the reads were assigned to a taxon. LCA analysis was considered more flexible for taxonomic classification, identifying more sequences related to *Bacteria*, *Archaea*, *Eukarya*, and *Viruses*. Both approaches generated similar results for bacteria.

In an attempt to decrease redundancies and obtain more complete genes, a restrict assembly was executed using Newbler assembler software version 2.5.3 with the "-rip" flag that allows the user to output each read in only one contig. Reads identified as "problematic" by the GS De Novo Assembler (Partial, Repeat, Outlier, or TooShort) or showing High-Quality Discrepancies (HQD) were filtered out. A new round of assembly and filtering was performed. These steps were repeated until the problematic and high-quality discrepant reads reached the cutoff point of 1% of the total number of reads filtered at the first assembly step. As a result, 423,310 reads remained as singletons while 2,623 contigs ≥500 bp were formed and the average contig size was 4,474 bp. Metagenome data reported in this paper have been deposited into the GenBank database (SRA051264) and on the MG-RAST server (4460151, 4481175, 4480501, and 4482672).

Analysis of Snail Crop-associated Microbiota

Analysis of community composition in filtered crop fluid confirmed an enrichment for prokaryotic populations, thus bypassing the difficulty of interpreting large amounts of repetitive and non-coding eukaryotic sequence data [45]. As expected, the non-filtered crop fluid sample contained more sequences related to *Eukaryota* and *Viruses* (Figure 2; Table S2). No host sequences were observed. The majority of eukaryotic sequences were from *Fungi*

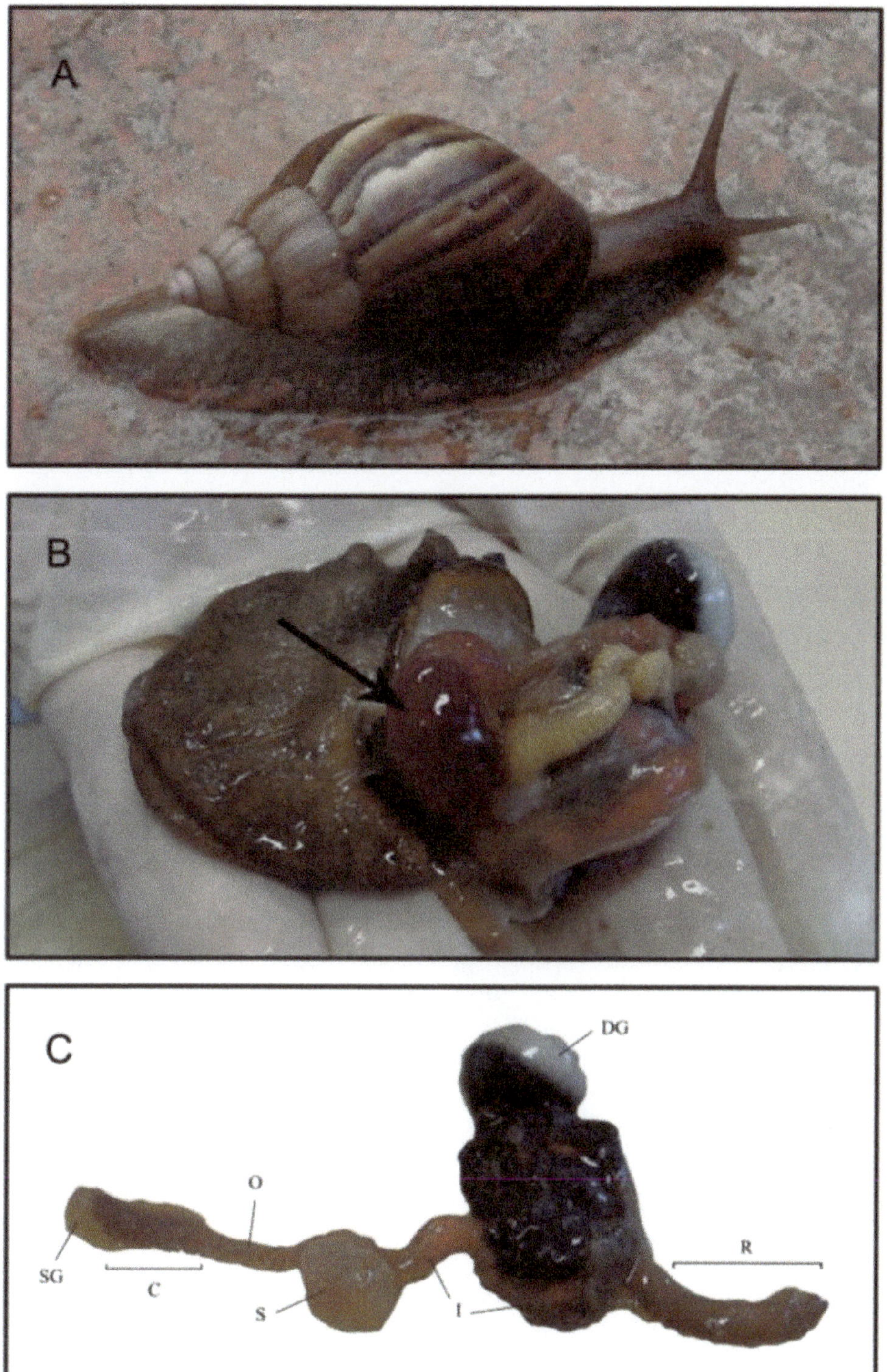

Figure 1. Digestive system of the snail. (A) Photograph of the African giant snail *Achatina fulica*. (B) Whole digestive system dissected. The black arrow indicates the crop filled with juice. DNA for the metagenome analysis was extracted from this fluid. (C) Spread out digestive system of the snail showing the salivary gland (SG), empty crop (C), esophagus (O), stomach (S), intestine (I), digestive gland (DG) and rectum (R).

(1,618), *Metazoa* (1,218), and *Viridiplantae* (806), which may represent plant DNA contamination (less than 0.088%). A small number of sequences related to the *Alveolata* (108) were identified. Although the role that fungi and protozoa play in cellulose digestion in rumen and lower termites is widely recognized [46–47], their exact function in snails remains unknown. Recently, it was shown that the host phylogeny is an important driver that shapes fungal community diversity and composition [48].

The majority of the reads that did match known viruses in the database were from phages that infect bacteria, mostly members of the order *Caudovirales* (Figure 2). Viruses regulate the composition of microbial communities in many environments [49]; however,

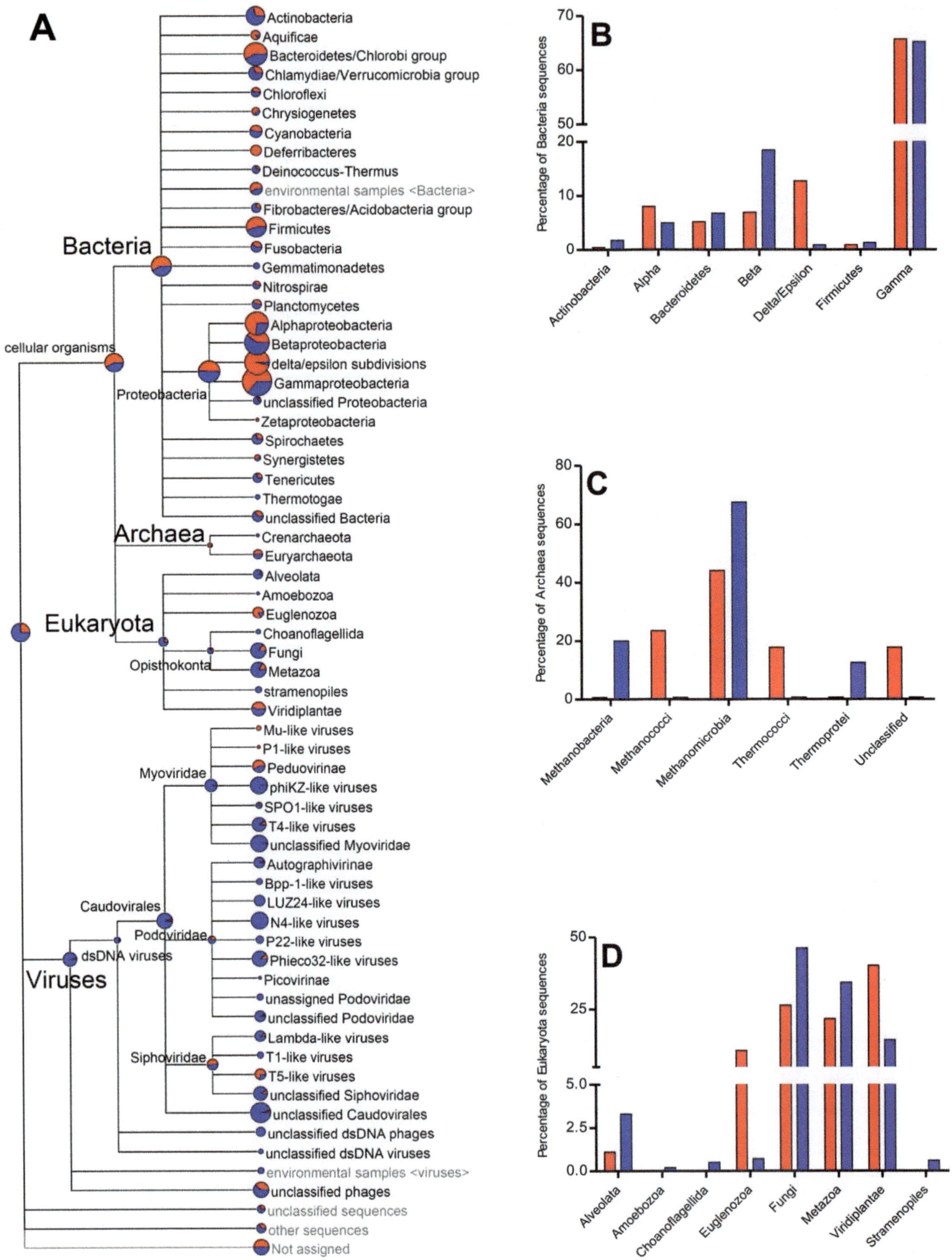
A
Bacteria
Archaea
Eukaryota
Viruses
cellular organisms
Proteobacteria
Opisthokonta
Myoviridae
Caudovirales
Podoviridae
dsDNA viruses
Siphoviridae
Actinobacteria
Aquificae
Bacteroidetes/Chlorobi group
Chlamydiae/Verrucomicrobia group
Chloroflexi
Chrysiogenetes
Cyanobacteria
Deferribacteres
Deinococcus-Thermus
environmental samples <Bacteria>
Fibrobacteres/Acidobacteria group
Firmicutes
Fusobacteria
Gemmatimonadetes
Nitrospirae
Planctomycetes
Alphaproteobacteria
Betaproteobacteria
delta/epsilon subdivisions
Gammaproteobacteria
unclassified Proteobacteria
Zetaproteobacteria
Spirochaetes
Synergistetes
Tenericutes
Thermotogae
unclassified Bacteria
Crenarchaeota
Euryarchaeota
Alveolata
Amoebozoa
Euglenozoa
Choanoflagellida
Fungi
Metazoa
stramenopiles
Viridiplantae
Mu-like viruses
P1-like viruses
Peduovirinae
phiKZ-like viruses
SPO1-like viruses
T4-like viruses
unclassified Myoviridae
Autographivirinae
Bpp-1-like viruses
LUZ24-like viruses
N4-like viruses
P22-like viruses
Phieco32-like viruses
Picovirinae
unassigned Podoviridae
unclassified Podoviridae
Lambda-like viruses
T1-like viruses
T5-like viruses
unclassified Siphoviridae
unclassified Caudovirales
unclassified dsDNA phages
unclassified dsDNA viruses
environmental samples <viruses>
unclassified phages
unclassified sequences
other sequences
Not assigned
B
Percentage of Bacteria sequences
70
60
50
20
10
0
Actinobacteria
Alpha
Bacteroidetes
Beta
Delta/Epsilon
Firmicutes
Gamma
C
Percentage of Archaea sequences
80
60
40
20
0
Methanobacteria
Methanococci
Methanomicrobia
Thermococci
Thermoprotei
Unclassified
D
Percentage of Eukaryota sequences
50
25
5.0
2.5
0.0
Alveolata
Amoebozoa
Choanoflagellida
Euglenozoa
Fungi
Metazoa
Viridiplantae
Stramenopiles

Figure 2. Analysis of composition of the snail microbial community. (A) Phylogenetic diversity of metagenomic sequences, computed by MEGAN4 based on a BLASTX using an e-value cut-off of 1e−5 comparison for filtered (red) and non-filtered (blue) sequences. The size of the circles is scaled logarithmically to represent the number of reads assigned to each taxon. (B) Bacteria, (C) Archaea, and (D) Eukaryota.

little is known about the ecological roles that they play in gastrointestinal systems. In the human gut, the viral population seems to be stable and persistant in individuals over time, suggesting that bacteria and viruses are in a mutually beneficial relationship [50].

The distribution of phylotypes and Environmental Gene Tags (EGTs) detected within the crop fluid sample primarily originated from the *Proteobacteria, Bacteroidetes*, and *Firmicutes* groups, consistent with previous SSU rDNA libraries of the snail crop [19]. At the genus level, *Pseudomonas* (37.5%), *Sulfurospirillum* (8.5%), and *Stenotrophomonas* (7.3%) were assigned as principal bacterial groups in the snail crop. Species from these genera are widely distributed in nature, including moist environments, water, soil, and on plants, including fruits and vegetables [51]. Therefore, they can use a varied range of nutrients and may assist the host in the degradation of lignocellulose material, detoxification of organic compounds, and synthesis of amino acids and vitamins.

Previous culture-dependent methodologies have identified cellulolytic *Pseudomonas* and *Stenotrophomonas* species in various invertebrates [52]. Moreover, previous culture-independent studies of bacterial species in the snail crop fluid, based on the 16S rRNA gene sequences, have demonstrated the presence of bacteria related to taxa found in the gut of herbivores, including *Pseudomonas, Clostridiaceae, Lactococcus, Bacteroides, Flavobacteriaceae, Mucilaginibacter, Citrobacter, Klebsiella, Aeromonas, Acinetobacter*, and sulfate-reducing bacteria related to *Sulforospirillum* [19]. In both the deepwater sea snail *Alviniconcha hessleri* [53] and the gutless marine oligochaete worm *Olavius algarvensis, Epsilonproteobacteria* occur as endosymbionts that serve as an energy source for the host and may participate in the removal of the end products of fermentation [54]; however, such a symbiosis has not yet been described for land snails.

Searches on rRNA databanks (RDP, SILVA, and Greengenes) resulted in no archaeal sequence detection, although BlastX analysis has revealed that genes related to *Euryarchaeota* are present within the snail crop, highlighting how little of this habitat has been described in the literature. Methanogenic archaea are known to be permanent residents in the digestive systems of ruminants, termites, and humans [55]. Although sulfate reduction and homoacetogenesis serve as alternate pathways for removing hydrogen, production of methane may contribute as hydrogen scavengers from the human distal gut [56–57].

In addition, sequences associated with *Crenarchaeota*, including the class *Thermoprotei*, have been detected (Figure 2). To the best of our knowledge, this is the first report of the detection of *Archaea, Fungi*, and *Viruses* within the crop of a snail. Compared with other metagenomes, the snail microbiota has high species richness (Figure S1), similar to the wallaby foregut and P3 luminal fluid from termite microbiomes. Principal coordinates analysis (PCoA) separated the snail microbiome across axes (Figure S2), indicating diverse and divergent community assemblages inhabiting the crop. In the scatter plot, the first two principal coordinates PC1 and PC2 explained 38.55% and 26.21% of data variation, respectively. PC1 separated snail crop-associated microbial communities from the panda, termite, and chicken microbiomes. PC2 separated from the wallaby and cow rumen environments. This finding suggests that *A. fulica* may represent an unexplored resource of uncharacterized microorganisms and genes.

Functional Metagenome Analysis

Snail metagenome assembly retrieved 363,365 predicted protein features. Of this total, 294,404 (81.0%) have been assigned an annotation using at least one protein database and 277,079 (94.1% of the annotated features) could be assigned to functional categories. The functional classification based on the COG database was obtained, and the abundance of the main pathways found on the *A. fulica* crop microbiome was compared with those of other metagenomes (Figures S3, S4).

Subsystems-based annotations (SEED) were performed to gain insights into the metabolic and physiological potential of the snail crop metagenome (Figure 3). In addition to the abundance of genes related to the main metabolic routes: carbohydrate, protein, and nucleic acid (Figure 3A; Figure S3), we found sequences associated with other important processes, such as iron acquisition, nitrogen, phosphorous, sulfur, potassium, cofactor and vitamins pathways, stress response and defense which are more abundant in the snail metagenome, as compared to other gut metagenomes (Figure S4).

The genetic repertoire of the *A. fulica* microbiome includes many sequences associated with protection against oxidative and osmotic stress, heat shock, and metabolism of aromatic compounds (Figure 3B; Figure S5), which could play an important role in the detoxification of phenolic and toxic compounds. Furthermore, we found a great variety of genes coding for antibiotic resistance as well as complete biosynthetic pathways for amino acids and vitamins (Figure 3).

Previous studies based on culture-independent high-throughput technologies have suggested that animals constitute complex biological systems in which mutualistic and symbiotic microbiota can influence host metabolism and complement important physiological functions [58–60]. In fact, microbiota that inhabit the gastrointestinal tract of animals considerably affect the growth and health of their hosts by influencing nutrition, pathogenesis, and immunity [61–62]. In this context, microbial communities within the snail crop could play an important role in detoxifying phenolic compounds and xenobiotics as well as in the synthesis of essential nutrients and defensive molecules. This relationship may be related to the high physiological adaptability and invasive properties of *A. fulica.*

We also analyzed the functional composition of the snail metagenome using similarity to a non-redundant protein database against KEGG metabolic pathways. This functional assignment revealed the main pathways represented on the microbiome (Figure S6A). A detailed examination of the carbohydrate metabolic pathways (Figure S6B) revealed a significant number (~7.5%) of putative genes belonging to the subgroup "starch and sucrose metabolism", including sequences coding for critical enzymes that cover all steps of cellulose deconstruction (Figure S6C).

Repertoire of Glycoside Hydrolase Functional Domains and Related Modules

To identify how the microbial community associated with *A. fulica* mediates plant fiber degradation, database searches for glycoside hydrolase (GH) domains were performed by HMMER software against the Pfam database. This analysis identified more than 4,000 EGTs from 93 different CAZy (Carbohydrate-Active EnZymes) families [33] and modules belonging predominantly to

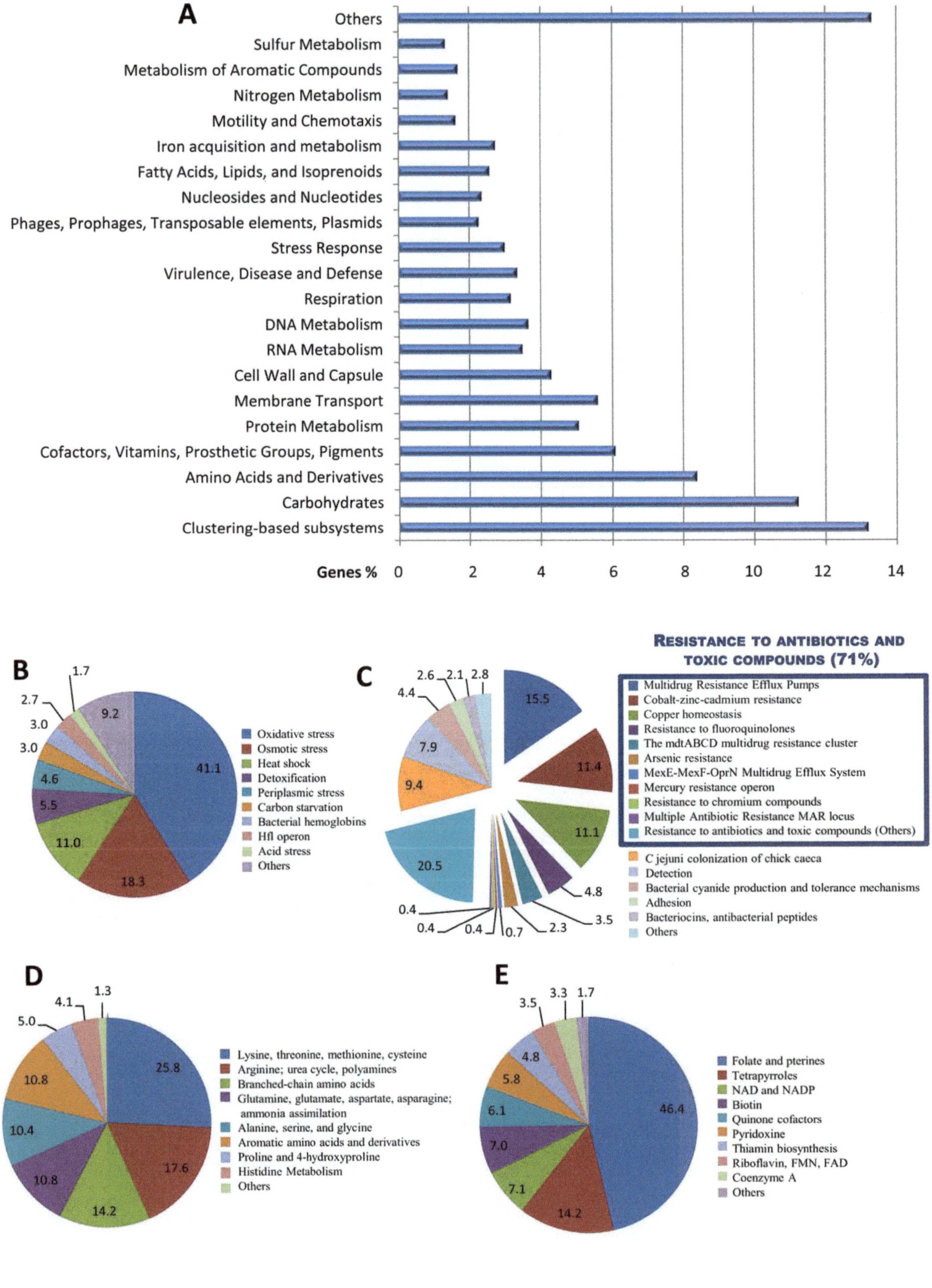
A
Others
Sulfur Metabolism
Metabolism of Aromatic Compounds
Nitrogen Metabolism
Motility and Chemotaxis
Iron acquisition and metabolism
Fatty Acids, Lipids, and Isoprenoids
Nucleosides and Nucleotides
Phages, Prophages, Transposable elements, Plasmids
Stress Response
Virulence, Disease and Defense
Respiration
DNA Metabolism
RNA Metabolism
Cell Wall and Capsule
Membrane Transport
Protein Metabolism
Cofactors, Vitamins, Prosthetic Groups, Pigments
Amino Acids and Derivatives
Carbohydrates
Clustering-based subsystems
Genes %
0
2
4
6
8
10
12
14
B
1.7
2.7
9.2
3.0
3.0
4.6
41.1
5.5
11.0
18.3
Oxidative stress
Osmotic stress
Heat shock
Detoxification
Periplasmic stress
Carbon starvation
Bacterial hemoglobins
Hfl operon
Acid stress
Others
C
2.6
2.1
2.8
4.4
15.5
7.9
11.4
9.4
11.1
20.5
4.8
0.4
3.5
0.4
0.4
0.7
2.3
RESISTANCE TO ANTIBIOTICS AND TOXIC COMPOUNDS (71%)
Multidrug Resistance Efflux Pumps
Cobalt-zinc-cadmium resistance
Copper homeostasis
Resistance to fluoroquinolones
The mdtABCD multidrug resistance cluster
Arsenic resistance
MexE-MexF-OprN Multidrug Efflux System
Mercury resistance operon
Resistance to chromium compounds
Multiple Antibiotic Resistance MAR locus
Resistance to antibiotics and toxic compounds (Others)
C jejuni colonization of chick caeca
Detection
Bacterial cyanide production and tolerance mechanisms
Adhesion
Bacteriocins, antibacterial peptides
Others
D
1.3
4.1
5.0
25.8
10.8
10.4
17.6
10.8
14.2
Lysine, threonine, methionine, cysteine
Arginine; urea cycle, polyamines
Branched-chain amino acids
Glutamine, glutamate, aspartate, asparagine; ammonia assimilation
Alanine, serine, and glycine
Aromatic amino acids and derivatives
Proline and 4-hydroxyproline
Histidine Metabolism
Others
E
3.3
1.7
3.5
4.8
5.8
6.1
46.4
7.0
7.1
14.2
Folate and pterines
Tetrapyrroles
NAD and NADP
Biotin
Quinone cofactors
Pyridoxine
Thiamin biosynthesis
Riboflavin, FMN, FAD
Coenzyme A
Others

Figure 3. Subsystems-based annotation (SEED) analysis of the snail crop microbiome. A) Percent of genes assigned by MG-RAST to each SEED subsystem. B) Percent of genes in the "Stress Response" subsystem, which includes sequences mainly associated with the protection against oxidative and osmotic stress, as well as heat shock. C) Genes assigned to the "Virulence, Disease and Defense" subsystem. Seventy-one percent of the sequences correspond to genes providing resistance to antibiotics and other toxic compounds, including multi-drug efflux pumps and genes for detoxification of a wide variety of heavy metals (cobalt, zinc, cadmium, chromium, copper and mercury). D) and E) Percent of genes in the subsystems "Amino Acids and Derivatives" and "Cofactors, Vitamins, Prosthetic Groups, Pigments", respectively.

Proteobacteria (Table S3; Figure S7). We found a wide diversity of GH catalytic modules: 2,534 EGTs belonging to 53 different GH families, which indicates an abundance of potential polysaccharide and plant cell wall biomass-degrading enzymes.

We also detected a total of 458 EGTs from 11 families of non-catalytic carbohydrate-binding modules (CBMs), which may promote the interaction of an enzyme with its target substrate, thereby increasing the efficiency of catalysis. This abundance reflects a high capacity to metabolize and degrade diverse lignocellulosic substrates. For example, modules that are likely to bind cellulose were found (CBM4-9-16-22 (PF02018)) [63], but the most frequently occurring module was CBM50 (152 members, with 18% of the sequences belonging to the *Pseudomonas* genus), which is also known as a LysM domain (PF01476). These are modules found attached to enzymes cleaving either chitin or peptidoglycan [64]. This domain is found in many of the enzymes involved in cell wall degradation and is also present in other proteins that are associated with bacterial cell walls. LysM-containing proteins have multiple functions, including modulating host–microbe interactions, such as signaling and recognition, between *Eukarya* and bacterial symbionts [64].

The most represented GH families were GH1 (294 members, with 73% of the sequences belonging to *Gammaproteobacteria*) and GH3 (219 members), which contains a widespread group of glycosidases that cleave non-reducing carbohydrates in oligosaccharides and the side chains of hemicelluloses and pectins. In addition, we found an abundance of GH13 (251 members), which is the most abundant family in metagenomes, such as the termite gut [7,65], and is the major GH family. This family includes enzymes that have a wide range of substrate specificities and catalytic activities, such as α-amylase, cyclodextrin glycosyltransferase (CGTase), branching enzymes, and cyclomaltodextrinase. The snail microbiome also contained a range of CAZy known to be involved in plant polymer degradation, including cellulases (GH5, GH9), β-galactosidases (GH2, GH27, GH35), chitinases (GH18, GH19), xylanases (GH26, GH43), and CE4 acetylxylan esterases that are required for the complete hydrolysis of xylan, α-xylosidases (GH31), α-mannosidases (GH38), and others targeting pectic polysaccharides, including GH43 and GH62 arabinofuranosidases, GH78 rhamnosidases, CE8 pectin methylesterases, GH28 rhamnogalacturonases, pectate lyases (PL1 and 3), and alginate lyase (PL5). In total, 21 families of glycosyltransferases (782 members), 4 families of carbohydrate esterases (212 members), and 4 families of polysaccharide lyases (94 members) were identified, representing the first and the largest microbial gene catalogue from a terrestrial snail. Moreover, 91 putative full-length genes with a significant match in the KEGG database were identified (Table S3). These include 43 candidate GHs, some of which are possibly involved in cellulose and hemicellulose degradation (such as five putative genes for β-glucosidases from the GH3 family, one putative protein from the GH5 family, and two putative cellulases from the GH8 family).

It is also worth noting the lack of specific CAZy family members, such as genes encoding for the GH7, GH12, GH44, and GH45 endoglucanases, which are essential in many cultured bacteria for cellulose solubilization. Cohesin and dockerin gene modules, which are often found associated with GHs, and are also known as cellulosome-associated modules, were not found in this metagenome. This may be a result of this particular microbiome structure, suggesting that there is still a deficiency in the information about the mechanisms involved in cellulose hydrolysis and dockerin-cohesin mediated complex assemblies in gut microbiomes.

Trends in the CAZy Family Frequencies

Comparing the snail GH profile with those of herbivores (giant panda, wallaby, and termite) and an omnivore (human), some interesting characteristics and patterns can be observed (Table 1).

First, the fraction of annotated carbohydrate active enzymes in the snail crop assigned to the oligosaccharide degradation category from the snail (36%) was very similar to that observed in samples from the giant panda, wallaby, and human. The profile of debranching enzymes from the snail metagenome (4%), while most similar to that of the termite and panda (and lower than that of the wallaby), showed higher relative abundance of GH78 rhamnosidases, which include enzymes with α-L-rhamnosidase activity. These enzymes are very interesting in terms of their suitability for various biotechnological applications within the food, pharmaceutical, and chemical industries and in the detoxification of rhamnose-conjugated plant secondary metabolites [66]. Nevertheless, the fraction of cellulases and endohemicellulases was higher in the snail than the levels observed in the giant panda and human but still lower than that of the termite indicating that despite its preference for fresh plants and fruits, the snail may be able to consume and degrade more recalcitrant materials with the help of its microbiota. These results suggest that the snail holobiont represents a novel and interesting model for studying the degradation of lignocellulose.

The dendrogram obtained by hierarchical clustering shows many clusters from identical or similar biomes (Figure 4). Upon analysis of the host-associated metagenomes, we found that the majority of samples from fungus-growing ants clustered together, including those from *Atta colombica* (ACOFUNT, ACOFUNB), *Cyphomyrmex longiscapus* (CLOFUN), and the *Trachymyrmex* (TRAFUN) genus. The microbiome of these herbivore insects is responsible for the degradation of plant biomass and simulates the role of the gut in other plant biomass degradation systems, such as those found in the termite, cow, panda, and wallaby [37,67].

A bigger cluster included the human fecal (HGUT7, HGUT8), mouse intestinal (MGUTL, MGUTOB), swine fecal (SWIN266, SWIN267), wild panda gut (PANDA1, PANDA2, PANDA5), and wallaby forestomach (WALLABY) microbiomes. In the same cluster, a termite gut microbiome (TERMITE1) was found. The wallaby metagenome was closer to the panda than to the human or swine metagenomes. The macropods are herbivores and possess unique adaptations to this diet, such as a forestomach where host-microbe associations allows the digestion of hemicellulose and cellulose [10]. Despite the fact that giant pandas do not have the long intestinal tract typical of other herbivores, it has been demonstrated that their gut microbiome contains putative cellulose-metabolizing symbionts [11]. Consequently, the

Table 1. Comparison of the glycoside hydrolase (GH) profiles targeting plant structural polysaccharides in the snail, termite, giant panda, wallaby, and human metagenomes.

CAZy Family	Known activity	SNAIL	TERMITE	WALLABY	PANDA	HUMAN
Cellulases						
GH5	cellulase	36	125	27	1	7
GH6	endoglucanase	4	0	0	0	0
GH7	endoglucanase	0	0	0	0	0
GH9	endoglucanase	15	43	5	0	0
GH44	endoglucanase	0	0	0	0	0
GH45	endoglucanase	0	6	0	0	0
GH48	endo-processive cellulases	2	0	0	0	0
Total		57 (2)	174 (16)	32 (4)	1 (0.5)	7 (1)
Endohemicellulases						
GH8	endo-xylanases	46	21	2	1	2
GH10	endo-1,4-β-xylanase	25	102	19	1	2
GH11	xylanase	1	19	0	0	0
GH12	endoglucanase & xyloglucan hydrolysis	0	0	0	0	0
GH26	β-mannanase & xylanase	11	20	8	0	1
GH28	galacturonases	69	15	10	0	3
GH53	endo-1,4-β-galactanase	9	20	11	4	11
Total		161 (6)	197 (18)	50 (6)	6 (2.5)	19 (3)
Cell wall elongation						
GH16	xyloglucanases & xyloglycosyltransferases	12	6	6	6	1
GH17	1,3-β-glucosidases	2	0	0	0	0
GH81	1,3-β-glucanase	1	0	0	0	0
Total		15 (1)	6 (1)	6 (1)	6 (3)	1 (0)
Debranching enzymes						
GH51	α-L-arabinofuranosidase	22	13	19	2	15
GH62	α-L-arabinofuranosidase	2	0	0	0	0
GH67	α-glucuronidase	5	6	1	2	1
GH78	α-L-rhmnosidase	73	7	46	1	13
Total		102 (4)	26 (2)	66 (8)	5 (2)	29 (5)
Oligosaccharide-degrading enzymes						
GH1	β-glucosidase & many other β-linked dimers	294	27	94	41	54
GH2	β-galactosidases & other β-linked dimer	66	32	39	4	29
GH3	mainly β-glucosidases	219	109	101	11	55
GH29	α-L-fucosidase	70	12	5	0	7
GH35	β-galactosidase	32	7	8	1	4
GH38	α-mannosidase	18	18	3	8	6
GH39	β-xylosidase	6	13	3	8	2
GH42	β-galactosidase	54	33	17	7	15
GH43	arabinases & xylosidases	185	63	72	13	34
GH52	β-xylosidase	0	3	0	0	0
Total		944 (36)	317 (28)	342 (39)	93 (41)	206 (36)

Data are presented with the GHs grouped according to their major function roles in the degradation of plant fiber, as classified in Allgaier *et al.*, 2010 [70]. The numbers in parentheses represent the percentages of these groups relative to the total number of GHs identified in the metagenomic datasets [2590 for snail, 872 for wallaby, 1117 for termite, 227 for panda (from 3 samples), and for human (from 2 samples)].

similarity between the wallaby and panda microbiomes is not surprising. Humans, mice, and swine share similar genetic complexity and omnivorous feeding behavior [7,68], which explains why their metagenomes were placed together in the major cluster. However, previous work has demonstrated that the cell wall and capsule profiles in the swine samples are more similar to those from the termite than those from the human gut [68]. This result was confirmed by our metagenomic comparison

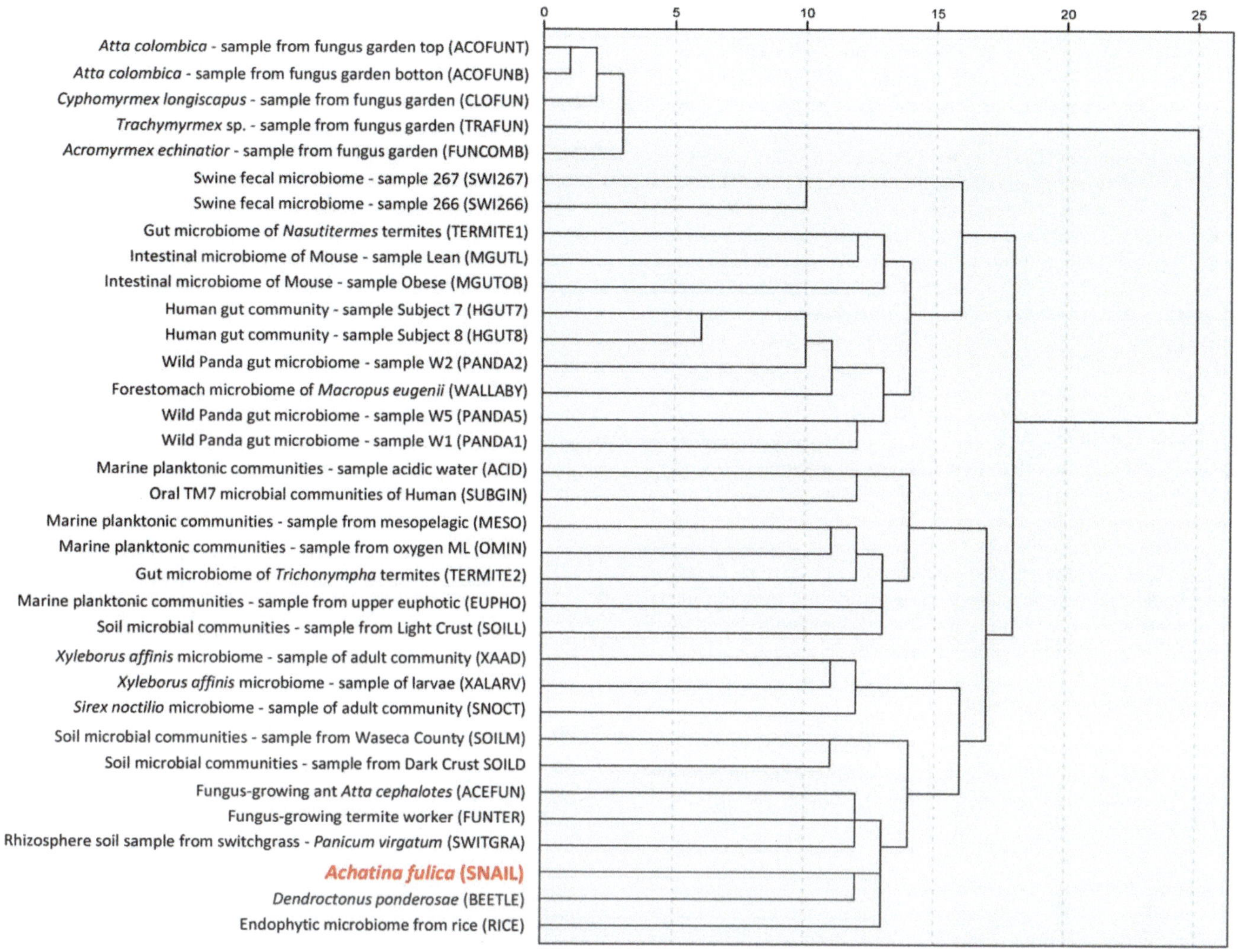

Figure 4. Dendrogram showing the relationships among metagenomes obtained through SPSS hierarchical clustering analysis. *Achatina fulica* crop microbiome (SNAIL) was compared with fungus garden microbial communities from *Atta colombica* in top (ACOFUNT) and bottom (ACOFUNB), *Atta cephalotes* (ACEFUN), *Cyphomyrmex longiscapus* (CLOFUN), *Trachymyrmex* (TRAFUN), and *Acromyrmex echinatior* (FUNCOMB); fecal microbiome of swine (SWI267 and SWI266); P3 luminal contents of *Nasutitermes* (TERMITE1); intestinal micobiome of lean (MGUTL) and obese (MGUTOB) mice; fecal microbiome of healthy adult humans (HGUT7 and HGUT8); feces from wild pandas (PANDA2, PANDA5, and PANDA1); forestomach from *Macropus eugenii* (WALLABY); Richmond acid mine drainage (ACID); oral microbiome from humans (SUBGIN); aquatic environment from mesopelagic (MESO), oxygen minimum layer (OMIN), and planktonic zones (EUPHO) in Hawaii; microbiome of *Trichonympha* from termites (TERMITE2); soil microbial communities from farm silage (SOILM), dark crust (SOILD), distinct crusts (SOILL), and switchgrass rhizosphere (SWITGRA); *Xyleborus affinis* from adult (XAAD) and larvae (XALARY); *Sirex noctilio* (SNOCT); fungus-growing termite (FUNTER); *Dendroctonus ponderosae* (BEETLE); and endophytic microbiome from rice (RICE).

using carbohydrate-active enzymes, indicating similar metabolic capabilities between these microbiomes.

Another group was formed from four aquatic metagenomes (MESO, EUPHO, OMIN, ACID) as well as soil (SOILL), human oral (SUBGIN), and a protist microbiome that lives in the intestine of the termite (TERMITE2). The oral microbial communities were close to the acidic water, showing the similarities among communities belonging to acidic environments. A smaller cluster was formed by *Sirex noctilio* (SNOCT), adult (XAAD) and larvae (XALARV) *Xyleborus affinis* microbiomes, which are arthropod host-associated.

Finally, a cluster including microbiota associated with Arthropoda (ACEFUN, FUNTER, BEETLE), two plant-host associated (RICE, SWITCHGRA), and the snail metagenome were identified and found to be similar. Two soil samples (SOILD, SOILM) formed an external group. Clustering analysis of CAZy profiles showed that the microbiota from the crop of the giant snail is closest to microbial communities from the beetle *Dendroctonus ponderosae*. Native to the forests of western North America, this beetle is the most important insect pest of pine and is responsible for the loss of millions of trees. These beetles attack trees by boring tunnels under the bark. Consequently, their microbiota may reveal important aspects of the cellulosic plant biomass conversion. Additionally, the similarity to the endophytic microbiome from rice and to microbial communities from the switchgrass (*Panicum virgatum*) rhizosphere may be due to the snail's feeding habits. In these environments, hydrolytic plant polymer-degrading enzymes have been demonstrated to be an important characteristic for the establishment of endophytes in their hosts, resulting in a high abundance of cellulases, xylanases, hydrolases, cellulose-binding proteins, and pectinases [69].

Concluding Remarks

Deconstructing plant cell walls to improve the efficiency of processing biomass from bagasse and other agriculture residues by enzymatic hydrolysis is considered the most promising environmentally friendly technology available for second-generation ethanol production. However, the heterogeneity and complexity of cell wall structural components and the presence of inhibitors of these enzymes make industrial scale-up a difficult task. Presented in this study is a comparative metagenome analysis of the first land snail crop microbiome from a highly invasive species, which has a wide-ranging diet and is capable of consuming different varieties of plants and substrates. The *A. fulica* holobiont represents a prosperous reservoir of novel GH genes and related modules, some of which have not been previously detected in other metagenomic datasets, representing a remarkable and unique complexity and diversity of genes. Moreover, the diverse metabolic capabilities of the snail metagenome may provide important physiological functions that assist the host in adapting successfully to different environments and feeding habitats. In these circumstances, we anticipate that future explorations of the land snail microbiome will contribute not only to our understanding of basic biology and host-microbiota co-evolution but also to areas of biotechnological relevance, such as discovery and bioengineering of lignocellulolytic enzymes, for use in biomass-to-bioethanol applications to develop a more eco-friendly biofuel.

Supporting Information

Figure S1 Rarefaction curve of annotated species richness. The curves represent the average number of different species annotations for subsamples of the complete dataset calculated in MGRAST.

Figure S2 Principal coordinates analysis (PCoA) for giant snail (this study), wallaby, giant panda, chicken, termite, and cow rumen metagenomes. The data were compared with RDP using a maximum e-value of 1e−5, a minimum alignment length of 50, normalized values, and bray-curtis distances calculated in MGRAST.

Figure S3 Metabolic comparison according to the clusters of orthologous groups' functional categories. The color scale corresponds to the relative number of putative genes within each metagenome.

Figure S4 Extended error bar of functional gene groups from *Achatina fulica* versus other gut metagenomes. Pair-wise comparisons were calculated for the snail metagenome versus (A) termite, (B) cow rumen, (C) panda, (D) wallaby, (E) chicken, and (F) human metagenomes are shown.

Figure S5 Functional analysis (SEED) for Metabolism of Aromatic Compounds category.

Figure S6 Functional analysis based on the KEGG database. A) Relative number of genes from the snail crop metagenome classified according to the KEGG main categories. As expected, the sequences related to carbohydrate and amino acid metabolism predominate, accounting for 17% and 15% of total coding sequences related to KEGG pathways, respectively. B) Examination of the relative gene diversity within the KEGG pathway for carbohydrate metabolism. C) KEGG sub-pathway for starch and sucrose metabolism. The color scale represents the number of genes (in logarithmic scale) found for each KEGG entry in the giant snail metagenome. Notice that a great variety of putative genes are related to cellulose degradation, including 49 endo-1,4-β-D-glucanases (entry 3.2.1.4), 3 types of 1,4-β-cellobiosidases (3.2.1.91), and 390 β-glucosidases (3.2.1.21). Furthermore, 27 sequences were classified as putative xylan-degrading enzymes (3.2.1.37, 1,4-β-xylosidases).

Figure S7 Contribution of organism taxa (Class level) to glycoside hydrolase (GH) hits on *Achatina fulica* metagenome. The color represents the proportion of each organism shown relative to the number of each GH hit. The scale is linear; red indicates 0, and blue indicates 100%.

Table S1 Food preferences of *Achatina fulica*.

Table S2 Summary of metagenomic data obtained from the *Achatina fulica* crop microbiome.

Table S3 Inventory of putative glycoside hydrolases (GH), polysaccharide lyases (PL), glycosyltransferases (GT), carbohydrate esterases (CE) and carbohydrate-binding modules (CBM) identified in the snail crop microbiome.

Acknowledgments

We would like to thank the staff of INMETRO, UFRJ, and LNCC for insightful discussions and comments.

Author Contributions

Conceived and designed the experiments: AMC OBM ESG WdS ATRV. Performed the experiments: AMC JJVC ALG. Analyzed the data: AMC JJVC MEC CET RBF AG SMNS YBS PJMSIB ALG ATRV. Contributed reagents/materials/analysis tools: OBM ESG WdS ATRV. Wrote the paper: AMC JJVC CET RBF MEC AG SMNS YBS PJMSIB ALG ATRV.

References

1. Goldemberg J (2007) Ethanol for a sustainable energy future. Science 315: 808–810.
2. Somerville C, Youngs H, Taylor C, Davis SC, Long SP (2010) Feedstocks for Lignocellulosic Biofuels. Science 329: 790–792.
3. Goldemberg J (2008) The Brazilian biofuels industry. Biotechnol Biofuels 1: 6–13.
4. Jenkins BM, Baxter LL, Miles-Jr TR, Miles TR (1998) Combustion properties of biomass. Fuel Proc Technol 54: 17–46.
5. Ragauskas AJ, Williams CK, Davison BH, Britovsek G, Cairney J, et al. (2006) The Path Forward for Biofuels and Biomaterials. Science 311: 484–489.
6. Rubin EM (2008) Genomics of cellulosic biofuels. Nature 454: 841–845.
7. Warnecke F, Luginbühl P, Ivanova N, Ghassemian M, Richardson TH, et al. (2007) Metagenomic and functional analysis of hindgut microbiota of a wood-feeding higher termite. Nature 450: 560–565.
8. Brulc JM, Antonopoulos DA, Miller ME, Wilson MK, Yannarell AC, et al. (2009) Gene-centric metagenomics of the fiber-adherent bovine rumen microbiome reveals forage specific glycoside hydrolases. Proc Natl Acad Sci USA 106: 1948–1953.
9. Hess M, Sczyrba A, Egan R, Kim TW, Chokhawala H, et al. (2011) Metagenomic discovery of biomass-degrading genes and genomes from cow rumen. Science 331: 463–467.

10. Pope PB, Denman SE, Jones M, Tringe SG, Barry K, et al. (2010) Adaptation to herbivory by the Tammar wallaby includes bacterial and glycoside hydrolase profiles different from other herbivores. Proc Natl Acad Sci USA 107: 14793–14798.
11. Zhu L, Wu Q, Dai J, Zhang S, Wei F (2011) Evidence of cellulose metabolism by the giant panda gut microbiome. Proc Natl Acad Sci USA 108: 17714–17719.
12. Karasov WH, Martínez del Rio C, Caviedes-Vidal E (2011) Ecological physiology of diet and digestive systems. Annu Rev Physiol 73: 69–93.
13. Thiengo SC, Faraco FA, Salgado NC, Cowie RH, Fernandez MA (2007) Rapid spread of an invasive snail in South America: the giant African snail, *Achatina fulica*, in Brasil. Biol Invasions 9: 693–702.
14. Albuquerque FS, Peso-Aguiar MC, Assunção-Albuquerque MJT (2008) Distribution, feeding behavior and control strategies of the exotic land snail *Achatina fulica* (Gastropoda: Pulmonata) in the northeast of Brazil. Braz J Biol 68: 837–842.
15. Biedermann W, Moritz P (1898) Beiträge zur vergleichenden Physiologie der Verdauung II. Ueber ein celluloselösendes Enzym im Lebersecret der Schnecke (*Helix pomatia*). Pflüg Arch 73: 219–287.
16. Charrier M, Daguzan J (1980) Food consumption: production and energy budget in *Helix aspersa* Müller (Gastropoda Pulmonata). Ann Nutr Alim 34: 147–166.
17. Tringe SG, Rubin EM (2005) Metagenomics: DNA sequencing of environmental samples. Nature Rev Gen 6: 805–814.
18. Oliveira-Filho AT, Fontes MAL (2000). Patterns of floristic differentiation among Atlantic Forests in Southeastern Brazil and the influence of climate. Biotropica 32: 793–810.
19. Cardoso AM, Cavalcante JJ, Vieira RP, Lima JL, Grieco MA, et al. (2012) Gut bacterial communities in the giant land snail *Achatina fulica* and their modification by sugarcane-based diet. PloS ONE 7: e33440.
20. Vieira RP, Gonzalez AM, Cardoso AM, Oliveira DN, Albano RM, et al. (2008) Relationships between bacterial diversity and environmental variables in a tropical marine environment, Rio de Janeiro. Environ Microbiol 10: 189–199.
21. Gomez-Alvarez V, Teal TK, Schmidt TM (2009) Systematic artifacts in metagenomes from complex microbial communities. ISME J 3: 1314–1317.
22. Chou HH, Holmes MH (2001) DNA sequence quality trimming and vector removal. Bioinformatics 17: 1093–1104.
23. Huson DH, Mitra S, Ruscheweyh HJ, Weber N, Schuster SC (2011) Integrative analysis of environmental sequences using MEGAN4. Genome Res 21: 1552–1560.
24. Altschul SF, Madden TL, Schäffer AA, Zhang J, Zhang Z, et al. (1997) Gapped BLAST and PSI-BLAST: a new generation of protein database search programs. Nucleic Acids Res 25: 3389–3402.
25. Huang Y, Gilna P, Li W (2009) Identification of ribosomal RNA genes in metagenomic fragments. Bioinformatics 25: 1338–1340.
26. Cole JR, Wang Q, Cardenas E, Fish J, Chai B, et al. (2009) The Ribosomal Database Project: improved alignments and new tools for rRNA analysis. Nucleic Acids Res 37: 141–145.
27. Overbeek R, Begley T, Butler RM, Choudhuri JV, Chuang HY, et al. (2005) The subsystems approach to genome annotation and its use in the project to annotate 1000 genomes. Nucleic Acids Res 33: 5691–5702.
28. Tatusov RL, Koonin EV, Lipman DJ (1997) A genomic perspective on protein families. Science 278: 631–637.
29. Meyer F, Paarmann D, D'Souza M, Olson R, Glass EM, et al. (2008) The metagenomics RAST server - a public resource for the automatic phylogenetic and functional analysis of metagenomes. BMC Bioinformatics 9: 386–394.
30. Parks DH, Beiko RG (2010) Identifying biologically relevant differences between metagenomic communities. Bioinformatics 26: 715–721.
31. Kanehisa M, Goto S, Sato Y, Furumichi M, Tanabe M (2012) KEGG for integration and interpretation of large-scale molecular data sets. Nucleic Acids Res 40: 109–114.
32. Almeida LG, Paixão R, Souza RC, Costa GC, Barrientos FJ, et al. (2004) A System for Automated Bacterial (genome) Integrated Annotation - SABIA. Bioinformatics 20: 2832–2833.
33. Cantarel BL, Coutinho PM, Rancurel C, Bernard T, Lombard V, et al. (2009) The Carbohydrate-Active EnZymes database (CAZy): an expert resource for Glycogenomics. Nucleic Acids Res 37: 233–238.
34. Finn RD, Clements J, Eddy SR (2011) HMMER web server: interactive sequence similarity searching. Nucleic Acids Res 39: 29–37.
35. Punta M, Coggill PC, Eberhardt RY, Mistry J, Tate J, et al. (2012) The Pfam protein families database. Nucleic Acids Res 40: 290–301.
36. Tringe SG, von Mering C, Kobayashi A, Salamov AA, Chen K, et al. (2005) Comparative metagenomics of microbial communities. Science 308: 554–557.
37. Suen G, Scott JJ, Aylward FO, Adams SM, Tringe SG, et al. (2010) An insect herbivore microbiome with high plant biomass-degrading capacity. PLoS Genet 6: e1001129.
38. Markowitz VM, Ivanova NN, Szeto E, Palaniappan K, Chu K, et al. (2008) IMG/M: a data management and analysis system for metagenomes. Nucleic Acids Res 36: 534–538.
39. Willner D, Thurber RV, Rohwer F (2009) Metagenomic signatures of 86 microbial and viral metagenomes. Environ Microbiol 11: 1752–1766.
40. Ghose KC (1963) The alimentary system of *Achatina fulica*. Trans Am Microsc Soc 82: 149–167.
41. Zbinden M, Pailleret M, Ravaux J, Gaudron SM, Hoyoux C, et al. (2010) Bacterial communities associated with the wood-feeding gastropod *Pectinodonta* sp. (Patellogastropoda, Mollusca). FEMS Microbiol Ecol 74: 450–463.
42. Lucena SA, Lima LS, Cordeiro LS Jr, Sant'anna C, Constantino R, et al. (2011) High throughput screening of hydrolytic enzymes from termites using a natural substrate derived from sugarcane bagasse. Biotechnol Biofuels 4: 51–60.
43. Saeed I, Tang SL, Halgamuge SK (2012) Unsupervised discovery of microbial population structure within metagenomes using nucleotide base composition. Nucl Acid Res 40: e34.
44. Ghai R, Rodriguez-Valera F, McMahon KD, Toyama D, Rinke R, et al. (2011) Metagenomics of the water column in the pristine upper course of the Amazon river. PLoS ONE 6: e23785.
45. Kunin V, Copeland A, Lapidus A, Mavromatis K, Hugenholtz P (1998) A Bioinformatician's Guide to Metagenomics. Microbiol Mol Biol Rev 72: 557–578.
46. Hungate RE (1975) The rumen microbial ecosystem. Ann Rev Ecol Syst 6: 39–66.
47. Breznak JA, Brune A (1994) Role of Microorganisms in the Digestion of Lignocellulose by Termites. Ann Rev Entomol 39: 453–487.
48. Liggenstoffer AS, Youssef NH, Couger MB, Elshahed MS (2010) Phylogenetic diversity and community structure of anaerobic gut fungi (phylum *Neocallimastigomycota*) in ruminant and non-ruminant herbivores. ISME J 4: 1225–1235.
49. Breitbart M, Rohwer F (2005) Here a virus, there a virus, everywhere the same virus? Trends Microbiol 13: 278–284.
50. Reyes A, Haynes M, Hanson N, Angly FE, Heath AC, et al. (2010) Viruses in the faecal microbiota of monozygotic twins and their mothers. Nature 466: 334–338.
51. Madigan MT, Martinko JM, Parker J (2003). Brock Biology of Microorganisms. 10th ed. Pearson Education, Inc. Upper Saddle River, NJ.
52. Huang S, Sheng P, Zhang H (2012) Isolation and Identification of Cellulolytic Bacteria from the Gut of *Holotrichia parallela* Larvae (*Coleoptera: Scarabaeidae*) Int J Mol Sci 13, 2563–2577.
53. Suzuki Y, Sasaki T, Suzuki M, Nogi Y, Miwa T, et al. (2005) Novel chemoautotrophic endosymbiosis between a member of the *Epsilonproteobacteria* and the hydrothermal-vent gastropod *Alviniconcha* aff. *hessleri* (*Gastropoda: Provannidae*) from the Indian Ocean. Appl Environ Microbiol 71: 5440–5450.
54. Dubilier N, Mülders C, Ferdelman T, de Beer D, Pernthaler A, et al. (2001) Endosymbiotic sulphate-reducing and sulphide-oxidizing bacteria in an oligochaete worm. Nature 411: 298–302.
55. Thauer RK, Kaster AK, Seedorf H, Buckel W, Hedderich R (2008) Methanogenic archaea: ecologically relevant differences in energy conservation. Nat Rev Microbiol 6: 579–591.
56. Stams AJ (1994) Metabolic interactions between anaerobic bacteria in methanogenic environments. Antonie Van Leeuwenhoek 66: 271–294.
57. Gill SR, Pop M, Deboy RT, Eckburg PB, Turnbaugh PJ, et al. (2006) Metagenomic analysis of the human distal gut microbiome. Science 312: 1355–1359.
58. Dethlefsen L, McFall-Ngai M, Relman DA (2007) An ecological and evolutionary perspective on human-microbe mutualism and disease. Nature 449: 811–818.
59. Li M, Wang B, Zhang M, Rantalainen M, Wang S, et al. (2008) Symbiotic gut microbes modulate human metabolic phenotypes. Proc Natl Acad Sci USA 105: 2117–2122.
60. Qin J, Li R, Raes J, Arumugam M, Burgdorf KS, et al. (2010) A human gut microbial gene catalogue established by metagenomic sequencing. Nature 464: 59–65.
61. Turnbaugh PJ, Ley RE, Mahowald MA, Magrini V, Mardis ER, et al. (2006) An obesity-associated gut microbiome with increased capacity for energy harvest. Nature. 444: 1027–1031.
62. Qu A, Brulc JM, Wilson MK, Law BF, Theoret JR, et al. (2008) Comparative metagenomics reveals host specific metavirulomes and horizontal gene transfer elements in the chicken cecum microbiome. PLoS ONE 3: e2945.
63. Bayer EA, Shoham Y, Lamed R (2006) in Prokaryotes, eds Dworkin M, Falkow S, Rosenberg E, Schleifer K-H, Stackebrandt E (Springer-Verlag, New York), 578–617.
64. Buist G, Steen A, Kok J, Kuipers OP (2008) LysM, a widely distributed protein motif for binding to (peptido)glycans. Mol Microbiol 68: 838–847.
65. Li LL, McCorkle SR, Monchy S, Taghavi S, van der Lelie D (2009) Bioprospecting metagenomes: glycosyl hydrolases for converting biomass. Biotechnol Biofuels 2: 10.
66. Manzanares P, Vallés S, Ramón D, Orejas M (2007) in *Industrial Enzymes*, eds Polaina J, MacCabe AP (Springer Verlag, Berlin-Heidelberg), 117–140.
67. Scott JJ, Budsberg KJ, Suen G, Wixon DL, Balser TC, et al. (2010) Microbial community structure of leaf-cutter ant fungus gardens and refuse dumps. PloS ONE 5: e9922.
68. Lamendella R, Domingo JW, Ghosh S, Martinson J, Oerther DB (2011) Comparative fecal metagenomics unveils unique functional capacity of the swine gut. BMC Microbiol 11: 103.
69. Sessitsch A, Hardoim P, Döring J, Weilharter A, Krause A, et al. (2010) Functional characteristics of an endophyte community colonizing rice roots as revealed by metagenomic analysis. Mol Plant Microbe Interact 25: 28–36.
70. Allgaier M, Reddy A, Park JI, Ivanova N, D'haeseleer P, et al. (2010) Targeted Discovery of Glycoside Hydrolases from a Switchgrass-Adapted Compost Community. PLoS ONE 5: e8812.

PERMISSIONS

All chapters in this book were first published in PLOS ONE, by The Public Library of Science; hereby published with permission under the Creative Commons Attribution License or equivalent. Every chapter published in this book has been scrutinized by our experts. Their significance has been extensively debated. The topics covered herein carry significant findings which will fuel the growth of the discipline. They may even be implemented as practical applications or may be referred to as a beginning point for another development.

The contributors of this book come from diverse backgrounds, making this book a truly international effort. This book will bring forth new frontiers with its revolutionizing research information and detailed analysis of the nascent developments around the world.

We would like to thank all the contributing authors for lending their expertise to make the book truly unique. They have played a crucial role in the development of this book. Without their invaluable contributions this book wouldn't have been possible. They have made vital efforts to compile up to date information on the varied aspects of this subject to make this book a valuable addition to the collection of many professionals and students.

This book was conceptualized with the vision of imparting up-to-date information and advanced data in this field. To ensure the same, a matchless editorial board was set up. Every individual on the board went through rigorous rounds of assessment to prove their worth. After which they invested a large part of their time researching and compiling the most relevant data for our readers.

The editorial board has been involved in producing this book since its inception. They have spent rigorous hours researching and exploring the diverse topics which have resulted in the successful publishing of this book. They have passed on their knowledge of decades through this book. To expedite this challenging task, the publisher supported the team at every step. A small team of assistant editors was also appointed to further simplify the editing procedure and attain best results for the readers.

Apart from the editorial board, the designing team has also invested a significant amount of their time in understanding the subject and creating the most relevant covers. They scrutinized every image to scout for the most suitable representation of the subject and create an appropriate cover for the book.

The publishing team has been an ardent support to the editorial, designing and production team. Their endless efforts to recruit the best for this project, has resulted in the accomplishment of this book. They are a veteran in the field of academics and their pool of knowledge is as vast as their experience in printing. Their expertise and guidance has proved useful at every step. Their uncompromising quality standards have made this book an exceptional effort. Their encouragement from time to time has been an inspiration for everyone.

The publisher and the editorial board hope that this book will prove to be a valuable piece of knowledge for researchers, students, practitioners and scholars across the globe.

LIST OF CONTRIBUTORS

David Chagné and Susan E. Gardiner
Plant and Food Research, Palmerston North Research Centre, Palmerston North, New Zealand

Ross N. Crowhurst and Roger P. Hellens
Plant and Food Research, Mount Albert Research Centre, Auckland, New Zealand

Michela Troggio, Alessandro Cestaro and Riccardo Velasco
IASMA Research and Innovation Centre, Foundation Edmund Mach, San Michele all'Adige, Trento, Italy

Mark W. Daveyn and Stijn Vanderzande
Laboratory for Fruit Breeding and Biotechnology, Department of Biosystems, Katholieke Universiteit Leuven, Heverlee, Leuven, Belgium

Barbara Gilmore and Nahla Bassil
USDA-ARS, National Clonal Germplasm Repository, Corvallis, Oregon, United States of America

Cindy Lawley
Illumina Inc., Hayward, California, United States of America

Satish Kumar
Plant and Food Research, Hawke's Bay Research Centre, Havelock North, New Zealand

Dorrie Main and Cameron Peace
Department of Horticulture and Landscape Architecture, Washington State University, Pullman, Washington, United States of America

Jasper D. Rees
Agricultural Research Council, Onderstepoort, South Africa

Amy Iezzoni
Department of Horticulture, Michigan State University, East Lansing, Michigan, United States of America

Todd Mockler
The Donald Danforth Plant Science Center, St. Louis, Missouri, United States of America

Larry Wilhelm
Oregon Health Sciences University, Portland, Oregon, United States of America

Eric Van de Weg
Plant Breeding, Wageningen University and Research Centre, Wageningen, The Netherlands

Santosh Kumar Gupta and Amit Kumar Rai
National Research Centre on Plant Biotechnology, Indian Agricultural Research Institute, New Delhi, India
Department of Biotechnology, Himachal Pradesh University, Summer-Hill, Shimla, India

Shamsher Singh Kanwar
Department of Biotechnology, Himachal Pradesh University, Summer-Hill, Shimla, India

Tilak R. Sharma
National Research Centre on Plant Biotechnology, Indian Agricultural Research Institute, New Delhi, India

Imri Ben-Israel, Habte Nida and Eyal Fridman
The Robert H. Smith Institute of Plant Sciences and Genetics in Agriculture, Faculty of Agriculture, Food and Environment, The Hebrew University of Jerusalem, Israel

Benjamin Kilian
Institute of Plant Genetics and Crop Plant Research (IPK), Gatersleben, Germany

Najmeh Sahebzadeh, Nur Azura Adam and Wei Hong Lau
Department of Plant Protection, Faculty of Agriculture. Universiti Putra Malaysia, Serdang, Selangor, Malaysia

Makhdzir Mardan
Department of Agriculture Technology, Faculty of Agriculture, Universiti Putra Malaysia, Serdang, Selangor, Malaysia

Abdul Manaf Ali
Department of Biotechnology, Faculty of Agriculture and Biotechnology, Universiti Sultan Zainal Abidin, Kuala Terengganu, Terengganu, Malaysia

Soon Guan Tan
Department of Cell and Molecular Biology, Faculty of Biotechnology and Biomolecular Sciences, Universiti Putra Malaysia, Serdang, Selangor, Malaysia

Xiao Yang and Bao-Rong Lu
Ministry of Education Key Laboratory for Biodiversity Science and Ecological Engineering, Department of Ecology and Evolutionary Biology, Fudan University, Shanghai, China

Feng Wang and Jun Su
Fujian Province Key Laboratory of Genetic Engineering for Agriculture, Fujian Academy of Agricultural Sciences, Fuzhou, China

Ruby Chandna, Rehna Augustine and Naveen C. Bisht
National Institute of Plant Genome Research, Aruna Asaf Ali Marg, New Delhi, India

Yulin Jia and Melissa H. Jia
Dale Bumpers National Rice Research Center, Agricultural Research Service, United States Department of Agriculture, Stuttgart, Arkansas, United States of America

Xueyan Wang
Department of Biology, College of Life Sciences, China Jiliang University, Hangzhou, Zhejiang, China

Guangjie Liu
Rice Research and Extension Center, University of Arkansas, Stuttgart, Arkansas, United States of America

Bhimana Gautami, Shyam N. Nigam, Amindala BhanuPrakash, Trushar M. Shah and Dave A. Hoisington
Center of Excellence in Genomics, International Crops Research Institute for the Semi-Arid Tropics (ICRISAT), Patancheru, India

Daniel Foncéka and Jean-Francois Rami
UMR Dwveloppement et Amélioration des plantes, Centre de coopération Internationale en Recherche Agronomique pour le Développement (CIRAD), Montpellier, France

Manish K. Pandey
Center of Excellence in Genomics, International Crops Research Institute for the Semi-Arid Tropics (ICRISAT), Patancheru, India
Department of Plant Pathology, University of Georgia (UGA), Tifton, Georgia, United States of America
Crop Protection and Management Research Unit, USDA-Agricultural Research Service, Tifton, Georgia, United States of America

Márcio C. Moretzsohn
Plant Genetics Lab, EMBRAPA Genetic Resources and Biotechnology, Brasilia, Brazil

Venkataswamy Sujay
Center of Excellence in Genomics, International Crops Research Institute for the Semi-Arid Tropics (ICRISAT), Patancheru, India
Department of Genetics and Plant Breeding, University of Agricultural Sciences (UAS-D), Dharwad, India

Hongde Qin
Department of Plant Pathology, University of Georgia (UGA), Tifton, Georgia, United States of America
Cash Crop Research Institute, Hubei Academy of Agricultural Sciences (HAAS), Wuhan, Hubei, China

Yanbin Hong Xiaoping Chen and Xuanqiang Liang
Crops Research Institute, Guangdong Academy of Agricultural Sciences (GAAS), Guangzhou, Guangdong, China

Issa Faye
Centre National de Recherche Agronomique (CNRA), Institut Sénégalais de Recherches Agricoles (ISRA), Bambey, Sénégal

Makanahally V. C. Gowda
Department of Genetics and Plant Breeding, University of Agricultural Sciences (UAS-D), Dharwad, India

Baozhu Guo
Crop Protection and Management Research Unit, USDA-Agricultural Research Service, Tifton, Georgia, United States of America

David J. Bertioli
Genetics Department, University of Brasilia, Brasilia, Brazil

Rajeev K. Varshney
Center of Excellence in Genomics, International Crops Research Institute for the Semi-Arid Tropics (ICRISAT), Patancheru, India
Crops Research Institute, Guangdong Academy of Agricultural Sciences (GAAS), Guangzhou, Guangdong, China
Theme- Comparative and Applied Genomics, CGIAR Generation Challenge Programme (GCP), CIMMYT, Mexico, Mexico

Jacopo Diamanti, Franco Capocasa, Francesca Balducci and Bruno Mezzetti
Dipartimento di Scienze Agrarie, Alimentari e Ambientali, Università Politecnica delle Marche, Ancona, Italy

Maurizio Battino
Dipartimento di Scienze Cliniche Specialistiche ed Odontostomatologiche, Università Politecnica delle Marche, Ancona, Italy

Jim Hancock
Department of Horticulture, Michigan State University, East Lansing, Michigan, United States of America

Rod J. Mahon and Bill James
Ecosystem Sciences Division, The Commonwealth Scientific and Industrial Research Organisation, Black Mountain Laboratories, Canberra, Australian Capital Territory, Australia

Sharon J Downes
Ecosystem Sciences Division, The Commonwealth Scientific and Industrial Research Organisation, Australian Cotton Research Institute, Narrabri, New South Wales, Australia

Roisin C. McGarry and Brian G. Ayre
Department of Biological Sciences, University of North Texas, Denton, Texas, United States of America

Cameron Peace, Dorrie Main and Stephen Ficklin
Department of Horticulture, Washington State University, Pullman, Washington, United States of America

Nahla Bassil and Barbara Gilmore
USDA-ARS, National Clonal Germplasm Repository, Corvallis, Oregon, United States of America

Umesh R. Rosyara, Travis Stegmeir, Audrey Sebolt and Amy Iezzoni
Department of Horticulture, Michigan State University, East Lansing, Michigan, United States of America

Cindy Lawley
Illumina Inc., Hayward, California, United States of America

Todd C. Mockler
The Donald Danforth Plant Science Center, St. Louis, Missouri, United States of America

Douglas W. Bryant
The Donald Danforth Plant Science Center, St. Louis, Missouri, United States of America
Intuitive Genomics, Inc., St. Louis, Missouri, United States of America

Larry Wilhelm
Oregon Health Sciences University, Portland, Oregon, United States of America

Mulualem T. Kassa, R. Varma Penmetsa, Noelia Carrasquilla-Garcia and Douglas R. Cook
Department of Plant Pathology, University of California Davis, Davis, California, United States of America

Birinchi K. Sarma
Department of Plant Pathology, University of California Davis, Davis, California, United States of America
Department of Mycology and Plant Pathology, Banaras Hindu University, Varanasi, India

Subhojit Datta
Department of Plant Pathology, University of California Davis, Davis, California, United States of America
Indian Institute of Pulses Research, Uttar Pradesh, India

Hari D. Upadhyaya and Rajeev K. Varshney
International Crops Research Institute for the Semi-Arid Tropics, Patancheru, Andhra Pradesh, India

Eric J. B. von Wettberg
Department of Biological Sciences, Florida International University, Miami, Florida, United States of America
Center for Tropical Plant Conservation, Fairchild Tropical Botanic Garden, Coral Gables, Florida, United States of America

Lewis H. Ziska and Martha B. Tomecek
Crop Systems and Global Change Laboratory, United States Department of Agriculture, Agricultural Research Service, Beltsville, Maryland, United States of America

David R. Gealy, Aaron K. Jackson and Howard L. Black
Dale Bumpers National Rice Research Center, United States Department of Agriculture, Agricultural Research Service, Stuttgart, Arkansas, United States of America

Maria Hurtado, Santiago Vilanova, Mariola Plazas, Pietro Gramazio and Jaime Prohens
Instituto de Conservación y Mejora de la Agrodiversidad Valenciana, Universitat Politècnica de València, Valencia, Spain

H. Hemal Fonseka
Horticultural Crop Research and Development Institute, Gannoruwa, Sri Lanka

Ramya Fonseka
Department of Crop Science, Faculty of Agriculture, University of Peradeniya, Peradeniya, Sri Lanka

Gaoyi Cao
College of Agriculture and Biotechnology, China Agricultural University, Beijing, China
Institute of Crop Sciences, Chinese Academy of Agricultural Sciences, Beijing, China

Yunjun Liu, Shengxue Zhang, Xuewen Yang, Rongrong Chen, Yan Liu and Guoying Wang
Institute of Crop Sciences, Chinese Academy of Agricultural Sciences, Beijing, China

Yuwen Zhang and Jianhua Wang
College of Agriculture and Biotechnology, China Agricultural University, Beijing, China

Wei Lu and Min Lin
Biotechnology Research Institute, Chinese Academy of Agricultural Sciences, Beijing, China

Rey Gaston Loor Solorzano
UMR AGAP, CIRAD, Montpellier, France
EET-Pichilingue, INIAP, Quevedo El Empalme, Ecuador

Olivier Fouet, Brigitte Courtois, Ange Marie Risterucci, Claire Lanaud and Xavier Argout
UMR AGAP, CIRAD, Montpellier, France

Arnaud Lemainque and Sylvana Pavek
Centre National de Génotypage, CEA Institut de Génomique, Evry, France

Michel Boccara
UMR AGAP, CIRAD, Montpellier, France
Cocoa Research Unit (CRU), University of the West Indies, St Augustine, Trinidad and Tobago

Freddy Amores
EET-Pichilingue, INIAP, Quevedo El Empalme, Ecuador

Ourania I. Pavli and George N. Skaracis
Department of Crop Sciences, Agricultural University of Athens, Athens, Greece

Anastasia P. Tampakaki
Department of Agricultural Biotechnology, Agricultural University of Athens, Athens, Greece

Daniel Fonceka, Ronan Rivallan, Hélène Vignes, Elodie Lacut, Fabien de Bellis, Jean-Christophe Glaszmann, Brigitte Courtois and Jean-François Rami
CIRAD, UMR AGAP, Montpellier, France

Hodo-Abalo Tossim and Ousmane Ndoye
ISRA/Ceraas, Thiès, Sénégal

Issa Faye
ISRA, Centre National de Recherche Agronomique, Bambey, Sénégal

Soraya C. M. Leal-Bertioli and José F. M. Valls
Embrapa Recursos Genéticos e Biotecnologia, Brasilia, Distrito Federal, Brazil

David J. Bertioli
Universidade de Brasília, Brasília, Distrito Federal, Brazil

Y. Tunginba Singh
Department of Botany, University of Delhi, Delhi, India
International Centre for Genetic Engineering & Biotechnology, New Delhi, India

Sudeshna Mazumdar-Leighton, Mahaswetta Saikia, Prashant Pant and Sochanngam Kashung
Department of Botany, University of Delhi, Delhi, India

Kartik Neog and Rajen Chakravorty
Central Muga and Eri Research and Training Institute, Ladoigarh, Assam, India

Suresh Nair
International Centre for Genetic Engineering & Biotechnology, New Delhi, India

Javaregowda Nagaraju
Center for DNA Fingerprinting & Diagnostics, Hyderabad, Andhra Pradesh, India

Cheerukeri Raghavendra Babu
Center for Management of Degraded Ecosystems, University of Delhi, Delhi, India

Li Huang, Huifang Jiang, Xiaoping Ren, Yuning Chen, Xinyan Zhao, Mei Tang, Jiaquan Huang and Boshou Liao
Key Laboratory of Biology and Genetic Improvement of Oil Crops, Ministry of Agriculture, Oil Crops Research Institute of the Chinese Academy of Agricultural Sciences, Wuhan, China

Yingjie Xiao
National Key Laboratory of Crop Genetic Improvement and National Center of Plant Gene Research (Wuhan), Huazhong Agricultural University, Wuhan, China

Hari D. Upadhyaya
International Crops Research Institute for the Semi-Arid Tropics, Patancheru, Andhra Pradesh, India

Alexander M. Cardoso, Janaína J. V. Cavalcante, Roberto B. Flatschart, Sandra M. N. Scapin, Youssef B. Sade and Paulo J. M. S. I. Beltrão
Instituto Nacional de Metrologia, Qualidade e Tecnologia, Duque de Caxias, Rio de Janeiro, Brazil

Maurício E. Cantão
Laboratório Nacional de Computação Científica, Petrópolis, Rio de Janeiro, Brazil

Empresa Brasileira de Pesquisa Agropecuária - CNPSA, Concórdia, Santa Catarina, Brazil

Claudia E. Thompson, Arnaldo Glogauer, Alexandra L. Gerber and Ana Tereza R. Vasconcelos
Laboratório Nacional de Computação Científica, Petrópolis, Rio de Janeiro, Brazil

Orlando B. Martins
Instituto Nacional de Metrologia, Qualidade e Tecnologia, Duque de Caxias, Rio de Janeiro, Brazil
Instituto de Bioquímica Médica, Universidade Federal do Rio de Janeiro, Rio de Janeiro, Rio de Janeiro, Brazil

Eloi S. Garcia
Instituto Nacional de Metrologia, Qualidade e Tecnologia, Duque de Caxias, Rio de Janeiro, Brazil
Instituto Oswaldo Cruz, Fundação Oswaldo Cruz, Rio de Janeiro, Rio de Janeiro, Brazil

Wanderley de Souza
Instituto Nacional de Metrologia, Qualidade e Tecnologia, Duque de Caxias, Rio de Janeiro, Brazil
Instituto de Biofísica Carlos Chagas Filho, Universidade Federal do Rio de Janeiro, Rio de Janeiro, Rio de Janeiro, Brazil

Index